METHODS IN MOLECULAR BIOLOGY™

Series Editor
John M. Walker
School of Life Sciences
University of Hertfordshire
Hatfield, Hertfordshire, AL10 9AB, UK

For further volumes:
http://www.springer.com/series/7651

DNA Recombination

Methods and Protocols

Edited by

Hideo Tsubouchi

University of Sussex,
Brighton, United Kingdom

Humana Press

Editor
Hideo Tsubouchi
MRC Genome Damage and Stability Centre
University of Sussex
Science Park Road, Falmer
Brighton, BN1 9RQ
United Kingdom
h.tsubouchi@sussex.ac.uk

ISSN 1064-3745 e-ISSN 1940-6029
ISBN 978-1-4939-5778-1 ISBN 978-1-61779-129-1 (eBook)
DOI 10.1007/978-1-61779-129-1
Springer New York Dordrecht Heidelberg London

Printed on acid-free paper

Humana Press is part of Springer Science+Business Media (www.springer.com)

Preface

Homologous recombination has been intensively studied in budding yeast. I think we are extremely lucky to find that homologous recombination is exceptionally robust in this organism, making it an ideal model to study this process. Historically, the availability of powerful genetics in this simple, unicellular organism has enabled the isolation of genes that play key roles in homologous recombination, and we have learnt a lot about homologous recombination using this organism. Homologous recombination is important in various aspects of DNA metabolism, including damage repair, replication, telomere maintenance, and meiosis. We also now know that key players in homologous recombination identified and characterized in yeast, such as proteins encoded by the genes belonging to the so-called RAD52 group, are well conserved among eukaryotic species, including humans. This offers promise that further in-depth characterization of homologous recombination using yeast will help provide the basic framework for understanding the universal mechanism(s) of homologous recombination conserved in eukaryotes. When asked to edit a book about methods for studying homologous recombination, I decided to include chapters that cover recent techniques that best utilize the advantages of the yeast system, with the belief that yeast will keep serving as a great model organism to study homologous recombination.

On the other hand, there is a group of genes involved in recombination that are apparently found only in higher eukaryotes, such as BRCA2, indicating the presence of an extra layer of mechanistic complexity in these organisms. Obviously, the most straightforward approach to study these mechanisms is to use models in which these particular mechanisms exist. From this point of view, chapters for studying recombination using higher eukaryotes have also been included.

Although we have gained significant understanding of the entity underlying homologous recombination, I have to say that we still do not know much about it when we see it as a "micro machine" that is incredibly efficient at finding similarity between two DNA molecules inside a cell. Obviously, a necessary step in the direction of understanding this process is to isolate the machine and let it work in a test tube. Understanding the design by studying the appearance and behavior of the machinery as a single molecule will be an important milestone toward understanding the mechanism of action of the machinery. Almost as important is to learn how the machinery behaves inside living cells. In recent years, this approach has flourished due to advances in microscopy and the availability of various fluorescent proteins. Techniques covering these topics have been included.

Yeast genetics has successfully provided a framework for the mechanism of homologous recombination. Now the question is, what can we do next to bring it to the next level of understanding? This is a question I ask myself, but I believe it is more or less a question for anyone who is enthusiastic about understanding this very fascinating phenomenon. I hope this protocol book will prove useful for this purpose. Finally, I would like to thank all the contributors who willingly agreed to share their expertise/knowledge. Needless to say, this book would not exist without their effort.

Hideo Tsubouchi

Contents

Contributors

CAROLINE ADELFALK • *Max-Planck-Institute for Molecular Genetics, Berlin, Germany*
ANDRÉS AGUILERA • *Centro Andaluz de Biología Molecular y Medicina Regenerativa, Universidad de Sevilla-CSIC, Sevilla, Spain*
JASVINDER S. AHUJA • *Department of Biological, Geological and Environmental Sciences, Center for Gene Regulation in Health and Disease, Cleveland State University, Cleveland, OH, USA*
VERA BATRAK • *Independent Scientist, Istra, Moscow Region, Russia*
GRACE C. BAZAN • *Biological Sciences, California Polytechnic State University, San Luis Obispo, CA, USA*
ARTEM BLAGODATSKI • *Institute of Protein Research, Russian Academy of Sciences, Russian Federation, Moscow, Russia*
HANNAH G. BLITZBLAU • *Whitehead Institute for Biomedical Research, Cambridge, MA, USA*
G. VALENTIN BÖRNER • *Department of Biological, Geological and Environmental Sciences, Center for Gene Regulation in Health and Disease, Cleveland State University, Cleveland, OH, USA*
ERIKA BRUNET • *Muséum National d'Histoire Naturelle, Paris, France*
JEAN-MARIE BUERSTEDDE • *Independent Scientist, Hildesheim, Germany*
DMITRY V. BUGREEV • *Department of Biochemistry and Molecular Biology, Drexel University College of Medicine, Philadelphia, PA, USA*
KERSTIN BYSTRICKY • *Laboratoire de Biologie Moléculaire Eucaryote (LBME), Université de Toulouse, Toulouse, France*
FRANCESCA CAVALLO • *Department of Public Health and Cell Biology, Section of Anatomy, University of Rome Tor Vergata, Rome, Italy*
STACY Y. CHEN • *Department of Obstetrics, Gynecology, and Reproductive Sciences, University of California, San Francisco, CA, USA*
FRANCESCA COLE • *Developmental Biology Program, Memorial Sloan-Kettering Cancer Center, New York, NY, USA*
VINCENZO COSTANZO • *Clare Hall Laboratories, London Research Institute, Hertsfordshire, UK*
JAN DROUAUD • *Institut Jean-Pierre Bourgin, UMR1318 INRA-AgroParisTech, Versailles Cedex, France; Institut National de Recherche, Agronomique, Centre de Versailles-Grignon Route de St-Cyr (RD10), Versailles Cedex, France*
NADINE ECKERT-BOULET • *Department of Biology, University of Copenhagen, Copenhagen, Denmark*
KIRK T. EHMSEN • *Department of Microbiology, University of California, Davis, CA, USA*
ANASTASIYA EPSTEIN • *Department of Biochemistry, New York University School of Medicine, New York, NY, USA*
SARAH FARMER • *MRC Genome Damage and Stability Centre, University of Sussex, Sussex, UK*

ILYA J. FINKELSTEIN • *Department of Biochemistry and Molecular Biophysics, Columbia University, New York, NY, USA*
JENNIFER C. FUNG • *Department of Obstetrics, Gynecology, and Reproductive Sciences, University of California, San Francisco, CA, USA*
BELÉN GÓMEZ-GONZÁLEZ • *Centro Andaluz de Biología Molecular y Medicina Regenerativa, Universidad de Sevilla-CSIC, Sevilla, Spain*
ERIC C. GREENE • *Department of Biochemistry and Molecular Biophysics, Columbia University, New York, NY; Howard Hughes Medical Institute, Chevy Chase, MD, USA*
JILL M. GRIMME • *US Army Engineer Research Development Center, Construction Engineering Research Laboratory, Champaign, IL, USA*
EDGAR HARTSUIKER • *North West Cancer Research Fund Institute, Bangor University, Bangor, UK*
YOSHITAMI HASHIMOTO • *Clare Hall Laboratories, London Research Institute, Hertsfordshire, UK*
WOLF-DIETRICH HEYER • *Department of Microbiology and Department of Molecular and Cellular Biology, University of California, Davis, CA, USA*
KENNETH J. HILLERS • *Biological Sciences, California Polytechnic State University, San Luis Obispo, CA, USA*
SEIKI HIRANO • *Weatherall Institute of Molecular Medicine, University of Oxford, Oxford, UK*
ANDREAS HOCHWAGEN • *Whitehead Institute for Biomedical Research, Cambridge, MA, USA*
NANCY M. HOLLINGSWORTH • *Department of Biochemistry and Cell Biology, Stony Brook University, New York, NY, USA*
SCOTT HOUGHTALING • *Department of Molecular Genetics and Microbiology, University of New Mexico School of Medicine, Albuquerque, NM, USA*
HIROSHI IWASAKI • *School and Graduate School of Bioscience and Biotechnology, Tokyo Institute of Technology, Tokyo, Japan*
MARIA JASIN • *Developmental Biology Program, Memorial Sloan-Kettering Cancer Center, New York, NY, USA*
HIROYUKI KITAO • *Department of Molecular Oncology, Kyushu University, Kyushu, Japan*
HANNAH L. KLEIN • *Department of Biochemistry, New York University School of Medicine, New York, NY, USA*
YOUNGHO KWON • *Department of Molecular Biophysics and Biochemistry, Yale University School of Medicine, New Haven, CT, USA*
IMEN LASSADI • *Laboratoire de Biologie Moléculaire Eucaryote, Université de Toulouse, Toulouse, France*
WING-KIT LEUNG • *MRC Genome Damage and Stability Centre, University of Sussex, Sussex, UK*
MICHAEL LISBY • *Department of Biology, University of Copenhagen, Copenhagen, Denmark*
JIE LIU • *Department of Microbiology, University of California, Davis, CA, USA*
HSIAO-CHI LO • *Department of Biochemistry and Cell Biology, Stony Brook University, New York, NY, USA*
WENJIAN MA • *Chromosome Stability Section, National Institute of Environmental Health Sciences (NIEHS), NIH, Research Triangle Park, NC, USA*

ANNA MALKOVA • *Biology Department, Indiana University Purdue University, Indianapolis, IN, USA*
ALEXANDER V. MAZIN • *Department of Biochemistry and Molecular Biology, Drexel University College of Medicine, Philadelphia, PA, USA*
OLGA M. MAZINA • *Department of Biochemistry and Molecular Biology, Drexel University College of Medicine, Philadelphia, PA, USA*
CHRISTINE MÉZARD • *Institut Jean-Pierre Bourgin, Versailles Cedex, France*
KUNTAL MUKHERJEE • *School of Biology, Georgia Institute of Technology, Atlanta, GA, USA*
YASUTO MURAYAMA • *Cancer Research UK, London Research Institute, London, UK*
KENTARO NABESHIMA • *Department of Cell and Developmental Biology, University of Michigan, Medical School, Ann Arbor, MI, USA*
WATARU NAKAI • *Chromosome Stability Section, National Institute of Environmental Health Sciences (NIEHS), NIH, Research Triangle Park, NC, USA*
KOJI NAKANISHI • *Developmental Biology Program, Memorial Sloan-Kettering Cancer Center, New York, NY, USA*
MARY ANN OSLEY • *Department of Molecular Genetics and Microbiology, University of New Mexico School of Medicine, Albuquerque, NM, USA*
MICHAEL A. RESNICK • *Chromosome Stability Section, National Institute of Environmental Health Sciences (NIEHS), NIH, Research Triangle Park, NC, USA*
MATTHEW J. ROSSI • *Department of Biochemistry and Molecular Biology, Drexel University College of Medicine, Philadelphia, PA, USA*
RODNEY ROTHSTEIN • *Department of Genetics and Development, Columbia University Medical Center, New York, NY, USA*
JOSÉ F. RUIZ • *Centro Andaluz de Biología Molecular y Medicina Regenerativa, Universidad de Sevilla-CSIC, Sevilla, Spain*
HARRY SCHERTHAN • *Bundeswehr Institute of Radiobiology, affiliated to the University of Ulm, Munich, Germany; Max-Planck-Institute for Molecular Genetics, Berlin, Germany*
HILARIE A. SEHORN • *Department of Genetics and Biochemistry, Clemson University, Clemson, SC, USA*
MICHAEL G. SEHORN • *Department of Genetics and Biochemistry, Clemson University, Clemson, SC, USA*
YING SHEN • *School of Biology, Georgia Institute of Technology, Atlanta, GA, USA*
JESSICA SNEEDEN • *Department of Microbiology, University of California, Davis, CA, USA*
MARIA SPIES • *Department of Biochemistry, Howard Hughes Medical Institute, University of Illinois, Urbana-Champaign, Urbana, IL, USA*
FRANCESCA STORICI • *School of Biology, Georgia Institute of Technology, Atlanta, GA, USA*
SAMANTHA STUCKEY • *School of Biology, Georgia Institute of Technology, Atlanta, GA, USA*
PATRICK SUNG • *Department of Molecular Biophysics and Biochemistry, Yale University School of Medicine, New Haven, CT, USA*
MINORU TAKATA • *Laboratory of DNA Damage Signaling, Department of Late Effects Studies, Kyoto University, Kyoto, Japan*
HIDEO TSUBOUCHI • *MRC Genome Damage and Stability Centre, University of Sussex, Brighton, UK*

TOYOKO TSUKUDA • *Department of Molecular Genetics and Microbiology, University of New Mexico School of Medicine, Albuquerque, NM, USA*
JIM WESTMORELAND • *Chromosome Stability Section, National Institute of Environmental Health Sciences (NIEHS), NIH, Research Triangle Park, NC, USA*
WILLIAM D. WRIGHT • *Department of Microbiology, University of California, Davis, CA, USA*
XIAO-PING ZHANG • *Department of Microbiology, University of California, Davis, CA, USA*
WEIXING ZHAO • *Department of Molecular Biophysics and Biochemistry, Yale University School of Medicine, New Haven, CT, USA*
XIUZHONG ZHENG • *Department of Biochemistry, New York University School of Medicine, New York, NY, USA*

Section I

Genetic and Molecular Biological Approaches with Yeast

Chapter 1

Methods to Study Mitotic Homologous Recombination and Genome Stability

Xiuzhong Zheng, Anastasiya Epstein, and Hannah L. Klein

Abstract

Spontaneous mitotic recombination occurs in response to DNA damage incurred during DNA replication or from lesions that do not block replication but leave recombinogenic substrates such as single-stranded DNA gaps. Other types of damages result in general genome instability such as chromosome loss, chromosome fragmentation, and chromosome rearrangements. The genome is kept intact through recombination, repair, replication, checkpoints, and chromosome organization functions. Therefore when these pathways malfunction, genomic instabilities occur. Here we outline some general strategies to monitor a subset of the genomic instabilities: spontaneous mitotic recombination and chromosome loss, in both haploid and diploid cells. The assays, while not inclusive of all genome instability assays, give a broad assessment of general genome damage or inability to repair damage in various genetic backgrounds.

Key words: Genomic instability, gene conversion, chromosome loss, mitotic recombination, cell division.

1. Introduction

Mitotic recombination and genome instability are outcomes of DNA damage and the cellular repair response. Many of the types of rearrangements and general instability that can be seen in yeast are typical of human cancer cells. Thus, yeast has become an excellent model system to detect genes essential for genome maintenance and to decipher the numerous pathways used to prevent genomic instability (1). There are several advantages to the yeast systems. First, double-strand break (DSB) repair genes that are essential in mammalian cells are frequently not essential in yeast, allowing the study of null mutants. Second, yeast can be grown

H. Tsubouchi (ed.), *DNA Recombination*, Methods in Molecular Biology 745,
DOI 10.1007/978-1-61779-129-1_1, © Springer Science+Business Media, LLC 2011

vegetatively as a haploid or a diploid. The haploid phase allows rapid genetic and physical detection of rearrangements and the easy use of recessive mutations. The diploid phase allows the study of haplolethal rearrangements such as chromosome loss. Third, it is relatively easy to conduct whole genome analyses of rearrangements by comparative genome hybridization to detect changes in gene copy number and chromosomal location. Fourth, many of the DNA recombination, repair, and damage checkpoint functions are highly conserved, so studies in yeast have direct applicability to mammalian cells. Last, many reporter systems have been developed to be quantitative so that rates can be determined and statistical comparisons can be made between strains with mutations in pathway components.

There have been several recent articles on methods to detect genomic instabilities such as mutations, repeat slippage, aneuploidy, and gross chromosomal rearrangements (1–6). Here we describe methods to detect mitotic gene conversion and chromosome loss as general markers for DNA lesions.

2. Materials

2.1. Media

Media for Petri plates are prepared in 2-l flasks or beakers, with each flask or beaker containing 1 l of medium, which is sufficient for about 30 plates. Unless otherwise stated, all components are autoclaved together for 20 min at 250°F (121°C) and 15 lb/square inch of pressure (103 kPa). The plates should be allowed to dry for 2–3 days at room temperature after pouring. Plates can be stored in sealed plastic bags for at least 3 months. The agar is omitted for liquid media. Liquid media can be prepared in smaller volumes for individual use:

1. Liquid and agar YPDA: 1% Bacto yeast extract, 2% Bacto peptone, 2% glucose, 2.5% Bacto agar, 1% adenine (2 ml), and distilled H_2O (1,000 ml). Store at room temperature.
2. YPGA: 1% Bacto yeast extract, 2% Bacto peptone, 3% glycerol, 2.5% Bacto agar, and 1% adenine (2 ml). Omit Bacto agar for liquid YPDA. Store at room temperature.
3. Liquid and agar synthetic complete (SC) and dropout media: SC is a medium in which the dropout mix contains all possible supplements (i.e., nothing is "dropped out"):

   Dropout media is a medium that contains all but one of the amino acid or base supplements listed below, for use with common strain auxotrophies: Bacto yeast nitrogen base without amino acids and ammonium sulfate, 2% glucose,

0.5% ammonium sulfate, 2.5% Bacto agar, dropout mix (49 ml), and distilled H_2O (1,000 ml).

Dropout mix: Dropout mix is a combination of the following ingredients minus the appropriate supplement: 1% adenine (2 ml), 1% arginine (3 ml), 1% histidine (3 ml), 1% isoleucine (3 ml), 2% leucine (3 ml), 1% lysine (3 ml), 1% methionine (3 ml), 5% phenylalanine (5 ml), 1% proline (3 ml), 10% serine (4 ml), 5% threonine (5 ml) (add threonine after autoclaving. This amino acid supplement is necessary only if the strains require threonine), 1% tryptophan (3 ml), 1% tyrosine (3 ml), 1% uracil (3 ml), and 1% valine (3 ml).

4. SC+ CAN plates: Make 1 l of SC-arginine dropout agar media, autoclave and cool the media down to 50–55°C, and supplement with L-canavanine sulfate salt (60 mg) (Sigma, C9758) diluted in water and filter sterilized. Mix well before pouring into Petri plates.
5. SC+ 5-FOA (fluoroorotic acid) plates: Make 1 l of SC agar media in two different flasks. In one flask, mix 500 ml dH_2O with 25 g Bacto agar, autoclave, and cool the media to 50–55°C. In the other flask, combine all of the dropout mix ingredients with 5-FOA (750 mg) (US Biological, F5050), filter sterilize, and prewarm to 50°C. Slowly pour the prewarmed dropout mix and 5-FOA solutions into the agar solution and mix well before pouring into Petri plates.

2.2. Strains

These strains can be modified to carry mutations in a particular gene of interest to test its role in genome stability:

1. Diploid chromosome loss assay:
   YWT-1 *MATa leu2-3, 112 his3-11, 15 ade2-1 ura3-1 trp1-1 can1-100 RAD5+*
   YWT-2 *MATα leu2-3, 112 his3-11, 15 ADE2+ ura3-1 trp1-1 CAN1+ RAD5+*
2. Diploid recombination assay:
   YWT-3 *MATa leu2-ecoRI his3-11, 15 ade2-1 ura3-1 trp1-1 can1-100 RAD5+*
   YWT-4 *MATα leu2-bstEII his3-11, 15 ade2-1 ura3-1 trp1-1 can1-100 RAD5+*
3. Haploid chromosome fragment loss assay:
   YWT-5 *MATa* CFV/D8B-tg *(URA3+ SUP11+) leu2-3, 112 his3-11, 15 ade2-1 ura3-1 trp1-1 can1-100 RAD5+*
4. Haploid gene conversion assay:
   YWT-6 *MATa leu2-ecoRI::URA3::leu2-bstEII his3-11, 15 ade2-1 ura3-1 trp1-1 can1-100 RAD5+*

3. Methods

3.1. Diploid Chromosome Loss Assay

To determine chromosome loss, recombination, and mutation rates, we perform fluctuation tests using the median method (7) (*see* **Fig. 1.1**). To make diploids, we cross YWT-1 and YWT-2 strains, pull about 36 zygotes (*can1-100 hom3-10/CAN1+ HOM3+*) on YPDA, and grow them at 30°C for 3 days (*see* **Note 1**). For each test, nine zygote colonies are used, and three separate tests are performed for each assay:

1. Pick nine colonies from the YPDA plate and disperse each into 1 ml sterile dH_2O.

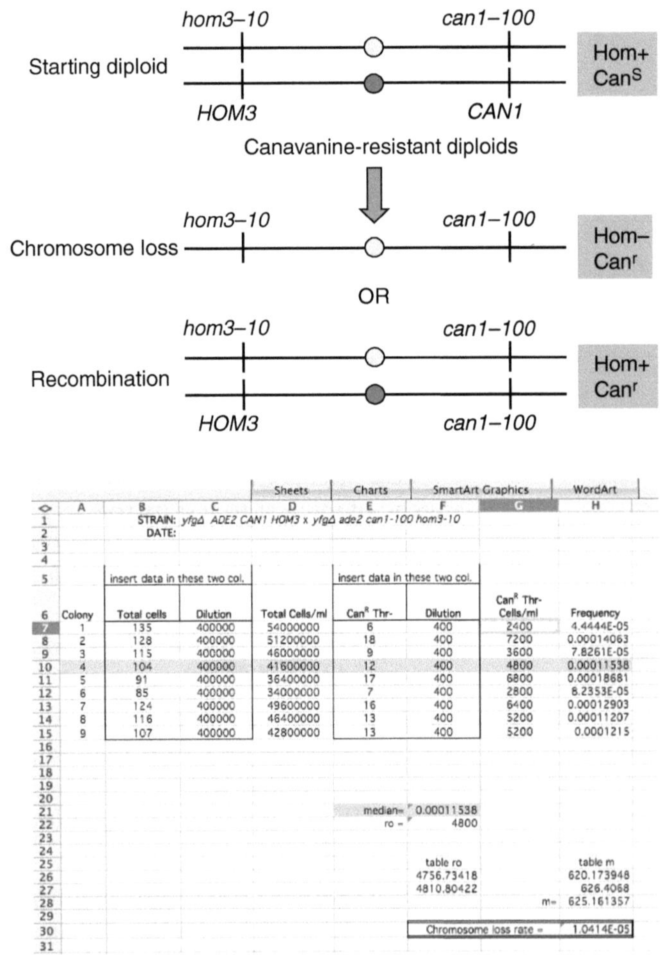

STRAIN: *yfgΔ ADE2 CAN1 HOM3* x *yfgΔ ade2 can1-100 hom3-10*
DATE:

	insert data in these two col.			insert data in these two col.			
Colony	Total cells	Dilution	Total Cells/ml	Can^R Thr-	Dilution	Can^R Thr- Cells/ml	Frequency
1	135	400000	54000000	6	400	2400	4.4444E-05
2	128	400000	51200000	18	400	7200	0.00014063
3	115	400000	46000000	9	400	3600	7.8261E-05
4	104	400000	41600000	12	400	4800	0.00011538
5	91	400000	36400000	17	400	6800	0.00018681
6	85	400000	34000000	7	400	2800	8.2353E-05
7	124	400000	49600000	16	400	6400	0.00012903
8	116	400000	46400000	13	400	5200	0.00011207
9	107	400000	42800000	13	400	5200	0.0001215
				median=	0.00011538		
				ro =	4800		
					table ro		table m
					4756.73418		620.173948
					4810.80422		626.4068
						m=	625.161357
					Chromosome loss rate =		1.0414E-05

Fig. 1.1. Schematic of the chromosome V markers and the selection for canavanine-resistant (Can^r) segregants are shown. Chromosome loss events are also threonine requiring (Hom$^-$), while recombination events are threonine prototrophic (Hom$^+$). Below the schematic an example of a fluctuation test spread sheet with the median frequency highlighted in *grey* is shown. YFG indicates your favorite gene.

2. Make 10-fold serial dilutions for each colony, up to 10^4 dilution.
3. Each plate is divided into quarters and 25 μl of each dilution is spread on one quadrant so that cultures from two diploids can be plated on one plate. Spread 25 μl of the 10^4 dilution from each diploid onto a SC plate and 25 μl of the 10^0 and 10^1 dilutions onto the SC+ CAN plate in order to get a reasonable number of colonies to count (somewhere between 10 and 100).
4. Incubate the plates at 30°C for 3 days and count the number of colonies that grow on SC+ CAN plates (N_{Can^r}) and the SC plates (N_{total}).
5. Replica-plate the SC+ CAN plates to SC-threonine plates and incubate at 30°C for one additional day.
6. Count the number of colonies that grow on the SC-threonine plates ($N_{\text{Can}^r\ \text{Thr}^+}$).
7. The number of colonies from chromosome loss events ($N_{\text{Can}^r\ \text{Thr}^-} = N_{\text{Can}^r} - N_{\text{Can}^r\ \text{Thr}^+}$) and the total number of colonies (N_{total}) for each diploid are entered into an Excel spreadsheet along with the dilution factor and event frequencies are calculated (*see* the example in **Fig. 1.1**). A rate is calculated from the median frequency using the equations (*see* **Note 2**) from the Lea and Coulson paper, which have been embedded into the Excel spread sheet. Chromosome loss events are detected by the above analysis, and other events such as recombination events consisting of crossovers and gene conversions, plus additional events ($N_{\text{Can}^r\ \text{Thr}^+}$), are not analyzed here, as they cannot be separately distinguished. Chromosome loss events can be verified by sporulation and dissection of the diploids, which will give two viable spores and two dead spores in each tetrad, or by CHEF gel analysis for chromosome copy number of the diploid segregant.

3.2. Diploid Recombination Assay (Gene Conversion)

To determine the recombination rate in diploids, we use diploids heterozygous at *LEU2* locus: *leu2-ecoRI/leu2-bstEII*. Diploids are obtained from zygotes, and we routinely perform three crosses, using different isogenic parental strains (usually three crosses for each assay):

1. Make diploids: cross YWT-3 and YWT-4 yeast strains with heterozygous alleles at the *LEU2* locus (*leu2-ecoRI/leu2-bstEII*) on the YPDA plate; pull nine or more zygotes for each of three crosses. Let the diploids grow for 3 days at 30°C (*see* **Note 1**).
2. Resuspend nine single diploid colonies each in 1 ml of dH_2O. Make 10-fold serial dilutions for each colony, up to the 10^4 dilution (*see* **Note 3**).

3. Spread 25 μl of 10^4 dilutions for each diploid onto the SC plates to calculate total number of cells per 1 ml and 25 μl of 10^0 and 10^1 dilutions onto the SC-leucine plates to calculate recombination rate (Leu2$^+$ colonies). Incubate for 3 days at 30°C (*see* **Note 4**).
4. Count the number of colonies on the SC plates and the number of Leu2$^+$ colonies on SC-leucine plates. The number of colonics for each diploid is entered into an Excel spreadsheet along with the dilution factor and event frequencies are calculated. The diploid gene conversion rate is calculated using the median method (7). The mean diploid gene conversion rate and standard deviation for each assay are calculated based on results from three tests.

3.3. Haploid Chromosome Fragment Loss Assay

Since haploid strains cannot lose a chromosome and remain viable, we monitor loss of a supernumerary chromosome fragment (*see* **Fig. 1.2**). As the fragment is smaller than a normal chromosome, it is less stable and is lost at a significant rate. Due to the high loss rate, the Lea and Coulson fluctuation test methods do not accurately measure the chromosome loss rate. Therefore we examine chromosome loss events that occur in one generation so that the loss frequency and the loss rate are identical, as described in a variation of this assay (8). The original chromosome fragment strain (YWT-5) was a gift from Dr. Symington. It contains a linear chromosome fragment (CF) vector (CFV/D8B-tg which contains the *URA3* and *SUP11* genes, *CEN4*, and an *ARS* element) derived as described (9). Appropriate haploid strains are made by crossing YWT-5 to a mutant strain of interest, followed by tetrad dissection and selection of spore colonies that are Ura$^+$Ade$^-$ white (due to partial suppression of the chromosomal *ade2-1* mutation by *SUP11*). Three different segregants of the same genotype are used for one assay:

1. Streak the YWT-5 strain onto a SC-uracil plate for 2–3 days for single colonies.
2. Pick up one single colony and grow in 5 ml liquid YPD overnight until OD_{600} = 0.5–0.6 (mid-log phase) (*see* **Note 5**).
3. Take 1 ml of culture, spin down at 3,000 rpm for 1 min.
4. Resuspend the cell pellet in 1 ml dH_2O (10^0 dilution).
5. Make 10-fold serial dilutions in 1 ml of dH_2O up to 10^4 dilution.
6. Spread 100 μl of 10^4 dilution onto each SC plate and spread all the 1 ml of 10^4 dilution using 10 plates in total.
7. Incubate plates at 30°C for 3 days. Four types of colonies grow: all white colonies that show no visible chromosome fragment loss, all red colonies that have lost the chromosome

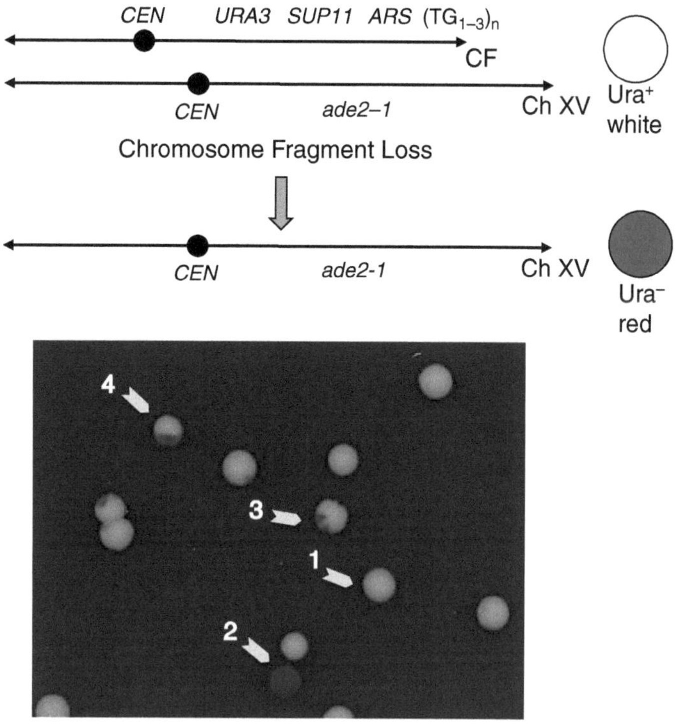

Fig. 1.2. Schematic of a chromosome fragment strain is shown with markers on the chromosome fragment and chromosome XV. Strains that have the chromosome fragment are *white*, as shown in colony 1 below. Strains that have lost the chromosome fragment are *red* seen as *dark grey* in the figure, as shown in colony 2 below. Strains that lose the chromosome fragment during division on the Petri plate are sectored for *red* (*grey*) and *white*, as shown in colonies 3 and 4 below. Chromosome fragment loss during the first cell division on the plate results in *red/white* (*grey/white*) half-sectors, as shown in colony 4 below. Chromosome fragment loss during later cell division on the place results in red (*grey*) sectors that are less than half the colony. The example shown in colony 3 has undergone two independent chromosome loss events to give two non-adjacent *red* (*grey*) sectors that are less than one-quarter of the colony.

fragment prior to plating, white colonies with red sectors that are less than half of the colony, indicating colonies that have experienced a chromosome loss after the first division on the plates, and colonies that are half red/white sectors, indicating chromosome fragment loss in the first division on the plates. These are the colonies of interest.

Count half-sector colonies and all viable colonies.

8. The chromosome fragment loss rate is determined by considering only the first cell division after plating and is calculated by dividing the total number of half-sectored colonies by the total number of colonies (white plus half-sectors plus partial sectors plus red):
   Chromosome fragment loss rate = number of half-sector colonies/total number of viable colonies.

9. The two-tailed Student's *t*-test is used to analyze significance between chromosome fragment loss rates.

3.4. Haploid Gene Conversion Assay

To determine the rates of intrachromosomal gene conversion, three different haploid strains (YWT-6) with the recombination system *leu2-ecoRI*:: *URA3*::*leu2-bstEII* are generated from crosses (*see* **Fig. 1.3**). Then each haploid strain is first grown on SC-uracil plates to ensure that the strain has the recombination reporter and then streaked on the YPDA plate for 2–3 days for single colonies.

For each test, nine colonies from one haploid strain are used and three individual haploid stains are used for one assay:

1. Pick nine colonies each from the YPDA plate and disperse into 1 ml sterile dH_2O (*see* **Note 3**).
2. Make 10-fold serial dilutions for each colony, up to the 10^4 dilution.
3. Each plate is divided into quarters and 25 μl of each dilution is spread on one quadrant so that cultures from two diploids can be spread on one plate. Spread 25 μl of the 10^4 dilution for each diploid onto the SC plate, 25 μl of the 10^0 and 10^1 dilutions (*see* **Note 4**) onto the SC-uracil-leucine plate, and 25 μl of the 10^1 and 10^2 dilutions (*see* **Note 4**) onto the SC + 5-FOA plate.

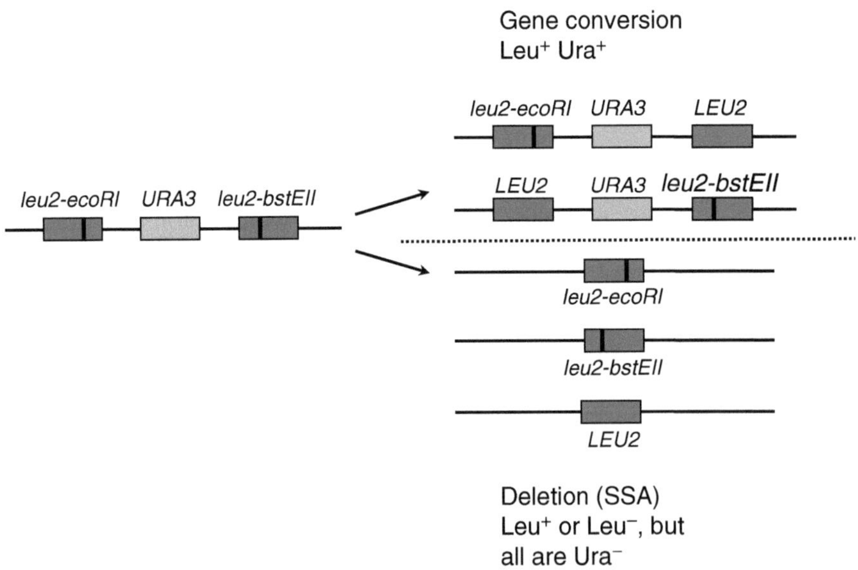

Fig. 1.3. Schematic for the intrachromosomal gene conversion assay is shown. Gene conversion events are detected as Leu$^+$ Ura$^+$ segregants, while deletion or single-strand annealing (SSA) events are Ura$^-$ and may be Leu$^+$ or Leu$^-$.

4. Incubate the plates at 30°C for 3 days and count the numbers of colonies that grow on the SC-uracil-leucine plates ($N_{Leu+Ura+}$), the SC + 5-FOA plates (N_{FOA^r}), and the SC plates (N_{total}).
5. The rates of intrachromosomal gene conversion are calculated from the frequencies of the Leu$^+$ Ura$^+$ mitotic segregants. $N_{Leu+Ura+}$ and N_{total} for each single colony are entered into an Excel spreadsheet along with the dilution factor and event frequencies are calculated. From the median frequency, a rate is calculated using the equations according to Lea and Coulson (7).
6. The rates of the recombination system *leu2-ecoRI::URA3::leu2-bstEII* events that are Ura3$^-$, considered to be single-strand annealing events, are calculated from the frequencies of the 5-FOA acid-resistant mitotic segregants. N_{FOA^r} and N_{total} for each single colony are entered into an Excel spreadsheet along with the dilution factor and event frequencies are calculated. From the median frequency, a rate is calculated using the equations according to Lea and Coulson (7).

3.5. Haploid Doubling Times

1. Strains are patched on the YPGA plate to ensure that there are no petite cells in the culture and grown for 1–2 days.
2. Cells from the YPGA plate are used to make cultures in liquid YPDA. The cultures are grown overnight at 30°C.
3. Each culture is resuspended at an OD_{600} of 0.05–0.07 and grown at 30°C during the day. The OD_{600} is taken every hour from 0 to 7 h.
4. The doubling time is calculated for log phase cells by converting the OD_{600} at 3 and 7 h into the number of cells. The time period is 240 min:

$$t_{doubling} = 240/\log_2(N_7/N_3)$$

5. The experiment is repeated three times. The mean doubling time and the standard deviation are then calculated.

4. Notes

1. Fresh diploids of 2–3 mm diameter in size are used for the assay. Most diploid strains will take 3 days to reach this size, but some mutant strains grow slower and will take 5 days to reach this size. Zygotes can be kept at 4°C for 2 additional days but not any longer, as some diploid strains, including

the W303 background, will sporulate on YPDA after several days and this will complicate chromosome loss rates which rely on the appearance of recessive markers.

2. The number of initial recombination events or chromosome loss events (m) is derived from the number of recombination or loss events (r) observed in median frequency sample. The equation is r/m–log m = 1.24. Rate = $m \times \ln 2/N$, where N is the total number of cells/ml in the median sample used to calculate m. The rate is measured as "events/cell/generation."
3. Sometimes the gene conversion rate is less than 10^{-6} and cannot be determined from suspension of a single colony in 1 ml dH_2O. In this case, nine single colonies are resuspended each in 5 ml liquid YPDA and grown overnight at 30°C to give more cells. The following morning, take 1 ml of each overnight culture, spin it down, resuspend in 1 ml dH_2O, and make 10-fold serial dilutions up to 10^5. Spread 25 μl of the 10^5 dilution for each diploid onto SC plate for the total number of colonies (N_{total}).
4. Two different dilution factors are used to get a reasonable range of colony numbers (10–100 colonies) for counting. Occasionally, the gene conversion or the deletion event being studied occurs early during growth of the colony, resulting in a large number of colonies growing on the selection plate, regardless of the dilution plated. These are called "Jackpot events." Since the Lea and Coulson method uses the median number (7), this will not affect the rate, but to facilitate calculations, we often enter a large number such as 1,000 into the Excel spread sheet and do not attempt to count the number of colonies growing on the selection plate.
5. To ensure that strains with different growth rates reach uniform OD_{600} following overnight incubation, single colonies are resuspended in YPDA and three serial dilutions are made. The cultures with the appropriate OD_{600} are used for the assay.

References

1. Kolodner, R.D., Putnam, C.D., and Myung, K. (2002) Maintenance of genome stability in *Saccharomyces cerevisiae*. *Science* **297**, 552–557.
2. Basrai, M.A. and Hieter, P. (1995) Is there a unique form of chromatin at the *Saccharomyces cerevisiae* centromeres? *Bioessays* **17**, 669–672.
3. Crouse, G.F. (2000) Mutagenesis assays in yeast. *Methods* **22**, 116–119.
4. Dion, B. and Brown, G.W. (2009) Comparative genome hybridization on tiling microarrays to detect aneuploidies in yeast. *Methods Mol Biol* **548**, 1–18.
5. Gordenin, D.A. and Resnick, M.A. (1998) Yeast ARMs (DNA at-risk motifs) can reveal sources of genome instability. *Mutat Res* **400**, 45–58.
6. Motegi, A. and Myung, K. (2007) Measuring the rate of gross chromosomal

rearrangements in *Saccharomyces cerevisiae*: a practical approach to study genomic rearrangements observed in cancer. *Methods* **41**, 168–176.
7. Lea, D.E. and Coulson, C.A. (1948) The distribution of the numbers of mutants in bacterial populations. *J Genet* **49**, 264–285.
8. Merker, R.J. and Klein, H.L. (2002) hpr1Delta affects ribosomal DNA recombination and cell life span in *Saccharomyces cerevisiae*. *Mol Cell Biol* **22**, 421–429.
9. Davis, A.P. and Symington, L.S. (2004) RAD51-dependent break-induced replication in yeast. *Mol Cell Biol* **24**, 2344–2351.

Chapter 2

Characterizing Resection at Random and Unique Chromosome Double-Strand Breaks and Telomere Ends

Wenjian Ma, Jim Westmoreland, Wataru Nakai, Anna Malkova, and Michael A. Resnick

Abstract

Resection of DNA double-strand break (DSB) ends, which results in 3′ single-stranded tails, is an early event of DSB repair and can be a critical determinant in choice of repair pathways and eventual genome stability. Current techniques for examining resection are restricted to model in vivo systems with defined substrates (i.e., HO-endonuclease targets). We present here a robust assay that can analyze not only the resection of site-specific DSBs which typically have "clean" double-strand ends but also random "dirty-ended" DSBs such as those generated by ionizing radiation and chemotherapeutic agents. The assay is based on our finding that yeast chromosomes with single-stranded DNA tails caused by resection are less mobile during pulsed-field gel electrophoresis (PFGE) than those without a tail. In combination with the use of a circular chromosome and enzymatic trimming of single-stranded DNA, resection of random DSBs can be easily detected and analyzed. This mobility-shift assay provides a unique opportunity to examine the mechanisms of resection, early events in DSB repair, as well as factors involved in pathway regulation.

Key words: DNA, double-strand break repair, resection, pulsed-field gel electrophoresis (PFGE), ionizing radiation, HO endonuclease, I-*Sce*I, mung bean nuclease, telomere.

1. Introduction

DNA double-strand breaks (DSBs) are among the most lethal and destabilizing DNA lesions that cells can encounter. They are induced by a variety of factors including ionizing radiation (IR), chemotherapeutic agents, endogenously arising reactive oxygen species, errors during replication such as fork collapse, as well as processing of closely spaced single-strand lesions (1). Two major

H. Tsubouchi (ed.), *DNA Recombination*, Methods in Molecular Biology 745,
DOI 10.1007/978-1-61779-129-1_2, © Springer Science+Business Media, LLC 2011

pathways have been identified to repair DSBs: non-homologous end joining (NHEJ) and homologous recombination (HR). DSB repair via the HR pathway is a multi-stage process using undamaged homologous DNA sequence as a template for accurate repair (2). An early step in this pathway involves resection of DSBs to produce 3′ single-stranded DNA tails that are critical for recombinational repair. The resected tails are utilized in strand invasion processes for priming repair synthesis and serve as a signal for checkpoint activation (3, 4).

Although long studied, mechanisms of resection have remained elusive, especially at the ends of random DSBs. To date, most studies on resection employ in vivo model systems with defined substrates such as DSBs induced by HO endonuclease or I-*Sce*I endonuclease (5). In these studies, the nuclease recognition site is placed at a defined location, and the cut is induced by the expression of a site-specific endonuclease (6–8). A direct approach for addressing resection involves a combination of restriction site analysis and probes to specific sequences at different distances from the DSB. Loss of restriction sites due to resection diminishes Southern blot hybridization signal (9, 10). Resection at a defined DSB can also be detected by using denaturing alkaline gels. In this case, the loss of restriction sites due to resection results in the formation of higher molecular weight bands that could be detected by sequence-specific probes. Finally, formation of ssDNA resection intermediates can be detected by slot blots which take advantage of the ssDNA binding to positively charged nylon membranes (whereas dsDNA cannot bind). The amount of ssDNA formed is determined by hybridization with strand-specific probes (11).

Both HO and I-*Sce*I recognize long nonpalindromic sequences and generate 4-bp staggered cuts with 3′-OH overhangs (12, 13). The DSB ends generated in this way are considered "clean" since they have 5′-P and 3′-OH groups suitable for ligation via end-joining processes or for priming DNA synthesis (14). However, most spontaneous or biologically relevant DSBs caused by environmental and therapeutic reagents such as IR, oxidative stress, and cancer drugs produce a variety of chemically modified termini or even protein–DNA adducts that cannot be directly ligated. These types of DSBs are referred to as "dirty" ends and require end processing by nucleases or other modifying enzymes to enable repair by HR or NHEJ (15). Analyzing the resection and repair of random "dirty" DSBs in vivo has been a challenge in the field. The appearance and repair of these types of DSBs can be determined qualitatively by the appearance of foci of proteins associated with DSB induction, such as H2AX chromatin modification, or foci appearing at various steps in repair (16, 17). However, there are few opportunities to address molecular events associated with random DSBs. Here we present a system capable of detecting resection at randomly induced DSBs as well as uncapped telomeres in addition to events at site-specific DSBs.

1.1. Large DNA Molecules with Single-Stranded Tail(s) Show Slower Mobility on PFGE

Pulsed-field gel electrophoresis (PFGE) is a widely used approach to monitor yeast chromosome changes since it permits very large DNA molecules to be resolved on agarose gels (for, e.g., *see* (18)). The system that we developed for the detection of resection is based on the finding that large chromosomal DNAs with single-stranded tails have significantly reduced mobility on PFGE. This mobility shift was observed in a study of the fate of radiation-induced DSBs in repair-deficient *rad50*, *rad51*, and *rad52* mutants of the yeast *Saccharomyces cerevisiae* (19). The repair was assessed by monitoring the fragmentation and restitution of full-size yeast chromosomes in nocodazole-arrested G2/M haploid yeast. Unexpectedly, *rad51* and *rad52* mutants showed a decrease in mobility of the smear of the chromosome fragments, initially interpreted as representing a low level of repair. There was no such PFGE mobility shift in the *rad50* mutant up to 4 h after irradiation. Further analysis that employed a circular chromosome and in vitro biochemical assays of the broken chromosome, as described below, demonstrated that the PFGE mobility change associated with the smear is due to the presence of single-stranded DNA (19).

1.2. Detecting Resection of Randomly Produced Single DSBs Using Circular Chromosome

The combination of PFGE along with an analysis of changes in circular chromosomes that have been broken provides the opportunity to study events at random DSBs (19). The use of a circular chromosome to detect a single DSB was initially developed in yeast by Game and colleagues (20). The principle of this method is that under most PFGE conditions a circular form of yeast chromosome is unable to move through the agarose matrix and is, therefore, retained in the loading well. However, the circle is converted into a full-length linear molecule by a single DSB, which enables the molecule to enter into the gel and give rise to a single band upon PFGE. The band is detectable either by Southern hybridization or by ethidium bromide. Since any single DSB on the circular chromosome leads to full-length linear DNA molecules of a uniform size, this approach provides the opportunity to address DSBs regardless of where they appear in a circular chromosome.

We recently found that the resection of IR-induced DSBs can be readily detected based on the shift in mobility of linearized circular chromosomes that have experienced a single DSB (19). **Figure 2.1** shows the "PFGE-shift" of the corresponding linear band that is seen in samples taken at various times after γ-irradiation (IR) of a recombination-deficient *rad52* mutant that is unable to repair DSBs. The yeast strain we constructed contained

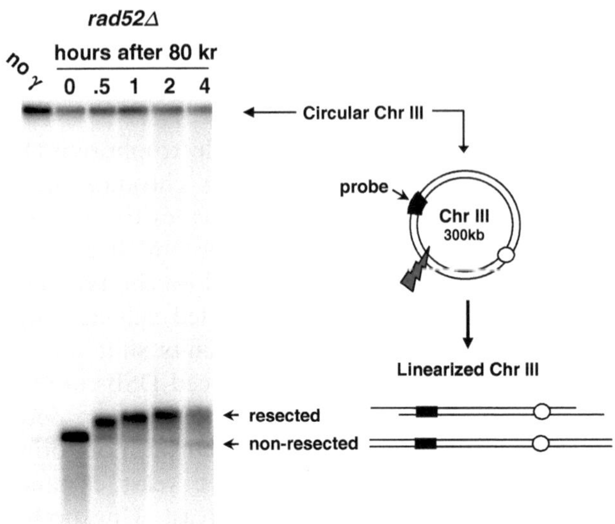

Fig. 2.1. PFGE-shift of circular chromosomes broken by IR-induced random DSBs. Nocodazole-arrested G2/M *rad52Δ* cells were irradiated with 80 krads, and post-IR incubation was done in YPDA medium. Cells were collected and prepared at the indicated times. Plug preparation and CHEF parameters are described in text. Chr III was detected by a probe targeting the *CHA1* gene. The circular (*unbroken*) form of Chr III is trapped in the well during PFGE. A single random DSB results in full-length linearized Chr III molecules that can migrate out of the well, forming a unique 300 kb band. The PFGE-shifted DNA corresponding to resected DNA reaches a plateau with "apparent" size of 430 kb. (This image is from 19.)

a circularized Chr III (~300 kb) (21). At "0" time after an 80 krad exposure an intense single band was detected (**Fig. 2.1**). The smear below this band corresponds to Chr III molecules with multiple DSBs. With time after post-irradiation incubation in YPDA, the DNA exhibited a shift that is clearly seen by 30 min after IR with further shift in PFGE mobility at 1 h reaching a plateau of ~430 kb *apparent size* by 4 h. We found that the increase in apparent size was actually due, paradoxically, to a loss in mass of the chromosomes due to resection, as described in the next section.

The resection is initiated uniformly and progresses at a comparable speed among the molecules examined based on the fairly sharp PFGE-shifted band at various times after irradiation. This also suggests that resection is not markedly affected by DNA sequence/structures. Nearly all the linearized molecules exhibited a shift by 1 h, independent of dose (19). The shift during post-irradiation incubation appears to occur even if the resected tail is a few hundred nucleotides based on the observation of shift in as little as 7.5 min after IR (19). The reasons for the shift remain to be established. The slower mobility of the resected DNA might be due to extension and contraction of single-stranded DNA (ssDNA) tails during PFGE providing stronger interactions than double-stranded DNA rods. It is also possible that secondary structure in the resected ssDNA contributes to its reduced migration.

This system based around PFGE-shift also has the potential to address the issue of whether resection of the two ends of the same DSB is coordinated or not. We found that the mobility shift of linear lambda DNA molecules with ssDNA tails generated in vitro at both ends moves much slower than molecules of the same length with ssDNA at only one end (19). This property provides a unique opportunity to address resection at both sides of a single randomly induced DSBs in circular molecules, where the two ends of the break are connected by the intervening intact DNA of the rest of the molecule. For example, for *rad50* mutants exposed to low IR doses, multiple PFGE-shift bands are detected that appear to be due to one- and two-end events (19). Since DSBs induced in linear chromosomes would result in the two ends becoming separated (the two fragments each bounded by a telomere at one end), it has not been possible until now to address events at both sides of the same DSB.

1.3. Measuring Resection Length Using a ssDNA-Degrading Enzyme

To establish that the PFGE-shift of the linearized molecules was due to resection, chromosomal DNA was treated with mung bean nuclease (MBN) in order to degrade the single-stranded tails. As shown in **Fig. 2.2** (using DNA from IR-exposed *rad52* cells), MBN treatment of the chromosomal DNAs within the plugs used for PFGE led to a reduction in the apparent MW of the Chr III linear molecules that showed PFGE-shift. This demonstrates that the PFGE-shift in radiation-broken chromosomes is due to the formation of ssDNA resulting from resection at the DSB ends. The mobility of the molecules at "0" time, when no resection is expected to occur, did not change with MBN treatment. The PFGE-shift in combination with MBN provides a sensitive method for measuring resection length and processing rate. In *rad52* cells treated with 80 krads, the resection rate was ~2 kb/h per DSB end. The opportunity to follow resection of random DSBs makes it possible to characterize the roles of different genetic components in DSB repair, especially the initial stage which is critical for signaling and repair pathway regulation.

1.4. Assessing Resection at Site-Specific DSBs and Telomeres

The resection-related PFGE-shift can be detected over a broad range of chromosome sizes that extends from tens of kilobytes (lambda DNA) to large chromosomes over 800 kb (e.g., yeast Chr II). The approach can also be employed to analyze resection at site-specific DSBs. We note that the induction of DSBs by ionizing radiation is "synchronous" in that they are induced simultaneously, unlike the enzymatically induced DSBs. Following induction of a single, DSB induced in a linear Chr III of G2/M yeast by HO endonuclease, we observed PFGE-shift with

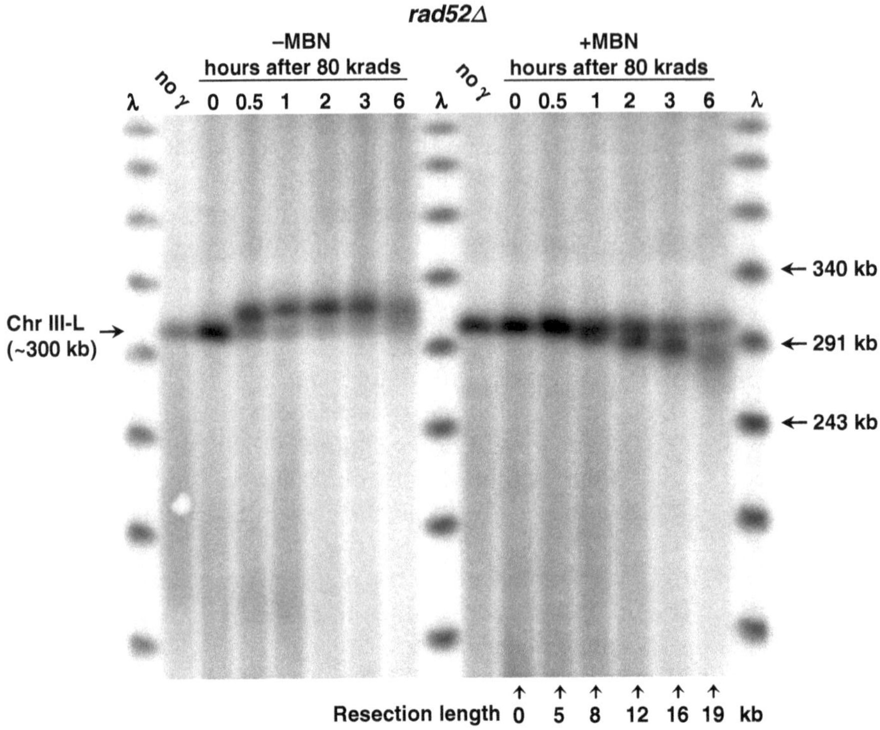

Fig. 2.2. PFGE-shift DNA is due to resection, based on mung bean nuclease treatment which can also be used to quantitate resection length. PFGE plugs from an experiment involving 80 krads to *rad52Δ* cells and post-irradiation incubation (such as that described in **Fig. 2.1**) were treated with MBN (+MBN lanes) or without MBN (−MBN lanes) and run on a CHEF gel. Chromosome bands after Southern blotting were detected by probing for the *LEU2* gene (*see* **Note 4**). The mung bean nuclease treatment (*right half* of image) abolished the PFGE-shift seen with untreated plugs (*left half* of image); the products ran at a faster rate than the unresected monomer in the 0 h lane. The numbers below each lane (*right half* of image) indicate the molecular weight change compared to the unresected linear Chr III band. The molecular weight of each band was calculated by comparing to positions identified in lambda DNA ladder (*first* and *last* lanes). (This image is from 19.)

kinetics similar to those for IR-induced DSBs under somewhat different PFGE conditions (19). The results obtained with an I-*Sce*I-induced DSB in Chr II (Nakai and Resnick, unpublished) using the PFGE procedures described here are presented in **Fig. 2.3**. Within 2 h after expression of I-*Sce*I, the two expected fragments (340 and 465 kb, respectively) were observed with the wild-type and the *rad50* null strains. PFGE-shift was detected in the ethidium bromide stained gels (and confirmed by Southern) for most of the broken molecules of the WT strain, but for less than half of the molecules in the *rad50* mutant.

The PFGE-shift phenomenon can also be used to distinguish events at uncapped telomeres of individual chromosomes. Using the temperature-sensitive mutant *cdc13-1*, which is deficient in telomere capping, we detected resection of telomeres at elevated temperatures as shown in **Fig. 2.4**. These findings are consistent with those of Maringele and Lydall (22, 23) using a very different

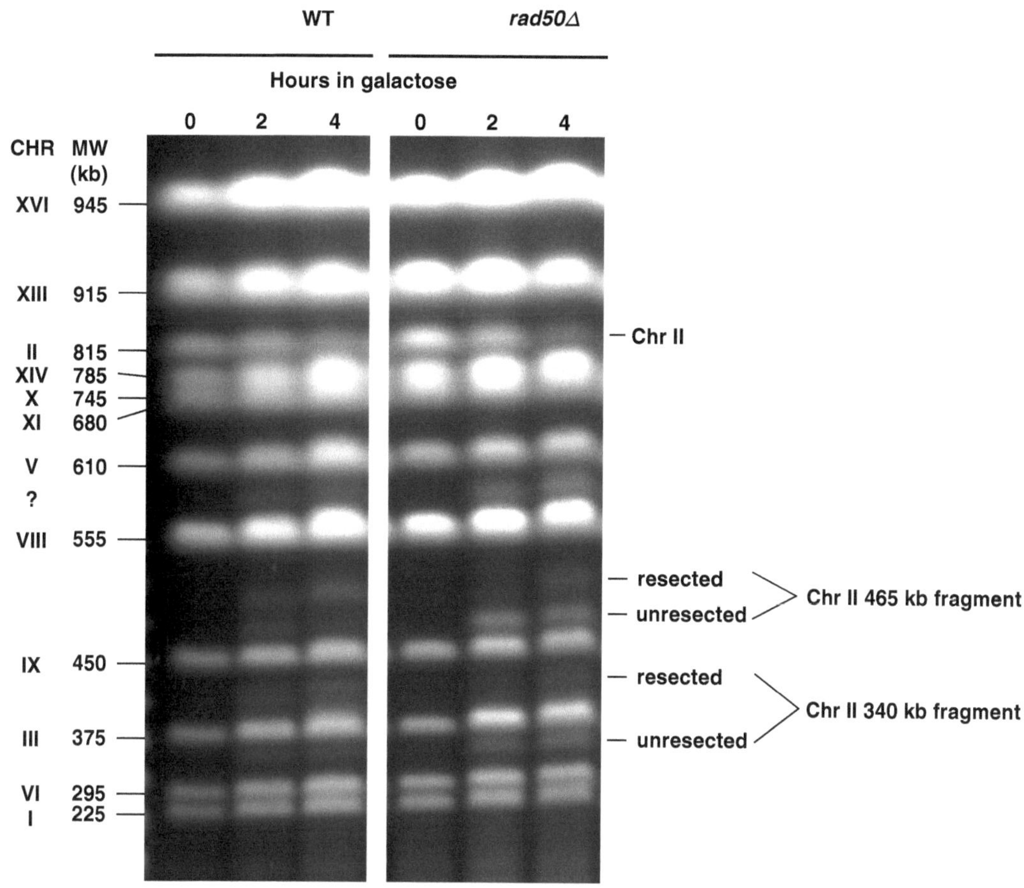

Fig. 2.3. PFGE-shift of chromosome fragments generated by an I-*Sce*I site-specific break is detected on ethidium bromide stained gels. The galactose-inducible I-*Sce*I endonuclease that cuts at a specific site engineered into Chr II was induced by transferring cells to galactose (*see* (23)). Samples were taken at 0, 2, and 4 h and analyzed by PFGE. The I-*Sce*I site is on Chr II (815 kb) and cuts the DNA into 340 and 465 kb fragments. The efficiency of I-*Sce*I cutting was 60% in WT and 80% in a *rad50*-null mutant at 4 h. Most of the fragments generated in WT cells within 4 h after transferring to galactose were shifted on PFGE (*left* image). However, for the *rad50* mutant, less than half of the molecules were shifted (*right half* of image). These results are consistent with those described by Westmoreland et al. (19) using an HO endonuclease acting at a different site and demonstrate with the PFGE-shift approach a role for the MRX complex in resection. (We note that in these experiments an unidentified fragment appeared between 555 and 610 kb as shown by the symbol "?" The origin of this cryptic target remains to be determined but the site of cutting is likely highly related to the I-*Sce*I site.) Experimental protocol: The experiment was performed at 30°C. Cells were grown overnight in YPDA medium, resuspended in YEP lactate medium (3.15% lactic acid, pH 5.5), and grown for an additional 18 h. The cells were then transferred to synthetic lactate medium (3.15% lactic acid, pH 5.5) containing 2% galactose. Cells were harvested at 0, 2, and 4 h and plugs were prepared for PFGE as described in the text.

approach that involves quantitative amplification of ssDNA (QAOS). Upon PFGE analysis, many chromosomes appeared as doublets. Based on Southern hybridization of Chr I (**Fig. 2.4**) there was, in fact, a doublet consisting of the original chromosome (230 kb) and an apparently larger version (~270 kb). This shift is considered to be due to the telomeres of this mutant becoming uncapped at 37°C and subject to resection by the repair

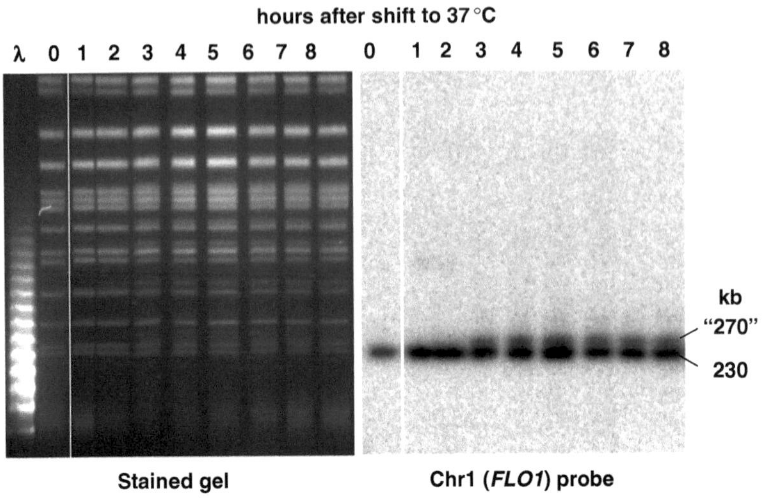

Fig. 2.4. PFGE-shift of uncapped telomeres. A temperature-sensitive *cdc13-1* strain, which is defective for telomere capping, was grown to stationary phase at permissive temperature, 23°C, then diluted 20-fold in fresh YPDA media at the nonpermissive temperature, 37°C, to induce telomere uncapping and subsequent 5′ to 3′ resection. (By 3 h, over 90% of the cells were arrested in G2.) Samples were collected at the indicated times following 37°C incubation. In the subsequent PFGE analysis, novel bands were observed at positions corresponding to molecular weights of ~40 kb above several of the chromosomal bands (*left image*). The shift in chromosomes was confirmed by Southern blot using a *FLO1* probe which is specific to Chr I (*right image*). This image is from (19). Likewise, shifts in Chromosomes II (813 kb), III (340 kb), V (576 kb), and VIII (565 kb) were also confirmed using chromosome-specific probes (data not shown). PFGE-shifts were not detected for cells incubated at the permissive temperature (data not shown). Although the image shown was obtained with a Beckman Geneline II TAFE system (no longer commercially available), we also have similar unpublished results with *cdc13-1* strains using CHEF. The TAFE running parameters were as follows: The first 18 h were run at constant current of 350 mA with 9 h of 60 s pulses, 3 h of 70 s pulses, 3 h of 80 s pulses, and 3 h of 90 s pulses. The remaining 6 h used 300 mA constant current and 4 min pulses.

system that deals with DSBs. Southern analysis of other chromosomes revealed that most (except Chr IV) exhibited a PFGE-shift (19). This approach for detecting resection at telomeres is expected to provide a useful tool for addressing mechanisms that maintain telomeres as well as the impact on genome stability of altered telomere metabolism.

2. Materials and Methods

2.1. Yeast Strains

All strains used here are haploids, although the approaches can be applied to diploid cells. Construction of strains containing circular Chr III (mwj49, mwj50, and derived yeast mutants) was described in (21). Construction of yeast strains for I-*Sce*I-induced DSBs (KS406 and derived mutants) was described in (24). Construction of strains containing the *cdc13-1 ts* mutation (DAG760) was described in (25).

2.2. Media and Solutions

2.2.1. Media for Yeast Cultures

1. YPDA: 1% yeast extract, 2% Bacto Peptone, 2% dextrose, and 60 μg/ml adenine sulfate, autoclave.
2. YEP lactate: 1% yeast extract, 2% Bacto Peptone, 3.7% lactic acid (pH 5.5), and 60 μg/ml adenine sulfate, autoclave.
3. Nocodazole stock solution: 10 mg/ml dissolved in DMSO; store at −20°C.

2.2.2. Solutions for PFGE and Southern Blotting

1. Cell suspension buffer: 10 mM Tris (pH 8.0), 100 mM EDTA, and 2 mM NaCl.
2. 2% low-melting agarose (LMP): 2% low-melting point agarose dissolved and melted in 10 mM Tris–HCl (pH 8.0), 100 mM EDTA.
3. Zymolyase: 1 mg/ml Zymolyase dissolved in 50% glycerol.
4. Agarose plug molds: see, for example, Bio-Rad, catalog no. 170-3622.
5. Proteinase K reaction buffer: 10 mM Tris (pH 8.0), 100 mM EDTA, 1.0% *N*-lauroyl sarcosine, 1 mg/ml proteinase K.
6. Plug washing buffer: 10 mM Tris, 50 mM EDTA (pH 8.0).
7. TBE 10X stock solution: 890 mM Tris base, 890 mM boric acid, 20 mM EDTA, pH 8.0.
8. TE buffer: 10 mM Tris, pH 7.4, 1 mM EDTA.
9. Mung bean nuclease (Promega, Madison, WI): stock solution 100 U/μl.
10. DNA detection: 10 mg/ml ethidium bromide solution or other DNA stains.
11. Southern blotting solutions. The following are used for Southern blotting: 0.25 N HCl; alkaline solution (0.4 N NaOH and 1.5 M NaCl); neutralizing buffer (0.5 M Tris–HCl and 1.5 M NaCl); 10× SSC (1.5 M NaCl, 0.15 M citrate, pH 7.0); Sigma PerfectHyb Plus hybridization buffer.

2.3. Probe to Detect Yeast Chromosome III

Chr III is detected by Southern blot with probes specifically targeting either the *CHA1* gene or the *LEU2* gene. The *CHA1* probe size is 279 bp, and the following primer pairs were used to amplify this fragment:

CHA1-5′: AACGGCCGTGATCTCTAATC

CHA1-3′: TCCAACGCTTCTTCCAAGTC

The *LEU2* probe size is 288 bp, and the following primer pairs were used to amplify:

LEU2-5′: TGTCAGAGAATTAGTGGGAGG

LEU2-3′: ATCATGGCGGCAGAATCAAT

2.4. Equipment and Other Materials

1. PFGE systems: transverse alternating field electrophoresis (TAFE) (Gene Line II apparatus from Beckman Instruments, Fullerton, CA, or equivalent) or contour-clamped homogeneous electric field (CHEF) (CHEF Mapper XA system from Bio-Rad, Hercules, CA, or equivalent).
2. Southern blotting apparatus and materials: UV crosslinker (Stratagene Stratalinker or equivalent); nylon membrane (Hybond N+, GE Healthcare or equivalent); Stratagene Prime-It RmT Random Primer Labeling Kit; ProbeQuant G-25 or G-50 Micro Columns; hybridization oven and bottles (260 × 40 mm); Whatman 3MM filter paper.

3. Methods

3.1. Cell Culture and Yeast Preparation

3.1.1. G2 Yeast Cell-Cycle Synchronization by Nocodazole

1. Growth: cells are grown logarithmically under aerobic conditions in liquid YPDA medium at 30°C to a concentration of 5–20 × 10^6 cells/ml.
2. Arrest at G2/M with nocodazole: nocodazole is added to a final concentration of 20 μg/ml and an additional 10 μg/ml every 1 h. Cells are incubated for 3 h at 30°C. Most cells are arrested in G2/M as determined microscopically by the presence of large budded cells and verification using flow cytometry.

3.2. Pulsed-Field Gel Electrophoresis (PFGE)

3.2.1. Preparation of Agarose-Embedded DNA (DNA Plug)

1. Prepare 2% low-melting agarose and keep it warm in a 55°C heat block.
2. Centrifuge ~1.2 × 10^8 cells and resuspend in cell suspension buffer at a total volume of 120 μl; add 20 μl Zymolyase (1 μg/μl), vortex and warm up to ~40–50°C using a heat block. Zymolyase should be added immediately prior to imbedding the cells in agarose (*see* **Note 1**).
3. Add 60 μl 2% agarose, quickly mix by gentle but thorough vortexing. Transfer the mixture to plug molds using sterile transfer pipettes (two plugs). Allow the agarose to solidify at room temperature or, to expedite this process, place the molds at 4°C for 10–15 min. (Note: this results in ~6 × 10^7 G2-arrested cells per 100 μl plug, which is the amount normally used in our experiments.)

4. Push the solidified agarose plugs into cell suspension buffer in a container such as multi-well tissue culture plate or conical centrifuge tube. Using ~1 ml for two plugs, incubate at 37°C for 1–2 h.
5. Remove cell suspension buffer and add 1 ml of Proteinase K reaction buffer for two plugs. Incubate the plugs overnight at 37°C without agitation.
6. Wash the plugs three to four times with plug washing buffer, 1 h for each wash at room temperature with gentle agitation.
7. Store plugs at 4°C. Depending on the type of DNA lesions induced, the plugs should be stable for a few weeks.

3.2.2. PFGE to Separate Yeast Chromosomes

The following protocol is for the preparation of a CHEF gel. The preparation of TAFE gels is similar and the running parameters for TAFE are provided in **Fig. 2.4**.

1. Preparation of gel casting stand with removable end plates (comes with the CHEF Mapper system) and comb. We found that a 3 mm thick preparative well comb (i.e., no teeth) is convenient for placing and organizing plugs during loading.
2. Melt 1% LE agarose (Seakem, Rockland, ME) in 0.5× TBE and pour into casting stand. While gel is solidifying, prepare 2.2 l 0.5× TBE running buffer and put into CHEF apparatus tank; cool to 14°C.
3. Take the DNA-containing agarose plug out of buffer; use a clean razor blade to cut out 1/4–1/2 size pieces (a thickness of ~2 mm); load into the bottom of a preparative well. Seal the well containing the plugs using 1% agarose and allow to set ~30 min at room temperature.
4. Install the gel from the casting stand into the PFGE electrophoresis tank according to CHEF Mapper instructions. Make sure the gel is not able to move or float during the electrophoresis. Equilibrate the gel placed in the tank with 14°C gel running buffer for 10 min before starting electrophoresis.
5. Run CHEF gel with appropriate conditions to separate the target DNA. For example, the following conditions can be used to separate all yeast chromosomes: 6 V/cm (120 V in the CHEF or DRII Bio-Rad units) at 14°C, 120° switch angle, switch time is ramped from 10 to 90 s over the 24 h run time.

3.3. Southern Blot and Hybridization

1. After electrophoresis, stain the gel for 60 min to overnight in 0.5× TBE with 1 μg/ml ethidium bromide. Destain in 0.5× TBE for 2–3 h and photograph the gel.

2. Rinse the gel briefly with water; add 0.25 N HCl to the tray containing the CHEF gel and gently shake forl 45–60 min.
3. Rinse gel briefly with water; treat with the alkaline solution for 30–60 min.
4. Neutralize with neutralization buffer for 30 min.
5. Cut a Hybond N+ membrane, wet first in water, and then soak for 10–15 min in 10× SSC.
6. Select a suitable method for Southern transfer. For example, use capillary method or a vacuum blotter according to the manufacturer's instructions.
7. Rinse the membrane with 10× SSC. Dry the membrane or UV-crosslink with Stratalinker (120 mJ/cm^2). Clearly mark the DNA side and top of the membrane.
8. Before hybridization, wet the membrane with 10× SSC and place it into a hybridization bottle.
9. Prehybridization: pour 15–20 ml of hybridization solution (e.g., PerfectHyb from Sigma) into the bottle (260 × 40 mm), add 10 μl of denatured salmon sperm DNA (10 mg/ml) per ml of hybridization buffer, and rotate at 68°C for 1 h in a hybridization oven.
10. Prepare radioactively labeled probe during prehybridization, using 50–100 ng of template DNA (preparation described in **3.4.1**). ^{32}P-labeled double-stranded DNA probe can be prepared by random priming using an appropriate commercial kit according to the manufacturer's instructions (e.g., Stratagene Prime-It RmT Random Primer Labeling Kit). Purify the radiolabeled probe using a gel filtration spin column (e.g., ProbeQuant G-50 or G-25 Micro Columns).
11. Denature the probe at 100°C for 10 min and quickly cool down in ice. Add denatured probe directly to the hybridization bottle with prehybridization solution. (No need to replace with fresh hybridization solution.) Rotate hybridization bottle at 68°C overnight (16–24 h).
12. Cold washes: discard the hybridization solution, put the membrane into a tray, add 300–400 ml of 2× SSC/0.1% SDS, and shake at ambient temperature for 30 min.
13. Stringent washes: add 400 ml of pre-warmed (68°C) 0.1× SSC/0.1% SDS into the tray, shake at 68°C for 20 min, two to three washes.
14. Wrap the blot with plastic wrap and expose to phosphor screen or film for 1–2 days.

3.4. Detection of DSBs and Resection by PFGE

3.4.1. Detection of Random DSBs Using Circularized Chromosome

1. Chr III is detected by Southern blot with probes that specifically target either the *CHA1* gene or the *LEU2* gene.
2. Amplify the *CHA1* or *LEU2* sequence by PCR using yeast genomic DNA and purify by agarose gel electrophoresis with an appropriate gel extraction kit. Use the purified DNA as template for secondary PCR with the same primers to prepare a large amount of probe for long-term use. Purify the second PCR product by gel extraction or PCR purification methods; dissolve in TE and store at –20°C.
3. After Southern transfer of DNA materials onto membrane, use the *CHA1* or *LEU2* probe for hybridization at a concentration of 5–10 ng probe/ml hybridization buffer (50–100 ng/hybridization tube). Autoradiographs can be analyzed by specific software such as Carestream MI.

3.4.2. Detection of Resection at Single, Random DSBs in Circularized Chromosomes by PFGE-Shift

The following protocol for DSB induction and repair is derived from studies that employed ionizing radiation. The method can be modified to detect random DSBs generated by other sources causing DNA damage such as chemotherapeutic reagents.

1. Harvest nocodazole-arrested G2 yeast by centrifugation (2,000×*g*, 2 min), wash once with water, and resuspend in ice-cold water at 5–10 × 10^7 cells/ml. Save 1.2 × 10^8 cells (for two DNA plugs as described below) to be used as the unirradiated control for PFGE.
2. Cell suspensions are kept on ice throughout the entire irradiation process. Irradiate cells at desired doses (we typically use 5–80 krads with a ^{137}Cs irradiator (J. L. Shepherd Model 431, 2.3 krads/min)) in plastic 50 ml tubes and vortex well every 10 krads exposure to assure good aeration.
3. Following irradiation, collect a volume corresponding to 1.2 × 10^8 cells, centrifuge and resuspend the pellet in ice-cold cell suspension buffer. These cells represent the time point "0" of DSB repair.
4. To address events during post-irradiation incubation, centrifuge the remaining cells and resuspend in YPDA with nocodazole (final concentration is 5–10 × 10^6 cells/ml) and incubate at 30°C with shaking. Since nocodazole is unstable in aqueous solution, add 10 μg/ml nocodazole every hour during incubation to maintain cells in G2/M.
5. At designated post-irradiation time points such as 30 min, 1 h, 2 h, collect cells to assess repair events. Cells (1.2 × 10^8 for two DNA plugs) are centrifuged and resuspended in ice-cold cell suspension buffer for DNA plug preparation as described above.

6. Run CHEF gel. The following parameters (or modify to other suitable parameters) can be used to detect DSBs (linear Chr III) and resected chromosomes (shifted band of linear Chr III): 6 V/cm, switch angle 120°, switch time 10–90 s with linear ramp, 24 h run time at 14°C with buffer recirculation (*See* **Note 2**).
7. Southern blot and hybridize with *CHA1* or *LEU2* probe.

3.4.3. Detection of Resection at Site-Specific DSBs Using PFGE-Shift

Procedures similar to those described above can be used to follow events at a site-specific DSB produced by galactose-induced HO endonuclease (19) or I-*Sce*I (Nakai and Resnick, unpublished, also *see* **Fig. 2.3**).

3.4.4. Detection of Resection at Telomeres Using PFGE-Shift

The following is an example of how to detect resection associated with unstable telomere ends using the temperature-sensitive yeast mutant *cdc13-1* which is defective for telomere capping (DAG760 described in (25)).

1. Grow *cdc13-1* cells to stationary phase for 3 days in YPDA medium at the permissive temperature, 23°C.
2. Dilute 20-fold into fresh YPDA and incubate at 37°C, a condition resulting in 5′ to 3′ resection at telomeres (22). Within 3 h, greater than 90% of the cells are arrested in G2.
3. Collect cells at different time points after shifting to 37°C along with control cells kept at 23°C. The cells are processed for PFGE and Southern blot analysis as described above.

3.4.5. Mung Bean Nuclease Digestion of DNA in PFGE Plugs to Identify Resection and Determine Length

Mung bean nuclease can be used to measure resection length. It removes the single-stranded resected ends that develop at DSBs. The nuclease generates blunt ends resulting in linear chromosomal DNAs with reduced length as exhibited by greater PFGE mobility. The resection length can be determined by comparing length after MBN treatment with the length at the time of DSB induction.

1. For MBN digestion of yeast plugs, cut the plug in half (50 μl) and put into a 96-well multi-well plate. Plugs are equilibrated with three changes (20 min) of 150 μl of TE at room temperature. The other half of the plug is used as a non-MBN control.
2. Remove the TE buffer and incubate with 40 U/ml of MBN in 150 μl of MBN reaction buffer for 20 min at room temperature with gentle shaking (*see* **Note 3**).
3. Quickly remove the MBN reaction solution and wash four times with ice-cold 50 mM EDTA to stop the reaction.
4. Preparation of CHEF gel, for sufficient separation of DNA at ~300 kb range, a long gel is preferred (using the 14 cm

(width) × 21 cm (length) casting stand from Bio-Rad, *see* **Note 2**). Load plugs (with and without MBN treatment) into the CHEF gel, using lambda DNA as the length marker. Run the gel with the following parameters: 6 V/cm (120 V in the CHEF and DRII Bio-Rad apparatuses), 120° switch angle, switch time 6–36 s, linear ramp, 48 h run time at 14°C with buffer recirculation.

5. Southern blot and hybridize with *LEU2* probe (*see* **Note 4**). The molecular weights associated with bands can be calculated using Kodak MI (version 4.0) software (Eastman Kodak Co., Rochester, NY) by comparing with positions in marker bands (lambda DNA ladder; New England Biolabs, Beverly, MA).

4. Notes

1. During plug preparation, the cell suspension after adding Zymolyase should be kept at 40–50°C as briefly as possible in order to minimize inactivation of enzymatic activity and avoid possible damage to DNA.
2. The 14 cm long by 21 cm wide gel can be used to visualize the PFGE-shift caused by resection. But for measuring resection length with mung bean nuclease, a 21 cm long gel should be used.
3. Mung bean nuclease should be used at low concentration (40 U/ml) and <30 min incubation to minimize the generation of nonspecific DSBs. The small amount of nonspecific activity at this low concentration does not interfere with the measurement of resection.
4. For measuring resection length, it is important to include appropriate DNA size standards on the same gel. In general, one needs to use two probes to visualize both the size marker and the chromosome in the autoradiograph of the Southern blot. We have found under our conditions of *LEU2* gene amplification, both the lambda DNA ladder (from NEB) and Chromosome III were detectable (*see* **Fig. 2.2**).

Acknowledgments

This work was supported by the Intramural Research Program of the NIEHS (NIH, DHHS) under project 1 Z01 ES065073 (MAR).

References

1. Ma, W., Panduri, V., Sterling, J.F., Van Houten, B., Gordenin, D.A., and Resnick, M.A. (2009) The transition of closely opposed lesions to double-strand breaks during long-patch base excision repair is prevented by the coordinated action of DNA polymerase delta and Rad27/Fen1. *Mol Cell Biol* **29**, 1212–1221.
2. Krogh, B.O., and Symington, L.S. (2004) Recombination proteins in yeast. *Annu Rev Genet* **38**, 233–271.
3. Zou, L. and Elledge, S.J. (2003) Sensing DNA damage through ATRIP recognition of RPA-ssDNA complexes. *Science* **300**, 1542–1548.
4. Haber, J.E. (2008) Evolution of models of homologous recombination. Genome Dynamics and Stability, Vol. **3** (Berlin/Heidelberg: Springer), pp. 1–64.
5. Mimitou, E.P. and Symington, L.S. (2009) DNA end resection: many nucleases make light work. *DNA Repair (Amst)* **8**, 983–995.
6. Kramer, K.M., Brock, J.A., Bloom, K., Moore, J.K. and Haber, J.E. (1994) Two different types of double-strand breaks in Saccharomyces cerevisiae are repaired by similar RAD52-independent, nonhomologous recombination events. *Mol Cell Biol* **14**, 1293–1301.
7. Plessis, A., Perrin, A., Haber, J.E. and Dujon, B. (1992) Site-specific recombination determined by I-SceI, a mitochondrial group I intron-encoded endonuclease expressed in the yeast nucleus. *Genetics* **130**, 451–460.
8. Haber, J.E. (2006) Transpositions and translocations induced by site-specific double-strand breaks in budding yeast. *DNA Repair (Amst)* **5**, 998–1009.
9. Ira, G., Pellicioli, A., Balijja, A., Wang, X., Fiorani, S., Carotenuto, W., Liberi, G., Bressan, D., Wan, L., Hollingsworth, N.M., et al. (2004) DNA end resection, homologous recombination and DNA damage checkpoint activation require CDK1. *Nature* **431**, 1011–1017.
10. Zhu, Z., Chung, W.H., Shim, E.Y., Lee, S.E. and Ira, G. (2008) Sgs1 helicase and two nucleases Dna2 and Exo1 resect DNA double-strand break ends. *Cell* **134**, 981–994.
11. Sugawara, N. and Haber, J.E. (2006) Repair of DNA double strand breaks: in vivo biochemistry. *Methods Enzymol* **408**, 416–429.
12. Colleaux, L., D'Auriol, L., Galibert, F. and Dujon, B. (1988) Recognition and cleavage site of the intron-encoded omega transposase. *Proc Natl Acad Sci USA* **85**, 6022–6026.
13. Kostriken, R. and Heffron, F. (1984) The product of the HO gene is a nuclease: purification and characterization of the enzyme. *Cold Spring Harb Symp Quant Biol*, **49** 89–96.
14. Daley, J.M., Palmbos, P.L., Wu, D., and Wilson, T.E. (2005) Nonhomologous end joining in yeast. *Annu Rev Genet* **39**, 431–451.
15. Wyman, C. and Kanaar, R. (2006) DNA double-strand break repair: all's well that ends well. *Annu Rev Genet* **40** 363–383.
16. Lisby, M., Barlow, J.H., Burgess, R.C. and Rothstein, R. (2004) Choreography of the DNA damage response: spatiotemporal relationships among checkpoint and repair proteins. *Cell* **118**, 699–713.
17. Rogakou, E.P., Pilch, D.R., Orr, A.H., Ivanova, V.S. and Bonner, W.M. (1998) DNA double-stranded breaks induce histone H2AX phosphorylation on serine 139. *J Biol Chem* **273**, 5858–5868.
18. Argueso, J.L., Westmoreland, J., Mieczkowski, P.A., Gawel, M., Petes, T.D. and Resnick, M.A. (2008) Double-strand breaks associated with repetitive DNA can reshape the genome. *Proc Natl Acad Sci USA* **105**, 11845–11850.
19. Westmoreland, J., Ma, W., Yan, Y., Van Hulle, K., Malkova, A. and Resnick, M.A. (2009) RAD50 is required for efficient initiation of resection and recombinational repair at random, gamma-induced double-strand break ends. *PLoS Genet* **5**, e1000656.
20. Game, J.C., Sitney, K.C., Cook, V.E. and Mortimer, R.K. (1989) Use of a ring chromosome and pulsed-field gels to study interhomolog recombination, double-strand DNA breaks and sister-chromatid exchange in yeast. *Genetics* **123**, 695–713.
21. Ma, W., Resnick, M.A. and Gordenin, D.A. (2008) Apn1 and Apn2 endonucleases prevent accumulation of repair-associated DNA breaks in budding yeast as revealed by direct chromosomal analysis. *Nucleic Acids Res* **36**, 1836–1846.
22. Maringele, L. and Lydall, D. (2002) EXO1-dependent single-stranded DNA at telomeres activates subsets of DNA damage and spindle checkpoint pathways in budding yeast yku70Delta mutants. *Genes Dev* **16**, 1919–1933.
23. Zubko, M.K., Maringele, L., Foster, S.S. and Lydall, D. (2006) Detecting repair intermediates in vivo: effects of DNA

damage response genes on single-stranded DNA accumulation at uncapped telomeres in budding yeast. *Methods Enzymol* **409**, 285–300.

24. Lobachev, K., Vitriol, E., Stemple, J., Resnick, M.A. and Bloom, K. (2004) Chromosome fragmentation after induction of a double-strand break is an active process prevented by the RMX repair complex. *Curr Biol* **14**, 2107–2112.
25. Yang, Y., Sterling, J., Storici, F., Resnick, M.A. and Gordenin, D.A. (2008) Hypermutability of damaged single-strand DNA formed at double-strand breaks and uncapped telomeres in yeast Saccharomyces cerevisiae. *PLoS Genet* **4**, e1000264.

Chapter 3

Characterization of Meiotic Recombination Initiation Sites Using Pulsed-Field Gel Electrophoresis

Sarah Farmer, Wing-Kit Leung, and Hideo Tsubouchi

Abstract

High levels of homologous recombination are induced during meiosis. This meiotic recombination is initiated by programmed formation of DNA double-strand breaks (DSBs) by a conserved meiosis-specific protein, Spo11. Meiotic DSBs are not formed at random along chromosomes but are formed in clusters known as recombination hot spots. To understand the regulation of this initiation step of meiotic recombination, determining the timing and location of meiotic DSBs is essential. In this chapter, we describe a method to detect genome-wide meiotic DSBs by using a combination of pulsed-field gel electrophoresis and Southern blotting.

Key words: Budding yeast, chromosomes, double-strand breaks, meiosis, pulsed-field gel electrophoresis, recombination, recombination hot spot, Spo11.

1. Introduction

Homologous recombination (HR) is essential for accurate segregation of chromosomes in meiosis (1, 2). HR plays two important roles in segregating homologous chromosomes at meiosis I. First, HR is used for homologous chromosomes to recognize each other. Second, HR between homologs leads to a fraction of crossovers which establish physical connections between them. This, along with sister-chromatid cohesion, provides tension at metaphase I by holding two homologs together, ensuring the faithful segregation of homologs at meiosis I.

H. Tsubouchi (ed.), *DNA Recombination*, Methods in Molecular Biology 745,
DOI 10.1007/978-1-61779-129-1_3,

The mechanism of meiotic recombination has been extensively characterized using budding yeast as a model system. HR is highly induced upon entry into meiosis and is initiated by programmed DSBs formed in early prophase I. These DSBs are not formed at random but are intensively formed at certain locations called recombination hot spots (3). The DSB ends are subject to exonucleolytic digestion in the 5′–3′ direction, leading to exposed 3′-ended single-stranded DNA (ssDNA) tails at their ends (4). This ssDNA is essential for the subsequent homology searching and strand exchange steps.

Meiotic DSBs are formed by a conserved meiotic protein, Spo11 (5, 6). Spo11 shows homology to the type II topoisomerase and after forming DSBs it stays covalently attached to their 5′-ends. Spo11 needs to be removed for subsequent resection to occur. The removal requires the Mre11/Rad50/Xrs2 complex and Sae2. In certain non-null mutants of *MRE11*, *RAD50*, and *XRS2*, and the null mutant of *SAE2*, DSBs accumulate with Spo11 attached to DSB ends.

Homology searching and strand exchange in meiotic recombination are mainly catalyzed by two RecA homologs, Rad51 and Dmc1 (7, 8). Dmc1 functions specifically in meiotic recombination, whereas Rad51 is involved in both mitotic and meiotic recombinations. In the absence of Dmc1, the function of Rad51 is blocked, leading to the accumulation of recombination intermediates before the strand-exchange steps (i.e., DSBs with 3′-tailed ssDNA).

Meiotic recombination initiation sites have been characterized by employing mutant backgrounds in which DSBs are not processed (e.g., *rad50S* and *sae2* null mutants), and thus DSB locations can be unambiguously determined. However, recent studies revealed that the amount and the distribution of DSBs differ, depending on the presence or the absence of resection at DSB ends; more DSBs are formed in the *dmc1* mutant, in which DSB ends are resected, than in *rad50S* or *sae2* null mutants, where Spo11 is still attached to DSB ends, blocking their subsequent resection (9, 10).

Pulsed-field gel electrophoresis combined with Southern blotting provides an effective way to identify meiotic DSB locations throughout the genome (11, 12). Pulsed-field gel electrophoresis is able to separate yeast chromosomes, whose size ranges from 100 kilobases to a few megabases. Formation of DSBs releases chromosome fragments that migrate faster on the gel. Thus, with a probe that recognizes one end of a chromosome, the location and magnitude of meiotic DSBs can be determined by Southern blotting. In this chapter, we describe a detailed protocol to determine the amount and locations of meiotic DSB, using meiotically synchronized budding yeast cell cultures.

2. Materials (*See* Note 1)

2.1. Meiotic Induction of Yeast Cells

1. YPADU: 1% Yeast extract, 2% Bacto peptone, 2% glucose, 0.3 mM adenine hemisulfate, 0.2 mM uracil (*see* **Note 2**). Autoclaved for 15 min at 121°C.
2. YPADU agar: YPADU, 2% agar. Autoclaved for 15 min at 121°C.
3. YPA: 1% Yeast extract, 2% potassium acetate, 2% Bacto peptone. Prepared and sterile filtered immediately prior to use (*see* **Note 3**).
4. Sporulation medium: 2% Potassium acetate (pH 6.5 with HCl). Autoclaved for 15 min at 121°C. Prewarmed to 30°C prior to use with SK1 strains.

2.2. Preparation of Sample Plugs

1. Plug buffer: 10 mM Tris–HCl (pH 7.5), 0.5 M EDTA (pH 8). 8 ml per sample (*see* **Note 4**).
2. Agarose solution: 1% (w/v) Low melting point agarose (InCert® agarose; Lonza), 125 mM EDTA (pH 8). 150 μl per sample.
3. Zymolyase solution: 0.8 mg/ml 100T zymolyase, 30 mM DTT, 125 mM EDTA. 75 μl per sample. Freshly prepared and kept on ice.
4. PK buffer: 1% (w/v) Sarkosyl (*N*-lauroylsarcosine sodium salt), 0.2% (w/v) proteinase K, 10 mM Tris–HCl (pH 7.5), 0.5 M EDTA (*see* **Note 4**). 1 ml per sample. Freshly prepared and kept on ice.

2.3. Pulsed-Field Gel Electrophoresis

1. TE-10: 1 mM Tris–HCl (pH 7.5), 0.1 mM EDTA. 15 ml per PFG.
2. 10× TBE stock: 10.8% (w/v) Tris base, 5.5% (w/v) sodium borate, 20 mM EDTA (pH 8).
3. PFG agarose mixture: 0.85% Invitrogen electrophoresis-grade agarose (*see* **Note 5**) in 0.5× TBE. 200 ml per PFG. Prepared, boiled, and kept at 50°C prior to use.
4. EtBr stain solution: 1 μg/ml Ethidium bromide. 200 ml per PFG. Mutagenic. Care should be taken with use and disposal should be undertaken according to workplace guidelines.

2.4. Southern Blotting

1. Depurination solution: 0.25 M Hydrochloric acid. 250 ml per PFG. Corrosive. Care should be taken with preparation, adding the stock solution of acid to premeasured water.

2. Denaturation solution: 0.4 M Sodium hydroxide. 1.5 l per PFG. Corrosive.
3. Blotting paper: 3 MM Whatman chromatography paper.
4. Positively charged nylon membrane: Hybond™-XL (GE Healthcare).
5. Labeling reaction: Rediprime™ II Random Prime Labelling System (GE Healthcare).
6. [α-^{32}P]dCTP (10 μCi/μl).
7. Sodium phosphate buffer (pH 7.2) 1 M stock solution: 9.71% (w/v) Na_2HPO_4, 4.36% (w/v) NaH_2PO_4.
8. Hybridization buffer: 7% (w/v) SDS, 0.5 M sodium phosphate buffer (pH 7.2), 1 mM EDTA. 30 ml per membrane. Prewarmed to 65°C before use. Care should be taken in preparation if powdered SDS is used.
9. Standard PCR reaction materials.
10. Gel purification kit, e.g., QIAquick® Gel Extraction Kit (Qiagen), used according to the manufacturer's instructions.
11. TE: 10 mM Tris–HCl (pH 8), 1 mM EDTA (pH 8).
12. Wash buffer: 1% SDS, 40 mM sodium phosphate buffer (pH 7.2), 1 mM EDTA. 800 ml per membrane. Prewarmed to 65°C before use. Care should be taken in preparation if powdered SDS is used.

3. Methods

3.1. Synchronous Meiotic Induction – SK1 Background

1. Diploid SK1 cells are streaked onto YPADU agar and incubated at 30°C for 2–3 days.
2. Single colonies (*see* **Note 6**) are cultured individually for 24 h (*see* **Note 7**) at 30°C in 10 ml YPADU in 100-ml flasks.
3. The saturated cultures are used to inoculate 100 ml fresh YPA in 1-l flasks (*see* **Note 8**) to absorbances (A_{595}) of 0.2. These premeiotic cultures are incubated for approximately 11 h, shaking vigorously at 30°C.
4. Premeiotic cultures measuring $2.0 < A_{595} < 4.0$ after 11 h and comprising >80% large, unbudded cells are selected for meiotic induction. Cells are rapidly washed twice in 50 ml distilled water, pre-equilibrated to 30°C, and finally resuspended in 100 ml of 30°C-pre-equilibrated sporulation medium in a 1-l flask (*see* **Note 8**) to an absorbance of 1.7. A 10 ml "time zero" sample is extracted (10 ml

meiotic culture equates to two PFG plugs) and the remainder of the meiotic cultures is incubated shaking vigorously at 30°C.

5. "Time zero" samples are washed twice in an equal volume of distilled water and once in 1 ml distilled water (*see* **Note 9**). Following a final wash in 1 ml plug buffer or distilled water (*see* **Note 10**) and transfer to 1.5- or 2-ml tubes, cell pellets are stored at –80°C.
6. Further time-point samples are taken as appropriate (*see* **Notes 11** and **12**). Cells are washed once in 1 ml distilled water or plug buffer, transferred to 1.5- or 2-ml tubes, pelleted, and stored at –80°C.

3.2. Meiotic Induction – BR Background

1. Diploid BR cells are streaked onto YPADU agar and incubated at 30°C for 2–3 days.
2. Single colonies are cultured individually overnight, shaking at 30°C in 4 ml YPADU.
3. 10 ml YPADU is added to the saturated cultures, which are incubated for 8 h (*see* **Note 13**), shaking vigorously at 30°C.
4. Cells are washed once in 50 ml distilled water and resuspended in 60 ml sporulation medium in 500-ml flasks (*see* **Note 14**). A "time zero" sample is extracted (10 ml meiotic culture equates to two PFG plugs) and the remainder of the meiotic cultures are incubated with vigorous shaking at 30°C.
5. "Time zero" samples are washed twice in an equal volume of distilled water and once in 1 ml distilled water (*see* **Note 9**). Following a final wash in 1 ml plug buffer or distilled water (*see* **Note 10**) and transfer to 1.5- or 2-ml tubes, cell pellets are stored at –80°C.
6. Further time-point samples are taken as appropriate (*see* **Note 15**). Cells are washed once in 1 ml plug buffer, transferred to 1.5- or 2-ml tubes, pelleted, and stored at –80°C.

3.3. Preparation of Agarose–Cell Plugs

1. Agarose solution is prepared and placed in a beaker of water, microwaved carefully until all agarose is dissolved, and is then kept at 50°C in a hotblock.
2. The base of the plug molds is sealed with tape and zymolyase solution is prepared and kept on ice.
3. Cell pellets are thawed on ice, resuspended in 66 μl (per 10 ml original sample) zymolyase solution, and allowed to come to room temperature.
4. 134 μl Agarose solution (per 10 ml original sample) is quickly mixed into the cell suspension by pipetting gently and is dispensed into two plug molds (*see* **Note 16**).

5. Plugs are allowed to solidify at room temperature or at 4°C (*see* **Note 17**). Excess agarose is scraped off the tops of the molds with a scalpel and the tape is peeled from the base. Plugs are expelled from molds into 1 ml plug buffer (per 2 plugs) in round-bottomed tubes by creating a seal over the exposed base of the plug with the rubber bulb from a small glass pipette, wetted with plug buffer, and squeezing the bulb.
6. Plugs are incubated in plug buffer at 37°C for 3 h. PK buffer is prepared and kept on ice.
7. Plug buffer is aspirated and replaced with 1 ml PK buffer (per 2 plugs) and plugs are incubated at 55°C overnight.
8. Plugs are washed four times for approximately an hour each time at 4°C in plug buffer and can then be stored in plug buffer at 4°C for months.

3.4. Pulsed-Field Gel Electrophoresis

1. 2.2 l of 0.5× TBE buffer is prepared from TBE 10× stock solution.
2. 200 ml PFG agarose mixture is boiled in a microwave until the agarose is completely melted, cooled to approximately 55°C, and poured into an assembled PFG gel cast and allowed to solidify.
3. While the PFG solidifies, the plugs to be run are equilibrated in 1 ml TE-10 for 30 min at room temperature.
4. The CHEF tank is filled with 2 l of 0.5× TBE buffer, the pump is operated at maximum (*see* **Note 18**), and the tank buffer cooled to 14°C.
5. The comb is removed from the PFG. Plugs are inserted into, and pushed to the bottom of, the wells with a slim spatula and a pipette tip is used to help dislodge bubbles (*see* **Note 19**).
6. The gel is submerged in the CHEF tank or other similar apparatus and the appropriate program is run (*see* **Note 20**).
7. The PFG is transferred into approximately 200 ml EtBr stain solution or enough to cover the gel, preferably in a lidded container which may be drained without tipping, and is agitated on a rotary platform for at least 30 min.
8. EtBr stain solution is drained and the PFG is washed twice rapidly in distilled water (*see* **Note 21**).
9. The ethidium bromide-stained DNA is visualized in a UV transilluminator (*see* **Note 22**).

3.5. Southern Blotting

1. The whole PFG is submerged in depurination solution for 15 min (*see* **Note 23**), agitating on a rotary platform.
2. Depurination solution is drained and replaced with denaturation solution for a further 15 min gentle agitation.
3. Step 2 is repeated (*see* **Note 24**).
4. During denaturation, the blotting base is constructed: a platform is set up over a pool of approximately 1 l denaturation solution and a large strip of blotting paper measuring 21 cm by length long enough to span the platform and extending at least 4 cm into the solution at either end (*see* **Note 25**) and a 21-cm × 15-cm piece are wet in denaturation solution and arranged on the platform.
5. The PFG is carefully flipped over and placed on the prepared platform so that the former top of the PFG is in contact with the blotting paper. Bubbles are expelled from under the PFG by smoothing clean, gloved fingers over its surface.
6. Cling film is stretched over the entire blotting base and a razor blade is used to cut out the piece in contact with the PFG surface, leaving a few millimeter of cling film overlapping the edges of the PFG (*see* **Note 26**). It does not matter if the blade cuts into the agarose. The central rectangle of cling film is removed to expose the area which is intended for blotting.
7. A piece of positively charged nylon membrane of the size of the PFG (thus slightly larger than the cling film-less window) is wet in denaturation solution and aligned on top of the PFG.
8. A piece of blotting paper of the size of the PFG is wet in denaturation solution and placed on top of the construction and gloved fingers are used to expel bubbles between the membrane and the PFG and between the membrane and the blotting paper.
9. Step 9 is repeated with a second piece of blotting paper.
10. Paper towels are stacked onto the assembled blot to a height of at least 10 cm and are topped by a flat weight.
11. The assembled blotting apparatus is left for 12–24 h, usually overnight.

3.6. Membrane Preparation for Probe Hybridization

1. A hybridization oven is brought to 65°C and 30 ml hybridization buffer is warmed to 65°C.
2. The blotting apparatus is deconstructed and the membrane is upturned (so that the surface previously in contact with the PFG faces upward) onto a fresh piece of blotting

paper and is UV irradiated at an energy of 120 J/cm^2 (*see* **Note 27**).

3. The membrane is rolled, with the cross-linked DNA surface inward, and inserted into a hybridization bottle containing 15 ml hybridization buffer at 65°C. The membrane is incubated, rotating at 65°C for at least 30 min.

3.7. Probe Preparation

1. Genomic DNA is amplified using a standard PCR reaction protocol with the appropriate oligonucleotide primer pair.
2. The completed reaction is run on a standard agarose gel and the separated PCR product band is excised and purified, for instance, using the Qiagen gel purification kit, according to the manufacturer's instructions.
3. Purified PCR product is diluted in TE to a final volume of 45 μl and a final concentration of about 25 ng/μl and is boiled for 5 min at 95°C and quenched on ice.
4. The denatured DNA is added to one aliquot of labeling reaction on ice and the labeling reaction mixture is gently pipetted up and down to resuspend (*see* **Note 28**).
5. 5 μl (50 μCi) [α-^{32}P]dCTP is added to the labeling reaction, which is pipetted up and down to mix and incubated at 37°C for approximately 20 min (*see* **Note 29**).
6. The labeling mix is boiled at 95°C for 3 min, quenched on ice, and spun down to the bottom of the tube briefly in a benchtop centrifuge. The labeled probe is kept on ice until use.

3.8. Hybridization and Imaging of Chromosomes

1. The hybridization buffer is carefully drained from the membrane, discarded, and replaced with the second 15 ml hybridization buffer, pre-equilibrated to 65°C. The labeled probe is added (*see* **Note 30**) and the hybridization bottle is rotated at 65°C, for 12–24 h, usually overnight.
2. Wash buffer is pre-equilibrated to 65°C.
3. Hybridization buffer is drained from the membrane and is safely discarded.
4. The hybridization bottle is half-filled with 65°C pre-equilibrated wash buffer and is upended several times.
5. Wash buffer is discarded and replaced with a second half-bottle volume of wash buffer. The hybridization bottle is rotated at 65°C for approximately 3 min.
6. Step 5 is repeated with a wash length of 30 min.
7. The membrane is removed from the bottle and is submerged in wash buffer in a tray for a brief period of up to a few minutes.

8. The membrane is wrapped in unwrinkled cling film, the edges of which are folded to create a water-tight pouch, and a phosphor screen (*see* **Note 31**) is exposed to the membrane for hours to days, usually overnight, depending on the radioactive decay of the [α-^{32}P] dCTP, prior to the experiment, the efficiency of the probe labeling and hybridization processes, and the desired signal strength (*see* **Note 32**).
9. The phosphor screen image is detected using a phosphoimager (*see* **Note 33**) and, if desired, quantified, e.g., using QuantityOne® (Bio-Rad) software (*see* **Note 34**). *See* **Fig. 3.1** for an example image.

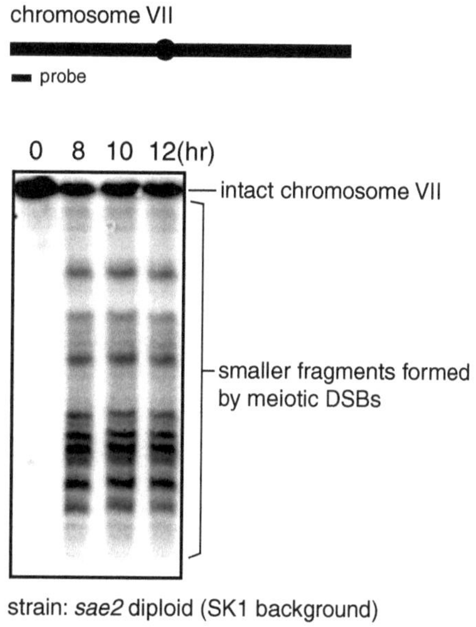

Fig. 3.1. An example of the visualization of a single budding yeast chromosome. Diploid *sae2* mutant was introduced into meiosis and samples taken at 0, 8, 10, and 12 h after introduction into meiosis. Chromosomal DNA was prepared, separated by pulsed-field gel electrophoresis, and chromosome VII was visualized by Southern blotting. The used probe recognizes the region from 14,960 to 16,234 on chromosome VII.

4. Notes

1. All buffers and solutions are made up in distilled water with a resistance of 18.2 MΩ/cm and are stored indefinitely at room temperature unless otherwise noted.
2. This preparation is our laboratory standard but, for SK1 cells, the addition of uracil is likely unimportant. Most BR

strains are *ade2 ura3* and grow significantly better in media supplemented with adenine and uracil.

3. We have found it to be extremely helpful in obtaining pre-meiotic cultures of the desired density for induction of synchronous meioses.
4. A slightly lower molarity of EDTA is permissible where a 0.5 M stock solution is being used to prepare the buffer.
5. This is the most economical agarose tested. Other types and brands of agarose may be employed but the gel percentage should be optimized if the degree of separation of chromosome bands is critical. 1% SeaKem® HGT or Gold (specifically formulated for PFGE) agaroses give roughly equivalent genomic "ladders" to 0.85% Invitrogen electrophoresis-grade agarose (but no observed upgrade in quality).
6. SK1 cells notoriously regularly produce "petite" colonies. These are usually smaller and whiter and should be avoided, as "petite" cells do not sporulate. If there is doubt as to the cells' mitochondrial function, it should be confirmed by growth on YP agar containing a non-fermentable carbon source such as glycerol or lactate.
7. This period is not critical; variation of several hours on either side can be tolerated.
8. A 100 ml culture will allow at least nine samples to be taken (each sample generating two PFG plugs). If more samples or plugs are desired, this volume can be scaled up. The volume of the flask should be 10 times the volume of the culture to allow adequate aeration.
9. We have observed artifactual smear-like signals in PFG "time zero" lanes which seem to be reduced with increased washing of these samples.
10. The EDTA in plug buffer should better preserve the DNA but we have observed no degradation using distilled water at this stage and furthermore, we have found cells washed only with distilled water easier to resuspend at later stages.
11. Due to the "clumpy" nature of SK1 cells, prior to samples being taken, cells "climbing" the sides of the flasks should be thoroughly resuspended.
12. SK1 meiotic cultures should sporulate synchronously (synchrony should always be verified, if critical, as the synchronous induction protocol can be "temperamental" (e.g., *see* **Note 3**)). DSBs should appear between 2 and 4 h (this can vary from experiment to experiment even using the same strain). Despite employing DSB-processing mutants, as described here, we have nevertheless found it

useful to determine the synchrony of meiotic cultures by taking samples for PFG analysis at 2, 3, and 4 h as well as at 5 h, when DSBs are usually accumulated. We routinely analyze 8 and 12 h samples to ensure that DSBs are fully accumulated.

13. This period is not critical; variation of an hour on either side can be tolerated.
14. A 60 ml culture will allow at least five samples to be taken (each sample generating two PFG plugs). If more samples or plugs are desired, this volume can be scaled up. The volume of the flask should be 10 times the volume of the culture to allow adequate aeration.
15. BR meiotic cultures do not sporulate synchronously but will usually be enriched for prophase I cells at 15 h and, when employing DSB-processing mutants, 24 h samples can be analyzed to ensure that DSBs are fully accumulated.
16. The plug volume tends to decrease slightly while solidifying, so dispense any extra cell-agarose mixture on top of the molds (excess will be removed in step 5 of **Section 3.3**).
17. Plugs will rapidly dehydrate, so do not allow to stand for long once solidified.
18. The pump needs to be operated only to the level required to maintain a tank buffer temperature of 14°C, but care must be taken that it does not pump so slowly that the refrigeration system freezes.
19. This step is technically difficult at first. The plugs are fragile and can disintegrate if not treated with care. Extra plugs made ready might be useful for a first attempt.
20. The program to be selected depends on the desired degree of separation of the chromosomes and whether an overall impression of the level of DSBs or an in-depth analysis of the DSBs formed on a single chromosome (population) is preferred. For typical analyses, the following initial and final switching times (seconds) are employed: 20 and 70 for chromosomes XII and IV; 5 and 30 for chromosomes I, VI, and III; 20 and 60 for the rest of chromosomes. Electric field and run time are 6 V/cm and 24 h, respectively.
21. Due to the mutagenic properties of ethidium bromide, care must be taken when handling and disposing of not only EtBr stain solution but also distilled water washes, and with all subsequent steps of the protocol.
22. A further stain period is possible if the signal is inadequate and, if there are time constraints, the gel may be stored either in EtBr stain solution or in 0.5× TBE at 4°C at least overnight.

23. This period is critical and must not be exceeded. Depurination nicks the DNA, which, following subsequent denaturation, generates short fragments of single-stranded DNA which can be more readily transferred to the membrane.
24. The two denaturation solution washes can vary in length but should total 30 min.
25. The large strip of blotting paper acts as a wick, allowing solution to be sucked upward, by capillary transfer action, through the PFG and membrane.
26. The overlap is required because of the importance of maintaining a barrier between the absorbent upper layers of the blot construction and the lower wick and sink so that the capillary transfer action does not bypass the DNA to be transferred.
27. Other protocols suggest methods which allow storage of the membrane at the stage, such as washing in 2× SSC followed by dry storage. While this can work, we have found the protocol more reliable when continuing straight to the hybridization stage.
28. For purposes of economy, one aliquot can be split between two (or even three) probes if desired. For two probes, one aliquot of labeling reaction is resuspended in 20 μl TE and divided between two screw cap tubes on ice. The PCR products are each boiled for 5 min at 95°C in 12.5 μl TE at a concentration of about 15 ng/μl in PCR tubes in a PCR machine and quenched on ice before being added to the labeling mix. (Subsequently, 2.5 μl (25 μCi) [α-^{32}P]dCTP is added to each.)
29. This step and subsequent steps should be carried out in areas and with equipment designated for "hot" work and exposure to radiation should be monitored according to workplace guidelines.
30. It is advisable not to pipette the probe directly onto the membrane. It may be added either to the hybridization buffer before it is poured into the hybridization bottle or part of the bottle where it can be mixed with the hybridization buffer before being directly in contact with the membrane (preferable, in order to minimize personal exposure, but sometimes difficult with large membranes).
31. Autoradiography can also be used to visualize the hybridization of the probe to the membrane but is less useful for quantification of the signal.
32. It is important that the signal is not saturated if it is intended for quantification, but re-exposure of the

membrane to the phosphor detector screen (and even re-washing of the blot if required) is simple.

33. A resolution of 100 μm is adequate for most purposes.
34. Quantification for comparison between lanes, at its simplest, can be the amount of DSB (i.e., the total lane signal minus the "uncut" parental band signal) expressed as a ratio of the total lane signal.

Acknowledgments

We would like to thank Prof. Alan Lehmann for critical reading of the manuscript. This work was supported by a Marie Curie Cancer Care transitional program grant.

References

1. Gerton, J.L. and Hawley, R.S. (2005) Homologous chromosome interactions in meiosis: diversity amidst conservation. *Nat Rev Genet* **6**, 477–487.
2. Roeder, G.S. (1997) Meiotic chromosomes: it takes two to tango. *Genes Dev* **11**, 2600–2621.
3. Petes, T.D. (2001) Meiotic recombination hot spots and cold spots. *Nat Rev Genet* **2**, 360–369.
4. Sun, H., Treco, D., and Szostak, J.W. (1991) Extensive 3′-overhanging, single-stranded DNA associated with the meiosis-specific double-strand breaks at the ARG4 recombination initiation site. *Cell* **64**, 1155–1161.
5. Bergerat, A., de Massy, B., Gadelle, D., Varoutas, P.C., Nicolas, A., and Forterre, P. (1997) An atypical topoisomerase II from Archaea with implications for meiotic recombination. *Nature* **386**, 414–417.
6. Keeney, S., Giroux, C.N., and Kleckner, N. (1997) Meiosis-specific DNA double-strand breaks are catalyzed by Spo11, a member of a widely conserved protein family. *Cell* **88**, 375–384.
7. Sehorn, M.G. and Sung, P. (2004) Meiotic recombination: an affair of two recombinases. *Cell Cycle* **3**, 1375–1377.
8. Masson, J.Y. and West, S.C. (2001) The Rad51 and Dmc1 recombinases: a non-identical twin relationship. *Trends Biochem Sci* **26**, 131–136.
9. Buhler, C., Borde, V., and Lichten, M. (2007) Mapping meiotic single-strand DNA reveals a new landscape of DNA double-strand breaks in *Saccharomyces cerevisiae*. *PLoS Biol* **5**, e324.
10. Blitzblau, H.G., Bell, G.W., Rodriguez, J., Bell, S.P., and Hochwagen, A. (2007) Mapping of meiotic single-stranded DNA reveals double-stranded-break hotspots near centromeres and telomeres. *Curr Biol* **17**, 2003–2012.
11. Game, J.C. (1992) Pulsed-field gel analysis of the pattern of DNA double-strand breaks in the *Saccharomyces* genome during meiosis. *Dev Genet* **13**, 485–497.
12. Zenvirth, D., Arbel, T., Sherman, A., Goldway, M., Klein, S., and Simchen, G. (1992) Multiple sites for double-strand breaks in whole meiotic chromosomes of *Saccharomyces cerevisiae*. *EMBO J* **11**, 3441–3447.

Chapter 4

Genome-Wide Detection of Meiotic DNA Double-Strand Break Hotspots Using Single-Stranded DNA

Hannah G. Blitzblau and Andreas Hochwagen

Abstract

The controlled fragmentation of chromosomes by DNA double-strand breaks (DSBs) initiates meiotic recombination, which is essential for meiotic chromosome segregation in most eukaryotes. This chapter describes a straightforward microarray-based approach to measure the genome-wide distribution of meiotic DSBs by detecting the single-stranded DNA (ssDNA) that transiently accumulates at DSB sites during recombination. The protocol outlined here has been optimized to detect meiotic DSBs in *Saccharomyces cerevisiae*. However, because ssDNA is a universal intermediate of homologous recombination, this method can ostensibly be adapted to discover and analyze programmed or damage-induced DSB hotspots in other organisms whose genome sequence is available.

Key words: ssDNA, meiosis, double-strand breaks, hotspots, microarray.

1. Introduction

In most eukaryotes, the proper segregation of homologous chromosomes during meiosis I depends on their physical linkage by crossover recombination. The first step in the process of forming crossovers is the introduction of Spo11-dependent DNA double-strand breaks (DSBs) on every chromosome (1). Each DSB is processed via strand resection to expose ssDNA, which then serves as a template for homology search and recombinational repair (2). Because the number and location of DSBs influence where and how many crossovers can form, their distribution across chromosomes is important to ensure proper chromosome assortment and viability of the resulting gametes.

H. Tsubouchi (ed.), *DNA Recombination*, Methods in Molecular Biology 745,
DOI 10.1007/978-1-61779-129-1_4, © Springer Science+Business Media, LLC 2011

Meiotic DSBs occur with high frequency in specific "hotspot" regions (3), and two methods have previously been described to measure DSB formation across chromosomes and genomes. First, Southern blot analysis has been used to detect DSB hotspots along restriction fragments and even whole chromosomes (4–6). This method can have very high spatial resolution; however, it is labor intensive and difficult to expand to a genome-wide scale. A second approach takes advantage of the fact that the Spo11 enzyme transiently forms a covalent bond with the DNA at a DSB site. Purification of Spo11-associated DNA enables the genome-wide detection of meiotic DSBs using microarrays (7) or high-throughput sequencing methods (8). However, this approach only detects Spo11-dependent DSBs and requires either epitope tags or antibodies against Spo11. Additionally, the *rad50S*-type mutations that improve Spo11 purification are known to change the distribution of DSB formation in budding yeast (9, 10).

As an alternative, we developed a microarray-based technique to detect the ssDNA that naturally accumulates at DSB hotspots. This method has the advantage that it can be used in wild-type, unperturbed cells, obviating the need for antibodies or epitope tags. Moreover, the repair mutations that trap ssDNA-containing DSBs, such as *dmc1Δ* or *rad52Δ*, have not been shown to affect DSB formation (9, 10). The analysis of mutants with persistent DSBs is useful because it enables cumulative DSB measurements and enhances the ssDNA signal of weaker or transient DSBs (9, 10). Finally, because ssDNA is a universal intermediate of homologous recombination, it should be straightforward to adapt this method to detect natural or induced DSB hotspots in other systems.

Our method utilizes the unique biochemical properties of ssDNA to specifically enrich and label DSB-associated sequences for microarray hybridization (**Fig. 4.1**). To detect meiotic DSB hotspots, cells are first synchronized in meiosis and total genomic DNA is carefully isolated and fragmented. At the strongest meiotic DSB hotspots, breaks are formed in only a small percentage of cells. Therefore, to gain sufficient signal for microarray detection, the ssDNA surrounding DSB sites must be enriched using benzoylated naphthoylated DEAE (BND) cellulose adsorption (11). Next, ssDNA regions are fluorescently labeled by carrying out a random priming reaction without denaturation of the template (12). Finally, enrichment of ssDNA is detected by comparative genomic hybridization (CGH) of the DSB-containing DNA with a control sample using high-density tiled microarrays. This approach allows for the specific and quantitative detection of meiotic DSB-associated ssDNA (9, 10).

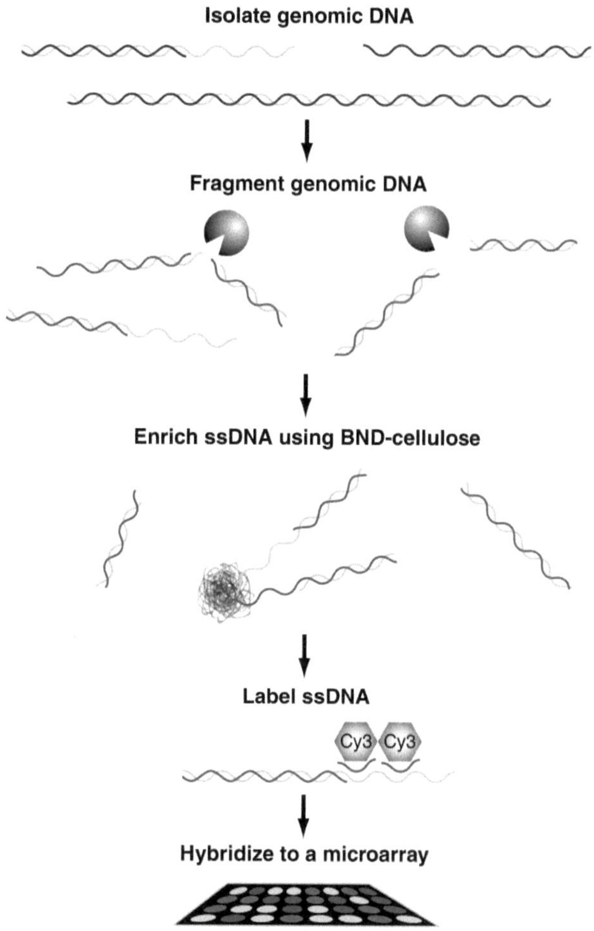

Fig. 4.1. Overview of the procedure used to detect DSB hotspots by measuring ssDNA enrichment. Genomic DNA is carefully isolated and then fragmented by restriction enzyme digestion. Next, the population of molecules containing ssDNA is enriched by batch adsorption to BND cellulose. Subsequently, the ssDNA regions are fluorescently labeled by carrying out a random priming reaction without denaturation of the template DNA in the presence of Cy3- or Cy5-dUTP. Finally, labeled probes are denatured and hybridized to a microarray to detect regions of ssDNA enrichment.

2. Materials

2.1. Cell Synchronization

1. YPG plates: 1% yeast extract, 2% peptone, 3% glycerol, 2% agar, 0.03 mg/ml adenine.
2. 4% YPD plates: 1% yeast extract, 2% peptone, 4% glucose, 2% agar, 0.03 mg/ml adenine.
3. Liquid YPD medium: 1% yeast extract, 2% peptone, 2% glucose, 0.03 mg/ml adenine.

4. Buffered YTA (BYTA) medium: 1% yeast extract, 2% bactotryptone, 1% potassium acetate, 50 mM potassium phthalate. Store at room temperature in the dark for several weeks.
5. Sporulation (SPO) medium: 0.3% potassium acetate, pH 7.0 with 250 μl of 5% acetic acid (v/v) per liter.

2.2. ssDNA Isolation

1. Ethanol, 70% (v/v)
2. Sorbitol buffer: 1 M sorbitol, 0.1 M EDTA, 20 mM Tris-HCl, pH 7.4
3. β-Mercaptoethanol
4. Zymolyase 100T (Associates of Cape Cod, Inc.): 10 mg/ml stock in 1 M sorbitol, store at –20°C
5. 10 mM Tris-HCl, pH 9.5
6. NDS: 0.5 M EDTA, 1% SDS, 10 mM Tris-HCl, pH 9.5 (*see* **Note 1**)
7. TE: 10 mM Tris-HCl, pH 7.5, 1 mM EDTA
8. Proteinase K: 14 mg/ml
9. Phenol:chloroform:isoamylalcohol (25:24:1)
10. Chloroform
11. RNase A solution: 30 mg/ml (Sigma), store at −20°C
12. 3 M sodium acetate, pH 5.2 with acetic acid
13. Ethanol, absolute

2.3. Genomic DNA Fragmentation

1. *Eco*RI restriction enzyme and 10X EcoRI reaction buffer (New England Biolabs)
2. Spermidine, >98% (Sigma)

2.4. ssDNA Enrichment Using BND Cellulose Adsorption

1. NET buffer: 1 M NaCl, 10 mM Tris-HCl, pH 7.5, 1 mM EDTA
2. 5 M NaCl
3. Caffeine elution buffer: NET + 1.8% caffeine (w/v). Put solution at 50°C to dissolve caffeine and then equilibrate to room temperature. This solution should be prepared fresh for each experiment.
4. Benzoylated naphthoylated DEAE–cellulose (Sigma, B-6385), 50% slurry in NET buffer, prepared as follows:
   (a) Weigh out 10 g BND cellulose into a 50 ml tube.
   (b) Wash BND cellulose five times by resuspending the resin in 5 M NaCl in a 50 ml total volume, spinning down for 2 min at 1,350×*g* in a bench top centrifuge and pouring off the supernatant.
   (c) Wash once with water in a 50 ml total volume.

(d) Wash twice with NET buffer in a 50 ml total volume.

(e) Adjust to 50% (v/v) BND cellulose in NET buffer.

(f) Store at 4°C for up to 1 year.

5. 2 ml round bottom microcentrifuge tubes
6. 15 ml conical tubes
7. 14 ml round bottom polypropylene tubes (*see* **Note 2**)

2.5. ssDNA Labeling and Microarray Hybridization

1. DNA Polymerase I, Large (Klenow) Fragment (50,000 units/ml) and 10X NEBuffer 2 (New England Biolabs)
2. Filter-sterilized TE: 10 mM Tris-HCl, pH 7.5, 1 mM EDTA
3. Random nonamer oligonucleotides: 867 μg/ml in filter-sterilized TE (25% of each nucleotide, IDT)
4. LowT dNTP mix: 2.4 mM each dATP, dCTP, dGTP, 1.2 mM dTTP, diluted in filter-sterilized TE
5. Cy3-dUTP and Cy5-dUTP (GE Healthcare), supplied as 1 mM stock solutions
6. 30,000 MWCO Amicon Ultra filter columns (Millipore)
7. 4x44K yeast whole genome tiled oligonucleotide microarrays (Agilent) or equivalent
8. 2X Hi-RPM hybridization buffer (Agilent) or equivalent
9. Slide hybridization chambers and gasket slides (Agilent)
10. 20X SSPE: 3 M NaCl, 200 mM NaH_2PO_4, 200 mM EDTA, pH 7.4 using NaOH. Filter sterilize and store at room temperature.
11. 20% *N*-lauroylsarcosine, sodium salt solution (Sigma)
12. Wash 1: 6X SSPE, 0.005% *N*-lauroylsarcosine. Filter sterilize and store at room temperature.
13. Wash 2: 0.6X SSPE, 0.005% *N*-lauroylsarcosine. Filter sterilize and store at room temperature.
14. Wash 3: Agilent stabilization and drying solution (*see* **Note 3**)
15. Agilent microarray scanner (or equivalent)

2.6. Microarray Detection of DSBs Using ssDNA Enrichment

1. Feature Extraction software (Agilent) or an equivalent program that can calculate Cy3 and Cy5 intensities from scanned microarray TIFF images.
2. R, a computer language and environment for statistical computing (v2.1.0, http://www.r-project.org), or an equivalent program that can be used to perform statistical analyses and visualize data and results.

3. A program for normalizing microarray data, such as the limma package (www.bioconductor.org) (13) for R (in the sample data set we used a similar but now obsolete R package, Statistics for Microarray Analysis (SMA) (http://www.stat.berkeley.edu/~terry/zarray/Software/smacode.html) (14)).

3. Methods

The protocol outlined below provides a step-by-step procedure for the isolation and labeling of meiotic ssDNA. When performing the initial experimental design, it is important to consider two key parameters that influence the ability to detect DSB-associated ssDNA: the relative abundance of DSBs (*see* **Note 4**) and the presence of non-break-associated ssDNA in the sample (*see* **Note 5**). Both should be taken into account when choosing the strain background and culture conditions. Moreover, preparing a proper control sample in parallel is critical for the quantitative detection of hotspots (*see* **Note 6**). We have provided a sample experiment in which we demonstrate one method to calculate ssDNA enrichment and identify DSB hotspots. In this experiment, we used a *dmc1Δ* strain, which accumulates meiotic DSBs, and an *spo11-Y135F* strain, which does not make meiotic DSBs. Samples were collected from each strain at 0 h before DSBs were formed and at 5 h after meiotic induction when the *dmc1Δ* cells had completed break formation. Biological replicate experiments were performed for each strain.

3.1. Synchronous Meiotic Time Course

Synchronization procedures vary between strain backgrounds. Below is a procedure that provides high synchrony for meiotic cultures of the SK1 background.

1. Remove cells from the –80°C stock onto a YPG plate and grow them overnight at 30°C.
2. Transfer cells onto a 4% YPD plate and grow them overnight at 30°C.
3. Inoculate a 12–15 ml liquid YPD culture for each strain and grow on a shaker for 24 h at room temperature to reach saturation.
4. Dilute the saturated cultures to $OD_{600} = 0.3$ in BYTA presporulation medium and grow on a shaker for 16 h at 30°C.
5. Collect cells from the BYTA cultures by centrifugation for 3 min at $1,350\times g$ in a bench top centrifuge. Wash cells with 2 vol. of sterile water.
6. To induce sporulation, resuspend the cells in SPO at $OD_{600} = 1.9$ in highly aerated flasks (maximum SPO culture

volume = 10% of flask volume). Incubate cultures at 30°C on a shaker.

7. At the appropriate time points (e.g., 0 and 5 h for the *dmc1Δ* and *spo11-Y135F* strains), harvest 25–50 ml of SPO culture by centrifugation using 50 ml conical tubes.
8. Resuspend the cell pellet in 25 ml of 70% ethanol and store in ethanol at –20°C.

3.2. Genomic DNA Extraction

Care must be taken when isolating genomic DNA to avoid the artificial creation of ssDNA during the purification procedure. Improper handling can both increase background and/or create artifacts. Two rules of thumb should preserve the original ssDNA content. First, samples should never be exposed to temperatures higher than 50°C to avoid heat denaturation of DNA duplexes. Second, random DNA shearing should be avoided. Therefore, samples should never be vortexed, but rather mixed thoroughly by inversion. Furthermore, wide-orifice pipette tips (which can be purchased or made by cutting off the end of regular pipette tips with a razor blade) should be utilized for **Sections 3.2**, **3.3**, and **3.4**.

1. Pellet the cells for 3 min at 1,350×*g* in a bench top centrifuge and discard the ethanol.
2. Wash the cells once with 10 ml of sorbitol buffer.
3. Resuspend the cells in 10 ml of sorbitol buffer containing 100 μl β-mercaptoethanol and 200 μl of Zymolyase stock by gently pipetting. Incubate at 37°C for 25 min to digest the cell walls.
4. Collect the spheroplasts by spinning for 4 min at 500×*g* in a bench top centrifuge and discard the supernatant.
5. Carefully resuspend the cells in 2 ml of 10 mM Tris-HCl, pH 9.5, by pipetting up and down using a 5 ml pipette. Transfer cells to a 15 ml conical tube.
6. Add 3 ml of NDS and 100 μl of proteinase K and incubate at 50°C for 1 h to digest proteins.
7. Add 2 ml of TE to increase the sample volume for phenol extraction.
8. Extract the DNA three times with 5 ml of phenol:chloroform:isoamyl alcohol. Invert tubes approximately 60 times per extraction to ensure thorough mixing. Centrifuge at 2,800×*g* for 10 min in a bench top centrifuge to separate the phases. It is normal for the aqueous phase to remain cloudy throughout these extractions. It should become clear during the next step (*see* **Note 7**).
9. Extract the DNA once with 5 ml of chloroform to remove traces of phenol and transfer the top phase to a new 15 ml tube.

10. Precipitate the DNA by adding 9 ml of absolute ethanol and pellet by centrifugation at 2,800 × *g* for 10 min at room temperature in a bench top centrifuge.
11. Wash the pellet once with 5 ml of 70% ethanol.
12. Drain the ethanol and dissolve the pellet in 3 ml of water by careful tapping.
13. Add 3.5 μl of RNase A and mix by inversion. Incubate samples at 37°C for 30 min to eliminate co-purified RNA (*see* **Note 8**).
14. Add 300 μl of sodium acetate and 7 ml of absolute ethanol to precipitate the DNA. Place tube at –20°C for 10 min. Pellet the DNA by centrifugation at 2,800 × *g* for 10 min in a bench top centrifuge.
15. Wash the pellet once with 5 ml of 70% ethanol.
16. Drain the ethanol, spin briefly, and remove any residual ethanol by aspiration. Dissolve the pellet immediately in 1 ml of TE. Store the sample at 4°C for up to several months (*see* **Note 9**).
17. Measure the concentration of genomic DNA using a spectrophotometer. The expected yield from 50 ml of cells is approximately 0.5–1 mg of genomic DNA.

3.3. Genomic DNA Fragmentation

1. Digest approximately 250 μg of DNA with 200 U of *Eco*RI restriction enzyme plus 2 μl of spermidine in a 2.5 ml reaction with 1X EcoRI buffer for 3–4 h at 37°C (*see* **Note 10**).
2. To precipitate the DNA, add 250 μl of sodium acetate and 5.5 ml of absolute ethanol to the digestion reaction. Place at –20°C for 10 min and then collect the precipitated DNA by centrifugation at 2,800 × *g* for 10 min in a bench top centrifuge.
3. Discard the supernatant and eliminate traces of ethanol with a pipette. Resuspend the pellet in 500 μl of TE and store the sample at 4°C.
4. Confirm the completion of the digest by analyzing 20 μl of the digestion reaction on a 0.7% agarose gel. Incompletely digested samples usually contain a bright band above the 12 kb band of the ladder.

3.4. ssDNA Enrichment Using BND Cellulose Adsorption

1. Prepare 3 ml of fresh caffeine elution buffer per sample.
2. Adjust *Eco*RI-digested samples to a final concentration of 1 M NaCl by adding 125 μl of 5 M NaCl.
3. For each sample, prepare a 500 μl bed volume of BND cellulose by placing 1 ml 50% BND cellulose slurry in a 2 ml round bottom tube using a wide-orifice pipette tip.

Briefly pellet the resin at full speed in a microcentrifuge and remove the NET buffer with a pipette.

4. Apply the entire *Eco*RI-digested genomic DNA sample to the prepared tube of BND cellulose and resuspend the resin by inverting and flicking the tube. Bind the ssDNA to the BND cellulose by rotating the suspension at room temperature for 5 min.
5. Pellet the BND cellulose for 30 s at full speed in a microcentrifuge. Remove the supernatant with a pipette and discard it.
6. Wash the resin five times with one bed volume (500 μl) NET buffer by inverting and flicking the tube to resuspend the resin. Pellet the resin and remove the supernatant as in the previous step. Discard the washes.
7. Elute the ssDNA by washing the BND cellulose five times with 600 μl of caffeine elution buffer and saving each elution. The five elutions (3 ml total) should be combined into a single 15 ml conical tube.
8. Spin each sample for 10 min at 1,350×*g* in a bench top centrifuge to remove any excess BND cellulose.
9. Carefully pour the supernatant into a 14 ml round bottom tube, which has been labeled on the side of the tube.
10. Add 6 ml of absolute ethanol and incubate at –20°C overnight to precipitate the eluted DNA.
11. Pellet the DNA by spinning for 10 min at 9,800×*g* in a Beckman JA25.50 or comparable rotor. Caps must be removed for the tubes to fit in the JA25.50 rotor adapters.
12. Wash the pellet once with 3 ml of 70% ethanol by spinning as described above. Dry the pellet completely.
13. Resuspend the pellet in 100 μl of TE and transfer the sample to a 1.5 ml microcentrifuge tube.
14. Spin the sample briefly at full speed in a microcentrifuge to remove the excess BND cellulose. Transfer the supernatant to a new 1.5 ml microcentrifuge tube for storage at 4°C.
15. Measure the OD_{260} to estimate the yield of enriched ssDNA. Typically, a total of about 20–25 μg of genomic DNA is recovered after BND cellulose adsorption for both 0 and 5 h samples.

3.5. ssDNA Labeling and Microarray Hybridization

The ssDNA regions are specifically labeled by carrying out a random priming reaction, without denaturing the genomic DNA. Because the 0 and 5 h BND-enriched ssDNA samples from each culture will be co-hybridized to a single microarray, one is labeled with Cy3 and the other with Cy5. Biological replicates are labeled as dye swaps; the 0 h sample is labeled with Cy3 for one experiment and Cy5 for the replicate.

1. For each sample, combine 20 μl (approximately 5 μg) of enriched ssDNA, 5 μl of random nonamer oligonucleotides, 3.5 μl of 10X NEBuffer 2, and 6.5 μl of water in a thermocycler tube or plate.
2. Heat the samples to 50°C in a thermocycler for 5 min to remove secondary structure in the ssDNA. Cool the samples to 4°C to allow annealing of the primers to the ssDNA.
3. For each sample, prepare 5 μl of extension mix containing 1.25 μl of water, 0.5 μl of 10X NEBuffer 2, 1 μl of lowT dNTP mix, 2 μl of Cy3- or Cy5-dUTP, and 0.25 μl (12.5 U) of Klenow DNA polymerase. Add the extension mix to the samples while they incubate at 4°C and mix well by pipetting.
4. Heat the samples to 37°C at a rate of increase of 0.1°C/s to allow extension of the primers. Incubate at 37°C for 1 h to complete the extension/labeling reaction. Store samples at 4°C in the dark until proceeding to the next step.
5. Remove unincorporated Cy3- and Cy5-dUTP by applying the sample to a Amicon Ultra column, as per manufacturer's instructions. Preload each column with 450 μl of filter-sterilized TE. Add the entire volume of the 0 and 5 h samples for each experimental array to a single column.
6. Spin the column at 14,000×*g* in a microcentrifuge for approximately 8 min to reduce volume to <100 μl.
7. Wash the sample two more times with 450 μl of filter-sterilized TE, followed by centrifugation as described in **Section 3.5**, step 6.
8. Make sure the final volume is reduced to a volume appropriate for microarray hybridization. For a 4x44K Agilent format, this is less than 56.5 μl.
9. Recover the labeled sample by flipping the column into a clean 1.5 ml microcentrifuge tube (provided) and spinning at 1,000×g for 3 min.
10. Adjust the volume to 56.5 μl with filter-sterilized TE (55 μl for hybridization and 1.5 μl for quality control).
11. Measure the Cy3- and Cy5-dUTP incorporation of 1.5 μl of each sample on a NanoDrop spectrophotometer using the microarray application and DNA setting. A typical labeling reaction should yield a total of 20–30 pmol each of Cy3 and Cy5 in each sample pair (*see* **Note 11**).

12. Boil the samples at 95°C for 5 min.
13. Immediately add 55 μl of 2X Hi-RPM hybridization buffer and mix each sample carefully by pipetting without creating bubbles.
14. Spin the samples at full speed for 1 min in a microcentrifuge to remove large bubbles and particulate matter.
15. Apply the entire sample to a single microarray and hybridize according to manufacturer's instructions. Agilent microarrays are hybridized at 65°C for 16–24 h in an Agilent rotating hybridization oven.
16. Set up slide washing chambers containing wash 1, wash 2 and wash 3. An additional open container of wash 1 is needed for opening the slide assembly.
17. Remove the array and gasket slide assembly from the hybridization chamber and submerge the slides under wash 1 in the open container. Immediately remove the array slide from the gasket slide by inserting forceps between the slides to release them.
18. Transfer the microarray slide to the chamber containing wash 1 for 1–5 min, then to wash 2 for 5 min, and finally to wash 3 for 30 s. Use either a stir bar or gentle manual agitation to fully clean the slides in each wash step. Remove the slide from wash 2 carefully so that the solution does not carry over to wash 3. Remove the slide from wash 3 very slowly, allowing the surface tension to gently remove all particulate matter from the surface of the microarray. If particulate matter is visible on the surface of the slide, repeat the wash 2 and wash 3 steps. Let the slides dry completely.
19. Scan the microarrays using an Agilent or equivalent scanner and appropriate laser power such that no microarray features have saturated signals in the Cy3 or Cy5 channel. The resulting data are stored in a split TIFF file that contains the Cy3 and Cy5 images for each slide.

3.6. Microarray Detection of DSBs Using ssDNA Enrichment

Following microarray hybridization and scanning, the raw image data are extracted to calculate ssDNA enrichment for all features ("spots") on the array, and DSB hotspots are identified. Reliable measurements require a number of controls and normalizations that are outlined below. For the sample data set, we performed the extraction and the subsequent calculations using the Agilent Feature Extraction program and R, although equivalent alternatives exist (*see* **Note 12**). The biological replicate experiments for the *dmc1Δ* and *spo11-Y135F* strains were hybridized to independent microarrays, so four total microarrays were analyzed.

1. Measure the fluorescence values and monitor the quality of each array hybridization using Feature Extraction. For each slide to be analyzed, select the TIFF image to be extracted and 'CGH' from the pull-down 'Protocols' menu. The program will automatically find and analyze the fluorescent levels in and around each feature on the array for both channels. Subsequently, several output files are produced, including quality control measures (*see* **Note 13**), a picture of each array, and a text file containing the results of the extraction. The text results file is used as the input for **Section 3.6**, step 2.
2. Calculate the adjusted log ratio of ssDNA enrichment for the 5 h sample versus 0 h sample for each array feature (*see* **Note 14**). Feature Extraction performs a log ratio calculation, which can be used directly from the imported results text file, or the limma function normalizeWithinArrays can be applied. For the sample data sets, the log ratio was calculated using the SMA package for R (13). The mean signal and mean background intensities of Cy3 and Cy5 for each array feature were imported into an R data file from the text results file. The SMA function stat.ma was applied to the data file to calculate log ratios for each feature on the array.
3. Perform scale normalizations for each set of biological replicate experiments. The sample data were normalized using the SMA function stat.norm.exp (the limma equivalent is the function normalizeBetweenArrays). This step normalizes the median absolute deviation of the log ratios for the individual experiments. The resulting scaled data sets are used for steps 4 and 5 of **Section 3.6**.
4. To visualize the results, plot the ssDNA enrichment for each array feature with respect to its chromosomal location. For the sample experiment, we plotted the ssDNA enrichment at 5 h versus the 0 h control for all points along chromosome 3 for the *dmc1Δ* and *spo11-Y135F* strains (**Fig. 4.2a**, black dots). To reduce the contribution of background noise, the results of the replicate experiments were averaged, and subsequently the log ratios were transformed into linear ratios to show fold enrichment. Consistent with the finding that >1 kb of ssDNA can be exposed at each DSB site (15), we observed clusters of adjacent features exhibiting specific ssDNA enrichment that were absent in the *spo11-Y135F* strain. These peaks of ssDNA enrichment in the *dmc1Δ* strain were confirmed by comparing the ssDNA profile from the microarray experiment to a Southern blot for full-length chromosome 3, which exhibited strong DSB hotspots at the same locations (**Fig. 4.2b**).

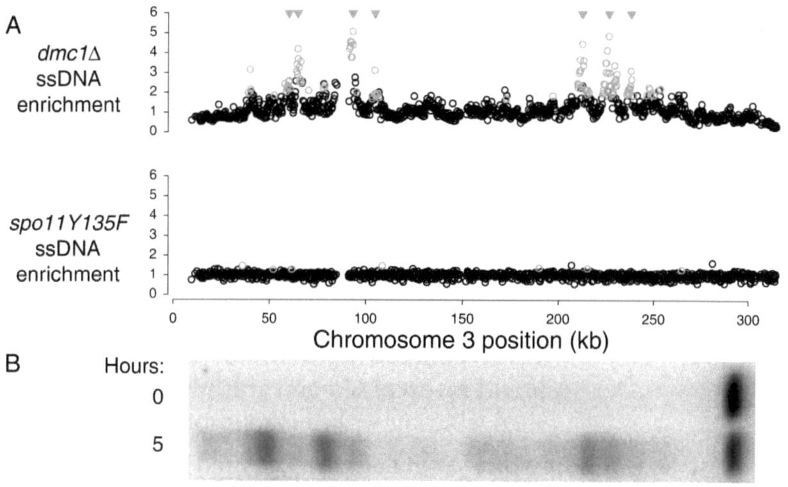

Fig. 4.2. Meiotic DSB hotspots predicted from the site-specific enrichment of ssDNA across the budding yeast genome. **a** DSB hotspot distribution on chromosome 3, as measured by ssDNA enrichment analysis. The mean ssDNA enrichment for biological replicate experiments is plotted with respect to position along chromosome 3 for the *dmc1Δ* (*top*) and *spo11-Y135F* (*bottom*) strains. Features that exhibited significant ssDNA enrichment ($p<0.125$) in both biological replicate experiments are indicated in *gray*, whereas all other features are drawn in *black*. *Inverted gray triangles* represent the positions of clusters of greater than three significantly enriched features, which denote the position of strong DSB hotspots predicted from the ssDNA enrichment. **b** The DSB hotspot distribution of chromosome 3 by Southern blot analysis. Samples were collected from the *dmc1Δ* strain at the indicated time points and DNA was separated by pulsed-field gel electrophoresis. A Southern blot for full-length chromosome 3 was carried out using a probe close to the left telomere.

5. To identify meiotic DSB hotspots using ssDNA enrichment, we applied several criteria to our ssDNA enrichment data.
   (a) A *p*-value cutoff was applied to determine all features significantly enriched above background in the 5 h sample versus the 0 h sample for each individual experiment. In this example, a cutoff of $p<0.125$ was applied to *p*-values that were determined using the pnorm function in R and assuming a single-tailed normal distribution of the data.
   (b) Only features that were reproducibly enriched in both of the individual replicate data sets were considered for further analysis (**Fig. 4.2a**, gray dots).
   (c) Because strong DSB hotspots should be surrounded by 1–2 kb of ssDNA, only peaks of ssDNA enrichment that contained greater than three contiguous features with a significant ssDNA enrichment signal were counted (**Fig. 4.2a**, inverted gray triangles).

This method identified a set of the strongest meiotic DSB hotspots that could be predicted with the highest confidence (*see* **Note 15**). No DSB hotspots were detected in the *spo11-Y135F* strain (**Fig. 4.2a**), indicating that this method does not detect specific ssDNA enrichment from DNA replication, telomeres, or spontaneous DNA damage repair (9).

4. Notes

1. The SDS in the NDS buffer precipitates after storage at room temperature. To resuspend the SDS, heat the buffer to 50°C immediately before use.
2. These tubes will be spun at high speed during the protocol. Polystyrene tubes cannot be used, as they can shatter in the centrifuge.
3. Wash 3 contains acetonitrile, which is highly toxic. Caution should be exercised and acetonitrile gloves should be used when working with this solution.
4. Choice of genetic background plays an important role in detecting DSB-associated ssDNA. As ssDNA is a transient intermediate in the repair process, a high level of synchrony in the experimental culture is crucial for ssDNA detection in wild-type cells. This can be achieved in efficiently sporulating backgrounds, such as SK1. Even in SK1, however, the signal-to-noise ratio was consistently higher in mutants that fail to complete the strand invasion step of homologous recombination, and thus prevent turnover and repair of ssDNA. The use of these mutants may be essential to detect ssDNA-associated DSB hotspots in other strains or organisms.
5. Cellular processes other than DSB formation, most notably DNA replication, can lead to the production of large amounts of ssDNA on all chromosomes (12). Under normal circumstances, replication-associated ssDNA occurs transiently and the synchrony between cells is insufficient to lead to local accumulation of signal. However, certain circumstances, such as the use of replication inhibitors or specific mutants that accumulate excess ssDNA at replication forks, may create abnormally high levels of ssDNA that could obscure the ssDNA signal at DSB hotspots.
6. The quantitative detection of ssDNA requires experimental samples to be compared to a control sample to normalize the data for biases generated by the method of sample preparation or microarray hybridization. The CGH method is a powerful tool for the quantitative analysis of DSB-associated ssDNA, because it enables the reliable measurement of small differences (less than twofold) in the enrichment of sequences relative to each other (16). For every ssDNA microarray experiment, we co-hybridize the experimental sample with a control DNA sample from the same strain collected at 0 h of the experimental time course, prior to meiotic DSB initiation. Alternatively, DNA from an isogenic but DSB-defective mutant (e.g., *spo11Δ*),

cultured in parallel to the experimental strain, can be used. It is critical that the control sample is treated identically to the experimental sample at every step of the protocol, to control for biases introduced by the ssDNA purification and labeling method, especially since other ssDNA is likely present in cells (*see* **Note 5**). Indeed, we consistently observe robust recovery and labeling of ssDNA from 0 h control samples.

7. When transferring the aqueous phase (top) to a new tube, strictly avoid the white interface. This can be aided by removing the aqueous phase with a disposable 5 ml plastic pipette.
8. Completion of this step is critical, because RNA can compete with ssDNA for binding to BND cellulose.
9. As with other applications sensitive to the intact nature of the DNA, purified DNA samples should never be frozen.
10. Do not let the digest proceed for longer or DNA degradation can occur.
11. Incorporation rates greater than 75 pmol of dye in either channel often lead to saturated signals on Agilent arrays. If this occurs, an appropriate portion of the labeled sample can be removed for hybridization, and the volume readjusted to 55 μl with filter-sterilized TE.
12. We used R to calculate ssDNA enrichment, identify hotspots, and visualize results, due to the ease of manipulation and comparison of large data sets in a Unix- and R-based environment. Additionally, the SMA or limma packages for R contain specific microarray normalization functions used to compare samples within or across separate microarray experiments. However, all of the calculations and graphs produced in **Section 3.6** could be performed using other available database and spreadsheet programs, such as Microsoft Excel. Because the Feature Extraction text results file is large and contains information that is not necessary for downstream processing, manipulation of these files is cumbersome in Excel. Therefore, a smaller file can be created for each microarray that contains only the relevant columns of data for every array feature (i.e., chromosome, position, description, log ratios, and significance), which can be used to perform simple calculations or visualize the data.
13. The Agilent Feature Extraction program provides multiple measures of quality control. The original TIFF can be visualized to monitor the quality of hybridization. During the extraction step, irregularities such as saturated or non-hybridizing features are detected. The appearance of a large number of irregularities often indicates insufficient

or saturated hybridization signals, dirt on the surface of the slide, or other hybridization problems that can interfere with data analysis. Because spike-in control samples are not used in the hybridization, error messages referring to the negative signal of control spots should be ignored.

14. To calculate a log ratio, most protocols first subtract local background, then mean normalize the Cy3 and Cy5 channels, and correct for dye bias at different intensities. This step is important to remove biases that are either inherent to the fluorescent dyes at different intensities or that can be introduced by differences in the amount of input DNA, the efficiency of labeling with Cy3 and Cy5, and cross-experiment variation. There are several programs that can be used to perform these calculations. We have used both the Agilent Feature Extraction program with 'CGH' or 'ChIP' settings and the SMA package in R to calculate the log ratios for ssDNA enrichment. The absolute values of the log ratios differ in each case, due to the specific data normalization methods used to calculate the log ratios. In spite of the different absolute values produced, all three of these methods enabled the detection of DSB hotspots and other prominent trends in the data. If an alternative method is used to calculate log ratios, the user should ensure that all of these normalization steps are performed. A good measure of the quality of data normalization is to plot the log ratio versus the log of the average intensity for each array feature (an M versus A plot), to ensure that the average log ratio is 0 across the range of all intensities.
15. Weaker hotspots can also be identified by using a less stringent *p*-value or by demanding that fewer contiguous features are enriched at a given site.

Acknowledgments

We would also like to thank Gerben Vader and Milan de Vries for technical discussions and critical reading of this protocol.

References

1. Keeney, S., Giroux, C.N., and Kleckner, N. (1997) Meiosis-specific DNA double-strand breaks are catalyzed by spo11, a member of a widely conserved protein family. *Cell* **88**, 375–384.
2. Bishop, D.K., and Zickler, D. (2004) Early decision; meiotic crossover interference prior to stable strand exchange and synapsis. *Cell* **117**, 9–15.

3. Petes, T.D. (2001) Meiotic recombination hot spots and cold spots. *Nat Rev Genet* **2**, 360–369.
4. Baudat, F., and Nicolas, A. (1997) Clustering of meiotic double-strand breaks on yeast chromosome III. *Proc Natl Acad Sci USA* **94**, 5213–5218.
5. Game, J.C. (1992) Pulsed-field gel analysis of the pattern of DNA double-strand breaks in the Saccharomyces genome during meiosis. *Dev Genet* **13**, 485–497.
6. Zenvirth, D., Arbel, T., Sherman, A., Goldway, M., Klein, S., and Simchen, G. (1992) Multiple sites for double-strand breaks in whole meiotic chromosomes of Saccharomyces cerevisiae. *EMBO J* **11**, 3441–3447.
7. Gerton, J.L., DeRisi, J., Shroff, R., Lichten, M., Brown, P.O., and Petes, T.D. (2000) Inaugural article: global mapping of meiotic recombination hotspots and coldspots in the yeast Saccharomyces cerevisiae. *Proc Natl Acad Sci USA* **97**, 11383–11390.
8. Pan, J., Sasaki, M., Kniewel, R., Murakami, H., Blitzblau, H.G., Tischfield, S.E., Zhu, X., Neale, M.J., Jasin, M., Socci, N.D., Hochwagen, A., and Keeney, S. (2011) A hierarchical combination of factors shapes the genome-wide topography of yeast meiotic recombination initiation. *Cell* **144**, 719–731.
9. Blitzblau, H.G., Bell, G.W., Rodriguez, J., Bell, S.P., and Hochwagen, A. (2007) Mapping of meiotic single-stranded DNA reveals double-strand-break hotspots near centromeres and telomeres. *Curr Biol* **17**, 2003–2012.
10. Buhler, C., Borde, V., and Lichten, M. (2007) Mapping meiotic single-strand DNA reveals a new landscape of DNA double strand breaks in Saccharomyces cerevisiae. *PLoS Biol* **5**, 2797–2808.
11. Huberman, J.A., Spotila, L.D., Nawotka, K.A., el-Assouli, S.M., and Davis, L.R. (1987) The in vivo replication origin of the yeast 2 microns plasmid. *Cell* **51**, 473–481.
12. Feng, W., Collingwood, D., Boeck, M.E., Fox, L.A., Alvino, G.M., Fangman, W.L., Raghuraman, M.K., and Brewer, B.J. (2006) Genomic mapping of single-stranded DNA in hydroxyurea-challenged yeasts identifies origins of replication. *Nat Cell Biol* **8**, 148–155.
13. Smyth, G.K. (2005) Limma: linear models for microarray data. In Bioinformatics and computational biology solutions using R and bioconductor, R.C. Gentleman, V.J. Carey, S. Dudoit, R. Irizarry, W. Huber, eds. (New York, NY: Springer), pp. 397–420.
14. Yang, Y.H., Dudoit, S., Luu, P., and Speed, T.P. (2001) Normalization of cDNA microarray data. In SPIE BiOS 2001. San Jose, CA.
15. Bishop, D.K., Park, D., Xu, L., and Kleckner, N. (1992) DMC1: a meiosis-specific yeast homolog of E. coli recA required for recombination, synaptonemal complex formation, and cell cycle progression. *Cell* **69**, 439–456.
16. Yabuki, N., Terashima, H., and Kitada, K. (2002) Mapping of early firing origins on a replication profile of budding yeast. *Genes Cells* **7**, 781–789.

Chapter 5

Detection of Covalent DNA-Bound Spo11 and Topoisomerase Complexes

Edgar Hartsuiker

Abstract

Topoisomerases can release topological stress and resolve DNA catenanes by a DNA strand breakage and re-ligation mechanism. During the lifetime of the DNA break, the topoisomerase remains covalently linked to the DNA and removes itself when the break is re-ligated. While the lifetime of a covalent topoisomerase–DNA complex is usually short, several clinically important cancer drugs kill cancer cells by inhibiting the removal of covalently linked topoisomerases. The topoisomerase-like protein Spo11 is responsible for meiotic double strand break formation. Spo11 is not able to remove itself and is removed by nucleolytic cleavage. This chapter describes a method which allows the reproducible and quantitative detection of proteins covalently bound to the DNA.

Key words: Topoisomerase I, topoisomerase II, Spo11, *Schizosaccharomyces pombe*, MRN complex, Tdp1.

1. Introduction

Various topoisomerases fulfil key functions within the cell through their ability to break and rejoin DNA. Type I topoisomerases (e.g. Top1) cleave one DNA strand, while type II topoisomerases (e.g. Top2) are able to cleave both DNA strands. In this way they can release torsional stress associated with DNA replication or transcription (both types I and II) or resolve DNA catenanes (type II only). During the lifetime of the DNA break the topoisomerase remains covalently bound to the DNA end through a transient phosphodiester bond between a tyrosine residue of the protein and the DNA, which is normally short-lived. The topoisomerase is released upon rejoining of the DNA strands (1).

H. Tsubouchi (ed.), *DNA Recombination*, Methods in Molecular Biology 745,
DOI 10.1007/978-1-61779-129-1_5, © Springer Science+Business Media, LLC 2011

Camptothecin (CPT) and etoposide derivatives (also called topoisomerase poisons) are clinically important cancer drugs that kill proliferating cells by inhibiting the release of Top1 and Top2 covalent complexes, respectively, thus interfering with replication and transcription (2, 3). For a (cancer) cell to resist treatment with these drugs, the topoisomerase must be removed and the remaining DNA break needs to be repaired. Various factors have been implicated in the removal of topoisomerases. Tdp1 (tyrosyl-DNA phosphodiesterase 1) is able to hydrolyse the phosphodiester bond between the tyrosine residue of Top1 and the 3′ phosphate of the DNA and thus remove the covalently linked topoisomerase (4). Recently, a protein which exhibits tyrosyl-DNA phosphodiesterase activity specific for 5′ phosphotyrosyl bonds, called Tdp2, has been identified (5). It has previously been proposed that, as an alternative to direct removal through cleavage of the tyrosyl-DNA phosphodiester bond by Tdp1, topoisomerases could be removed through nucleolytic cleavage, releasing the protein together with a short oligonucleotide (6). Neale et al. (7, 8) have developed and used a method that allows detection of a nucleolytic release product to show that the topoisomerase-like protein Spo11, which is responsible for meiotic DSB formation and is not able to remove itself from the DNA, is removed by nucleolytic cleavage.

To study Rec12^{Spo11} removal in *Schizosaccharomyces pombe*, I have developed a method that allows quantification of covalently bound Rec12^{Spo11}–DNA complexes in vivo (9), which is also suitable to detect covalently bound Top1 and Top2 (10). This method (the DNA-linked protein detection or DLPD assay) was originally developed with the aim to quantify covalently bound Rec12^{Spo11} in meiotic *S. pombe* cells and is based on a procedure (11) that was used to identify Spo11 as the protein responsible for meiotic DSB formation (*see* **Note 1**). The crucial step in this protocol is to disrupt non-covalent interactions, and thus remove all non-covalently bound protein from the DNA, using the chaotropic agent guanidine hydrochloride (GuHCl) in combination with a detergent at 65°C. In the next step, the non-covalently bound proteins are separated from the DNA fraction (containing the covalently bound Rec12^{Spo11}) using CsCl step gradient centrifugation (12). After centrifugation, DNA-containing fractions are removed from the gradient and loaded on a slot blot, and the covalently bound Rec12^{Spo11} is detected using a specific antibody (*see* **Fig. 5.1** for outline of procedure). Apart from detecting Rec12^{Spo11} (9), I have also adapted this procedure to detect Top1 and Top2 covalent complexes in *S. pombe* and study their Rad32^{Mre11}-dependent nucleolytic removal (10). This procedure has also been successfully used for the detection of Top1 covalent complexes in human cells (13).

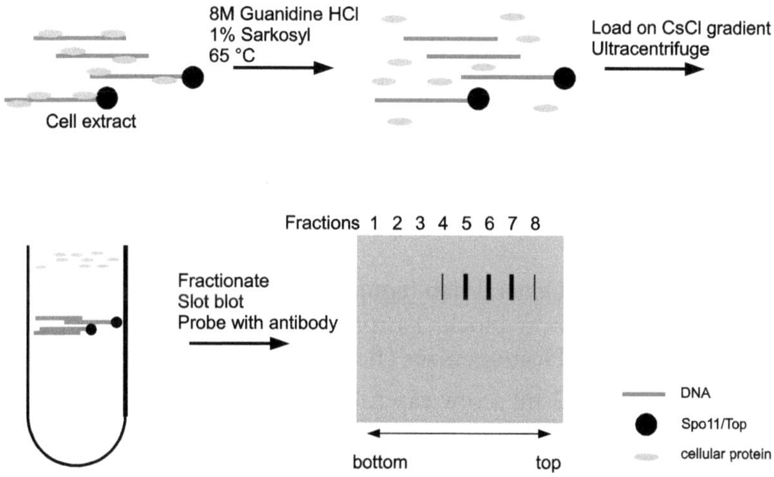

Fig. 5.1. Schematic outline of the DLPD assay. Cells are lysed in lysis buffer (containing the chaotropic agent GuHCl and sarkosyl) and incubated at 65°C. The non-covalently bound proteins are separated from the DNA fraction (containing the covalently bound protein) using CsCl step gradient centrifugation. After centrifugation, DNA-containing fractions are fractionated and loaded on a slot blot, and the covalently bound protein is detected using a specific antibody.

2. Materials

2.1. Preparation and Treatment of S. pombe Cultures

1. General lab equipment for culturing yeast: temperature-controlled shaking incubators, swing-out bench top centrifuge, and haemocytometer.
2. Strains containing a *pat1-114* mutation (*see* 14) and *rec12-6HA:kanMX6* for meiotic time courses and detection of covalently bound Rec12^{Spo11} (e.g. strain EH611 *h*$^-$ *smt0 ade6-M26 pat1-114 rec12-6HA:kanMX6* as described in (9)). Use a *top2-HA:kanMX6*-containing strain for detection of covalently bound Top2 (e.g. strain EH817 *h*$^+$ *leu1-32 ura4-D18 top2-HA:kan* as described in (10)). Top1 is detected using a specific antibody (*see* **Section 2.7**, Step 2) and a wild-type (WT) strain can thus be used.
3. Media (15): yeast extract (YE), per litre: 5 g yeast extract, 30 g glucose, 100 mg each of supplements (e.g. adenine, uracil, histidine, leucine, arginine), as required; autoclave. EMM2, per litre: 3 g potassium hydrogen phthalate, 2.2 g Na_2HPO_4, 5 g NH_4Cl, 20 g glucose, 20 ml 50x salt stock (per litre: 52.5 g $MgCl_26H_2O$, 0.735 g $CaCl_22H_2O$, 50 g KCl, 2 g Na_2SO_4), 1 ml 1,000x vitamin stock (per litre: 1 g pantothenic acid, 10 g nicotinic acid, 10 g inositol, 10 mg biotin), 0.1 ml 10,000x mineral stock (per litre: 5 g boric acid, 4 g $MnSO_4$, 4 g $ZnSO_47H_2O$, 2 g $FeCl_26H_2O$, 0.4 g molybdic acid, 1 g KI, 0.4 g $CuSO_45H_2O$, 10 g citric acid); filter sterilise or autoclave. Ready-mixed EMM2 powder is

available from ForMedium (www.formedium.com). EMM2 minus nitrogen: NH_4Cl is omitted (and added back from a 20% NH_4Cl in H_2O solution to initiate meiosis).

4. Drugs: CPT (Sigma), make up a stock solution of 10 mM in DMSO. TOP-53 (etoposide analogue, Taiho Pharmaceuticals, Japan) (16), make up a stock solution of 10 mg/ml in DMSO.

2.2. Preparing the Cell Extract

1. General lab equipment: microcentrifuge, water bath, or heat block. Fastprep-24 (MP Biomedicals) or Precellys 24 Homogenizer (Bertin) is used for lysing *S. pombe* cells.
2. 2 ml screw cap tubes suitable for use in the Fastprep-24 (or Precellys Homogenizer).
3. Lysis buffer: 8 M GuHCl, 30 mM Tris, pH 7.5, 10 mM EDTA, and 1% sarkosyl. Adjust pH to 7.5 with 10 M NaOH. This solution is made freshly. Over time the GuHCl precipitates out of the solution and should be re-dissolved by heating to 65°C.
4. Glass beads (e.g. Sigma G9268).

2.3. Preparing the CsCl Gradients

1. A refractometer can be used to check the refractive index (RI) of the CsCl stock solutions.
2. Polyallomer centrifuge tubes (order no. 326819, Beckman).
3. Prepare the following CsCl stock solutions in H_2O: 1.45 g/ml density: dissolve 60.90 g CsCl in 100 ml H_2O. RI should be 1.3764; 1.50 g/ml density: 68.48 g CsCl in 100 ml H_2O, RI 1.3815; 1.72 g/ml density: 98.04 g CsCl in 100 ml H_2O, RI 1.4012; 1.82 g/ml density: 111.94 g CsCl in 100 ml H_2O, RI 1.4104.
4. Ultracentrifuge with 6 × 5 ml swing-out rotor (e.g. AH650 rotor (Sorvall) or SW55Ti (Beckman)).

2.4. DNA Quantification

1. A suitable fluorometer capable of excitation at 480 nm and measuring emission at 520 nm (e.g. TBS-380 Mini-Fluorometer (Turner) or Qubit (Invitrogen)).
2. 50 μg/ml stock solution of RNaseA (DNase-free, Sigma).
3. Quant-iT PicoGreen dsDNA reagent (Invitrogen).
4. DNA standard (e.g. Lambda DNA (New England Biolabs)).

2.5. Gradient Fractionation

1. A retort stand with clamp, suitable to hold centrifuge tube.
2. Silicone tubing, inner diameter 1 mm, outer diameter 3 mm, attached to a hypodermic needle (21G × 1.5″).
3. Peristaltic pump suitable for 3 mm (outer diameter) tubing.

2.6. Slot Blotting

1. A slot blotter (e.g. PR 648 slot blot filtration manifold unit (Hoeffer)).
2. Nitrocellulose membrane for slot blotter (e.g. Amersham Hybond ECL (GE Healthcare)).
3. Blotting paper (e.g. 3 MM paper (Whatman)).
4. UV crosslinker (e.g. Stratalinker (Stratagene)).

2.7. Detection of Covalent Complexes

1. General lab equipment and reagents for probing membranes with antibodies and detection: platform shaker, film developer, blocking solution (3% non-fat dry milk, 0.1% Tween-20 (e.g. Sigma P1379) in PBS), washing solution (PBS + 0.1% Tween-20), ECL detection agent (e.g. Amersham ECL Western Blotting Detection Reagents (GE Healthcare)), light-sensitive film for detection (e.g. Amersham Hyperfilm ECL (GE Healthcare)), X-ray film cassette.
2. Antibodies (*see* **Note 2**): Mouse monoclonal anti-HA antibody (sc-7392, Santa Cruz), use 1:2,000. A specific antibody against the *S. pombe* Top1 sequence FSKREDVPIEKLFSK, nine amino acids downstream of the active tyrosine, was raised in rabbit and affinity purified by Eurogentec. This antibody was used 1:2,000. Secondary HRP-conjugated antibodies (e.g. polyclonal rabbit anti-mouse HRP (PO260, Dako) or polyclonal swine anti-rabbit HRP (PO217, Dako)).

3. Methods

3.1. Preparation and Treatment of *S. pombe* Cultures

While the procedures described here are optimised for the detection of covalent complexes in *S. pombe*, they can easily be adapted for other organisms. Mammalian cells can simply be lysed by resuspension in lysis buffer, after which the protocol can be continued from **Section 3.2**, Step 5 (13).

3.1.1. Meiotic Cultures

For basic *S. pombe* methods, *see* (15). This section describes the preparation of synchronous meiotic *S. pombe* cultures using the temperature-sensitive *pat1-114* mutant (*see* **Note 3**) and is based on previously described procedures (14, 17). In short, *S. pombe* cells are grown at 25°C in EMM2 media, and then transferred to EMM2 minus nitrogen to arrest the cells in G1. Meiosis is induced by a temperature shift to 34°C and addition of nitrogen. For four samples, 100 ml of meiotic culture is enough; the procedure can be scaled up as required.

1. Day 1. Set up a pre-culture of *S. pombe pat1-114* cells in 10 ml EMM2 (15), grow overnight (or till saturation) at 25°C in a shaking incubator.

2. Day 2. Inoculate overnight pre-culture in 100 ml EMM2 in an Erlenmeyer flask, such that the culture reaches a density of 5×10^6 cells/ml at a convenient time the next day (doubling time of WT cells in EMM2 at 25°C is approximately 5 h). Incubate at 25°C in a shaking incubator.
3. Day 3. Count the cells in a haemocytometer. When the cell density reaches 5×10^6 cells/ml, centrifuge culture (5 min 2,000×*g* in a swing-out centrifuge). Discard supernatant.
4. Wash cell pellet: resuspend in 100 ml H_2O, centrifuge (5 min 2,000×*g*). Discard supernatant.
5. Resuspend in original volume (*see* **Note 4)** of EMM2 without nitrogen (this medium may contain 10 mg/l adenine, *see* **Note 3**). Incubate at 25°C in a shaking incubator for 20 h.
6. Day 4. To start meiosis, add 2.5 ml NH_4Cl from a 20% stock solution and shift the temperature to 34°C (*see* **Note 5**). Take 25 ml samples for further processing (*see* **Note 6**). Continue with **Section 3.2**.

3.1.2. Mitotic Cultures and Treatment with Topoisomerase Poisons

For two samples, 100 ml of culture is enough and the procedure can be scaled up as required.

1. Day 1. Set up a pre-culture of *S. pombe* cells in yeast extract (YE, *see* (15)), grow overnight (or till saturation) at 30°C in a shaking incubator.
2. Day 2. Inoculate overnight pre-culture in 100 ml YE in an Erlenmeyer flask, such that the culture reaches a density of 5×10^6 cells/ml at a convenient time the next day (doubling time of WT cells in YE at 30°C is approximately 2.5 h). Incubate at 30°C in a shaking incubator.
3. Day 3. Count the cells in a haemocytometer. When the cell density is 5×10^6 cells/ml, add CPT at a final concentration of 50 μM or TOP-53 at a final concentration of 100 μg/ml. Incubate for 15 min (*see* **Note** 7) at 30°C in a shaking incubator. Take 25 ml samples for further processing (*see* **Note 6**).

3.2. Preparing the Cell Extract

1. Centrifuge the cells for 5 min at 2,000×*g* in a swing-out rotor. Discard the supernatant and resuspend cells in 1 ml lysis buffer (*see* **Note 8**). Transfer to 2 ml screw cap tubes.
2. Centrifuge for 30 s in a microcentrifuge at 16,000×*g*, discard supernatant, resuspend in 750 μl lysis buffer.
3. Add two Eppendorf lids full of glass beads (approximately 1.2 g) to the tubes, close the tubes, and freeze in liquid nitrogen (*see* **Note 9**).
4. Thaw the cells on ice. Lyse the cells in a Fastprep machine, 45 s at maximum speed (*see* **Note 10**).

5. Incubate for 15 min at 65°C.
6. Centrifuge 5 min at 16,000×*g*.
7. Transfer 600 μl to a new microcentrifuge tube (*see* **Note 11**), add 500 μl lysis buffer, and mix (*see* **Note 12**).
8. Centrifuge 15 min at 16,000×*g*. After centrifugation, the supernatant will be loaded on a CsCl step gradient. While waiting for the centrifugation, prepare the CsCl gradients as described under **Section 3.3**.

3.3. Preparing the CsCl Gradients

1. Add 1 ml of CsCl 1.82 g/ml to each ultracentrifuge tube.
2. Very carefully layer 1 ml of CsCl 1.72 g/ml on top of the first layer using a cut-off pipette tip. Repeat this for the 1.50 g/ml and 1.45 g/ml CsCl solutions.
3. Load 1 ml of cell extract in lysis buffer (from **Section 3.2**, Step 8) on the step gradients.
4. Centrifuge for 24 h at approximately 107,000×*g* (at maximum radius) at 25°C in a swing-out rotor (30,000 RPM in a Sorvall AH650).

3.4. DNA Quantification

Ensuring that equal amounts of DNA are loaded on the slot blot is essential to reproducibly detect small differences in the amount of covalently bound Spo11/topoisomerase between different mutants or experimental conditions (e.g. *see* (10)). Due to the presence of many contaminants in the cell extract, quantifying DNA by absorbance measurement at 260 and 240 nm is far from accurate. Therefore, the DNA concentration is quantified fluorometrically using PicoGreen, which is relatively insensitive to most contaminants found in cell extract.

1. Incubate the remaining cell lysate (from **Section 3.2**, Step 8) at 65°C for 5 min (*see* **Note 13**).
2. Centrifuge 2 min at 16,000×*g*.
3. Add 10 μl of the supernatant to 90 μl of TE containing 0.5 μg/ml RNase A (*see* **Note 14**).
4. Incubate for 3 h, or overnight, at 37°C.
5. Centrifuge 2 min at 16,000×*g* to remove any insoluble material.
6. Add 50 μl of the supernatant to 50 μl of a 1:200 dilution of PicoGreen in TE, mix, and incubate for 2–5 min at room temperature. Prepare a blank control (50 μl TE) and DNA standard (50 μl 100 ng/ml λ DNA in TE) in parallel (*see* **Note 15**).
7. Calibrate a fluorometer using the blank control and DNA standard and measure the DNA concentration in the samples (typical concentration approximately 2.5 μg/ml).

3.5. Gradient Fractionation

1. Remove the tubes from the ultracentrifuge.
2. Clamp a centrifuge tube containing the gradient in a retort stand.
3. Fit silicone tubing into peristaltic pump.
4. Pierce tube with a needle (attached to silicone tubing) at an angle of 45° (*see* **Fig. 5.2**); move needle to a horizontal position, making sure that the opening is at the bottom of the tube facing upwards. Support the needle such that it stays horizontal (e.g. supported on the lid of a Pyrex bottle).
5. Using the peristaltic pump, slowly (± 5 ml/min) pump the gradient out of the tube and collect 0.5 ml fractions in labelled and numbered microcentrifuge tubes, using the 0.5 ml mark on the tube (*see* **Note 16**).

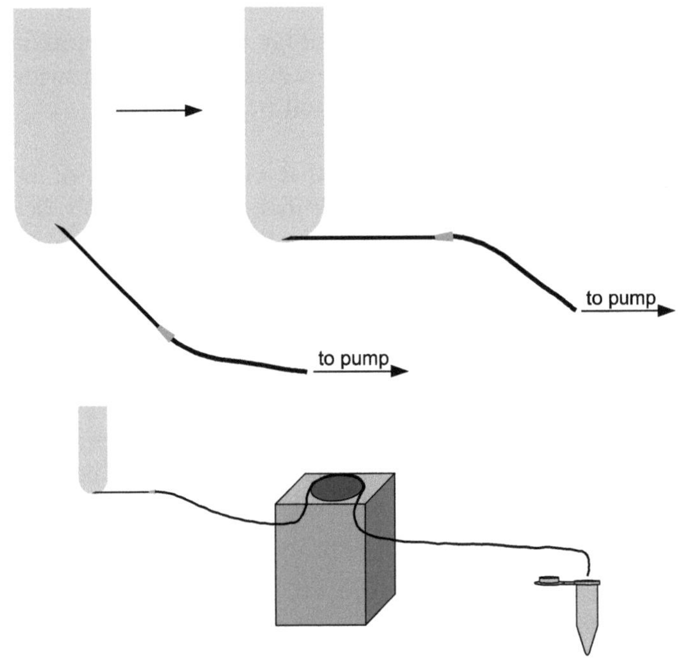

Fig. 5.2. Diagram explaining the fractionation procedure. The ultracentrifuge tube (fitted in a suitable retort stand) is pierced with a needle (attached to silicone tubing) at an angle of 45°. The needle is moved to a horizontal position such that the bevelled opening is at the bottom of the tube facing upwards. Fractions are pumped out of the tube using a peristaltic pump and collected in microcentrifuge tubes.

3.6. Slot Blotting

The volumes of the fractions loaded on the slot blot are adjusted to ensure equal loading.

1. Calculate the volumes needed to load equal amounts of DNA for each experimental condition. A total of 200 μl of the fractions coming from the cell lysate with the lowest

DNA concentration will be loaded (approximately 500 ng). Adjust the volume for the other fractions according to their DNA concentration (*see* **Note 17**).

2. Load the samples on the membrane using a slot blotter according to the manufacturer's instructions (*see* **Note 18**).
3. Once the samples have been sucked through the membrane, disassemble the slot blotter and dry the membrane face up on a piece of blotting paper.
4. Crosslink the DNA to the membrane using a Stratalinker (auto-crosslink, 120,000 μJ, 254 nm; this step might not be necessary).

3.7. Detection of Covalent Complexes

1. Block the membrane for 30 min in blocking solution on a shaker.
2. Remove the blocking solution and incubate the membrane in the appropriate dilution of primary antibody in blocking solution on a shaker overnight at 4°C or for 2 h at room temperature (*see* **Note 19**).
3. Remove the antibody solution and briefly rinse the membrane three times in a small volume of washing solution.
4. Wash the membrane three times 10 min in at least 100 ml of PBS + 0.1% Tween-20.
5. Incubate the membrane in the appropriate dilution of secondary antibody in blocking solution on a shaker overnight at 4°C or for 1 h at room temperature (*see* **Note 19**).
6. After discarding the antibody, wash the membrane three times 10 min in at least 100 ml of PBS + 0.1% Tween-20.
7. After removing the final wash, rinse the membrane in a mix of 1 ml each of the ECL reagents, remove the membrane from the reagents, remove excess liquid using absorbent paper wipes, and place in between two acetate sheets.
8. Put the assembly in an X-ray film cassette, face up, and place a film on top.

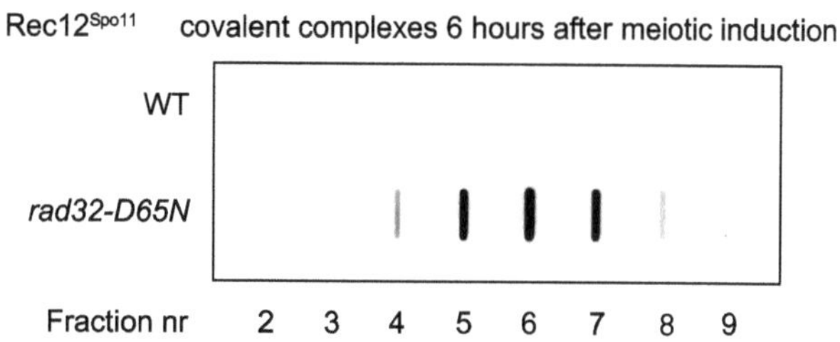

Fig. 5.3. Example of detection of covalent Rec12^{Spo11} complexes 6 h after initiation of meiosis. While in WT cells Rec12^{Spo11} has been removed, *rad32-D65N* nuclease-deficient cells are unable to remove the complex.

9. Expose for 2–3 min and develop film, adjust exposure for subsequent film if necessary. An example of what the result should look like is shown in **Fig. 5.3** (*see* **Note 20**).

4. Notes

1. When I developed the DLPD assay, I was unaware that a similar procedure, called the ICE Bioassay, had been previously described (18). While the principles on which these assays are based are similar, they differ considerably in technical detail. Lysing conditions as used in the DLPD assay are much harsher than in the ICE Bioassay, possibly contributing to a reduction of non-specific background signal. Also, the DNA quantification procedure as described in this chapter is essential to achieve equal loading and allows reproducible detection of small differences between mutants or experimental conditions, as demonstrated in (10).
2. Good antibodies are a prerequisite for success. Some antibodies, while suitable for use on Western blots, show a high non-specific background in the DLPD assay. Preferably, use monoclonal or affinity-purified polyclonal antibodies.
3. Please note that various artefacts have been reported for *pat1-114*-synchronised meiosis (14, 19). Alternatively, synchronous meiosis can be performed as described in (20). Efficient nitrogen starvation as used to synchronise *pat1-114* meiosis can only be performed in the absence of the common supplements uracil, histidine, leucine, and arginine, therefore the strain should be prototrophic for these substances. As nitrogen starvation works in the presence of 10 mg/l adenine, the strain can be auxotrophic for adenine.
4. This volume can be adjusted to correct for a deviation from the optimum cell density of 5×10^6 cells/ml. Please note that cell densities significantly higher (or lower) than 5×10^6 when shifted to EMM2 without nitrogen might negatively affect the degree of meiotic synchrony.
5. The maximum temperature at which *S. pombe* meiosis can be performed is 34°C. An orbital shaking water bath is ideal to shift the temperature of the culture quickly to 34°C. Alternatively, shake the Erlenmeyer flask containing the culture in a large volume of warm water. Insert a thermometer in the culture to monitor the temperature.

6. Optionally, add sodium azide to a final concentration of 0.1% and EDTA to a final concentration of 5 mM to stop biological processes and inhibit nuclease activity.
7. The effect of these drugs is instantaneous, and covalent complexes can be seen after as little as 1 min of treatment (10).
8. Optionally, add CPT or TOP-53 to the same concentration as used in the treatment of the cultures to prevent spontaneous resolution of the covalent complexes.
9. Quickly freezing the cells in liquid nitrogen is important to reduce the time at which a temperature-sensitive mutant is at permissive temperature during the cooling procedure. This might not be necessary for WT or mutants which are not temperature sensitive, but provides a convenient break point in the protocol.
10. If you do not have access to a Fastprep-24 (or Precellys 24) machine you can break the cells by vortexing for several minutes (15).
11. Try to minimise the transfer of glass beads. Alternatively, to separate the extract from the glass beads, the bottom of the tube can be pierced with a 25 Gauge syringe needle. Place the pierced tube on top of another 2 ml screw cap tube and place the assembly in a 15 ml centrifuge tube. Centrifuge for 4 min at 2,000×*g*. The extract should be in the bottom tube, while the beads should have remained in the top tube.
12. This is done to bring the volume up to 1.1 ml; 1 ml will be loaded on the gradient, after which 100 μl remains which can be used for subsequent DNA quantification.
13. This is to dissolve any precipitated GuHCl.
14. Large amounts of RNA interfere with the DNA measurement.
15. The volumes and DNA standard concentration used here are specific for the Turner TBS-380 fluorometer and might need to be adapted for other fluorometers.
16. Normally, only fractions 1–8 are collected, as fractions 9 and 10 contain free proteins, lipids, and other cellular components.
17. To calculate the volumes, the following formula gives you the amount in microlitre: $(A/B) \times 200$, where A is the DNA concentration of the sample with the lowest concentration and B is the DNA concentration of the sample to be loaded.

18. Also, *see* **Note 16**. Fractions 9 and 10 (and sometimes fraction 8) often block the membrane and might take a very long time to go through.
19. Optimal conditions (e.g. dilution, length of incubation) differ between antibodies and need to be determined empirically.
20. To compare amounts of covalent complexes between mutants or different experimental conditions, signal strengths can be quantified from film. To adjust for non-linearity between signal strength and film blackening, especially for weak signals, make a twofold dilution series of the fraction with the strongest signal (usually fraction 6) and load this on the slot blotter. This can be used to create a standard curve to allow correction of non-linearity and to determine relative protein amounts. After the film has been scanned using a standard flatbed scanner (make sure to avoid saturation), signals can be quantified using ImageJ (http://rsb.info.nih.gov/ij).

References

1. Champoux, J.J. (2001) DNA topoisomerases: structure, function, and mechanism. *Annu Rev Biochem* **70**, 369–413.
2. Pommier, Y. (2004) Camptothecins and topoisomerase I: a foot in the door. Targeting the genome beyond topoisomerase I with camptothecins and novel anticancer drugs: importance of DNA replication, repair and cell cycle checkpoints. *Curr Med Chem Anticancer Agents* **4**, 429–434.
3. Baldwin, E.L., and Osheroff, N. (2005) Etoposide, topoisomerase II and cancer. *Curr Med Chem Anticancer Agents* **5**, 363–372.
4. Pouliot, J.J., Yao, K.C., Robertson, C.A., and Nash, H.A. (1999) Yeast gene for a Tyr-DNA phosphodiesterase that repairs topoisomerase I complexes. *Science* **286**, 552–555.
5. Cortes Ledesma, F., El Khamisy, S.F., Zuma, M.C., Osborn, K., and Caldecott, K.W. (2009) A human 5′-tyrosyl DNA phosphodiesterase that repairs topoisomerase-mediated DNA damage. *Nature* **461**, 674–678.
6. Connelly, J.C., and Leach, D.R.F. (2004) Repair of DNA covalently linked to protein. *Mol Cell* **13**, 307–316.
7. Neale, M.J., Pan, J., and Keeney, S. (2005) Endonucleolytic processing of covalent protein-linked DNA double-strand breaks. *Nature* **436**, 1053–1057.
8. Neale, M.J., and Keeney, S. (2009) End-labeling and analysis of Spo11-oligonucleotide complexes in Saccharomyces cerevisiae. *Methods Mol Biol* **557**, 183–195.
9. Hartsuiker, E., Mizuno, K., Molnar, M., Kohli, J., Ohta, K., and Carr, A.M. (2009) Ctp1CtIP and Rad32Mre11 nuclease activity are required for Rec12Spo11 removal, but Rec12Spo11 removal is dispensable for other MRN-dependent meiotic functions. *Mol Cell Biol* **29**, 1671–1681.
10. Hartsuiker, E., Neale, M.J., and Carr, A.M. (2009) Distinct requirements for the Rad32(Mre11) nuclease and Ctp1(CtIP) in the removal of covalently bound topoisomerase I and II from DNA. *Mol Cell* **33**, 117–123.
11. Keeney, S., Giroux, C.N., and Kleckner, N. (1997) Meiosis-specific DNA double-strand breaks are catalyzed by Spo11, a member of a widely conserved protein family. *Cell* **88**, 375–384.
12. Shaw, J.L., Blanco, J., and Mueller, G.C. (1975) Simple procedure for isolation of DNA, RNA and protein fractions from cultured animal cells. *Anal Biochem* **65**, 125–131.
13. El-Khamisy, S.F., Hartsuiker, E., and Caldecott, K.W. (2007) TDP1 facilitates repair of ionizing radiation-induced DNA single-strand breaks. *DNA Repair (Amst)* **6**, 1485–1495.

14. Bähler, J., Schuchert, P., Grimm, C., and Kohli, J. (1991) Synchronized meiosis and recombination in fission yeast: observations with pat1-114 diploid cells. *Curr Genet* **19**, 445–451.
15. Forsburg, S.L., and Rhind, N. (2006) Basic methods for fission yeast. *Yeast* **23**, 173–183.
16. Utsugi, T., Shibata, J., Sugimoto, Y., Aoyagi, K., Wierzba, K., Kobunai, T., Terada, T., Ohhara, T., Tsuruo, T., and Yamada, Y. (1996) Antitumor activity of a novel podophyllotoxin derivative (TOP-53) against lung cancer and lung metastatic cancer. *Cancer Res* **56**, 2809–2814.
17. Cervantes, M.D., Farah, J.A., and Smith, G.R. (2000) Meiotic DNA breaks associated with recombination in S. pombe. *Mol Cell* **5**, 883–888.
18. Subramanian, D., Furbee, C.S., and Muller, M.T. (2001) ICE bioassay. Isolating in vivo complexes of enzyme to DNA. *Methods Mol Biol* **95**, 137–147.
19. Chikashige, Y., Kurokawa, R., Haraguchi, T., and Hiraoka, Y. (2004) Meiosis induced by inactivation of Pat1 kinase proceeds with aberrant nuclear positioning of centromeres in the fission yeast Schizosaccharomyces pombe. *Genes Cells* **9**, 671–684.
20. Bähler, J., Wyler, T., Loidl, J., and Kohli, J. (1993) Unusual nuclear structures in meiotic prophase of fission yeast: a cytological analysis, *J. Cell Biol* **121**, 241–256.

Chapter 6

Molecular Assays to Investigate Chromatin Changes During DNA Double-Strand Break Repair in Yeast

Scott Houghtaling, Toyoko Tsukuda, and Mary Ann Osley

Abstract

Multiple types of DNA damage, including bulky adducts, DNA single-strand breaks, and DNA double-strand breaks (DSBs), have deleterious effects on the genomes of eukaryotes. DSBs form normally during a variety of biological processes, such as V–D–J recombination and yeast mating type switching, but unprogrammed DSBs are among the most dangerous types of lesion because if left unrepaired they can lead to loss of genetic material or chromosomal rearrangements. The presence of DSBs leads to a DNA damage response involving activation of cell cycle checkpoints, recruitment of repair proteins, and chromatin remodeling. Because many of the proteins that mediate these processes are evolutionarily conserved, the budding yeast, *Saccharomyces cerevisiae*, has been used as a model organism to investigate the factors involved in the response to DSBs. Recent research on DSB repair has focused on the barrier that chromatin represents to the repair process. In this article, we describe molecular techniques available to analyze chromatin architecture near a defined DSB in budding yeast. These techniques may be of value to experimentalists who are investigating the role of a novel protein in DSB repair specifically in the context of chromatin.

Key words: DNA double-strand break repair, yeast, chromatin, nucleosome remodeling.

1. Introduction

1.1. DSB Repair in Yeast

DSBs can be repaired by non-homologous end joining (NHEJ) or homologous recombination (HR) (*see* **Fig. 6.1**) (1). NHEJ is used in the G1 phase of the cell cycle and HR functions during S/G2 phases when sister chromatids are available as templates for repair. During NHEJ in yeast, DSBs are recognized by the Ku hetero-dimer (Ku70-Ku80), processed by the Mre11/Rad50/Xrs1 (MRX) complex, and ligated by DNA

H. Tsubouchi (ed.), *DNA Recombination*, Methods in Molecular Biology 745,
DOI 10.1007/978-1-61779-129-1_6, © Springer Science+Business Media, LLC 2011

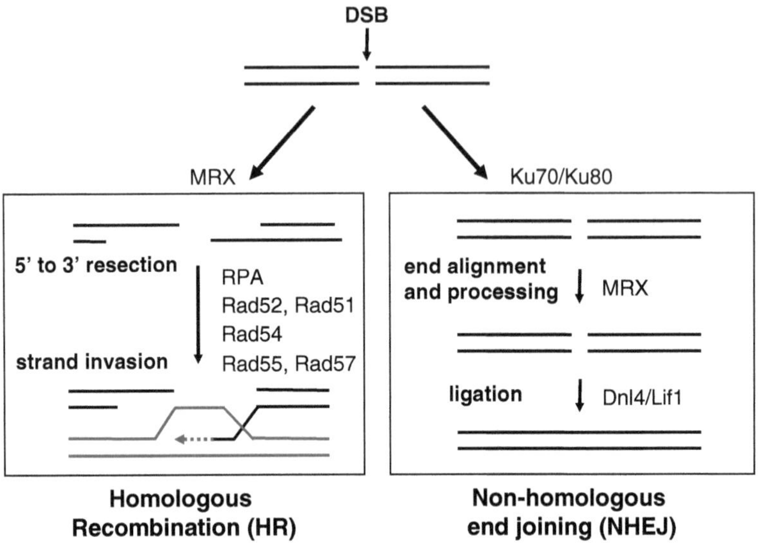

Fig. 6.1. DSB repair by HR or NHEJ. The DNA strand interactions and key repair factors associated with DSB repair by NHEJ and HR pathways are indicated (see text for additional details).

ligase IV. The NHEJ pathway requires little strand modification and results in direct rejoining of DNA ends (2, 3), but NHEJ is often an error-prone process that involves the deletion or the insertion of bases at the DSB.

Because DSB repair by HR uses homologous DNA as a template, it is usually an error-free process that maintains genomic integrity. In yeast, HR repair begins with end resection that is regulated by the MRX complex to reveal single-stranded DNA (ssDNA) at the DSB (4). This ssDNA overhang is bound by RPA, followed by Rad52 and Rad51. The Rad51-coated ssDNA then searches for a region of homology on a homologous chromosome or a sister chromatid, and DNA strand extension occurs from the invading 3′-end of the Rad51 filament. Rad54, a member of the Snf2 family of helicases, is a central component of HR that works in concert with Rad51, and evidence suggests that it functions both before and after synapsis of ssDNA with homologous duplex DNA (5).

1.2. Chromatin Dynamics During DSB Repair

Repair of a DSB by either HR or NHEJ takes place in a chromatin context. The basic repeating unit of chromatin is the nucleosome, which consists of 147 base pairs of DNA wrapped approximately two times around a histone octamer composed of two H2A–H2B dimers and an H3–H4 tetramer (6). Nucleosomes assemble into arrays of increasingly dense structures, and the compaction of DNA into chromatin presents an obstacle to proteins that act during many DNA transactions, including DSB repair.

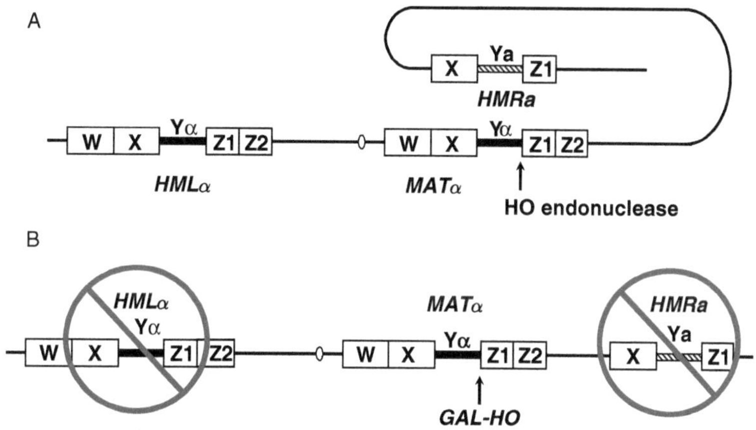

Fig. 6.2. The yeast *MAT* locus. (**a**) Schematic of the *S. cerevisiae MAT* locus on chromosome III. The *MATα* locus produces two regulatory transcripts from the Yα region. During late G1 phase of the cell cycle, the HO endonuclease creates a DSB at *MATα*, which then switches to *MATa* by gene conversion from a transcriptionally silent copy located at *HMRa*. W, X, and Z represent homology blocks present at the *MAT* and *HM* loci. (**b**) For analysis of events that occur in the presence of a persistent DSB at *MATα*, the two silent *HM* loci have been deleted and the *HO* gene has been placed under control of the *GAL10* promoter. The *MAT* DSB can be formed at all phases of the cell cycle by inducing the *GAL-HO* gene with galactose.

Multiple factors that remodel chromatin during DSB repair have been identified in yeast (7). In the context of the defined DSB model described in **Fig. 6.2**, these factors are recruited to the DSB, where they promote specific alterations to chromatin. The first group of factors includes modifying enzymes that mediate specific post-translational modifications of histones, which provide binding sites for repair and other remodeling proteins or alter chromatin folding (8). The Mec1/Tel1-mediated phosphorylation of histone H2A on its C terminus (equivalent to γH2A.X in vertebrates) (9, 10) is one of the earliest and most extensive chromatin modifications to appear at DSBs, closely followed by NuA4-dependent histone acetylation (11, 12). A second group of factors includes ATP-dependent nucleosome-remodeling complexes that disrupt contacts between histones and DNA, thus allowing the repositioning or the removal of nucleosomes (13). Many of these nucleosome-remodeling complexes, including INO80, SWR, RSC, and Swi-Snf, have been found to play key roles in DSB repair in yeast and likely function by allowing access of repair proteins at the DSB (14–19). While most of these factors promote efficient HR repair, SWR and INO80 also function during NHEJ repair to facilitate recruitment of end-joining proteins such as Ku80 (20). Many of the efforts in defining the roles of nucleosome-remodeling enzymes in DSB repair have focused on a specific locus where a DSB can be formed in the absence of a donor template for repair. However, roles for RSC,

Swi–Snf, INO80, and Rad54 at donor templates during HR have also been reported (5, 14, 21). However, we still do not understand how these remodelers affect chromatin structure at a donor locus. Just how the activities that remodel chromatin structure are integrated with the events associated with DSB repair is an area of intense investigation in a number of laboratories.

2. Materials

2.1. Strain Information

The assays described below utilize yeast strains in which a defined DSB can be created at the mating type or *MAT* locus with high efficiency through galactose-mediated induction of the HO endonuclease that specifically targets a unique sequence at this locus (*see* **Fig. 6.2**). The most commonly used strain, JKM179 (*MATα Δho Δhml::ADE1 Δhmr::ADE1 ade1 leu2,3-112 lys5 trp1::hisG ura3-52 ade3::GAL10-HO*), carries deletions of the two silent mating type loci (*HML* and *HMR*) used as HR donor sequences and is typically used for studies of chromatin changes that occur at a DSB in the absence of HR (22). To monitor chromatin changes that occur during HR, the HR competent strain, XW756 (*HMLα HMRa MATα-BamH1 lys5 trp1 ade1 leu2 ura3-52 ade3::GAL-HO*), is used (21). This strain carries both silent mating type loci, which provide donor templates for HR during *MAT* switching. Both strains have been engineered to contain a genomic insertion of an N-terminal, Flag epitope-tagged *HTB1* gene at the endogenous *HTB1* locus to facilitate measurement of H2B levels by chromatin immunoprecipitation (21, 23). These strains can be further manipulated to delete DNA repair or chromatin remodeling genes or to place selected epitope tags at the N or C terminus of any factor of interest using standard yeast genetic tools (24–26).

2.2. Cell Growth, Formaldehyde Fixation, Quenching, Washing, and Cell Lysis

1. Yeast growth media (GLGYP): 3% glycerol, 2% lactic acid, 0.05% glucose, 1% yeast extract, 2% peptone (pH 5.5).
2. Galactose (20% stock).
3. Glucose (20% stock).
4. Methanol-free formaldehyde (37% stock) (Polysciences, Inc., Cat. # 04018).
5. Glycine (2.5 M stock).
6. 1× TBS: 0.05 M Tris (pH 8.0), 0.138 M NaCl, 0.0027 M KCl.
7. Glass beads, acid washed (425–600 μm) (Sigma, Cat. # G8772-500G).

2.3. Analysis of Gross Chromatin Changes Near the MAT DSB by MNase Digestion and Southern Blot Analysis

1. Zymolyase 20T (25 mg/ml stock in zymolyase buffer) (Seikagaku Biosciences, Cat. # 120491).
2. Zymolyase buffer: 1 M sorbitol, 50 mM Tris (pH 7.4) with freshly added 2- mercaptoethanol (10 mM final concentration).
3. Micrococcal nuclease (MNase) (Worthington, Cat. # 4797) (2 units/μl stock).
4. MNase buffer: 1 M sorbitol, 50 mM NaCl, 10 mM Tris (pH 7.4), 5 mM $MgCl_2$, 1 mM $CaCl_2$, 0.075% NP40 with freshly added 1 mM 2-mercaptoethanol, and 500 μM spermidine.
5. EDTA (500 mM stock).
6. Proteinase K (Sigma, Cat. # p4850).
7. Phenol:chloroform:isoamyl alcohol (25:24:1) saturated with 10 mM Tris (pH 8), 1 mM EDTA.
8. 100% Ethanol.
9. 70% Ethanol.
10. RNase A (25 mg/ml stock).
11. Agarose (molecular biology grade).
12. TAE running buffer: 0.04 M Tris–acetate, 0.001 M EDTA.
13. Membrane for Southern blotting.
14. Radiolabeled DNA probe.

2.4. Analysis of Nucleosome Positioning by Indirect End Labeling

1. S-buffer: 1 M sorbitol, 50 mM Tris (pH 7.4).
2. 2-Mercaptoethanol (14.3 M stock).
3. 10% SDS.
4. Lyticase (Sigma, Cat. # L5263): 100 units/μl in 50% glycerol, 50% 50 mM Tris (pH 7.5), stored at –20°C.
5. MNase buffer: 1 M sorbitol, 15 mM Tris (pH 8), 1 mM $MgCl_2$, 50 mM NaCl, with PMSF added to 0.5 mM prior to use.
6. $CaCl_2$ (100 mM stock).
7. MNase (Worthington): 15 units/μl stock in water stored at –20°C.
8. Termination solution: 250 mM EDTA, 5% SDS, 50 mM Tris (pH 8).
9. Proteinase K (10 mg/ml stock).
10. Phenol:chloroform:isoamyl alcohol (25:24:1) saturated with 10 mM Tris (pH 8), 1 mM EDTA.
11. Chloroform:isoamyl alcohol (24:1).
12. 100% Ethanol.

13. 70% Ethanol.
14. TE: 10 mM Tris (pH 8), 1 mM EDTA.
15. RNase A (25 mg/ml stock stored at –20°C).
16. Qiagen spin column (Qiagen, Cat. # 28106).
17. BspEI restriction enzyme and appropriate buffer.
18. Agarose (molecular biology grade).
19. TBE running buffer: 89 mM Tris, 89 mM boric Acid, 2 mM EDTA (pH 8).
20. Membrane for Southern blotting.
21. Radiolabeled DNA probe.

2.5. Analysis of Nucleosome Positioning by qPCR Scanning

1. FA lysis buffer: 50 mM HEPES–KOH (pH 7), 140 mM NaCl, 1 mM EDTA, 1% Triton, 0.1% Na deoxycholate (filter sterilized and stored at 4°C).
2. Protease inhibitor cocktail (PI) (50× stock) (Sigma, Cat. # P2714).
3. PMSF (100 mM stock).
4. Acid-washed glass beads (size 425–600 μm; Sigma, Cat. # G8772).
5. MNase buffer: 15 mM Tris (pH 7.8), 10 mM NaCl, 1.4 mM $CaCl_2$, 0.2 mM EDTA.
6. EDTA (500 mM stock).
7. Adjusting buffer: 75 mM HEPES (pH 7.5), 200 mM NaCl, 1.5% Triton X-100, 0.15% Na deoxycholate (filter sterilized and stored at 4°C).
8. Elution buffer: 10 mM Tris (pH 8.5), 1% SDS, 1 mM EDTA.
9. Pronase (Sigma, Cat. # P6911).
10. $CaCl_2$ (1 M stock).
11. Qiagen spin column (Qiagen, Cat. # 28106).
12. SYBR Green PCR master mix (ABI or Fermentas).

2.6. Analysis of Nucleosome Occupancy by ChIP

1. ANTI-FLAG M2 affinity gel (Sigma, A220).
2. FA lysis buffer: 50 mM HEPES–KOH (pH 7.5), 140 mM NaCl, 1 mM EDTA, 1% Triton, 0.1% Na deoxycholate (filter sterilized and stored at 4°C).
3. Wash buffer #1: 0.05 M HEPES–KOH (pH 7.5), 0.5 M NaCl, 0.001 M EDTA, 1% Triton, 0.1% Na deoxycholate (filter sterilized and stored at 4°C).
4. Wash buffer #2: 0.25 M LiCl, 0.5% NP-40, 0.5% NA deoxycholate, 10 mM Tris, 1 mM EDTA (filter sterilized and stored at 4°C).

5. Elution buffer: 10 mM Tris (pH 7.4), 1% SDS, 1 mM EDTA.
6. Pronase (20 mg/ml stock).
7. $CaCl_2$ (1 M stock).
8. Qiagen spin column (Qiagen, Cat. # 28106).
9. SYBR Green PCR master mix (ABI or Fermentas).

3. Methods

DSB repair occurs in the context of a chromatin structure that must be remodeled to allow for efficient repair. Many of the chromatin remodeling enzymes first identified in the regulation of transcription are also required for efficient repair of DSBs (13). Repair factors that affect the efficiency of DSB repair could function either directly on DNA or at a step that involves the post-translational modification of histones or remodeling of nucleosomes. A number of molecular assays can be utilized to analyze chromatin architecture during repair of DSBs. In the following sections, we present assays for determining the occupancy and positioning of nucleosomes at the *MAT* DSB (*see* **Fig. 6.2**). These assays are presented in order of increasing resolution. The first assay allows for analysis of chromatin structure across a region encompassing many nucleosomes and the last assay allows for analysis of individual nucleosomes at defined positions.

3.1. Analysis of Gross Chromatin Changes Near the MAT DSB by MNase Digestion and Southern Blot Analysis

The chromatin organization at *MAT* after HO cleavage can be analyzed using micrococcal nuclease (MNase) digestion of chromatin followed by Southern blot analysis. This assay allows for a qualitative comparison of the gross chromatin state at *MAT* between different conditions or between different yeast mutants during repair of the DSB. For example, it has been used to investigate the role of the nucleosome-remodeling complex, INO80, in chromatin reorganization during DSB repair at *MAT* (19). MNase cleaves linker DNA between nucleosomes, and regions with well-positioned nucleosomes show greater protection from MNase digestion, whereas regions that have relatively poorly positioned nucleosomes are digested more readily. The method described below has been adapted from previously described protocols (27, 28). This and the other methods outlined in this chapter are typically performed in the donorless strain, JKM179, in which a persistent DSB forms that cannot be repaired due to deletion of the HR repair templates, *HMRa* and *HMLα* (*see* **Fig. 6.2**) (23). This strain serves as the wild-type control when comparing effects of deletions of genes encoding repair or chromatin remodeling factors. Importantly, the same techniques can also be applied to examine chromatin changes at *MAT* during HR repair in the switchable strain XW756 (*see* **Note 1**):

1. Grow 2–4 l of JKM179 cells in GLGYP medium to mid-log phase (Abs_{600} ~0.6–0.8). Retain 500 ml as an uninduced control and add galactose to a final concentration of 2% to induce *HO*. At 1 h intervals, remove 500 ml of culture. Collect cells by centrifugation (5,000 rpm for 5 min at room temperature) and wash once with 50 ml of room temperature water to remove media.
2. Resuspend cell pellet in 6 ml of zymolyase buffer in a 15-ml conical tube. Add zymolyase 20T to a final concentration of 0.25 mg/ml. Lyticase can also be used for the treatment of cells to generate spheroplasts (*see* **Section 3.2**). Treat cells with zymolyase for 30 min with gentle rolling at 30°C. Monitor spheroplast formation (*see* **Note 2**). Collect nuclei by centrifugation at 4,000 rpm at 4°C for 10 min. Keep the pellet on ice.
3. Resuspend nuclei in 2 ml of MNase buffer. Divide samples into six 300 μl aliquots and transfer to 1.5-ml microcentrifuge tubes. Perform MNase digestion of the samples at 37°C using a fixed concentration (10–15 units) of enzyme over time (0, 1, 2, 4, 8, and 16 min). Stop the reaction by adding EDTA (25 mM final concentration) and SDS (0.5% final concentration) (*see* **Note 3**).
4. Add 2 μl proteinase K and incubate samples at 50°C for 2 h.
5. Purify DNA by sequential phenol–chloroform extraction, chloroform extraction, and ethanol precipitation with a final wash in 70% ethanol.
6. Resuspend the DNA in 40 μl sterile water and add 2 μl RNase A. Incubate at 37°C for 1–2 h.
7. Resolve the purified DNA on a long (20 cm) 1.5% agarose TAE gel.
8. Transfer the DNA to an appropriate membrane and perform Southern blot analysis with a radiolabeled 800-bp DNA probe corresponding to sequences ~200 bp to the right of the HO cut site (19).

In JKM179 wild-type cells, a ladder of bands representing positioned nucleosomes surrounding *MAT* will appear increasingly diffuse over time following HO induction. A mutant strain that has a defect in nucleosome remodeling may have a delay in the appearance of this diffuse pattern or show no change in the ladder (19).

3.2. Analysis of Nucleosome Positioning by Indirect End Labeling

A higher resolution method for measuring nucleosome positioning at *MAT* involves the mapping procedure known as indirect end labeling (29–32). This assay provides improved resolution over the above method and has the benefit of being able to map the precise position of specific nucleosomes before and after induction of a DSB. In this assay, spheroplasts are digested with MNase, followed by digestion of purified DNA with a restriction enzyme flanking the HO cut site at *MAT* (BspEI or BanII). Southern blot analysis is performed using a radiolabeled probe that abuts the restriction enzyme recognition site. The precise position of nucleosomes can be inferred by measuring band positions relative to this site (*see* **Fig. 6.3**). Changes in band position can be compared before and after DSB formation to analyze changes in nucleosome position. Importantly, this assay is suitable for monitoring nucleosome dynamics near the DSB at *MAT* only during a short time interval after DSB formation because over longer periods (greater than 1 h) there is interference from extensive end resection that occurs in the donorless strain. In the modified protocol described below, formaldehyde fixed cells are

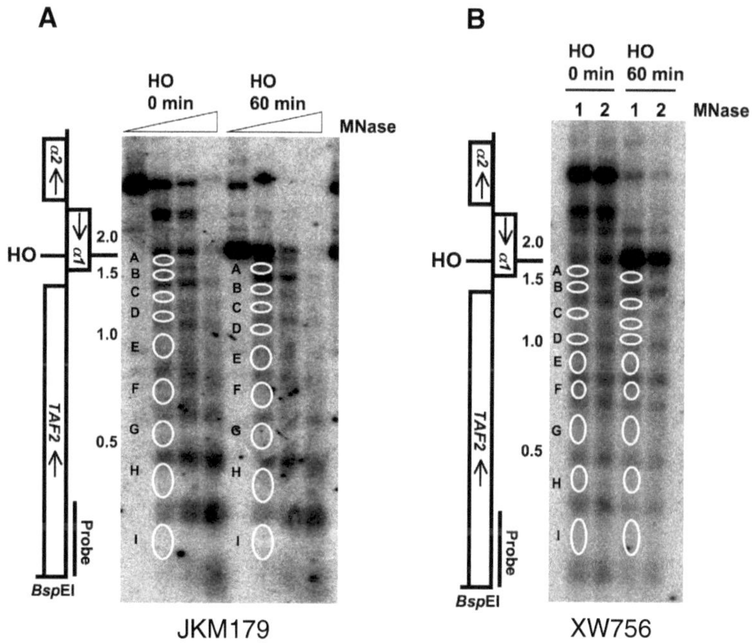

Fig. 6.3. Measurement of nucleosome positioning at *MAT*α by indirect end labeling. (**a**) A donorless (JKM179) or (**b**) switchable (XW756) strain was induced for 60 min with galactose to create a DSB at the *MAT*α locus, or left untreated (0 min), and chromatin was fixed with formaldehyde. Spheroplasts were prepared and incubated with MNase over time. Following DNA purification, samples were digested with BspEI, electrophoresed on a 1.5% agarose gel, and transferred to a nylon membrane. Southern blot analysis was performed with a short radiolabeled probe that abuts the BspEI site. The positions of 3–4 nucleosomes shift adjacent to the *MAT* cut site in both strains.

utilized so that histone/DNA interactions can be captured at the moment that cross-linking occurs:

1. Grow JKM179 cells in 1 L GLGYP media overnight to mid-log phase (Abs_{600} ~0.6–0.8) (*see* **Note 1**).
2. Add galactose to a final concentration of 2% to induce HO endonuclease.
3. At time 0, prior to HO induction, and at 1 h after induction, fix cells with formaldehyde (final concentration 1%) for 15 min at room temperature with shaking. Quench fixation by adding glycine to a final concentration of 0.125 M for 5 min at room temperature with shaking. Collect cells by centrifugation (5,000 rpm for 5 min at room temperature) and wash with ice-cold TBS. Weigh the cell pellet. Freeze cell pellets on dry ice and store at –80°C.
4. Thaw cell pellet on ice and wash with 5 ml S-buffer. Centrifuge at 4,500 rpm for 2 min at room temperature to collect cells.
5. Resuspend cell pellet in S-buffer (4 ml/g of pellet). Add 1/200th volume of 2-mercaptoethanol (stock 14.3 M) and incubate for 20 min at room temperature with shaking.
6. Retain 10 μl of cells and dilute to 1 ml with 0.1% SDS to obtain an Abs_{600} reading for an untreated sample.
7. Add 1,000 units of lyticase per gram of cell pellet. Monitor the cells for spheroplast formation (*see* **Note 2**).
8. When 80–90% of the cells are converted to spheroplasts, immediately wash twice with 0.5 ml of cold S-buffer (centrifuge at 4,500 rpm for 3 min at 4°C) by gently resuspending spheroplasts with a spatula or a glass rod (do not vortex). Remove the supernatant and measure the weight of the spheroplast pellet. The pellet can be stored at –80°C or it can be immediately used in step 9.
9. Gently resuspend spheroplasts in 1 ml of MNase digestion buffer for each gram of spheroplast cell pellet from step 8 and transfer to a 1.5-ml microcentrifuge tube. Centrifuge for 1 min at 11,500 rpm in a microcentrifuge. Repeat washes and centrifugation twice.
10. Aliquot 0.5 ml of washed spheroplasts to a fresh 1.5-ml microcentrifuge tube and prewarm tube at 37°C for 2 min.
11. Add 1.5 units of MNase (Worthington) and incubate at 37°C. Start reaction by addition of 5 μl of 100 mM $CaCl_2$. Remove 90 μl aliquots at 1, 2, 4, 8, and 16 min and mix with 10 μl termination solution and place on ice.
12. Add 2 μL proteinase K and incubate at 50°C for 2 h. Incubate at 65°C for 5 h (or overnight) to decross-link chromatin.

13. Purify DNA by phenol–chloroform extraction, chloroform extraction, and ethanol precipitation, with a final wash in 70% ethanol.
14. Resuspend DNA in 200 μl TE.
15. Add 1 μl of 100 mg/ml RNase and incubate at 37°C for 1–2 h.
16. Purify DNA by a spin column following manufacturer's directions.
17. Digest DNA to completion with BspEI and separate DNA on a 19 cm × 18.5 cm 1.5% agarose TBE gel run at 150 V for 4 h at 4°C (*see* **Note 4**).
18. Blot gel to appropriate membrane and hybridize with a 100–150-bp radiolabeled probe abutting the restriction enzyme recognition site.

Indirect end labeling was used to map the positions of nucleosomes at the *MAT* locus before and after HO induction to attribute a role for the chromatin remodeling complex, RSC, in the repositioning of several nucleosomes next to the DSB (16, 33). This repositioning of nucleosomes occurs at *MAT* in both the donorless (JKM179) and switchable (XW756) strains (*see* **Fig. 6.3**).

3.3. Analysis of Nucleosome Positioning by qPCR Scanning

The ability of nucleosomes to protect DNA from MNase digestion can be combined with quantitative PCR (qPCR) to provide more enhanced nucleosome positioning information across specific regions. This method provides improved resolution over the methods described above but is more costly due to the large number of qPCR reactions that must be performed. In this technique, formaldehyde-fixed chromatin is digested with MNase so that the yield is nearly 100% mono-nucleosomes (*see* **Fig. 6.4** and **Note 3**). Primers for qPCR are designed across the region of interest such that each ~100-bp amplicon overlaps with the neighboring amplicon (33). The density of primer pairs allows for a pattern of "peaks and valleys" to emerge when DNA quantities at specific locations are plotted as a function of genomic position. The "peaks" correspond to DNA that is relatively resistant to MNase and indicate genomic positions that are well occupied by nucleosomes. In contrast, the "valleys" represent regions that have been digested by MNase and indicate linker regions that are unoccupied by nucleosomes or where nucleosomes are not well positioned. The protocol outlined below has been modified from those published previously (33–35):

1. Grow 250 ml of JKM179 cells in GLGYP medium to mid-log phase (Abs_{600} ~0.6–0.8). Immediately before galactose induction, collect 50 ml for a time-zero sample. Add formaldehyde to a final concentration of 1% and shake at

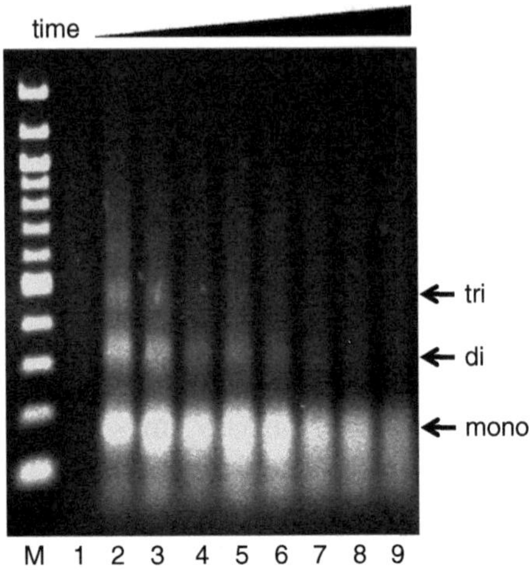

Fig. 6.4. MNase digestion of yeast chromatin to generate mono-nucleosomes. Exponentially growing XW756 cells were fixed with formaldehyde and a crude chromatin fraction was prepared. Chromatin was digested with MNase, decross-linked, and purified DNA was separated on a 1.5% agarose gel. The lane marked M contains a 100-bp molecular weight marker with lower bands of 100 and 200 bp. *Lane 1* is undigested chromatin in which the high molecular weight DNA is not visible. *Lanes 2–9* were incubated for increasing amounts of time with a fixed concentration of MNase. *Lane 6* contains a sample that was digested to nearly 100% mono-nucleosomes.

125 rpm for 15 min at room temperature. Add glycine to a final concentration of 0.125 M and incubate with shaking for 5 min at room temperature to quench cross-linking. Collect cells by centrifugation (5,000 rpm for 5 min at room temperature) and wash twice with 20 ml of ice-cold 1× TBS.

2. Add galactose to the remaining culture at a final concentration of 2% to induce HO and harvest 50 ml samples at 1 h intervals.
3. Harvest the fixed cells by centrifugation (5,000 rpm for 5 min at room temperature) and wash twice with ice-cold TBS. Remove all the liquid with a sequencing gel pipette tip and quick freeze cell pellets in dry ice before storage at –80°C.
4. Thaw cell pellets on ice, resuspend in 0.5 ml FA lysis buffer + 1 mM PMSF + 1× PI, and transfer to a pre-chilled 1.5-ml microcentrifuge tube. Add 0.4 mg of acid-washed glass beads to the cell suspension.
5. Lyse cells by vortexing at 4°C for 15 min at maximum speed using a multihead microcentrifuge tube adaptor.
6. Drain the whole cell lysate into a pre-chilled 15-ml conical tube by puncturing the bottom of the microcentrifuge tube and centrifuging at 1,000 rpm for 1 min. Alternatively, the

lysate can be removed from the glass beads by pipetting with a DNA sequencing gel loading tip.

7. Centrifuge the cell lysate at 14,000 rpm at 4°C for 10 min to collect the insoluble, chromatin-enriched fraction. Resuspend the chromatin fraction in 0.4 ml MNase buffer.
8. Use an amount of the chromatin fraction corresponding to ~7.5 total Abs_{600} units (~100 μl) and adjust to a final volume of 200 μl in MNase buffer. Solubilize chromatin by adding 10–15 units of MNase and incubate at 37°C for 15 min. Stop the reaction by adding 20 μl of 0.5 M EDTA and place the tube on ice (*see* **Note 3**).
9. Add 420 μl of adjusting buffer plus 1 mM PMSF and 1× PI. Add 360 μl of FA lysis buffer plus 1 mM PMSF and 1× PI for a final volume of 1.0 ml. Centrifuge at 14,000 rpm for 10 min at 4°C and transfer supernatant to a new microcentrifuge tube. This sample can be stored at –80°C. It is also suitable to use for chromatin immunoprecipitation (ChIP) to analyze histone occupancy as described below.
10. Transfer 50 μl of this sample to a 0.5-ml microcentrifuge tube and add 50 μl of elution buffer. Add 10 μl of pronase and 1 μl of 1 M $CaCl_2$ followed by incubation at 42°C for 2 h and 65°C for 12 h. This step can be performed in a PCR thermal cycler.
11. Purify DNA by a spin column following manufacturer's directions.
12. Perform qPCR using SYBR green fluorescent dye with overlapping sets of primer pairs (*see* **Notes 5** and **6**).

This method of nucleosome mapping generates a pattern of "peaks and valleys," corresponding to relatively well-positioned nucleosomes at the *MAT* locus before the DSB is induced. Using this technique, Shim et al. demonstrated that the chromatin remodeling complex, RSC, mobilizes and repositions three nucleosomes at the *MAT* locus following DSB induction in the donorless strain JKM179 (33). This method can also be used to analyze nucleosome positioning near the *MAT* DSB when HR repair occurs in a strain in which donor templates are present (XW756).

3.4. Analysis of Nucleosome Occupancy by ChIP

Analysis of nucleosome occupancy by ChIP complements the nucleosome scanning method described above by providing precise information on histone localization. ChIP followed by qPCR has been used to map the relative abundance of specific histones at defined positions relative to the *MAT* DSB in both donorless and switchable strains (21, 35, 36). While strains have been constructed that include epitope-tagged versions of specific histones (e.g. Flag-H2B), antibodies directed to specific histones have also been used reliably in ChIP experiments. A commonly used antibody directed against the C terminus of histone H3 is commercially available to detect this histone species (Abcam #1791). The choice of the histone to monitor is important. The histone octamer that makes up each nucleosome consists of an H3–H4 tetramer and two H2A–H2B dimers. Monitoring H2B by ChIP provides information on the relative occupancy of the H2A–H2B dimers, whereas monitoring H3 provides details on the relative occupancy of the tetramer, and thus the entire nucleosome. It is possible that loss of H2A–H2B dimers could occur, while H3–H4 tetramers are retained. Therefore, monitoring both H2B and H3 occupancy is considered to be the best approach. While we have focused on nucleosome occupancy and positioning here, the monitoring of histone variants or modified histones may also be of interest. The presence of distinct post-translational modifications of histones at the DSB, such as H3/H4 acetylation or H2A phosphorylation, can be analyzed using a number of commercially available antibodies against these modified histones (15, 19). In the method described below, H2B occupancy is monitored using a strain carrying an integrated copy of a *Flag-HTB* gene and chromatin is solubilized by sonication:

1. Grow 250 ml of JKM179 cells in GLGYP medium to mid-log phase (Abs_{600} ~0.6–0.8) (*see* **Note 1**). Immediately before galactose induction, collect 50 ml for a zero-time point. Add formaldehyde to a final concentration of 1% and shake at 125 rpm for 15 min at room temperature (*see* **Note 7**). Add glycine to a final concentration of 0.125 M and incubate with shaking for 5 min at room temperature to quench cross-linking. Collect cells by centrifugation (5,000 rpm for 5 min at room temperature) and wash twice with 20 ml of ice-cold 1× TBS.
2. Add galactose to the remaining culture to a final concentration of 2% to induce HO and harvest 50 ml samples at 1 h intervals. Harvest the fixed cells by centrifugation (5,000 rpm for 5 min at room temperature) and wash twice with ice-cold TBS. Quick-freeze cell pellets in dry ice and store at –80°C.
3. Thaw cell pellets on ice, resuspend in 0.5 ml FA lysis buffer + 1 mM PMSF + 1× PI, and transfer to a pre-chilled 1.5-ml microcentrifuge tube. Add 0.4 mg of acid-washed glass beads to the cell suspension.
4. Lyse cells by vortexing at 4°C for 15 min at maximum speed.
5. Drain the whole cell lysate into a pre-chilled 15-ml conical tube by puncturing the bottom of the microcentrifuge tube and centrifuging at 1,000 rpm for 1 min. Alternatively, the lysate can be removed from the glass beads by pipetting with a DNA sequencing gel loading tip.

6. Centrifuge the cell lysate at 14,000 rpm at 4°C for 10 min to collect the insoluble, chromatin-enriched fraction (*see* **Note 8**). Resuspend the pellet in 0.5 ml FA lysis buffer + 1 mM PMSF + 1× PI.
7. Solubilize the chromatin by sonication six times for 10 s each with 30% output on a Branson Sonifier 250. Sonicate samples to generate DNA fragments centered around 600 bp (*see* **Note 9**).
8. Add 2.5 Abs_{600} units of chromatin-enriched lysate to a 1.5-ml microcentrifuge tube on ice. Add 0.5 ml cold FA lysis buffer + 1 mM PMSF + 1× PI. Retain 10% of this sample and set aside to purify as input DNA. Add 80 μl ANTI-FLAG M2 Affinity gel (washed three times in 1× FA lysis buffer per manufacturer's instructions) and rotate at 4°C overnight.
9. Collect beads by centrifugation at 4,000 rpm for 2 min and wash sequentially in FA lysis buffer, wash buffer #1, wash buffer #2, and TE.
10. Resuspend beads in 0.25 ml elution buffer and incubate at 65°C for 15 min. Vortex for 5 s and centrifuge at 4,000 rpm for 4 min. Remove eluate to a new 1.5-ml microcentrifuge tube. Add pronase (2 mg/ml final concentration) and $CaCl_2$ (5 mM final concentration). Incubate at 42°C for 2 h and at 65°C overnight to reverse formaldehyde cross-linking.
11. Purify DNA from both the retained input sample and the immunoprecipitated sample by a spin column following manufacturer's directions and store DNA at –20°C.
12. Quantify the amount of input DNA and immunoprecipitated DNA using qPCR with SYBR Green dye (*see* **Notes 6** and **10**).

ChIP has been used to investigate a multitude of changes that occur in chromatin near DSBs. Modification of histones, replacement of canonical histones with histone variants, and occupancy of individual nucleosomes can be monitored. Analysis of histone dynamics by ChIP in the donorless strain, JKM179, revealed that entire nucleosomes in a defined region adjacent to the DSB at *MAT* are lost following DSB formation (19).

4. Notes

1. The methods described can utilize either the donorless strain (JKM179) or the switchable strain (XW756) (*see* **Fig. 6.2**). If chromatin changes at *MAT* during repair of the DSB are to be analyzed in the switchable strain, glucose

should be added to a final concentration of 2% just prior to collection of the 1 h time point to repress HO expression, thereby allowing for HR repair to occur. In addition to monitoring chromatin changes near the DSB at *MAT*, the switchable strain can also be used to monitor chromatin events at the donor locus (21).

2. To monitor spheroplast formation, read the Abs_{600} of 10 μl of cells diluted in 1 ml of 0.1% SDS at 30 min intervals until the Abs_{600} is ~10% of the untreated sample. The Abs_{600} will drop as spheroplasts are lysed in the presence of SDS. In order to avoid over-digestion, stop zymolyase or lyticase digestion when the percentage of spheroplasts reaches 80–90%. Spheroplast formation can also be monitored by light microscopy of the diluted cells. Spheroplasts will appear "ghost-like" and large and will eventually lyse, leading to cellular debris.
3. The amount of starting sample and the concentration of MNase must be determined empirically on small-scale samples before proceeding with large-scale MNase digestion. The appropriate amount of digestion can be determined by examining purified DNA from MNase digestion on a 1.5% TBE agarose gel to ensure that a pattern of mono-nucleosomes but not higher molecular weight di- or tri-nucleosomes has been achieved (*see* **Fig. 6.4**). Monosome-length DNA (~150 bp) can also be excised from an agarose gel, followed by DNA purification.
4. To analyze nucleosomes on the right/distal side of the HO cut site, digest DNA to completion with BspEI. To analyze nucleosomes on the left/proximal side of the HO cut site, digest to completion with BanII. Use a short, radio-labeled DNA fragment (~100–150 bp) that abuts the particular restriction enzyme recognition site (BspEI or BanII) to probe the Southern blot.
5. If comparison across different samples is required (for example, at different time points after DSB induction), the samples must be normalized to a region of the genome unaffected by DSB formation, for example, the *POL5* or *ACT1* gene.
6. To obtain accurate quantification of DNA, DNA standards must be included for each set of primer pairs. A large quantity of standard genomic DNA is prepared and 10-fold serially diluted from 1 to 1/10,000 (0.01–100 ng/μl). qPCR is performed with each primer pair, and a standard curve is generated from the critical threshold (Ct) value for each DNA concentration. The amount of DNA present at each site where a specific primer pair amplifies a unique amplicon

is then derived from the standard curve. As a rule of thumb, experimental DNA is diluted 1/10. However, if the fluorescent signal from an experimental sample does not fall within the linear range of the standard curve, the sample concentration should be adjusted. Multiple cycles of freezing and thawing can result in DNA degradation and should be avoided. Once diluted, experimental DNA and standards should be stored at 4°C. Primer pairs can be designed using Primer Express software from ABI.

7. Fixation for 15 min is sufficient to monitor histone levels by ChIP. However, longer fixation times may be required when monitoring certain histone modifications or histone variants, and the time of fixation should be determined empirically by IP-Western blot analysis using appropriate antibodies.
8. Centrifugation at this step will yield a pellet that is enriched in cross-linked chromatin by eliminating the supernatant that contains soluble protein. This step is not required but can lead to a reduction in background signal during immunoprecipitation. This enrichment step is recommended when using low-affinity antibodies or when the target protein is in low abundance.
9. One important caveat in adapting the ChIP technique is the choice of method to generate fragmented chromatin. Sonication will generally produce fragmented DNA on the order of 500–750 bp, while MNase digestion to mononucleosomes will be on the order of 150–175 bp. In our experience, sonication is suitable for most histone ChIP experiments. However, we have found that MNase digestion is necessary to analyze nucleosome occupancy at the donor template (*HMRa*) during *MAT* switching (21).
10. Primer pairs have been designed to detect nucleosome dynamics near the *MAT* DSB at ~0.3, ~0.5 and ~1.8 kb from the HO cut site (19, 21). If MNase digestion is used to solubilize chromatin prior to ChIP, it is important that primer pairs do not span more than a single nucleosome. It is useful to refer to a high-resolution genome-wide map of nucleosome positioning to identify primer positions relative to nucleosome position at *MAT* (37).

Acknowledgments

Supported by grants NIH CA118357 to M.A.O. and NIH F32 CA125955 to S.H.

References

1. Haber, J.E. (2000) Partners and pathways repairing a double-strand break. *Trends Genet* **16**, 259–264.
2. Daley, J.M., Palmbos, P.L., Wu, D., and Wilson, T.E. (2005) Nonhomologous end joining in yeast. *Annu Rev Genet* **39**, 431–451.
3. Lewis, L.K., and Resnick, M.A. (2000) Tying up loose ends: nonhomologous end-joining in Saccharomyces cerevisiae. *Mutat Res* **451**, 71–89.
4. Mimitou, E.P., and Symington, L.S. (2009) DNA end resection: many nucleases make light work. *DNA Repair (Amst)* **8**, 983–995.
5. Heyer, W.D., Li, X., Rolfsmeier, M., and Zhang, X.P. (2006) Rad54: the Swiss Army knife of homologous recombination? *Nucleic Acids Res* **34**, 4115–4125.
6. Luger, K., Mader, A.W., Richmond, R.K., Sargent, D.F., and Richmond, T.J. (1997) Crystal structure of the nucleosome core particle at 2.8 A resolution. *Nature* **389**, 251–260.
7. Krogh, B.O., and Symington, L.S. (2004) Recombination proteins in yeast. *Annu Rev Genet* **38**, 233–271.
8. Downs, J.A. (2007) Chromatin structure and DNA double-strand break responses in cancer progression and therapy. *Oncogene* **26**, 7765–7772.
9. Rogakou, E.P., Pilch, D.R., Orr, A.H., Ivanova, V.S., and Bonner, W.M. (1998) DNA double-stranded breaks induce histone H2AX phosphorylation on serine 139. *J Biol Chem* **273**, 5858–5868.
10. Shroff, R., Arbel-Eden, A., Pilch, D., Ira, G., Bonner, W.M., Petrini, J.H., Haber, J.E., and Lichten, M. (2004) Distribution and dynamics of chromatin modification induced by a defined DNA double-strand break. *Curr Biol* **14**, 1703–1711.
11. Bird, A.W., Yu, D.Y., Pray-Grant, M.G., Qiu, Q., Harmon, K.E., Megee, P.C., Grant, P.A., Smith, M.M., and Christman, M.F. (2002) Acetylation of histone H4 by Esa1 is required for DNA double-strand break repair. *Nature* **419**, 411–415.
12. Utley, R.T., Lacoste, N., Jobin-Robitaille, O., Allard, S., and Cote, J. (2005) Regulation of NuA4 histone acetyltransferase activity in transcription and DNA repair by phosphorylation of histone H4. *Mol Cell Biol* **25**, 8179–8190.
13. Osley, M.A., Tsukuda, T., and Nickoloff, J.A. (2007) ATP-dependent chromatin remodeling factors and DNA damage repair. *Mutat Res* **618**, 65–80.
14. Chai, B., Huang, J., Cairns, B.R., and Laurent, B.C. (2005) Distinct roles for the RSC and Swi/Snf ATP-dependent chromatin remodelers in DNA double-strand break repair. *Genes Dev* **19**, 1656–1661.
15. Downs, J.A., Allard, S., Jobin-Robitaille, O., Javaheri, A., Auger, A., Bouchard, N., Kron, S.J., Jackson, S.P., and Cote, J. (2004) Binding of chromatin-modifying activities to phosphorylated histone H2A at DNA damage sites. *Mol Cell* **16**, 979–990.
16. Kent, N.A., Chambers, A.L., and Downs, J.A. (2007) Dual chromatin remodeling roles for RSC during DNA double strand break induction and repair at the yeast MAT locus. *J Biol Chem* **282**, 27693–27701.
17. Liang, B., Qiu, J., Ratnakumar, K., and Laurent, B.C. (2007) RSC functions as an early double-strand-break sensor in the cell's response to DNA damage. *Curr Biol* **17**, 1432–1437.
18. Morrison, A.J., Highland, J., Krogan, N.J., Arbel-Eden, A., Greenblatt, J.F., Haber, J.E., and Shen, X. (2004) INO80 and gamma-H2AX interaction links ATP-dependent chromatin remodeling to DNA damage repair. *Cell* **119**, 767–775.
19. Tsukuda, T., Fleming, A.B., Nickoloff, J.A., and Osley, M.A. (2005) Chromatin remodelling at a DNA double-strand break site in Saccharomyces cerevisiae. *Nature* **438**, 379–383.
20. van Attikum, H., Fritsch, O., and Gasser, S.M. (2007) Distinct roles for SWR1 and INO80 chromatin remodeling complexes at chromosomal double-strand breaks. *EMBO J* **26**, 4113–4125.
21. Tsukuda, T., Lo, Y.C., Krishna, S., Sterk, R., Osley, M.A., and Nickoloff, J.A. (2009) INO80-dependent chromatin remodeling regulates early and late stages of mitotic homologous recombination. *DNA Repair (Amst)* **8**, 360–369.
22. Lee, S.E., Moore, J.K., Holmes, A., Umezu, K., Kolodner, R.D., and Haber, J.E. (1998) Saccharomyces Ku70, mre11/rad50 and RPA proteins regulate adaptation to G2/M arrest after DNA damage. *Cell* **94**, 399–409.
23. Sugawara, N., and Haber, J.E. (2006) Repair of DNA double strand breaks: in vivo biochemistry. *Methods Enzymol* **408**, 416–429.
24. Baudin, A., Ozier-Kalogeropoulos, O., Denouel, A., Lacroute, F., and Cullin, C. (1993) A simple and efficient method for direct gene deletion in Saccharomyces cerevisiae. *Nucleic Acids Res* **21**, 3329–3330.

25. Gietz, R.D., and Woods, R.A. (2002) Transformation of yeast by lithium acetate/single-stranded carrier DNA/polyethylene glycol method. *Methods Enzymol* **350**, 87–96.
26. Janke, C., Magiera, M.M., Rathfelder, N., Taxis, C., Reber, S., Maekawa, H., Moreno-Borchart, A., Doenges, G., Schwob, E., Schiebel, E., and Knop, M. (2004) A versatile toolbox for PCR-based tagging of yeast genes: new fluorescent proteins, more markers and promoter substitution cassettes. *Yeast* **21**, 947–962.
27. Fleming, A.B., and Pennings, S. (2001) Antagonistic remodelling by Swi-Snf and Tup1-Ssn6 of an extensive chromatin region forms the background for FLO1 gene regulation. *EMBO J* **20**, 5219–5231.
28. Lee, W., Tillo, D., Bray, N., Morse, R.H., Davis, R.W., Hughes, T.R., and Nislow, C. (2007) A high-resolution atlas of nucleosome occupancy in yeast. *Nat Genet* **39**, 1235–1244.
29. Nedospasov, S.A., and Georgiev, G.P. (1980) Non-random cleavage of SV40 DNA in the compact minichromosome and free in solution by micrococcal nuclease. *Biochem Biophys Res Commun* **92**, 532–539.
30. Ravindra, A., Weiss, K., and Simpson, R.T. (1999) High-resolution structural analysis of chromatin at specific loci: Saccharomyces cerevisiae silent mating-type locus *HMRa*. *Mol Cell Biol* **19**, 7944–7950.
31. Weiss, K., and Simpson, R.T. (1998) High-resolution structural analysis of chromatin at specific loci: Saccharomyces cerevisiae silent mating type locus *HMLα*. *Mol Cell Biol* **18**, 5392–5403.
32. Wu, C. (1980) The 5′ ends of Drosophila heat shock genes in chromatin are hypersensitive to DNase I. *Nature* **286**, 854–860.
33. Shim, E.Y., Hong, S.J., Oum, J.H., Yanez, Y., Zhang, Y., and Lee, S.E. (2007) RSC mobilizes nucleosomes to improve accessibility of repair machinery to the damaged chromatin. *Mol Cell Biol* **27**, 1602–1613.
34. Kuo, M.H., and Allis, C.D. (1999) In vivo cross-linking and immunoprecipitation for studying dynamic protein:DNA associations in a chromatin environment. *Methods* **19**, 425–433.
35. Tsukuda, T., Trujillo, K.M., Martini, E., and Osley, M.A. (2009) Analysis of chromatin remodeling during formation of a DNA double-strand break at the yeast mating type locus. *Methods* **48**, 40–45.
36. Chen, C.C., Carson, J.J., Feser, J., Tamburini, B., Zabaronick, S., Linger, J., and Tyler, J.K. (2008) Acetylated lysine 56 on histone H3 drives chromatin assembly after repair and signals for the completion of repair. *Cell* **134**, 231–243.
37. Jiang, C., and Pugh, B.F. (2009) A compiled and systematic reference map of nucleosome positions across the Saccharomyces cerevisiae genome. *Genome Biol* **10**, R109.

Chapter 7

Analysis of Meiotic Recombination Intermediates by Two-Dimensional Gel Electrophoresis

Jasvinder S. Ahuja and G. Valentin Börner

Abstract

During meiosis, programmed double strand breaks give rise to crossover and non-crossover recombination products. Meiotic recombination products are formed via several branched intermediates, including single end invasions and double Holliday junctions. Two-dimensional gel electrophoresis provides a sensitive and specific approach for detecting branched recombination intermediates, determining their genetic requirements, and enriching intermediates for further analysis. Here, we describe analysis of branched recombination intermediates in the yeast *Saccharomyces cerevisiae* by two-dimensional gel electrophoresis. We also provide an introduction to meiotic time-course procedures, stabilization of branched DNA molecules by interstrand crosslinking, extraction of genomic DNA from meiotic cultures, and quantitative analysis of two-dimensional gel blots.

Key words: Joint molecules, meiosis, recombination, two-dimensional gel electrophoresis, double Holliday junction, single end invasion.

1. Introduction

Meiotic recombination is initiated by the formation of double strand breaks (DSBs) on one of two homologous chromosomes ("Mom" and "Dad"). DSBs undergo 5′ resection to generate single-stranded 3′-ends (1). Following resection, DSB ends sequentially invade the intact homologous recombination partner (homolog), giving rise to several species of branched recombination intermediates collectively referred to as joint molecules. The first prominent joint molecule, the single end invasion (SEI), arises when one resected DSB end asymmetrically invades

H. Tsubouchi (ed.), *DNA Recombination*, Methods in Molecular Biology 745,
DOI 10.1007/978-1-61779-129-1_7,

the intact allelic DNA on the homolog (2). The other DSB end is subsequently captured by the single end invasion giving rise to double Holliday junctions which involve fully ligated parental DNA molecules connected by a pair of closely spaced Holliday junctions (3). Double Holliday junctions are resolved as crossovers (4, 5).

Apart from this prominent pathway that generates crossovers, several alternative recombination pathways exist which play minor roles during wild-type meiosis in *Saccharomyces cerevisiae*, but can become more prominent under certain conditions. First, recombination may occur not only between homologous chromosomes, but also between sister chromatids, with intersister double Holliday junctions as the main detectable recombination intermediates (6). Second, repeated strand invasion of SEIs with recombination partners other than the cognate DSB end generates long, multichromatid joint molecules encompassing more than two Holliday junctions (7).

Two-dimensional gel analysis has been used extensively to identify and monitor the kinetics of joint molecules in wild-type and mutant situations (4, 5, 7, 8). Branched recombination structures are stabilized by introducing covalent interstrand crosslinks, preventing branch migration, and loss of Holliday junctions. Following extraction of genomic DNA from meiotic cells, gel electrophoresis in the first dimension separates restriction fragments according to molecular weight only, while electrophoresis in the second dimension is performed under conditions that exaggerate effects of molecular shape on electrophoretic mobility, with branched molecules migrating slower than linear DNA fragments of equal length.

Meiotic joint molecules have been investigated in detail at a few recombination hotspots that carry restriction site polymorphisms for physical analysis and are linked to nutritional markers for genetic analysis (3, 4). The *HIS4LEU2* hotspot construct discussed here has been optimized for analysis in a single strain of several key meiotic recombination intermediates including DSBs and joint molecules, as well as crossover and non-crossover products. A particular version of this hotspot, called *HIS4::LEU2-(BamHI)/his4-X::LEU2-(NgoMIV)-URA3*, provides several advantages over earlier versions of the same hotspot, including a predominant, central DSB site equally active on both homologs ("Mom" and "Dad"), as well as flanking restriction polymorphisms (*Xho*I) that distinguish between "Mom" and "Dad," generate appropriately sized fragments for two-dimensional gel analysis, and rarely undergo size changes due to recombination-associated gene conversion (**Fig. 7.1**) (9). A central restriction site polymorphism (*Bam*HI/*Ngo*MIV) that comprises differences at only a few base pairs permits monitoring timing and frequency of gene conversion proximal to the

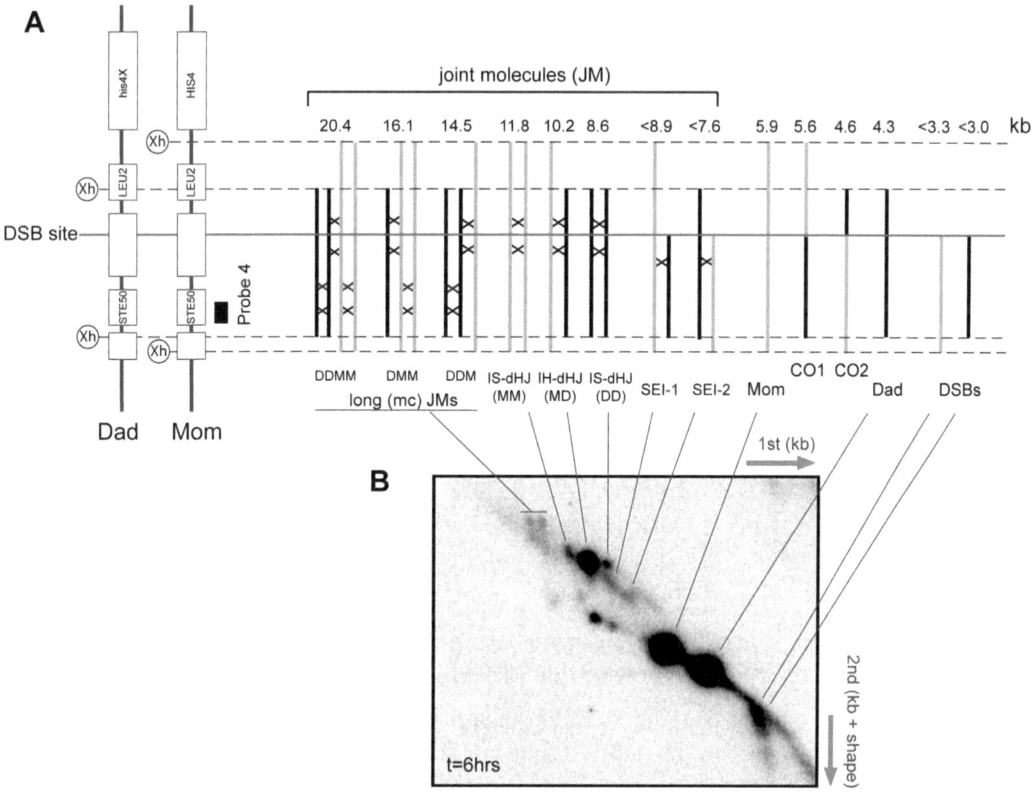

Fig. 7.1. Detection of joint molecules at the *HIS4LEU2* recombination hotspot by two-dimensional gel analysis. (**a**) Map of the two *HIS4LEU2* alleles on homologous chromosomes with restriction site polymorphisms for restriction enzyme *XhoI* (*circles*). Relevant gene names and the position of "Probe 4" are indicated. Mom (*grey*) and Dad (*black*) refer to the two parental restriction fragments. CO-1 and CO-2 are reciprocal crossover products that can be resolved by one-dimensional gel electrophoresis (5). SEI-1 and SEI-2 are prominent single end invasion species. IH-dHJ, interhomolog double Holliday junction; IS-dHJ, intersister double Holliday junction; mc-JMs, multichromatid joint molecules. The predicted chromatid composition of joint molecules is indicated with *M* (Mom) or *D* (Dad) (7). (**b**) Image of two-dimensional gel hybridized with Probe 4. A *pch2Δ* mutant sample at *t*=6 h exhibiting transient accumulation of joint molecules is shown for clarity (8).

DSB site. The SK1 strain background carrying this version of the hotspot exhibits high spore viability and undergoes meiosis with high synchrony, an important prerequisite for detection of short-lived recombination intermediates.

2. Materials

2.1. Culture Conditions and DNA Crosslinking

All growth media, including YPG, YPD, YPA, and SPM, should be prepared with double distilled water, referred to as water in the text. Growth media, SPM, and distilled water are sterilized by autoclaving for 20 min on liquid cycle setting.

1. YPG solid: 1% (w/v) Bacto yeast extract (BD, Franklin Lakes, NJ), 2% (w/v) Bacto peptone (BD, Franklin Lakes, NJ), 2% glycerol, 2% (w/v) agar (EMD Chemicals, Gibbstown, NJ).
2. YPD liquid: 1% (w/v) Bacto yeast extract (BD, Franklin Lakes, NJ), 2% (w/v) Bacto peptone (BD, Franklin Lakes, NJ), 2% (w/v) D-glucose.
3. YPD solid: same as liquid, in addition add 2% (w/v) agar (EMD Chemicals, Gibbstown, NJ).
4. YPA liquid: 1% (w/v) Bacto yeast extract (BD, Franklin Lakes, NJ), 2% (w/v) Bacto peptone (BD, Franklin Lakes, NJ), 1% (w/v) potassium acetate (Fisher, Pittsburgh, PA).
5. SPM liquid: 0.5% (w/v) potassium acetate (Fisher, Pittsburgh, PA), 0.02% (w/v) D-raffinose (Acros Organics, Morris Plains, NJ), 75 μl/l antifoam 204 (Sigma).
6. 2x DAPI fix: 80% ethanol, 100 mM sorbitol, 0.5 mM EDTA.
7. DAPI stock solution: 1 μg/μl 4′,6-diamidino-2-phenylindole (Thermo Scientific).
8. 10x crosslinking stock solution: 1 mg/ml trioxalen (Sigma) in 100% ethanol. Store at –20°C in an aluminum foil-packaged vial. Vortex vigorously for 5 min before each use. Trioxalen solution is "flakey," i.e., crystals only go partially into solution in 100% ethanol and crosslinking work solution.
9. Crosslinking work solution: Before each use, crosslinking stock solution is diluted 1:10 in 50 mM EDTA and 50 mM Tris (pH 8.0) buffer. The crosslinking work solution should be kept on ice during the time course, and trioxalen needs to be resuspended before addition to cell samples.
10. 35 × 10 mm cell culture dish (Corning).
11. Long-wave UV transilluminator (365 nm emission maximum).
12. Blak-Ray Ultraviolet Meter J221 (UVP, Upland, CA).
13. 200 proof (100%) ethanol for molecular biology applications.

2.2. Genomic DNA Extraction

All solutions for gel preparation, washing, and blotting should be prepared using filtered, deionized water with a resistivity of 18.2 MΩ cm at 25°C, referred to as NP (Nanopure) water in the text.

1. Zymolyase buffer (100 ml): 1 M sorbitol (50 ml of 2 M sorbitol), 50 mM potassium phosphate, pH 7.5 (4.2 ml 1 M K_2HPO_4, 0.8 ml 1 M KH_2PO_4), 5 mM EDTA (1 ml

of 0.5 M EDTA, pH 8.0), make up volume with NP water (44 ml), sterile filter, aliquot, and store at –20°C.

2. Zymolyase work solution: Prior to use add 1% (v/v) β-mercaptoethanol and 100 μg/mL zymolyase (US Biological, Marblehead, MA) to zymolyase buffer and vortex for 5 min to get zymolyase into solution.
3. 20% (w/v) SDS: SDS powder is toxic. Prepare by weighing a 100 g vial of SDS (e.g., VWR International), adding NP water, and dissolving powder in a final volume of 500 ml.
4. Lysis stock solution: 100 mM Tris–HCl (pH 8.0), 50 mM EDTA (pH 8.0), and 0.5% SDS. Make fresh on day of DNA extraction, keep at room temperature.
5. Proteinase K stock solution: 10 mg/ml proteinase K (Invitrogen) in 100 mM Tris–HCl (pH 8.0) and 50 mM EDTA (pH 8.0), make on day of DNA extraction, keep on ice.
6. 5 M potassium acetate (KAc), without pH adjustment.
7. TE: 10 mM Tris (pH 8.0), 1 mM EDTA.
8. RNase A work solution: TE plus 10 μg/ml RNase A (Invitrogen).
9. Phenol–chloroform–isoamyl alcohol (25:24:1) (Fisher, Pittsburgh, PA) is buffered at pH 8.0 with Tris–HCl, stabilized with 0.1% (w/v) 8-hydroxyquinoline as antioxidant/coloring agent, stored at –20°C, and thawed as needed.
10. Restriction enzyme *Xho*I (New England Biolabs).
11. 96% ethanol/150 mM NaAc: Mix 48 ml 100% ethanol with 2 ml 50% (w/v) NaAc without pH adjustment.
12. Loading dye: 6x loading dye: 0.25% (w/v) bromophenol blue, 0.25% (w/v) xylene cyanol, 15% (w/v) Ficoll 400 in water; mix thoroughly, sterile filter, and store at 4°C.

2.3. Two-Dimensional Gel

1. For the first dimension gel, we use a standard submarine electrophoresis apparatus with 26 cm long gel trays and wells at 2.5 cm. Teeth of the gel comb are 4.5 mm wide and 1.5 mm thick. The second dimension of two-dimensional electrophoresis is performed in a 26 × 19.5 cm gel tray.
2. Agarose: Seakem Gold Agarose (Lonza, Basel, Switzerland) for first dimension, Ultrapure agarose (Invitrogen) for second dimension.
3. 10x TBE stock solution: 890 mM Tris base (108 g/l), 20 mM EDTA (40 ml of 0.5 M EDTA), 890 mM boric acid (55 g/l), add NP H_2O to 1 l, stir, aliquot, and autoclave.
4. Prepare sufficient 1x TBE running buffer for two-dimensional gel tank (~3 l/gel) and equilibrate at 4°C.

5. Ethidium bromide: 1% stock solution (33,333x; Fisher, Pittsburgh, PA).

2.4. Southern Blot

1. Hybond-N nylon membrane (GE Health Care).
2. Depurination solution: 0.25 N HCl. Store at room temperature for up to 1 month.
3. Denaturation buffer: 0.5 N NaOH, 1 M NaCl. Store at room temperature for up to 3 months.
4. Neutralization buffer: 1.5 M Tris–HCl pH 7.4, 1.5 M NaCl. Adjust to pH 7.4 with concentrated hydrochloric acid. Store at room temperature for up to 3 months.
5. 20x SSC (sodium chloride/sodium citrate): 3 M NaCl (175 g/l), 0.3 M trisodium citrate (88 g/l), add NP water to 800 ml, adjust pH to 7.0 with 1 N HCl, adjust to 1 l, aliquot, and autoclave.
6. Nucleic acid transfer buffer: 10x SSC.
7. 1 M sodium phosphate (pH 7.2; 1 l): 34.14 g $NaH_2PO_4 \cdot 1H_2O$ (monobasic), 193 g $Na_2HPO_4 \cdot 7H_2O$ (dibasic), NP water to 1 l, confirm pH, aliquot into bottles, autoclave, and store for up to 3 months at room temperature.
8. Glass plates to put across Pyrex pan (28 × 28 cm).
9. Cut large Whatman 3 MM paper (3030-917) to 21 × 46 cm as bridge for Southern blot.
10. Small Whatman 3 MM paper (3030-866) (20 × 25 cm).
11. Paper towels 4,000/cs (Kimberly-Clark, catalogue # 01700).
12. All gel washes are carried out in Pyrex glass pans large enough to fit the large two-dimensional gel tray.
13. UV crosslinker (e.g., Stratalinker, Stratagene).

2.5. Southern Blot Hybridization

1. Southern probe for hotspot *HIS4LEU2*: "probe 4" corresponds to nucleotides 63,086–63,640 of *S. cerevisiae* chromosome III.
2. Prime-It RmT Random Primer Kit (Stratagene).
3. [α-^{32}P]dCTP (6,000 Ci/mmol) (Perkin Elmer).
4. ProbeQuant G-50 Micro Column (GE Healthcare).
5. Hybridization solution: 7% (w/v) SDS, 0.25 M sodium phosphate (pH 7.2), 0.25 M NaCl, 1 mM EDTA.
6. Wash solution: 0.1 × SSC, 0.1% (w/v) SDS.
7. Phosphoimaging system, e.g., Typhoon or Fuji.
8. Quantitation software, e.g., Quantity One (Biorad).

3. Methods

A yeast culture that efficiently and synchronously initiates meiosis is key for detection of joint molecules which reach peak at ~2% of total DNA at a given locus in synchronous wild-type cultures (5). UV-activated psoralen crosslinking is used to introduce interstrand crosslinks at an average distance of 500 bp (A. Schwacha, personal communication), thereby preventing Holliday junction branch migration and loss of Holliday junctions as well as of other branched recombination intermediates. Procedures for genomic DNA extraction from meiotic cells are similar to those widely used for DNA extraction from mitotic cells, yet consistent yields of genomic DNA are much more difficult to achieve from meiotic cultures. It is therefore important to closely follow the outlined protocol for time-course setup, DNA extraction, and restriction digest prior to two-dimensional gel analysis. Alternative approaches for DNA preparation and/or stabilization of branched structures have been described (10).

3.1. Strain Construction, Time Course, and Psoralen Crosslinking

1. The desired mutant allele is generated in an isogenic SK1 strain background carrying "Mom" and "Dad" versions of the *HIS4LEU2* meiotic recombination hotspot (e.g., NHY1296) (9) (*see* **Notes 1** and **2** for details on strain construction).
2. Mate strains of opposite mating type for <6 h at 30°C on YPD plates and freeze the mating mix in 25% glycerol at −80°C.
3. *Day 1 (evening)*: Patch mating mix on YPG plate and incubate overnight (i.e., <16 h; diploid SK1 strains enter meiosis if left on YPG for longer periods). Include an isogenic wild-type culture for every time course.
4. *Day 2 (morning)*: Streak for single colonies on YPD plates and incubate at 30°C for ~56 h.
5. *Day 4 (afternoon)*: Inoculate individual diploid colonies into 4 ml liquid YPD in a glass culture tube and incubate for 26 h at 30°C on a roller drum at maximum speed. Inoculate at least four cultures per strain to accommodate for possible haploid colonies or poor growth (*see* **Note 3**).
6. *Day 5 (morning)*: Vortex YPD cultures at least three times between morning and inoculation of YPA (below). Cells from saturated diploid SK1 cultures tend to stay in suspension while haploid cells flock out and fall to the bottom of the tube within <1 min.
7. *Day 5:* Pre-warm 150 ml YPA in 2 l Erlenmeyer flasks to 30°C; 13.5 h before start of time course, dilute the YPD

overnight culture at dilutions between 1:100 and 1:200. Dilutions should be performed from the same YPD culture (*see* **Note 4**).

8. Shake vigorously (300 rpm) at 30°C for 13.5 h.
9. *Day 6 (time course):* Measure OD_{600} of YPA cultures. Use YPA cultures with OD_{600} =1.2–1.6 (*see* **Note 5**).
10. Spin at 3,200×*g* for 5 min, resuspend cell pellet with equal volume of sterile SPM, repeat spin, resuspend in 150 ml SPM, transfer to 2l Erlenmeyer flask and shake at 300 rpm at appropriate temperature.
11. At each time point, prior to cell sampling, swirl flask and resuspend all cells sticking to flask wall. Consistency in taking samples is important to achieve consistent yield of DNA.
12. At each time point, 12 ml of SPM culture is transferred into a 15 ml centrifuge tube. Medium is removed via centrifugation. Take time points e.g. at 0, 2.5, 4, 5, 6, 7, 8.5, 10, and 12 h or adjust sampling times depending on mutant kinetics.
13. For analysis of nuclear divisions, one volume of cells should be mixed with one volume of 2x DAPI fix to monitor nuclear divisions; 200 μl of cells is sufficient for multiple rounds of DAPI staining and counting of nuclear divisions. Cell samples mixed with DAPI fix can be stored at –20°C for up to 12 months.
14. Resuspend pellet from 12 ml of SPM culture in 1.5 ml crosslinking work solution and transfer into 3 cm culture dish. Resuspend with a P1000 filter tip. Never vortex samples for genomic DNA extraction.
15. Culture dishes are placed on a long-wave UV transilluminator (365 nm) for 10 min, and cells are resuspended three times during the incubation by swirling culture dishes (*see* **Note 6**).
16. Using filter tips resuspend cells thoroughly and transfer to 1.5 ml microfuge tube.
17. Spin at 3,600×*g* and pour off trioxalen-containing supernatant (trioxalen has to be discarded as toxic waste). Collect cell pellets on ice and store in a cold room until time course is completed, but no longer than 24 h.
18. After 24 h in SPM, identify the culture that has undergone meiosis most synchronously. DAPI staining of nuclei can be used for wild-type and mutant strains that undergo divisions. For mutant strains defective for meiotic progression, assessing spindle status by tubulin staining or assessing premeiotic replication by FACS can be used (11).

3.2. Meiotic DNA Extraction and Restriction Digest

1. Genomic DNA is isolated from the two most synchronous cultures for each genotype. Cohorts of no more than eight samples should be processed to ensure consistent treatment (*see* **Note 7**).
2. Following storage on ice, cell pellets are centrifuged again at 3,600 × *g* and excess crosslinking solution is pipetted off.
3. Using P1000 filter tips, cell pellets are resuspended in 0.5 ml zymolyase work solution by slowly pipeting up and down. Samples are incubated for spheroplasting at 37°C for 30 min, inverting tubes at least three times during incubation.
4. Spheroplasted cells are centrifuged for 5 min at 7,000 rpm in a tabletop centrifuge and the supernatant is removed with a pipet, leaving as little liquid as possible.
5. For every cohort of eight tubes, a master mix of lysis work solution with proteinase K is prepared, pre-mixing 5.5 ml lysis stock solution with 200 μl 10 mg/ml proteinase K stock solution.
6. Add 570 μl of lysis work solution to each cell pellet. Resuspend the viscous spheroplast pellet, setting P1000 to 400 μl and pipeting up and down slowly with a filter tip.
7. Incubate in a 65°C water bath for 45 min. Vigorously flick tubes during the first 10 min of incubation until pellets are completely resuspended. Continue flicking tubes during the incubation. The solution remains opaque, but no particles or streaks should be visible at the end of incubation.
8. Let samples cool on ice, add 150 μl 5 M KAc, mix by repeated inverting, keep on ice for 15 min, and spin in a tabletop centrifuge at maximum speed for 20 min at 4°C.
9. Transfer 650 μl of supernatant into 700 μl pre-aliquoted 100% ethanol, avoiding the pellet. If the pellet becomes loose, centrifuge again and process fewer samples at a time. Invert the mixture of supernatant and ethanol >5 times. A sizable precipitate should be visible in all samples except at time points up to 3 h which tend to yield less DNA due to their pre-G2 DNA content.
10. Spin for 5 min at room temperature and discard supernatant completely, but do not dry the pellet.
11. Add 500 μl RNase A work solution to pellet and store overnight at 4°C.
12. On the next day, flick tubes until pellet separates from bottom of tube, incubate at 50°C for 10 min, flick tubes

frequently until pellet is completely dissolved. Incubate for another 30 min at 37°C, flicking each tube >3 times.

13. In a fume hood, add 500 μl phenol–chloroform–isoamyl alcohol (25:24:1), invert tubes 10 times (do not vortex), and spin for 20 min at 13,500 rpm in a tabletop centrifuge. Remove tubes promptly after centrifuge has stopped, transfer aqueous phase to 600 μl pre-aliquoted isopropanol using a P200.
14. Invert tube several times, the precipitate will be smaller compared to the ethanol precipitation. Spin at 13,500 rpm in a tabletop centrifuge.
15. Discard supernatant, rinse pellet with ~100 μl of 70% (v/v) ethanol by setting P200 to larger volume and taking off all supernatant. Place tubes with open lids into heating block at 42°C, covering tubes loosely with Saran wrap.
16. After pellet has dried completely, add 40 μl of TE and allow genomic DNA to dissolve overnight at 4°C.
17. On the next day, flick tubes vigorously until pellet is completely in solution. Collect liquid via a 1 min spin in a tabletop centrifuge at maximum speed.
18. Digest 10 μl genomic DNA from a meiotic culture with 50 U *Xho*I in 80 μl reaction volume and incubate for 16 h at 37°C.
19. Stop digest by adding three volumes (240 μl) of 96% ethanol/150 mM NaAc, invert, spin 10 min at 4°C, discard supernatant, wash with 70% ethanol, dry completely in 42°C heat block, add 30 μl TE (50 mM Tris, pH 8.0, 1 mM EDTA), and allow to dissolve overnight at 4°C.
20. Stir by flicking tube, add 8 μl 6x loading dye, and mix thoroughly.

3.3. Two-Dimensional Gel Run

1. Prior to performing two-dimensional gel analysis, half of the restriction digest (~18 μl) should be run on a 0.6% one-dimensional gel without ethidium bromide and analyzed by Southern blot analysis to ascertain consistent amounts of DNA in all samples and a complete genomic digest (5). For one-dimensional gel analysis at the *HIS4LEU2* hotspot, *Xho*I-restricted DNA is separated on a 26 cm gel at 1.6 V/cm for 26 h at room temperature and blotted as described for two-dimensional gels.
2. To perform two-dimensional gel analysis, pour a 0.4% (w/v) Seakem Gold agarose/1x TBE gel *without ethidium bromide* in a 26 cm long gel box. Pour gel at room temperature on a leveled surface and cover with glass or plastic plate while agarose solidifies.

3. Flick genomic digest and spin for 2 min at maximum speed in a tabletop centrifuge immediately before loading the gel.
4. Load 18 μl of sample, leaving one lane empty between samples.
5. Run at 0.75 V/cm at room temperature for 16.5 h. Do not run in cold room as this negatively affects resolution.
6. After gel run, transfer gel into 1x TBE containing 0.3 μg/ml ethidium bromide and agitate gently for 45 min at room temperature.
7. Fill two-dimensional gel box in cold room with 1x TBE/0.3 μg/ml ethidium bromide, pre-cooled to 4°C.
8. Set up long-wave (365 nm) UV transilluminator in a room that can be darkened. Tape up a large (19.5 cm × 26 cm) gel tray and set it up next to the UV transilluminator, with the position where the comb would normally be inserted pointing away from you. Direct and indirect UV causes damage to eyes and skin. Wear a UV protective face shield, gloves, and cover your forearms.
9. Put gel tray with ethidium bromide-stained first dimension gel on UV transilluminator and orient such that wells are on your left. To trim lanes to 8.5 cm, slide wells over left edge of tray and cut with a razor blade 2 cm below the wells. In the same way, trim gel by sliding it over the right edge of gel tray. Discard wells and other pieces of gel that you have cut off.
10. Slide the 8.5 cm gel fragment onto Saran wrap-covered UV transilluminator, darken the room, switch on UV transilluminator, and cut along both sides of each ethidium bromide-stained lane, lowering the front edge of the razor blade after the back edge into the agarose. A new blade should be used for every lane as razor blades tend to get blunt.
11. Transfer each gel slice with a similarly sized piece of semi-flexible plastic (e.g., X-ray film) into the taped two-dimensional gel tray, starting with the earliest time point. Place earliest time point in the top left corner (i.e., at the end where the comb would normally be inserted) and proceed in Z order to later time points, leaving at least 6 cm space between agarose slices. Two 8.5 cm strips can be layered next to each other, and up to four rows of slices fit on one standard tray. If DSBs need to be detected on the two-dimensional gel, six rather than eight slices should be used per tray, as DSBs run into the gel slice below if eight time points are analyzed.
12. In a microwave, boil appropriate volume (450 ml) of 0.8% (w/v) agarose (Ultrapure agarose, Invitrogen) in 1x TBE.

Interrupt heating after 2 min and stir flask to suspend agarose evenly. Visually inspect agarose solution for streaks, and boil again if streaks are present. Add 0.3 μg/ml ethidium bromide and stir slowly on a magnetic stirrer until agarose reaches ~60°C.

13. In a cold room, on a leveled surface, slowly pour agarose from one edge into the taped-up two-dimensional tray with the gel slices until agarose just covers gel slices. Cover gel with Plexiglas lid while it solidifies. Straighten gel slices with pipet tip if they have shifted while pouring the gel.
14. After the gel has solidified (>45 min), remove tape, securing the gel with one hand so it does not slide off the tray. Lower gel tray with two-dimensional gel into a large gel tank previously filled with refrigerated 1x TBE/0.3 μg/ml ethidium bromide, so that gel is supported by buffer. Cover with more 1x TBE/0.3 μg/ml ethidium bromide.
15. To inject loading dye into gel, use P20 pipet set to 10 μl, fill with 6x loading dye, poke hole into agarose between upper pair of cast-in gel slices from first dimension, and inject loading dye.
16. Perform electrophoresis at 3.2 V/cm (150 V) for 330 min. The bromophenol blue dye should run about 15 cm into the gel. Monitor gel run with hand-held UV lamp.
17. Following electrophoresis, take picture on UV transilluminator. A signal should be visible >1 cm above the arc representing the endogenous 2 μm plasmid (12).

3.4. Southern Blotting

1. The protocol for Southern analysis described here uses transfer to neutral nylon membrane with 10x SSC as transfer buffer. Alkaline transfer to positively charged nylon membrane reduces the sensitivity in our hands.
2. Following second dimension electrophoresis, transfer gel tray with two-dimensional gel into a large Pyrex glass pan, submerse in NP water, and agitate gently on a shaker for 30 min. Repeat this step to wash out all TBE buffer. Washing gel in a large excess volume of water improves depurination efficiency ensuring quantitative transfer of large DNA molecules.
3. Transfer tray with gel into depurination solution and agitate gently for 20 min or until bromophenol blue turns yellow.
4. Following depurination, rinse by submersing tray with gel briefly in Pyrex pan with NP water.
5. Transfer tray with gel into neutralization buffer and agitate gently for 30 min.

6. Following a brief rinse in NP water, transfer tray with gel into nucleic acid transfer buffer (10x SSC) and agitate gently for 30 min.
7. During incubation, pour new transfer buffer into large Pyrex pan, put glass plate across middle of pan, pre-wet 23 × 46 cm 3 MM Whatman paper by holding it at diagonally opposite corners and lowering it into transfer buffer, and lay across glass plate.
8. Similarly pre-wet small Whatman paper. Lay on top of bridge. Repeat this step with a second small Whatman paper.
9. Roll 5 ml serological pipet across Whatman paper to remove trapped air bubbles.
10. To invert two-dimensional gel, slide agarose gel from tray onto similarly sized Plexiglas plate. Cover with a second Plexiglas plate. Reach underneath the lower plate with one hand (thumb pointing away from you), and reach with the other hand across the gel (thumb pointing toward you), grip firmly and invert.
11. Slide gel from Plexiglas on pre-wetted small Whatman 3 MM paper. Remove air bubbles by rolling 5 ml serological pipet across gel.
12. Wearing gloves, cut nylon membrane to size of two-dimensional gel (19.5 cm × 26 cm) and label back of membrane in top left corner.
13. Pre-wet nylon membrane in 10x SSC and place on gel, starting at one corner of gel. If you have to shift the membrane after placing it on the gel, start over with a new membrane.
14. Cover entire blotting setup with Saran wrap to minimize evaporation. Remove trapped air bubbles with 5 ml serological pipet. Cut along edge of nylon membrane with razor blade. Remove cut out centerpiece of Saran wrap.
15. Cover nylon membrane with two pre-wetted small Whatman 3 MM papers (*see* step 8).
16. Pile 5 cm of dry paper towels on top of small Whatman papers.
17. Layer glass plate on top of paper towels and distribute four bottles with a total weight of ~0.5 kg evenly on the glass plate.
18. Allow transfer overnight.
19. Following transfer, dismantle blot wearing gloves, remove membrane, neutralize by briefly incubating in 50 mM

sodium phosphate (pH 7.2), and UV-crosslink membrane in UV crosslinking incubator in auto-crosslink mode (120,000 $\mu J/cm^2$). Store membrane between two pieces of Whatman 3 MM paper in a slider bag at room temperature or for long-time storage at –20°C.

3.5. Hybridization and Exposure

1. Use proper procedures while working with radioactivity.
2. Membranes should always be handled with gloves, avoiding creasing or bending (this will result in streaks of signal, obscuring the real signal), as well as stretching (this will distort signals).
3. Preheat hybridization solution and hybridization bottles to 65°C.
4. For prehybridization, pour 30 ml hybridization solution into large hybridization bottle. Roll UV-crosslinked nylon membrane and transfer into hybridization bottle. Rotate at 65°C in rotisserie for at least 30 min. Check that membrane unfolds in the rotisserie and does not stick upon itself during prehybridization or hybridization.
5. Following the supplier's instructions, resuspend pelleted buffer–nucleotide mix from PrimeIT kit RmT (Stratagene) with 30 ng "Probe 4" in 37 μl of water and boil at 100°C for 5 min in heating block filled with water.
6. Spin briefly and add 10 μl of alpha-^{32}P-dCTP (6,000 Ci/mmol) and 3 μl Magenta polymerase.
7. Incubate at 37°C for 20 min.
8. To reduce non-specific background, remove unincorporated nucleotides using a spin column (GE) following the supplier's instructions.
9. Using Geiger counter, ensure presence of strong radioactive signal in eluate.
10. After collecting probe from spin column, transfer into 1.5 ml tube, poke a hole in lid, and denature probe at 100°C for 5 min in heating block.
11. Add the denatured probe to 30 ml hybridization solution preheated to 65°C in a 50 ml Falcon tube. Do not attempt to mix. It is important to maintain the temperature. If the temperature drops substantially, the probe will re-anneal to itself and fail to bind to its target.
12. Discard prehybridization solution.
13. Pour mixture of labeled probe plus hybridization solution into hybridization bottle with membrane. Incubate overnight in rotisserie at 65°C.
14. Following overnight hybridization for >12 h, discard hybridization solution in radioactive waste, rinse with

60 ml wash solution preheated to 65°C, discard wash solution, and transfer membrane into 2 l plastic container in which membrane can unfold completely.

15. Wash for 20 min in shaking water bath at 65°C with 1 l wash solution preheated to 65°C. Discard wash solution and perform two more identical washes.
16. Following washes, remove nylon membrane from wash solution, pack between two layers of Saran wrap, squeeze out excess wash solution by rolling serological pipet across membrane, and fold in edges of Saran wrap.
17. Tape membrane into exposure cassette, expose for >30 h to erased phosphoimager imaging plate, and scan imaging plate at 100 μm/pixel resolution (files are ~20 Mb).

3.6. Quantitative Analysis

1. Quantitative analysis of two-dimensional gel Southern blots should be performed using software that allows drawing round volumes with a close fit to all signals, e.g., Quantity One™ (Biorad) or ImageJ. To ensure maximum signal detection within the linear range of the detection system, an exposure with a few (<10) saturated pixels at the center of the parental fragments ("Mom," "Dad" in **Fig. 7.2**) should be used. Comparable saturation of "Mom" and "Dad" signal indicates equal transfer of longer and shorter restriction fragments.
2. For quantitation, a mask is generated for a time point of maximum joint molecule levels (*see*, e.g., **Fig. 7.2**). At intermediate detection threshold, draw large circles around the two parental signals, as well as an equally sized background volume. Keep lowering detection threshold until

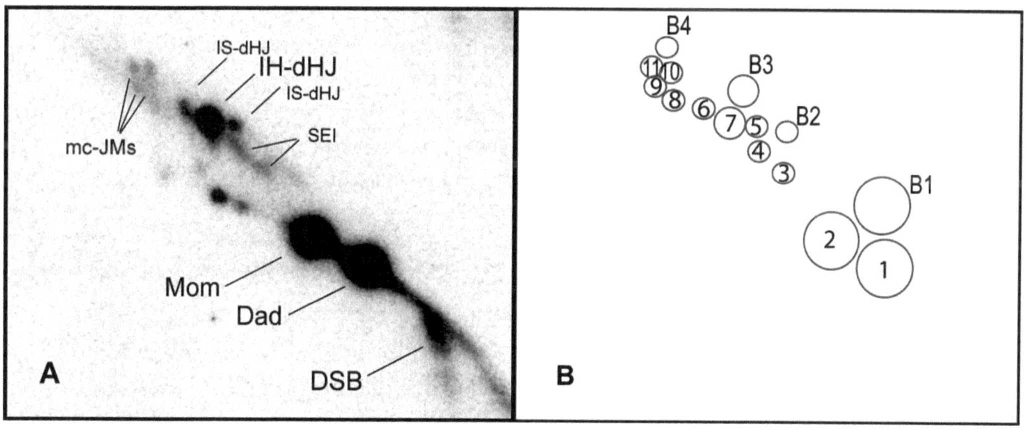

Fig. 7.2. Quantitation of two-dimensional gel. (**a**) Image of two-dimensional gel hybridized with Probe 4. (**b**) Example for a quantitation mask. In addition to 11 species of DNA molecules, 4 background regions (*B1–B4*) used for quantitation are also indicated. The background regions for corresponding signals are indicated in parenthesis: 1, 2 (*B1*); 3–6 (*B2*); 7 (*B3*); 8–11 (*B4*).

joint molecule signals above the arc become visible. For weak signals, switch to the sigmoid input–output setting (image output manipulation in quantitation software does not affect the actual counts). Separate background volumes are generated for every volume size and gel region.

3. With all visible signals circled, select all volumes, shift the entire mask to the earliest time point (**Fig. 7.2**). Copy–paste the mask to each time point on the same blot. Adjust position using parental signals as anchoring points. Enlarge each signal to maximum size on computer screen to ensure optimum fit.
4. Export numeric quantitation data to MS Excel worksheet and utilize Excel capabilities for automatic calculation to express branched signals as a percentage of the total signal (i.e., signals 1 through 11 minus appropriate background) intensity within the boxes, expressing joint molecules as "% of total DNA."
5. Joint molecules analyzed in detail include (from lower to higher molecular weight) two prominent SEIs, Dad–Dad intersister double Holliday junction, interhomolog double Holliday junction, Mom–Mom intersister double Holliday junction, and several species of multichromatid long joint molecules (**Fig. 7.1**). The respective species have been identified based on expected molecular weight, strand composition analysis, and/or electron microscopy (3, 7, 12).

4. Notes

1. Competent diploid cultures are transformed with a PCR-generated deletion construct marked e.g. with antibiotic resistance (13). A strain heterozygous for the appropriate mutant construct is sporulated, tetrads are dissected, spores with appropriate marker combinations (*HIS4*, *ura3* or *his4*, *URA3*) are identified, and the status of restriction sites at the hotspot is ascertained by restriction analysis with *Xho*I, as well as double digestion with *Xho*I/*Ngo*MIV and *Xho*I/*Bam*HI, respectively (9).
2. In mutant strains exhibiting aberrant intermediate levels and/or kinetics of joint molecules, absolute levels of double Holliday junctions can be determined in a strain background also deficient for pachytene exit factor *NDT80*. Absence of *NDT80* results in pachytene arrest with concomitant accumulation of double Holliday junction intermediates (4, 10).

3. Diploid colonies in the SK1 strain background exhibit a smooth morphology and margarine-like consistency, versus haploid colonies which exhibit a rough surface.
4. Pregrowth in YPA is notoriously difficult to control, and dilutions need to be optimized prior to each series of experiments. To achieve efficient and synchronous sporulation, we set up multiple parallel YPA and at least two SPM cultures for each genotype.
5. OD measurements vary between spectrophotometers; empirically identify OD_{600} that gives most synchronous time course and sufficient DNA for two-dimensional analysis.
6. The UV intensity of a UV transilluminator should be calibrated frequently to 0.6 mW/cm^2 using a Blak-Ray UV meter. UV lamps should be exchanged when the intensity deviates.
7. The genomic DNA extraction described here requires a few practice runs until it works reliably. When it works, a given amount of culture provides consistent amounts of genomic DNA. DNA concentrations are usually consistent and do not need to be measured in this case.

Acknowledgments

Work in the Börner lab is supported by NIH NIGMS grant GM080715.

References

1. Neale, M.J., and Keeney, S. (2006) Clarifying the mechanics of DNA strand exchange in meiotic recombination. *Nature* **442**, 153–158.
2. Hunter, N., and Kleckner, N. (2001) The single-end invasion: an asymmetric intermediate at the double-strand break to double-Holliday junction transition of meiotic recombination. *Cell* **106**, 59–70.
3. Schwacha, A., and Kleckner, N. (1995) Identification of double Holliday junctions as intermediates in meiotic recombination. *Cell* **83**, 783–791.
4. Allers, T., and Lichten, M. (2001) Differential timing and control of noncrossover and crossover recombination during meiosis. *Cell* **106**, 47–57.
5. Börner, G.V., Kleckner, N., and Hunter, N. (2004) Crossover/noncrossover differentiation, synaptonemal complex formation, and regulatory surveillance at the leptotene/zygotene transition of meiosis. *Cell* **117**, 29–45
6. Schwacha, A., and Kleckner, N. (1994) Identification of joint molecules that form frequently between homologs but rarely between sister chromatids during yeast meiosis. *Cell* **76**, 51–63.
7. Oh, S.D., Lao, J.P., Hwang, P.Y., Taylor, A.F., Smith, G.R., and Hunter, N. (2007) BLM ortholog, Sgs1, prevents aberrant crossing-over by suppressing formation of multichromatid joint molecules. *Cell* **130**, 259–272.

8. Börner, G.V., Barot, A., and Kleckner, N. (2008) Yeast Pch2 promotes domainal axis organization, timely recombination progression, and arrest of defective recombinosomes during meiosis. *Proc Natl Acad Sci USA* **105**, 3327–3332.
9. Lao, J.P., Oh, S.D., Shinohara, M., Shinohara, A., and Hunter, N. (2008) Rad52 promotes postinvasion steps of meiotic double-strand-break repair. *Mol Cell* **29**, 517–524.
10. Oh, S.D., Jessop, L., Lao, J.P., Allers, T., Lichten, M., and Hunter, N. (2009) Stabilization and electrophoretic analysis of meiotic recombination intermediates in Saccharomyces cerevisiae. *Methods Mol Biol* **557**, 209–234.
11. Hochwagen, A., Wrobel, G., Cartron, M., Demougin, P., Niederhauser-Wiederkehr, C., Boselli, M.G., et al. (2005) Novel response to microtubule perturbation in meiosis. *Mol Cell Biol* **25**, 4767–4781.
12. Bell, L., and Byers, B. (1979) Occurrence of crossed strand-exchange forms in yeast DNA during meiosis. *Proc Natl Acad Sci USA* **76**, 3445–3449.
13. Goldstein, A.L., and McCusker, J.H. (1999) Three new dominant drug resistance cassettes for gene disruption in Saccharomyces cerevisiae. *Yeast* **14**, 1541–1553.

Chapter 8

Mapping of Crossover Sites Using DNA Microarrays

Stacy Y. Chen and Jennifer C. Fung

Abstract

Crossovers (COs) play an essential role in promoting successful chromosome segregation during meiosis. Crossing over generates chiasmata, which are physical bridges between homologs that provide the appropriate tension to properly align chromosomes on the meiosis I spindle. Homolog pairs that fail to cross over can result in meiosis I nondisjunction, leading to aneuploid gametes. Therefore, the number and distribution of crossovers are tightly regulated to ensure that each chromosome pair receives at least one CO. Here, we describe a DNA microarray-based method to map CO distribution genome-wide, on a cell-by-cell basis, allowing for rapid and accurate analysis of multiple aspects of CO control.

Key words: Meiosis, recombination, crossover, noncrossover, direct allelic scanning, crossover interference, crossover homeostasis, S96, YJM789.

1. Introduction

Meiosis is the beginning stage of sexual reproduction during which one diploid parent undergoes two rounds of cellular division to produce four haploid progeny (1, 2). Recombination between homologous chromosomes during the first meiotic division is essential for successful chromosome segregation. Meiotic recombination leads to the formation of crossovers (COs) and noncrossovers (NCOs) (3). Crossing over creates chiasmata, which are interhomolog associations that provide the necessary tension to correctly align homologs on the meiosis I spindle. Defects in crossing over lead to meiosis I nondisjunction, resulting in the production of aneuploid gametes (4).

To ensure that each pair of homologous chromosomes receives at least one CO, the spatial distribution and the number

H. Tsubouchi (ed.), *DNA Recombination*, Methods in Molecular Biology 745,
DOI 10.1007/978-1-61779-129-1_8, © Springer Science+Business Media, LLC 2011

of COs are highly orchestrated. Some examples of CO control include CO interference and CO homeostasis. In most eukaryotes, CO interference regulates the spatial positioning of COs along a chromosome such that a CO event in one region reduces the likelihood of another one occurring nearby (5–7). This results in a nonrandom and more evenly spaced distribution of COs across the genome where the strength of interference diminishes as a function of distance.

CO homeostasis controls the number of COs in a single meiosis, whereby the normal level of COs is maintained despite fluctuations in the overall number of recombination-initiating events (8). Martini et al. observed that when the overall recombination-initiating events are reduced, CO levels are maintained at the expense of NCOs. CO homeostasis may function to reduce the occurrence of nonexchange chromosomes by ensuring that a sufficient number of COs are made.

One major difficulty in understanding CO control in meiosis has been the lack of an efficient and accurate method for determining CO distribution genome-wide and on a cell-by-cell basis. Here, we describe a microarray-based approach for mapping CO distribution using the method of direct allelic variation scanning of the genome that has been adapted to analyze multiple aspects of CO control (*see* **Note 1**) (9, 10). This method identifies sequence polymorphisms between two strains of yeast *Saccharomyces cerevisiae* – S96 and YJM789. Using the polymorphic markers, the parental origin of the meiotic progeny at each of the detectable sequence polymorphic loci is determined. The reciprocal CO events (and a subset of NCOs and gene conversions) can be mapped by following the inheritance pattern of allelic markers in the four haploid progeny strains. Multiple aspects of the CO landscape can thus be analyzed, including the genome-wide interference level, which can be calculated using the distribution of distances between adjacent COs and the gamma distribution function (11, 12), as well as CO homeostasis, which can be determined by the correlation between the number of COs and NCOs for each meiotic event (10).

2. Materials

2.1. Isolation of Four-Spore Viable Tetrads

1. *S. cerevisiae* strains: S96 (*MAT*a *ho lys5*) and YJM789 (*MAT*α *ho::hisG lys2 cyh*).
2. YPAD plates: dissolve 20 g dextrose, 20 g bactopeptone, 10 g yeast extract, and 20 g agar in water to a final volume of 940 ml. Sterilize by autoclaving. Add 50 ml of 10 mM

sterile adenine solution and 10 ml of 20 mM sterile uracil solution. Pour 20 ml into each Petri plate. Allow media to cool and solidify at room temperature.

3. Amino acid mix: 2.4 g adenine, 21.0 g arginine, 6.0 g glutamic acid, 2.6 g histidine, 2.4 g inositol, 31.2 g isoleucine, 15.8 g leucine, 5.4 g lysine, 9.0 g methionine, 4.8 g phenylalanine, 6.6 g serine, 7.2 g threonine, 4.8 g tryptophan, 1.2 g tyrosine, 1.4 g uracil, and 7.2 g valine.
4. Sporulation plates: dissolve 2 g yeast extract, 1 g dextrose, 20 g potassium acetate, 1 g amino acid mix, and 20 g agar in water to a final volume of 1 l. Sterilize by autoclaving. Pour into Petri plate as described for YPAD plates.
5. Zymolyase 100T (Seikagaku Corporation, Tokyo, Japan) stock solution is prepared at 10 mg/ml in 5% (w/v) dextrose solution and is stored in single-use aliquots at –20°C.
6. Ascus digestion solution is freshly prepared each time from the zymolyase stock solution to a working solution of 0.05 mg/ml in 1 M sorbitol.

2.2. Allele-Specific Primer Extension Colony PCR of Four-Spore Viable Tetrads

1. Overnight yeast cultures of all spores from four-spore viable tetrads which had been grown on a YPAD plate.
2. 0.02 M NaOH solution.
3. PCR primers (**Table 8.1**; synthesized by Integrated DNA Technologies, Carolville, IA): resuspended in sterile ddH_2O to 100 μM stock solution. Store at –20°C.
4. *Taq* PCR Core Kit (Qiagen, Germantown, MD), specifically: *Taq* DNA polymerase (5 units/μl), CoralLoad PCR buffer (10x), Q-solution (5x), and dNTP mix (10 mM of each dNTP). Store at –20°C.
5. DNA HyperLadder^{TM} IV (Bioline, Taunton, MA). Store at 4°C.
6. 1x Tris/acetate/EDTA (TAE) buffer: 40 mM Tris–acetate, pH 8.5, 2 mM $Na_2EDTA{\cdot}2H_2O$.
7. Agarose gel: 1.5% (w/v) agarose, 1x TAE buffer, 0.5 μg/ml ethidium bromide.

2.3. Isolation of Yeast Genomic DNA

1. YPAD media are prepared in a similar procedure as YPAD plates, omitting the agar.
2. TE: 10 mM Tris–Cl, pH 8.0, 1 mM EDTA, pH 8.0. Store at room temperature.
3. Buffer Y1 (yeast lysis buffer): 1 M sorbitol, 100 mM EDTA, 14 mM β-mercaptoethanol. Store at 2–8°C.

Table 8.1
Allele-specific primer sequences. Shown are primer sequences for 23 primer sets

Chr.	Coordinate (kbp)	Primer type	Primer sequence (5′ to 3′)
II	673	Common	GGCTATTGATGCGATAAATAAAGGC
		S96-specific	TTGGTTCTACGATACTGGGTGAC
		YJM789-specific	TTCCACATATCTTTGAAAAGAGTCGTA
III	82	Common	GCCGAGAGTATCACTGATTCAAGG
		S96-specific	CGCTGTTAGGTGGCTTTTTTACAGTA
		YJM789-specific	CACTTTCAGTCCCTTTTTCCTCCT
IV	344	Common	GTAATTCTACCTAGCCCACCAC
		S96-specific	GCATATCGTATGATTCGACTACAGACG
		YJM789-specific	CTTATCTAAGCTGATACCAGGGTATA
IV	1261	Common	CGGATACCAAGATTGTCATACTCACTAAAG
		S96-specific	CTTAATGGGTATGAATATATTCTTGTTTATTCTCC
		YJM789-specific	GGTGAATGTAAAATTAATACGGCGGTAAC
V	458	Common	GCGATAATTGACCTTTTCCAAGGAC
		S96-specific	GGTCCCTTATAAACGTATGAAGTGTAG
		YJM789-specific	GTTTCTTAGGCAATCTAGTAATGTTG
VI	239	Common	CATATGTATACACATATACATATCTGTACATACTC
		S96-specific	GATAGCTGCCCATCGAAATACGTTT
		YJM789-specific	GATTATAGATACCCACGACTGGTTGAAA
VII	773	Common	GGGTGATAATACATACTCCCCATC
		S96-specific	GTTGGGATTCCATTGTTAATAACACTAG
		YJM789-specific	CATGGAAAACCGGATTTCTAGGAAGGAAG
VIII	359	Common	GGTGAATAATGAAGATTGGGTGAATAATTTG
		S96-specific	GTGATAATACACTACTAATGTGACTACTAGTAGAC
		YJM789-specific	GCTGTGATAATTATTCATAGAAATATTACAGAGCATA
VIII	413	Common	CGCAAGACTTTCTTCACCAATACTTTG
		S96-specific	CATTTACTTCACTTCGTAGCAATGTTAAG
		YJM789-specific	GGCATGCATACTGGGACGT
IX	98	Common	GGCCAATGAGCAAAAATTTAGGC
		S96-specific	CAAATTGGAAGCAAAGAGAAAGGTTTC
		YJM789-specific	CCTCCCCGTTACAGTTTAGACTG
IX	191	Common	CTCGAAAGTGCTACCCACTGC
		S96-specific	GGGACGAAAAGAGCAGCTGTATTAACG
		YJM789-specific	GGGTTTATTACTTCAGGGAACTTTCTGGTT
X	137	Common	GAAATAGTAATCCCAACGCACTCATCCGC
		S96-specific	CTTCTGAAAATAATCTTGAAATGGCATGATATGAATCTA
		YJM789-specific	GGTGAACAGGTGCATTTTGAGAAGA

Table 8.1 (continued)

Chr.	Coordinate (kbp)	Primer type	Primer sequence (5′ to 3′)
X	148	Common	GTAATGACTATACGTATAAGGAAAATTAAGAAAAGGC
		S96-specific	CACCACAACAAGCTATGCTATAC
		YJM789-specific	GGTGCTATCAGTAAAAGGAAGGAGAACAT
X	516	Common	GTAAATCAGTATAGTAATGTCCTTCGGATGG
		S96-specific	GGATGTACCTAAAATACAGCAAACAAAGCGTT
		YJM789-specific	CACGCAAGCCATCACCCGATA
X	622	Common	CCATCCAGAGTATACATCGAAGG
		S96-specific	CACTTCAATCCTTTCAAGAAGACATAT
		YJM789-specific	GAAGAATCTTTGAAGACTGGTAATCCT
X	627	Common	CTGTGAACCTTAGAAATCCTCTATGC
		S96-specific	CGTCCAACCTGCCCATCACCCT
		YJM789-specific	AATGATGAGATAATTAACCCAACAGCCGG
XI	394	Common	CGTGTGGCTGCCTCTAAGAATTAAACTTC
		S96-specific	CCATTGATCATTTGCACAAATCATTGAAC
		YJM789-specific	GCTTCGCTCAATAAAAAAAGATCTTCATCGG
XI	624	Common	GAGGAGTTCAACAATGAACTGC
		S96-specific	ATGAATCCTTTTGGGCAGGATT
		YJM789-specific	AGTTTTTCACCGGAAAGTAACGGAATA
XI	320	Common	GTATAAGTGCATACTAACATACTGTGTACGTAC
		S96-specific	GACATGAACGACGTTTTGGGAAAAATAAC
		YJM789-specific	CTAAGAGAAGATTCGGGTTTTAATTTAAGGTT
XII	574	Common	GTTGAAGCACTGCCTCCAG
		S96-specific	GATCGAAGGAAACTAAAAGAGGTTTGATGTCAG
		YJM789-specific	GCGCCAAACAAGGGATGG
XII	780	Common	CATGGAGGCTAGACATGACTAATG
		S96-specific	CAGTCGATCTCTTGCCCTAG
		YJM789-specific	CCTTTTGTTCAATGGCAGAATTTCTATGCA
XIII	216	Common	GACCGCTATGCGTCTGATGT
		S96-specific	CAGCTGATAAAGAACACTGATCATGACA
		YJM789-specific	CCTTTTGGATCTTCTGTCTTTGAGCT
XIII	802	Common	CCAGCAGGGAAGCCATTAAATAG
		S96-specific	CTAGGTGAGTAGACTAACCGATCC
		YJM789-specific	GTATTTGAGAAGGGGGTTTAACACTAACA

Genomic coordinates are approximated in kilobase pair. Each primer set assesses SNP genotype at two SNP positions. Three primers are designed for each primer set: a common primer, a YJM789-specific primer (which anneals to the first SNP), and a S96-specific primer (which anneals to the second SNP). Primer sequences are given in the 5′- to 3′-direction. Mismatches internal to the 3′-end of the primer, when present, are underlined. The 3′-terminal nucleotide of each allele-specific primer is the position of the SNP and matches only one of the two possible SNP sequences.

4. Zymolyase 100T is dissolved at 10 mg/ml in 5% (w/v) dextrose solution and is freshly made each time.
5. Buffer G2 (digestion buffer): 800 mM guanidine hydrochloride, 30 mM Tris–Cl, pH 8.0, 30 mM EDTA, pH 8.0, 5% Tween-20, 0.5% Triton X-100. Store at 2–8°C or room temperature.
6. RNase A is dissolved at 100 mg/ml in 0.01 M sodium acetate, pH 5.2, followed by heating at 100°C for 15 min. Allow solution to cool slowly to room temperature before adding 0.1 volume of 1 M Tris–Cl, pH 7.4, to adjust the pH. Store in aliquots at –20°C.
7. Proteinase K (Invitrogen, Carlsbad, CA) is dissolved at 20 mg/ml in sterile ddH_2O and is freshly made each time.
8. Buffer QBT (equilibration buffer): 750 M NaCl, 50 mM MOPS, pH 7.0, 15% isopropanol, 0.15% Triton X-100. Store at 2–8°C or room temperature.
9. Genomic-tip 500/G (Qiagen, Valencia, CA).
10. Buffer QC (wash buffer): 1.0 M NaCl, 50 mM MOPS, pH 7.0, 15% isopropanol. Store at 2–8°C or room temperature.
11. Buffer QF (elution buffer): 1.25 M NaCl, 50 mM Tris–Cl, pH 8.5, 15% isopropanol. Store at 2–8°C or room temperature.
12. Isopropanol.
13. Glass microcapillary pipettes (10 μl) (VWR International, West Chester, PA): Pipettes are sealed on one end by flaming.
14. 70% (v/v) ethanol.

2.4. Fragmentation of DNA Using Deoxyribonuclease I

1. Deoxyribonuclease I (DNase I), amplification grade, 1 U/μl (Invitrogen).
2. 10x One-Phor-All Buffer *Plus* (discontinued item from GE Healthcare, Chalfont Saint Giles, UK): 100 mM Tris–acetate, pH 7.5, 100 mM magnesium acetate, 500 mM potassium acetate. Store at 4°C.
3. 25 mM $CoCl_2$ solution (in package contents of the terminal transferase used for biotinylation in **Section 2.5**).
4. 10,000x SYBR® Green I Nucleic Acid Gel Stain (Invitrogen). Store at 4°C and shield from light.
5. Agarose gel: 2% (w/v) UltraPure agarose 1,000 (Invitrogen), 1x TAE buffer, with SYBR® Green I Nucleic Acid Gel Stain. Shield from light.
6. TAE buffer: 40 mM Tris–acetate, pH 8.5, 2 mM $Na_2EDTA \cdot 2H_2O$.

7. DNA HyperLadder™ IV (Bioline, Taunton, MA). Store at 4°C.
8. Loading buffer: 7.5% (v/v) glycerol solution.

2.5. Biotinylation of DNA Fragments and Microarray Analysis

1. Bio-N^6-ddATP, 1 mM (Enzo Life Sciences, Inc., Farmingdale, NY).
2. Terminal transference, recombinant, 400 U/μl (Roche, Indianapolis, IN).
3. 12x MES stock solution (1.22 M MES, 0.89 M [Na+]): Dissolve 70.4 g MES free acid monohydrate and 193.3 g MES sodium salt in 800 ml sterile ddH_2O. Mix and adjust volume to 1 l. pH should be between 6.5 and 6.7 without adjustments. Sterile filter using 0.2 μm filter. Store at 4°C and shield from light. Discard if solution becomes yellow.
4. 2x hybridization buffer (200 mM MES, 2 M [Na+], 40 mM EDTA, 0.02% Tween-20): Mix 8.3 ml 12x MES stock solution, 17.7 ml 5 M NaCl, 4.0 ml 0.5 M EDTA, pH 8.0, 0.1 ml 10% Tween-20, and add 19.9 ml sterile ddH_2O to bring it to a final volume of 50 ml. Store at 4°C and shield from light.
5. Herring Sperm DNA, 10 mg/ml (Promega, Madison, WI).
6. Acetylated BSA, 20 mg/ml (Invitrogen).
7. Control Oligo B2, 3 nM (Affymetrix, Santa Clara, CA).
8. GeneChip® Yeast Genome S98 Array (Affymetrix, Santa Clara, CA).
9. Prepare wash buffer A, wash buffer B, Streptavidin Phycoerythrin (SAPE) stain and antibody solutions according to the *GeneChip® Expression Analysis Technical Manual* (13).
10. GeneChip® Hybridization Oven 645 (Affymetrix, Santa Clara, CA).
11. GeneChip® Fluidics Station 450 (Affymetrix, Santa Clara, CA).
12. GeneChip® Scanner 3000 (Affymetrix, Santa Clara, CA).
13. Affymetrix® Microarray Suite Software (Affymetrix, Santa Clara, CA).

2.6. Data Analysis Using Allelescan and CrossOver Software

1. MATLAB® 6.5 with the Statistics Toolbox™ (MathWorks, Natick, MA).
2. Allelescan (Davis Lab, Stanford University).
3. Python 2.5 or higher (http://www.python.org).
4. CrossOver (Fung Lab, University of California, San Francisco, CA).

3. Methods

S96 and YJM789 haploid parental strains are mated and four meiotic progeny are isolated via tetrad dissection by selecting for those which are four-spore viable. Four-spore viable tetrads are prescreened for possible errors in the selection procedure or abnormal genome-wide missegregation using the allele-specific primer extension colony PCR. Tetrads that show a normal 2:2 segregation of parental alleles in the majority of SNP (single-nucleotide polymorphism) primer sets tested by colony PCR are selected for further microarray analysis.

Genomic DNA is isolated from the parental strains, S96 and YJM789 haploids, and their meiotic progeny, the four-spore viable tetrads. Genomic DNA is fragmented using DNase I and end-labeled with biotin-*N*6-ddATP using terminal deoxynucleotidyl transferase before hybridizing to Affymetrix GeneChip® Yeast Genome S98 Array using the GeneChip® Hybridization Oven 645. The arrays are stained with R-phycoerythrin–streptavidin and amplified with biotinylated antistreptavidin antibody using GeneChip® Fluidics Station 450 and scanned using the laser confocal scanner, GeneChip® Scanner 3000.

Microarray experiment data are analyzed using the software Allelescan, a microarray analysis platform that analyzes genomic DNA hybridization data and identifies sequence polymorphisms between samples from two distinct genetic backgrounds using their differential hybridization signal intensities (14). Meiotic progeny generated from the two parental backgrounds can be genotyped at each of the polymorphic markers and a segregation profile is generated for the four-spore tetrad.

CrossOver is developed as a downstream analysis tool to process the segregation profile from Allelescan in order to determine locations of meiotic recombination events. In addition, CrossOver performs various analyses that address questions of particular interest to meiotic recombination. CrossOver can compute crossover densities for each chromosome, occurrence of nonexchange chromosomes, inter-crossover distances, CO-to-centromere and CO-to-closest telomere distances, gene conversion tract lengths, correlation coefficients of the number of COs and NCOs for each meiosis, and parameters of the gamma distribution function for inter-crossover distances (10).

3.1. Isolation of Four-Spore Viable Tetrads

1. Streak out *S. cerevisiae* yeast strains S96 and YJM789 from frozen stock onto YPAD plates and grow overnight at 30°C. Select and patch a few single colonies from each parent to proceed with and mate. Yeast mating is most efficient when parent cells are from fresh cultures.

2. Use sterile toothpicks to mix a small, but equal amount of S96 and YJM789 parent cells on a YPAD plate, creating a patch of cells of about 5 mm in diameter. Allow cells to mate at 30°C for 4–6 h.
3. Transfer the mixture of S96 and YJM789 cells from the YPAD plate to a sporulation plate using a sterile toothpick. Incubate cells at 30°C for 3–5 days until sufficient numbers of tetrads have formed. Cultures with fewer than 5% tetrads are difficult to dissect.
4. Prepare fresh ascus digestion solution from the zymolyase stock solution. To prepare cultures for dissection, resuspend a small dab of cells (about 1 mm in diameter) from the sporulation plate in 50 μl freshly prepared ascus digestion solution. Incubate for 10 min at 30°C. Add 100 μl of sterile water and mix gently.
5. To prepare a dissection plate, hold a fresh YPAD plate at a 45° angle and gently spot 15 μl of zymolyase-treated cells along a line down the center of the plate. Allow the liquid solution to dry on plate.
6. Dissect tetrads on a yeast dissection microscope. Incubate dissected plates at 30°C. After 3 days at 30°C, colonies should be of sufficient size to determine viability. Only four-spore viable tetrads are selected for further analysis (*see* **Note 2**).

3.2. Allele-Specific Primer Extension Colony PCR of Four-Spore Viable Tetrads

1. **Table 8.1** shows 23 sets of allele-specific PCR primers which display strong allele specificity in allele-specific primer extension PCR. Each primer set assesses SNP genotype at two different SNP positions, approximately 200 bp apart. Three primers are designed for each primer set: (1) a common forward primer that anneals to both the S96 and the YJM789 template; (2) a YJM789-specific reverse primer that anneals to the first SNP approximately 200 bp from the common forward primer; and (3) a S96-specific reverse primer that anneals to the second SNP approximately 400 bp from the common forward primer (*see* **Fig. 8.1**). Allele-specific primers are designed to match only one of the two possible SNP allele sequences at the 3′-terminal nucleotide, allowing efficient amplification of the matched SNP nucleotide, but not the mismatched allele. To increase allele specificity, a single-base mismatch is sometimes introduced to both allele-specific primers 3 or 4 bases inward from the 3′-end of the primer, causing further destabilization for primers that may have annealed to the wrong allele (15). Primer sequences within each primer set were selected to have similar melting temperatures, of approximately 54°C. The resulting PCR reaction generates

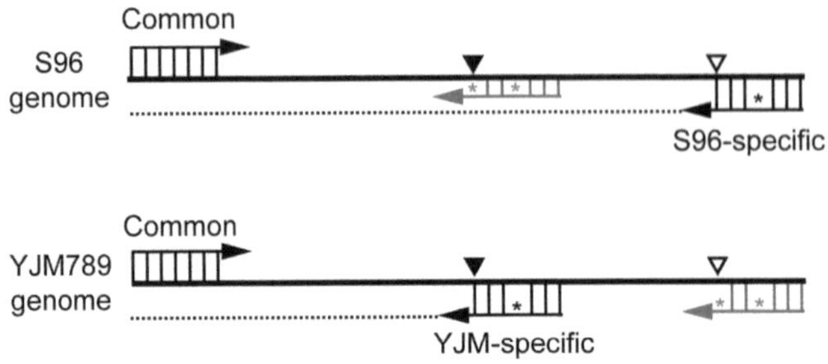

Fig. 8.1. A schematic of the design for allele-specific primer extension PCR. The common primer anneals to both the S96 and the YJM789 genome. SNP sites are engineered at the 3′-terminal nucleotide of each allele-specific primer, leading to amplification of only one of the two SNP alleles. Allele-specific primers are designed at two separate SNP positions (indicated by ▼ and ▽), approximately 200 bp apart. The resulting PCR yields two allele-specific bands with a 200 bp difference in size (shown in *dotted line*), which can then be visualized on the agarose gel. A single nucleotide internal mismatch is engineered in the allele-specific primers to enhance specificity by further destabilizing allele primers that may have annealed to the wrong allele (15). Positions of mismatch are denoted by an *asterisk* (∗). Primers containing mismatch at the 3′-terminal nucleotide do not successfully amplify and are illustrated in *gray*.

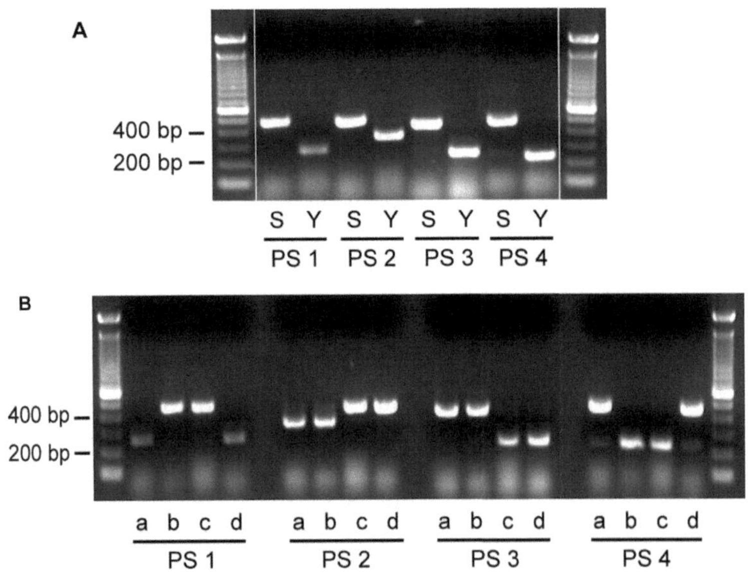

Fig. 8.2. Allele-specific primer extension colony PCR for S96 and YJM alleles. (**a**) Allele-specific primer extension PCR results for S96 (S) and YJM789 (Y) parental haploid strains. PCR primers are designed so that the S96 allele-specific band is approximately 200 bp longer than the YJM789 allele-specific band. PCRs of four primer sets (PS) are shown. (**b**) Allele-specific primer extension PCR is performed for a four-spore tetrad using the same four primer sets as shown for the parents. Four spores are indicated by *a*, *b*, *c*, and *d*. PCR products from all four primer sets show 2:2 segregation of the SNP alleles.

a YJM789-specific product of roughly 200 bp and a S96-specific product about 400 bp in length, which can easily be resolved from each other on a 1.5% agarose gel (*see* **Fig. 8.2**).

2. To set up allele-specific primer extension colony PCR for one full four-spore tetrad and one primer set (*see* **Note 3**), take a small dab of overnight yeast culture from each spore of the tetrad using a sterile pipette tip. Resuspend cells from each spore in a separate PCR tube containing 5 μl 0.02 M NaOH solution. Transfer cell mixture to a PCR machine and boil for 10 min at 99°C. Cool to 4°C.
3. Select the primer set to test with by PCR and its corresponding individual primers. Mix equal amounts of 100 μM common forward primer, 100 μM S96-specific reverse primer, and 100 μM YJM789-specific reverse primer to create a master primer mix consisting of all three primers (*see* **Note 4**).
4. Once the cells from Step 2 have cooled to 4°C, add the following to each of the four-spore cell mixtures: 10 μl 5x Q-solution, 5 μl 10x CoralLoad PCR buffer, 1 μl 10 mM dNTP mix, 1 μl of master primer mix from Step 3, 25.5 μl sterile ddH_2O, and 0.5 μl *Taq* DNA polymerase to a final volume of 50 μl.
5. Run the following PCR program: initial denaturing step at 94°C for 5 min, followed by 35 cycles of denaturing at 94°C for 30 s, annealing at 54°C for 30 s, and extending at 72°C for 1 min, and a final extension at 72°C for 10 min.
6. Prepare a 1.5% TAE agarose gel. Run 5 μl of each of the finished PCR reaction alongside 5 μl of HyperLadder^{TM} IV (or any DNA ladder that includes the 200–400 bp range) until the YJM789-specific band (~200 bp) is resolved from the S96-specific band (~400 bp). Assess the S96 and YJM789 allele segregation of the tetrad (*see* **Note 3** and **Fig. 8.2**).

3.3. Isolation of Yeast Genomic DNA

1. Yeast genomic DNA is isolated using the Qiagen Genomic-tip 500/G. The following procedures are adapted from the *Qiagen Genomic DNA Handbook* (16). Using a sterile toothpick, make a circular patch of yeast culture approximately 1 in. in diameter on a YPAD plate. Grow at 30°C overnight.
2. In a 1 l flask, inoculate the entire patch of overnight yeast culture in 150 ml YPAD liquid media. Grow culture for ~18 h on platform shaker at 30°C to a cell density of approximately 3×10^8 cells/ml (*see* **Note 5** for alternative inoculation method).
3. Harvest 100 ml of culture by centrifuging at 3,000–5,000×*g* for 5 min. Discard the supernatant.

4. Resuspend cell pellet in 12 ml TE buffer and transfer cells to a 50 ml conical tube. Centrifuge again at 3,000–5,000 × *g* for 5 min to remove residual YPAD media. Discard the supernatant (*see* **Note 6**).
5. Resuspend cell pellet in 12 ml of Buffer Y1. Vortex to resuspend cells thoroughly. Add 1 ml of zymolyase solution. Rotate on roller drum at 30°C for 1–1.5 h.
6. Following zymolyase digestion, centrifuge cells at 5,000 × *g* for 20 min at 4°C. During centrifugation, add 30 μl of RNase A solution to 15 ml of buffer G2 and prepare fresh proteinase K solution. Slowly decant supernatant after centrifugation to avoid disturbing the pellet. Discard supernatant.
7. Add 15 ml of buffer G2 (with RNase A) to the spheroplast pellet from Step 6. Resuspend pellet thoroughly by pipeting. A homogeneous suspension is critical for efficient lysis of the spheroplasts.
8. Add 400 μl of proteinase K solution and mix gently by inverting. Incubate at 50°C for at least 1 h and centrifuge at 5,000 × *g* for 20 min at 4°C. During centrifugation, place a Qiagen Genomic-tip 500/G over a waste collector tube using a tip holder and equilibrate Genomic-tip by adding 10 ml of Buffer QBT.
9. Gently pour supernatant into a fresh 50 ml conical tube and discard the pellet. Vortex for exactly 8 s at top speed. Add 10 ml of Buffer QBT to the supernatant and vortex again for two more seconds to mix. Pour mixture into the equilibrated Genomic-tip and allow it to slowly drip through the column by gravity. *See* **Note 7** if the column becomes clogged.
10. Wash the Genomic-tip by adding 30 ml of Buffer QC. Repeat wash. While waiting for the wash buffer to drip through, prewarm Buffer QF in 50°C water bath.
11. To collect eluate, place Genomic-tip over a clean 50 ml conical tube using a tip holder provided by the manufacturer. Elute DNA with 15 ml of prewarmed Buffer QF.
12. Precipitate DNA by adding 10.5 ml of room temperature isopropanol to the eluate. Gently invert the tube 10–15 times until white web-like precipitated DNA appears.
13. Using a sealed glass microcapillary pipette, gently spool the precipitated DNA and transfer to a 1.5 ml microcentrifuge tube containing 200 μl of 70% ethanol. Nutate DNA for 5 min before spinning for a few seconds in a microcentrifuge at top speed.

14. Gently remove the supernatant with a pipette. Briefly air-dry the pellet for less than 5 min. Take caution to not overdry the pellet. Overdried pellets become difficult to redissolve later.
15. Depending on pellet size, add 100–150 μl of TE, pH 8.0, to the pellet. The optimal target DNA concentration is around 1 μg/μl. Dissolve the DNA overnight on a nutator at room temperature.
16. Next day, place DNA in a 50°C water bath for 1 h to aid in dissolving the DNA. Flick the tube occasionally to help resuspension. If the pellet remains undissolved or if the DNA solution appears murky, *see* **Note 8**.
17. Measure DNA concentration with a spectrophotometer such as NanoDrop™. Take two readings to ensure consistency in DNA concentration. Widely different readings indicate the presence of undissolved DNA. Adjust DNA concentration to around 1 μg/μl with TE. Refer to **Note 8** for resuspending undissolved DNA. Store DNA sample at –20°C or precede to DNase I digestion.

3.4. Fragmentation of DNA Using Deoxyribonuclease I

1. Prepare a boiling water bath for deactivation of DNase I. Alternatively, set a heat block or program a PCR machine to 100°C.
2. Prepare appropriate dilutions of DNase I (*see* **Note 9**).
3. For each DNase I reaction, dilute 15 μg of genomic DNA in sterile ddH_2O to a volume of 36.8 μl, then add 4.5 μl of 10x One-Phor-All Buffer *Plus* and 2.7 μl of 25 mM $CoCl_2$ solution to a total volume of 44 μl.
4. Add 1 μl of diluted DNase I to the reaction tube. Thoroughly mix the tube with gentle flicks. (If you will be using a boiling water bath to deactivate DNase I, place a lid clamp on the tube at this step.) Immediately transfer the tube to a 37°C water bath and incubate for 5 min.
5. Place the sample in the boiling water bath or in a 100°C heat block for 10 min to deactivate DNase I and to convert dsDNA to ssDNA. Snap cool DNA sample on ice for 10 min immediately after boiling to retain DNA in single-stranded state. Quick spin the sample to collect any condensation that may have gathered on the sides of the tube.
6. Prepare a 2% TAE agarose gel with SYBR® Green I nucleic acid stain. Dilute SYBR® Green I stock solution 1:10,000 in the melted agarose solution just prior to pouring the gel. Shield from light.
7. Combine 1 μl of each digestion sample with 2 μl of 7.5% glycerol loading buffer. Load all 3 μl onto agarose gel along-

side 3 μl of HyperLadder™ IV. Run agarose gel until the range 30–100 bp is resolved.

8. If multiple DNase I digests were performed for a genomic DNA sample, select the best DNase I digest with which to proceed. Genomic DNA fragments should be under 100 bp but over 30 bp. *See* **Fig. 8.3** for an ideal fragmentation pattern (also *see* **Note 10**). Repeat DNase I fragmentation with genomic sample that does not show desirable fragmentation in any of the digests. Adjust DNase I concentration accordingly. Alternatively, one can consider increasing or reducing fragmentation time to achieve the desirable degree of fragmentation.

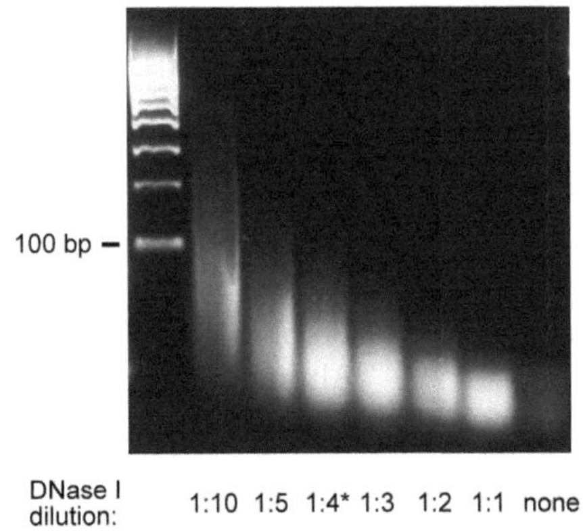

Fig. 8.3. Fragmentation pattern of genomic DNA. Multiple dilutions of DNase I were used in fragmenting genomic DNA samples. A total of 15 μg of genomic DNA was incubated with various dilutions of DNase I for 5 min. Shown are 1 μl of each digestion sample resolved on a 2% TAE agarose gel stained with SYBR Green I nucleic acid stain. In this sample, digestion using the 1:4 dilution of DNase I, indicated with an *asterisk*, reveals the most ideal fragmentation pattern.

3.5. Biotinylation of DNA Fragments and Microarray Analysis

1. Add 1.5 μl biotin-*N*6-ddATP and 1.5 μl terminal transferase to the DNase I-digested sample. Incubate at 37°C for 1.5 h to biotinylate DNA fragments.
2. In a boiling water bath, boil sample for 12 min and snap cool on ice for 10 min.
3. To prepare sample for hybridization, add 150 μl 2x hybridization buffer, 3 μl Herring sperm DNA (10 mg/ml), 7.5 μl acetylated BSA (20 mg/ml), 5 μl control oligonucleotide B2, and 87.5 μl of water to a final volume of 300 μl. Transfer sample to a 2 ml screw top tube.
4. Find a local genomics core facility that provides service for processing the Affymetrix GeneChip® microarrays. Alternatively, contact Affymetrix for array processing services available in your area (17).

5. Follow standard protocol procedures for hybridization, washing and staining, and scanning of the GeneChip® Yeast Genome S98 Array as described in *GeneChip® Expression Analysis Technical Manual* (13), with the following exceptions:
   a. In preparing samples for hybridization, incubate samples at 99°C for 10 min and then transfer to ice for 5 min before centrifuging at top speed for 3 min.
   b. Load 210 μl of sample onto GeneChip® Yeast Genome S98 Array. Avoid taking any particulates that may have settled at the bottom of the tube.
   c. Incubate array in hybridization oven at 42°C, 60 rpm for 18 h.
   d. Use fluidics protocol "EukGE-WS2v4_450."
6. After the array has been scanned, experiment data are generated by the Affymetrix® Microarray Suite Software in files with the following extensions: .exp, .dat, .cel, and .chp. Only .cel files, which contain probe location and signal intensity data, are needed to proceed with the downstream analysis.

3.6. Data Analysis Using Allelescan and CrossOver Software

1. Install MATLAB® and the Statistics Toolbox™ onto your computer.
2. Copy the Allelescan software folder onto your computer.
3. Run allelescan.m file using MATLAB®.
4. Following the instructions in the *Allelescan Users Manual*, create a new project, identify locations of sequence polymorphisms among samples, genotype one four-spore tetrad, and determine the segregation inheritance pattern of the tetrad (14). *See* figure 3 in Chen et al. for a sample segregation profile (10). Save the dump_segregation.txt file for further analysis using the CrossOver analysis software.
5. Install Python onto your computer.
6. Copy the CrossOver software folder onto your computer.
7. Following the instructions in the readme.txt file located inside the "doc" folder of CrossOver, copy the dump_segregation.txt file from Allelescan into the "segfiles" folder in CrossOver and change the filename according to the readme.txt file. Run CrossOver following the documentation given in the readme.txt file. CrossOver computes the location of meiotic recombination products, such as COs, NCOs, and gene conversions, as well as various aspects of meiotic recombination, such as crossover densities, occurrence of nonexchange chromosomes, inter-crossover distances, gene conversion tract lengths, and chromatid

interference. Also computed are the correlation coefficients between the number of COs and NCOs, which is an indicator of CO homeostasis, and the parameters of the gamma distribution function for inter-crossover distances, which are indicators of CO interference.

4. Notes

1. This chapter presents a method which utilizes the Affymetrix GeneChip® Yeast Genome S98 Array. For an alternative approach using the Affymetrix custom tiling array, see (18).
2. Approximately 45% of all dissected tetrads for the S96/YJM789 diploid strain are four-spore viable tetrads (10). Dissect enough tetrads to obtain the target number of four-spore viable tetrads.
3. Here we describe the allele-specific PCR procedure to test one primer set in all four spores of a tetrad. To test multiple primer sets for the same tetrad, increase the initial resuspension volume for each spore accordingly. Our lab generally tests eight different primer sets for each tetrad prior to microarray analysis. Only tetrads that display a 2:2 segregation of S96 and YJM789 alleles in at least seven primer sets tested will be chosen for downstream microarray analysis.
4. Store the unused master primer mix at –20°C for future use. However, repeated freezing and thawing will slowly degrade the master primer mix. Aliquot the master primer mix into smaller quantities to reduce the number of freeze and thaw cycles.
5. Alternatively, inoculate a 5 ml starter culture in YPAD liquid media from a single colony of fresh yeast culture. Grow at 30°C for 6–8 h. Measure the OD of the 5 ml starter culture. Calculate the volume of cells needed to set up a 500 ml culture at an OD of 0.005. Inoculate culture in YPAD liquid media in a 2.8 l flask. Grow for ~14 h on platform shaker at 30°C. In the next step, harvest all 500 ml of yeast culture by centrifugation.
6. Overloading the Genomic-tip with an excess of yeast culture leads to clogging of the tip and underloading results in low DNA yield. Visually inspect the size of the cell pellet after the TE wash. We find that a cell pellet of 5 ml reliably yields a generous amount of DNA.
7. Highly concentrated genomic DNA lysates may clog the column, leading to decreased flow rate. Gentle and slow

positive pressure may be applied to facilitate flow. When applying positive pressure, do not exceed the recommended flow rate of 20–40 drops/min.

8. If parts of the DNA pellet remain stubbornly undissolved, or if the DNA solution appears murky due to precipitated salt, spin DNA in a tabletop microcentrifuge at top speed for 5–10 min. Transfer the supernatant to a clean tube and measure the DNA concentration using a spectrophotometer. To recover additional DNA from the pellet, judiciously add small amounts of TE to see if it would progressively aid in dissolving DNA. Take caution to not dilute DNA to a final concentration of less than 500 ng/μl.
9. Because DNase I activity may vary between different batches, we recommend testing a range of DNase I dilutions to find the dilution that gives the most desirable digestion pattern. To start off, try 1:4, 1:3, and 1:2 dilutions of DNase I in sterile ddH_2O. Since the quality of genomic DNA can also affect the efficiency of fragmentation by DNase I, we recommend fragmenting every genomic DNA sample with at least two dilutions of DNase I, those which showed the best fragmentation patterns in the test sample. Please note that long-term storage of diluted DNase I in water is not recommended. DNase I should be freshly diluted before use to ensure consistent enzymatic activity. As an alternative to using varying dilutions of DNase I, one can also vary the incubation time for DNase I digestion and determine a time that yields the most desirable digestion pattern.
10. Oligo length of the probe set for the Affymetrix GeneChip® Yeast Genome S98 Array is 25 bp. Overdigestion of genomic DNA leads to nonspecific hybridization on the microarray and high background, while underdigestion results in low signal intensity. Therefore, it is crucial to fragment samples within 30–100 bp to ensure ideal hybridization.

Acknowledgments

We thank Carol Anderson for optimizing an alternative yeast inoculation method for genomic DNA extraction. We thank Ashwini Oke for her assistance in performing the DNase I digestion for this publication. We also thank Mike Pollard for critical reading of the manuscript. S.Y.C. is supported by a Genentech Fellowship. J.C.F. is supported by the American Cancer Society Research Scholar Award (RSG CCG 110688).

References

1. Roeder, G.S. (1997) Meiotic chromosomes: it takes two to tango. *Genes Dev* **11**, 2600–2621.
2. Zickler, D., and Kleckner, N. (1999) Meiotic chromosomes: integrating structure and function. *Annu Rev Genet* **33**, 603–754.
3. Whitby, M.C. (2005) Making crossovers during meiosis. *Biochem Soc Trans* **33**, 1451–1455.
4. Hassold, T. (2007) The origin of human aneuploidy: where we have been, where we are going. *Hum Mol Genet* **16**, 203–208.
5. Hillers, K.J. (2004) Crossover interference. *Curr Biol* **14**, R1036–R1037.
6. Jones, G.H. (1984) The control of chiasma distribution. *Symp Soc Exp Biol* **38**, 293–320.
7. Muller, H. (1916) The mechanism of crossing over. *Am Nat* **50**, 193–434.
8. Martini, E., Diaz, R.L., Hunter, N., and Keeney, S. (2006) Crossover homeostasis in yeast meiosis. *Cell* **126**, 285–295.
9. Winzeler, E.A., Richards, D.R., Conway, A.R., Goldstein, A.L., Kalman, S., McCullough, M.J., McCusker, J.H., Stevens, D.A., Wodicka, L., Lockhart, D.J., and Davis, R.W. (1998) Direct allelic variation scanning of the yeast genome. *Science* **281**, 1194–1197.
10. Chen, S.Y., Tsubouchi, T., Rockmill, B., Sandler, J.S., Richards, D.R., Vader, G., Hochwagen, A., Roeder, G.S., and Fung, J.C. (2008) Global analysis of the meiotic crossover landscape. *Dev Cell* **15**, 401–415.
11. McPeek, M.S., and Speed, T.P. (1995) Modeling interference in genetic recombination. *Genetics* **139**, 1031–1044.
12. Zhao, H., Speed, T.P., and McPeek, M.S. (1995) Statistical analysis of crossover interference using the chi-square model. *Genetics* **139**, 1045–1056.
13. Affymetrix. (2005) GeneChip expression analysis technical manual with specific protocols for using the GeneChip hybridization, wash, and stain kit.
14. Richards, D. (2004) Allelescan users manual. http://genomics.stanford.edu/Allelescan_user_manual.pdf
15. Okimoto, R., and Dodgson, J.B. (1996) Improved PCR amplification of multiple specific alleles (PAMSA) using internally mismatched primers. *Biotechniques* **21**, 20–22, 24, 26.
16. Qiagen. (2001) Qiagen Genomic DNA handbook.
17. Affymetrix. http://www.affymetrix.com.
18. Mancera, E., Bourgon, R., Brozzi, A., Huber, W., and Steinmetz, L.M. (2008) High-resolution mapping of meiotic crossovers and non-crossovers in yeast. *Nature* **454**, 479–485.

Chapter 9

Using the Semi-synthetic Epitope System to Identify Direct Substrates of the Meiosis-Specific Budding Yeast Kinase, Mek1

Hsiao-Chi Lo and Nancy M. Hollingsworth

Abstract

Recent studies have shown that the meiosis-specific kinase, Mek1, plays a key role in promoting recombination between homologous chromosomes during meiosis in budding yeast by suppressing recombination between sister chromatids, as well as playing a role in the meiotic recombination checkpoint. Understanding how Mek1 regulates recombination requires the identification of direct substrates of the kinase. We have applied the semi-synthetic epitope method developed by Shokat and colleagues to Mek1. This method uses an analog-sensitive version of Mek1, GST-Mek1-as, in conjunction with an ATPγS analog, for kinase assays that detect only those proteins that are directly phosphorylated by Mek1. This method may be applicable to any kinase for which an analog-sensitive version is available. In addition, it provides a non-radioactive alternative for kinase assays with wild-type kinases.

Key words: Meiosis, Mek1, kinase assays, Rad54, semi-synthetic epitope, yeast.

1. Introduction

In the past several years, protein kinases have been found to play key roles in a variety of meiotic processes, including the initiation of premeiotic DNA synthesis (Ime2, Cdc28) and recombination (Cdc28 and Cdc7), the promotion of recombination between homologous chromosomes instead of sister chromatids (Mek1/Mre4), resolution of recombination intermediates (Cdc5), mono-orientation of sister kinetochores at the first meiotic division (Cdc5 and Cdc7), and onset of the first meiotic division (Cdc28, Cdc7, and Ime2) (1–9). Understanding the molecular mechanisms by which these kinases work requires

H. Tsubouchi (ed.), *DNA Recombination*, Methods in Molecular Biology 745,
DOI 10.1007/978-1-61779-129-1_9, © Springer Science+Business Media, LLC 2011

the identification of the direct substrates of these kinases and the phenotypic analysis of phosphorylation site mutants. A major problem inherent in kinase studies is specificity, since all protein kinases do essentially the same reaction – i.e., transfer the gamma phosphate from ATP onto serine, threonine, or tyrosine residues in their targets. This specificity issue has been addressed by the development of analog-sensitive (*as*) kinases, in which the ATP-binding pocket of a kinase of interest is enlarged by mutation (10, 11). Addition of inhibitors that are too bulky to fit into the ATP-binding pockets of unmodified kinases results in the specific inactivation of the *as* kinase. In addition, use of ATP analogs in conjunction with an *as* kinase results in specific phosphorylation of target proteins (e.g., 12). These target proteins can be detected using the semi-synthetic epitope method (13) (**Fig. 9.1**). Use of an ATPγS analog in the kinase reaction results in thiophosphorylation of substrate proteins specifically by the *as* kinase. This thiophosphorylation is then converted into an affinity tag by an alkylation reaction using *p*-nitrobenzyl mesylate (PNBM). The resulting thiophosphate ester is then detected on immunoblots using a commercially available rabbit monoclonal antibody referred to as the α-hapten antibody.

Mek1 is a meiosis-specific kinase that regulates meiotic recombination in a variety of ways (6, 12). Using the semi-synthetic epitope system we have shown that Mek1 itself as well as the recombination protein, Rad54, are direct targets of the kinase (14) (**Fig. 9.2**). Although Mek1 is involved in suppressing Rad51-mediated strand invasion of sister chromatids as well as the meiotic recombination checkpoint, the substrates involved in these processes have not yet been identified (14, 15). The semi-synthetic epitope approach provides a relatively easy assay

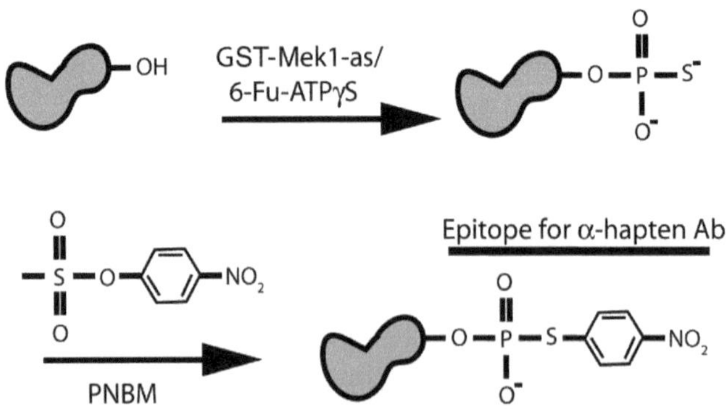

Fig. 9.1. The semi-synthetic epitope system. Substrates of interest are incubated with GST-Mek1-as and the ATP analog 6-Fu-ATPγS. Only the enlarged ATP-binding pocket of GST-Mek1-as can accommodate the bulky ATP analog, resulting in specific thio-phosphorylation of substrates. This thio-phosphorylation is converted to an affinity tag by alkylation using PNBM that can be detected using α-hapten antibodies.

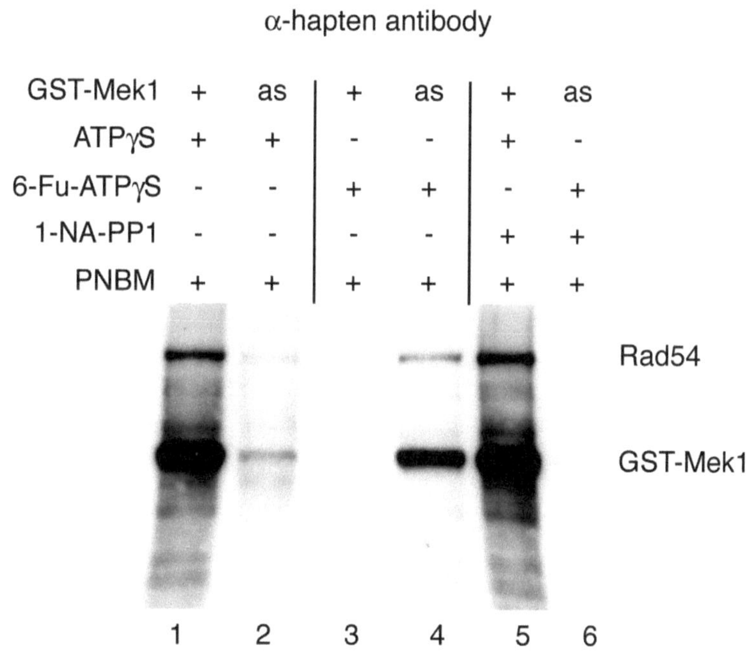

Fig. 9.2. Using the semi-synthetic epitope system to detect direct phosphorylation of GST-Mek1-as and Rad54 by GST-Mek1-as. GST-Mek1 and GST-Mek1-as were pulled down from meiotic extracts of NH520/pLW1 and NH520/pLW3, respectively. One microgram of bacterially purified Rad54 protein was included in each reaction. Where indicated, 15 μM of the 1-NA-PP1 inhibitor was included. Note that GST-Mek1-as utilizes ATPγS poorly compared to GST-Mek1 (Compare *lanes 1* and *2*), as was previously observed with ATP (6). In contrast, GST-Mek1-as, but not GST-Mek1, exhibited phosphorylation of GST-Mek1-as and Rad54 when the analog, 6-Fu-ATPγS, was used for the kinase assays (compare *lanes 3* and *4*). The GST-Mek1/ATPγS reaction was unaffected by the addition of inhibitor, while 1-NA-PP1 greatly reduced the phosphorylation observed in the reactions containing GST-Mek1-as/6-Fu-ATPγS (compare *lanes 5* and *6*). This blot was exposed to film for 1 s.

for testing whether candidate proteins are direct targets of Mek1. In addition, this protocol may be adapted for use with other *as* kinases such as Cdc5-as, Ime2-as, Cdc7-as, and Cdc28-as, to name a few (1, 5, 16). Finally, the semi-synthetic epitope system can be used with wild-type kinases for non-radioactive kinase assays (**Fig. 9.2**, lane 1).

For GST-Mek1-as, the semi-synthetic epitope method can be broken down into four parts: (1) generating a culture of meiotic cells containing activated GST-Mek1-as kinase, (2) pulling down GST-Mek1-as from soluble extracts using glutathione-sepharose, (3) using the beads in kinase assays containing a substrate of interest and the ATP analog, 6-Fu-ATPγS, followed by alkylation of the thio-phosphorylated proteins, and (4) probing the phosphorylated proteins on an immunoblot using the α-hapten antibodies.

2. Materials

2.1. Yeast Strains and Sporulation

1. SK1 diploid strain NH520 transformed with high copy number *GST-Mek1-as* plasmid, pLW3 (6)
   *MAT***a** *leu2::hisG his4-X dmc1Δ hoΔ lys2 ura3 Mek1Δ*
   MATα leu2::hisG his4-B dmc1Δ hoΔ lys2 ura3 Mek1Δ
   /2μ GST-Mek1-as URA3 (*see* **Note 1**).
2. SD-ura plates: 2% agar, 0.7% yeast nitrogen base without amino acids, 2% glucose, 1 g uracil dropout powder (*see* **Note 2**).
3. YEP + glycerol plates: 2% agar, 2% bactopeptone, 1% yeast extract, 3% glycerol.
4. Liquid YEPD medium: 2% bactopeptone, 1% yeast extract, 2% glucose.
5. Liquid YPA medium: 2% bactopeptone, 1% yeast extract, 2% potassium acetate.
6. Sporulation (Spo) medium: 2% potassium acetate.

2.2. Yeast Extracts and Glutathione Precipitation

1. Lysis buffer (LB): Make fresh the same day using ice-cold water and keep on ice. 50 mM Tris–HCl, pH 7.5, 10 mM EDTA, pH 8.0, 300 mM NaCl, 1 mM dithiothreitol, 10 mM NaF (Sigma, diluted from 1 M stock made up in water and stored at 4°C), 10 mM $Na_4P_2O_4$ (Sigma, diluted from 100 mM stock made up in water and stored at 4°C), 1 mM PMSF (Sigma, diluted from 100 mM stock made up in ethanol and stored at room temperature), 1 μg/ml leupeptin, 1 μg/ml aprotinin, 1 μg/ml pepstatin [USB Corporation, all diluted from 1 mg/ml stocks made up in either water (leupeptin and aprotinin) or ethanol (pepstatin) and frozen at –20°C in 1 ml aliquots]. Add PMSF immediately before use as it is unstable in aqueous solutions.
2. Glutathione-sepharose (Amersham Biosciences): The day of the experiment, glutathione-sepharose is equilibrated with LB. For each yeast cell pellet from 100 ml sporulating culture, use 40 μl of a 1:1 slurry of beads:liquid. Transfer slurry to a 1.7 ml microfuge tube and mark the volume on the tube. Add 1 ml LB, spin for 10 s at 2,000×*g*, remove supernatant by aspiration, leaving ~100 μl behind. Repeat wash two times with 1 ml LB. After the final wash, remove LB to the mark, so that the 1:1 slurry is regenerated.
3. Glass beads, 0.5 mm (#11079105), Biospec Products, Inc. Use directly from bottle. Place on ice at the beginning of the procedure to cool beads down.

4. BioRad Protein Assay reagent (#500-0001). Store at 4°C.
5. 30 gauge × 1/2 inch needle and 1 ml syringe.

2.3. Kinase Assays and Alkylation

1. ATP (Sigma): Make 100 μM stock in water, aliquot, and store at –80°C.
2. 6-Fu-ATPγS (BioLog #F008): Comes as a 10 mM solution. Aliquot 6 μl/microfuge tube and store at –80°C. The nucleotide should not be reused after thawing.
3. Kinase buffer: 50 mM Tris–HCl, pH 7.5, 200 mM NaCl, 10 mM $MgCl_2$.
4. *p*-Nitrobenzyl mesylate (PNBM) (Epitomics #3700-1): Add 420 μl DMSO to 5 mg powder in bottle to generate a 50 mM stock. Make 30 μl aliquots and store at –20°C. The PNBM can be refrozen and reused.
5. 1-(1,1-Dimethylethyl)-3-(1-naphthalenyl)-1*H*-pyrazolo[3,4-*d*]pyrimidin-4-amine (1-NA-PP1) (Tocris Bioscience #3063): Resuspend 10 mg in 3.15 ml dimethylsulfoxide (DMSO) to generate 10 mM stock. Store at –20°C. This stock can be refrozen and reused. Dilute to 375 μM in DMSO on the day of the experiment.
6. 5X protein sample buffer: 310 mM Tris–HCl, pH 6.8, 15% SDS, 25% 2-mercaptoethanol, 25% glycerol (w/v), and 0.25% bromophenol blue.

2.4. Protein Gels and Immunoblots

1. 8% SDS-polyacrylamide gel. (a) Resolving gel: For one mini-gel make 5 ml solution using 2.3 ml water, 1.3 ml 30% acrylamide/bis (29:1) (BioRad), 1.3 ml 1.5 M Tris–HCl, pH 8.8, 50 μl 10% SDS, 50 μl 10% ammonium persulfate (APS) (BioRad). APS can be stored at 4°C for up to a week. Polymerization is initiated by the addition of 3 μl *N,N,N,N′*-tetramethyl-ethylenediamine (TEMED) (BioRad) and should only be added immediately prior to pouring the gel. (b) Stacking gel: For a single gel, make 2 ml using 1.4 ml water, 330 μl 30% acrylamide/bis (29:1), 250 μl 1 M Tris–HCl, pH 6.8, 20 μl 10% SDS, 20 μl 10% APS, and 2 μl TEMED.
2. Benchmark Prestained Protein Ladder (Invitrogen).
3. Whatman 0.45 MM Opitran BA-S85 Reinforced nitrocellulose (VWR).
4. Whatman 3 MM filter paper (Fisher Scientific).
5. Running buffer: 0.3% Tris base, 0.1% SDS, 1.44% glycine. Can be stored at room temperature indefinitely as either 10X or 1X solution.
6. Transfer buffer: 1.47% glycine, 0.3% Tris base, 20% methanol. Can be stored indefinitely at room temperature as a 1X solution.

7. TBST: 20 mM Tris–HCl, pH 7.5, 250 mM NaCl, 0.1% Tween-20. This solution can be made in a large volume (e.g., 2 l) as a stock that is stored at room temperature.
8. Blocking buffer TBST plus 5% non-fat milk. Dissolve 2.5 g milk powder in 50 ml TBST (*see* **Note 3**).
9. Primary antibody: Thiophosphate ester rabbit monoclonal antibody (the α-hapten antibody) (Epitomics, #2686-1).
10. Secondary antibody: Goat anti-rabbit IgG (H+L) HRP antibody (Epitomics #3053-1).
11. Amersham ECL Plus Western Blotting Detection System (GE Healthcare RPN2132).
12. Amersham Hyperfilm ECL (GE Healthcare 28-9068-39).

3. Methods

3.1. Sporulation

1. Using a sterile toothpick, streak out NH520/pLW3 onto SD-ura plates to select for single colonies that contain the plasmid. Invert plate and incubate at 30°C for 2–3 days.
2. Inoculate a single colony into 5 ml YEPD in a 15 ml test tube. Incubate on a roller at 30°C overnight. Patch cells from the same colony onto YEP + glycerol plates to check for *petite* colonies (*see* **Note 4**).
3. At 5:00 pm the next day, dilute 1.2 and 2.0 ml of overnight culture into 600 ml YPA in two 2.8 l flasks, respectively. Incubate at 30°C shaking at 250 rpm for 16 h (*see* **Notes 5** and **6**).
4. At 9:00 am the next day, blank the spectrophotometer with 1 ml YPA and read the absorbance at a wavelength of 660 nm of 1 ml of undiluted culture. The OD_{660} reading should be between 1.2 and 1.4. Using the conversion chart in **Table 9.1**, convert the OD_{660} to cell density and multiply with the total volume of YPA to determine the total number of cells in the culture. Calculate the volume of Spo medium necessary to give a cell density of 3×10^7 cells/ml. Aliquot this volume minus 5 ml into a 2.8 l flask (*see* **Note 7**).
5. Divide cells between two 500 ml bottles and pellet in a centrifuge at 3,000×*g* for 5 min. Resuspend each pellet in 5 ml water, combine, transfer to a 15 ml test tube and pellet again in a tabletop centrifuge. Resuspend pellet in 5 ml Spo medium and add to flask with Spo medium. Place on shaker (250 rpm) at 30°C for 5 h (*see* **Note 8**).

Table 9.1
Conversion of optical density$_{660}$ (OD_{660}) values to cell density

OD_{660}	Cells/ml ($\times 10^7$)
0.80	0.63
0.85	0.69
0.90	0.76
0.95	0.83
1.00	0.92
1.05	1.02
1.10	1.12
1.15	1.24
1.20	1.35
1.25	1.48
1.30	1.61
1.35	1.76
1.40	1.91
1.45	2.06
1.50	2.24
1.55	2.42
1.60	2.61

6. Divide the sporulating culture into 100 ml aliquots in 250 ml bottles and pellet in a centrifuge. Resuspend each pellet in 10 ml water, transfer to a 15 ml test tube, and spin in the tabletop centrifuge. Pour off supernatant and resuspend the cells in 1 ml 25% glycerol. Transfer to 1.7 ml graduated microfuge tubes (Posi-Click from Denville Scientific) and store at –80°C.

3.2. Yeast Extracts and Glutathione Precipitation

This protocol uses a frozen cell pellet from 100 ml sporulating culture.

1. Thaw pellet on ice for approximately 10 min. If necessary, melting can be accelerated by gentle flicking of the tube.
2. Everything should be kept as cold as possible to reduce the chance of proteolysis. Pellet cells by spinning in a 4°C microfuge for 1 min at 3,300×*g*.
3. To wash the cells, resuspend pellet in 0.5 ml LB and transfer to a 14 ml round bottom graduated Falcon tube (2059, Fisher Scientific) containing 4.5 ml LB (*see* **Note 9**).

4. Pellet 1 min in a tabletop centrifuge. Discard supernatant and resuspend in 1 ml LB. Measure 1 ml glass beads using a graduated microfuge tube and add to cells (the total volume should be ~2 ml).
5. Vortex 10× 1 min at top speed with 1 min rests on ice in-between.
6. Spin 1 min in the tabletop centrifuge to remove bubbles. Using a pipetman, transfer the supernatant to a new 1.7 ml microfuge tube. Measure the volume using the graduations on the side of the tube and add 25% Triton X-100 to a final concentration of 1%. Mix gently by flicking. Incubate on ice for 10 min. During this time equilibrate the glutathione-sepharose with lysis buffer (*see* **Note 10**).
7. Pellet insoluble material by spinning the extract in a 4°C microfuge for 10 min at 16,200×*g*.
8. To measure the protein concentration, make a 1:10 dilution of the soluble extract in LB. For each extract, add 800 μl water and 200 μl BioRad Protein Assay reagent to a microfuge tube and add 1 μl of the diluted extract. Blank the spectrophotometer with 1X BioRad Protein Assay reagent at OD_{595} and compare the absorbance value for each extract to a standard curve previously generated using known quantities of a standard protein such as bovine serum albumin. Multiply by 10 to calculate the protein concentration of the extract. The concentration should be at least 20 mg/ml (*see* **Note 11**).
9. Transfer the entire soluble extract (~800–900 μl) to a microfuge tube containing 40 μl 1:1 glutathione-sepharose slurry equilibrated in LB, being careful not to transfer any insoluble material.
10. Rock tube for 1.5 h at 4°C to allow the GST-Mek1-as to bind the glutathione-sepharose.
11. Pellet beads by spinning in a microfuge for 30 s at 3,300×*g*. It is difficult to see the beads at this point so leave ~200 μl extract behind when removing the supernatant by either aspiration or with a pipetman.
12. Wash beads by resuspending them in 1 ml lysis buffer and spinning as in Step 11. Remove supernatant and repeat for a total of three times, then wash twice with 1 ml ice-cold kinase buffer. After the final wash, remove any residual liquid using a 1 ml syringe attached to a 30 1/2 gauge needle.

3.3. Kinase Assays and Alkylation

1. Kinase reactions contain between 24 and 28 μl and include 22 μl GST-Mek1-as-bound bead slurry, 1 μl 100 μM ATP, 1 μl 10 mM 6-Fu-ATPγS, 1–4 μl substrate, and 1 μl

375 μM 1-NA-PP1 (if appropriate). Determine the number of desired reactions and multiply by 22 μl to determine the volume of kinase buffer required to resuspend the kinase-bound beads (*see* **Note 12**). For four reactions, use 88 μl kinase buffer. Transfer 22 μl bead:buffer slurry to three 1.7 ml microfuge tubes, leaving 22 μl in the original microfuge tube. Before pipetting the slurry, mix well with the pipette tip as the beads tend to settle at the bottom of the tube. Add 0.2–1.0 μg of putative substrate to each reaction.

2. For reactions containing inhibitor, add 1 μl 375 μM 1-NA-PP1 (final concentration is 15 μM) and wait for 5 min at room temperature (*see* **Note 13**).
3. Add 6 μl 100 μM ATP to the 6 μl aliquot of 10 mM 6-Fu-ATPγS. To start the kinase reaction, add 2 μl of this mixture to the bead slurry + substrate and stir gently with pipette tip. Incubate at 30°C for 30 min.
4. To alkylate the proteins, add 1.3 μl 50 mM PNBM for a final concentration of 2.5 mM. Mix well with pipette tip and leave at room temperature for 2 h (*see* **Note 14**).
5. Stop the alkylation reaction by adding 6 μl of 5X protein sample buffer and incubate the tubes at 95°C for 5 min. Before loading gel, spin the samples for 10 s at 16,200×*g* in a microfuge.

3.4. Protein Gels and Immunoblots

1. This protocol assumes the use of a Hoeffer Mini-gel apparatus but other apparatuses may also be used. Take a square glass plate and put a 1 mm spacer on either side. Place a notched plate on top. Holding the plates and spacers together, tap the bottom of the sandwich gently on the bench top to make sure the plates and spacers are flush and then clamp onto a Hoeffer gel casting apparatus. Placing a piece of parafilm underneath the bottom of the plates helps prevent leaks.
2. Acrylamide is a neurotoxin and therefore gloves should be worn while pouring the gel. To make the resolving gel, combine all of the solutions except the TEMED using disposable pipettes in a 15 ml plastic tube and mix well. Add TEMED, mix, and using a Pasteur pipette, immediately fill up three-fourths of the volume between the glass plates. Using a squirt bottle, layer ethanol on top of the acrylamide. Leave at room temperature for at least 30 min to polymerize. Check the residual acrylamide in the plastic tube to confirm that the polymerization reaction has occurred. After polymerization, this tube can be discarded in the regular trash.

3. To pour the stacking gel, combine all the solutions except the TEMED in a 15 ml plastic tube. Pour ethanol off the resolving gel. Blot away residual ethanol using 3 MM filter paper. Add TEMED and fill to the top of the gel using a Pasteur pipette. Immediately insert comb and let polymerize for at least 15 min.
4. To run the gel, remove the plates from the gel casting apparatus and assemble onto gel tank with the notched plate facing inward and the notch on top. Fill upper and lower reservoirs with running buffer, being sure to submerge the teeth of the comb. Place a stencil on the outer glass plate to indicate the positions of the wells and remove comb.
5. Load 10 μl benchmark protein markers and 15–25 μl kinase reaction into separate wells using 1–200 μl microcapillary tips from Nortech (RSE-01R-204). Adding 15 μl of 1X protein sample buffer to empty lanes helps the gel run more evenly.
6. Run gel at 100 V (constant voltage) for 2 h. The pink marker should not run off the gel (*see* **Note 15**).
7. To stop gel, turn off power supply and remove gel sandwich from the apparatus. To disassemble, remove the spacers and use one spacer to leverage the notched gel plate off the gel. The gel should remain stuck to the unnotched plate. Cut off the stacking gel with a scalpel.
8. Measure the dimensions of the gel and cut four pieces of 3 MM filter paper and one piece of nitrocellulose to those dimensions.
9. Wet the nitrocellulose and pieces of filter paper in transfer buffer. Place one piece of wet filter paper on top of the gel and smooth out any bubbles by rolling a glass rod over the paper. Using a flat thin spatula, separate a corner of the gel from the glass plate and then, holding the gel and paper together, pull gently to peel the gel away from the plate.
10. With the gel side up, place the filter paper holding the gel onto a second piece of wet filter paper on a smooth surface. Layer the nitrocellulose and the remaining pieces of filter paper on top of the gel for a sandwich that is composed from bottom to top of the following: two pieces of filter paper, gel, nitrocellulose membrane, two pieces of filter paper. Remove bubbles using the glass rod each time a layer is added.
11. Proteins are transferred onto nitrocellulose using a Trans-blot SD Semi-Dry Transfer Cell apparatus from BioRad.

Invert the sandwich, so that the nitrocellulose membrane is under the gel, place on the transfer apparatus, and connect the lid. Attach the transfer apparatus to a power supply such that current flows from top to bottom in a negative to positive direction, so that the SDS-bound negatively charged proteins travel from the gel onto the membrane. For a mini-gel, use 70 mA (constant current) for 1 h.

12. Remove the lid and disassemble the sandwich using forceps to peel the nitrocellulose membrane off the gel. If transfer has occurred properly, the prestained standards should be visible on the membrane. Place the membrane in a small plastic container that is slightly larger than the filter. Cover the membrane with blocking buffer (5–10 ml) and rotate gently on a platform shaker at room temperature for 1 h (*see* **Note 16**).
13. Pour off the blocking buffer. Rinse the membrane by adding ~10 ml of TBST, briefly swirling the container by hand, and pouring the TBST into the sink. For the primary antibody incubation, add 5 ml of blocking buffer and a 1:10,000 dilution (0.5 μl) of α-hapten antibodies. Shake overnight (usually between 12 and 14 h) at 4°C.
14. Pour off blocking buffer with antibody and wash the membrane by adding ~10 ml TBST and shaking for 10 min at room temperature. Repeat for a total of three washes.
15. For the secondary antibody, add 5 ml of blocking buffer with a 1:5,000 dilution (1 μl) of goat anti-rabbit IgG HRP antibody and shake at room temperature for 1 h.
16. Wash membrane three times with TBST for 10 min each, shaking at room temperature. After the final wash, drain the TBST from the membrane. Cut a piece of parafilm that is larger than the membrane and place the membrane face up on the parafilm, so that the prestained molecular weight markers are visible.
17. In a test tube, mix 2 ml of ECL Plus kit Solution A with 50 μl Solution B. Using a pipette gently apply the mixture to the top of the membrane (*see* **Note 17**). Incubate at room temperature for 5 min.
18. Pick up the membrane with a forceps and remove any excess liquid by touching the edges of the membrane gently onto a paper towel.
19. For assaying the chemiluminescent signal, cover the membrane with plastic. Go to a dark room and put the membrane on film (*see* **Note 18**).
20. Develop film with an X-ray film developer (*see* **Note 19**).

4. Notes

1. Mek1 is activated in response to meiotic DSBs by autophosphorylation and therefore it is necessary to isolate the kinase from meiotic cells (17). Mek1 is constitutively active in *dmc1Δ*-arrested cells (6). Therefore purifying Mek1 from *dmc1Δ* diploids increases the number of cells containing active kinase, as does using a high copy plasmid to overexpress *GST-Mek1-as*. In vitro autophosphorylation of GST-Mek1-as serves as a good internal positive control for the kinase reaction. Using strains derived from the SK1 background is useful for two reasons: (1) sporulation is highly efficient with wild-type cells typically exhibiting >90% asci and (2) the *dmc1Δ* arrest is very tight (18, 19). Yeast strains and plasmids are available from the Hollingsworth lab.
   Although this method uses GST-Mek1-as, it may be applicable to any *as* kinase for which an ATPγS analog can be identified and the kinase can be precipitated from an extract. An excellent description of the analog-senstive kinase approach can be found in (20), including which amino acids to mutate and how to analyze the resulting mutants. In addition to the 6-Fu-ATPγS analog, 6-phenethyl-ATPγS (6-PhEt-ATPγS) and 6-benzyl-ATPγS (6-Bn-ATPγS) are also commercially available from BioLog (http://www.biolog.de). Since different *as* kinases exhibit preferences for different analogs, all of the analogs should be tried to see which one works best.
   The semi-synthetic epitope system can also be used with wild-type kinases by substituting ATPγS for the analog. Although these reactions lack the specificity provided by *as* kinases and therefore may have some additional background due to co-precipitating kinases that can also use ATPγS (**Fig. 9.2**, compare lanes 1 and 4), they have the advantage over traditional kinase assays in being non-radioactive.
2. To make –ura dropout powder, combine the following in a blender: 5 g adenine-HCl, 5 g tryptophan, 5 g histidine-HCl, 5 g arginine-HCl, 5 g methionine, 7.5 g tyrosine, 7.5 g leucine, 7.5 g isoleucine, 7.5 g lysine-HCl, 7.5 g valine, 7.5 g threonine, 7.5 g serine, 12 g phenylalanine. Blend for 5 × 1 min bursts using the "mix" setting. Pour into sterile bottle and use 2 g/l.
3. The blocking buffer should be made fresh for each experiment as the milk may go sour with time. Do not add sodium azide as this will inhibit the horseradish peroxidase

reaction used to detect the antibody. Keep at 4°C during the overnight incubation.

4. One characteristic of the SK1 background is that it gives rise to *petite* (respiration deficient) colonies. Because sporulation requires respiration, cells with defective mitochondria are unable to undergo meiosis. Every colony should therefore be checked for its ability to grow on a non-fermentable carbon source such as glycerol before sporulating the cells.
5. The synchrony and efficiency of sporulation are maximized if log phase cells at a density of 3×10^7 cells/ml are used. Because strains may exhibit different growth rates depending upon the starting inoculum, two different dilutions are used for a standard number of hours to increase the chances that the cells will be at the appropriate density at the start of the experiment.
6. The most time-consuming part of this protocol is sporulating the cells. We therefore sporulate several hundred milliliters of cells at a time and store the meiotically arrested cells in 100 ml aliquots at –80°C. To achieve the correct cell density, at least two times as much YPA is needed as Spo medium. OD_{660} readings between 1.2 and 1.4 from 600 ml of YPA will produce from 270 to 380 ml sporulating culture (**Table 9.1**).
7. Good aeration is critical for sporulating yeast cells. Therefore the flask volume should always be at least five times the volume of Spo medium.
8. Cell are incubated in Spo medium for 5 h to give the cells a chance to enter the meiotic program and arrest with unrepaired DSBs due to the meiotic recombination checkpoint. The time in Spo medium may be increased up to 8 h if necessary.
9. Mek1 is activated by phosphorylation of its activation loop (17). It is important to include phosphatase inhibitors in the lysis buffer to prevent loss of this phosphorylation and therefore kinase activity.
10. Detergent is used to dissolve membranes and is added after vortexing to prevent foaming.
11. If making more than one extract, the amount of protein used for the GST-Mek1-as precipitation can be equalized by varying the volume used for the pulldown.
12. The amount of GST-Mek1-as pulled down from 100 ml sporulating cell pellet is more than enough for four reactions. If more reactions are necessary, the amount of kinase/reaction can be reduced by resuspending the beads

in a larger volume of kinase buffer before distributing 22 μl into tubes for kinase reactions.

13. Addition of inhibitor should block the kinase reaction and serves as a good control for the specificity of the reaction (**Fig. 9.2**, compare lanes 4 and 6).
14. The kinase buffer must have less than 0.5 mM reductant (e.g., DTT) because otherwise the PNBM will be consumed. If necessary, reactions can be quenched by the addition of EDTA to final concentration twice that of the magnesium concentration.
15. This protocol is designed to detect proteins greater than 50 kD. If smaller substrates are being tested, a higher percentage acrylamide gel may be necessary and the length the gel is run should be adjusted accordingly.
16. Incubating the membrane in milk prior to the addition of antibody decreases non-specific binding of the antibody to the membrane.
17. The HRP reaction occurs using a small volume (~2 ml) of ECL Plus reagents. This volume can be placed directly on the membrane if the membrane is on a hydrophobic surface such as parafilm or saran wrap. The mixture must be added carefully so that the surface tension is not broken, or the mixture will spill off the membrane.
18. The membrane may be placed on a piece of plastic wrap that is then folded over to cover both sides. Alternatively, we slit the side of a hybridization bag and insert the membrane inside to eliminate creases that frequently arise when using plastic wrap. To conserve film, the membrane can be pressed down onto different areas of the same film for exposures ranging from 1 s to 5 min before the film is developed.
19. The ECL Plus kit generates both a chemiluminescent signal that is transient (gone after ~30 min) and a stable chemifluorescent signal. The chemiluminescent signal can be detected by exposing blots to X-ray film while the chemifluorescent signal can be detected and quantitated using a phosphoimager.

Acknowledgments

We thank Jasmina Allen, Kevan Shokat, and Beatrice Wang for help with ideas and reagents in the early development of this protocol. Patrick Sung generously provided bacterially purified

Rad54 protein. Aaron Neiman provided helpful comments on the manuscript. This work was supported by an NIH grant to N. M. H. (R01 GM50717).

References

1. Benjamin, K.R., Zhang, C., Shokat, K.M., and Herskowitz, I. (2003) Control of landmark events in meiosis by the CDK Cdc28 and the meiosis-specific kinase Ime2. *Genes Dev* **17**, 1524–1539.
2. Lee, B.H., and Amon, A. (2003) Role of Polo-like kinase CDC5 in programming meiosis I chromosome segregation. *Science* **300**, 482–486.
3. Matos, J., Lipp, J.J., Bogdanova, A., Guillot, S., Okaz, E., Junqueira, M., Shevchenko, A., and Zachariae, W. (2008) Dbf4-dependent CDC7 kinase links DNA replication to the segregation of homologous chromosomes in meiosis I. *Cell* **135**, 662–678.
4. Pak, J., and Segall, J. (2002) Regulation of the premiddle and middle phases of expression of the *NDT80* gene during sporulation of *Saccharomyces cerevisiae*. *Mol Cell Biol* **22**, 6417–6429.
5. Sourirajan, A., and Lichten, M. (2008) Polo-like kinase Cdc5 drives exit from pachytene during budding yeast meiosis. *Genes Dev* **22**, 2627–2632.
6. Wan, L., de los Santos, T., Zhang, C., Shokat, K., and Hollingsworth, N.M. (2004) Mek1 kinase activity functions downstream of *RED1* in the regulation of meiotic DSB repair in budding yeast. *Mol Biol Cell* **15**, 11–23.
7. Wan, L., Niu, H., Futcher, B., Zhang, C., Shokat, K.M., Boulton, S.J., and Hollingsworth, N.M. (2008) Cdc28-Clb5 (CDK-S) and Cdc7-Dbf4 (DDK) collaborate to initiate meiotic recombination in yeast. *Genes Dev* **22**, 386–397.
8. Ahmed, N.T., Bungard, D., Shin, M.E., Moore, M., and Winter, E. (2009) The Ime2 protein kinase enhances the disassociation of the sum1 repressor from middle meiotic promoters. *Mol Cell Biol* **29**, 4352–4362.
9. Henderson, K.A., Kee, K., Maleki, S., Santini, P., and Keeney, S. (2006) Cyclin-dependent kinase directly regulates initiation of meiotic recombination. *Cell* **125**, 1321–1332.
10. Bishop, A.C., Buzko, O., and Shokat, K.M. (2001) Magic bullets for protein kinases. *Trends Cell Biol* **11**, 167–172.
11. Bishop, A.C., Ubersax, J.A., Petsch, D.T., Matheos, D.P., Gray, N.S., Blethrow, J., Shimizu, E., Tsien, J.Z., Schultz, P.G., Rose, M.D., Wood, J.L., Morgan, D.O., and Shokat, K.M. (2000) A chemical switch for inhibitor-sensitive alleles of any protein kinase. *Nature* **407**, 395–401.
12. Ubersax, J.A., Woodbury, E.L., Quang, P.N., Paraz, M., Blethrow, J.D., Shah, K., Shokat, K.M., and Morgan, D.O. (2003) Targets of the cyclin-dependent kinase Cdk1. *Nature* **425**, 859–864.
13. Allen, J.A., Li, M., Brinkworth, C.S., Paulson, J.L., Wang, D., Hubner, A., Chou, W.-H., Davis, R.J., Burlingame, A.L., Messing, R.O., Katayama, C.D., Hedrick, S.M., and Shokat, K.M. (2007) A semisynthetic epitope for kinase substrates. *Nat Methods* **4**, 511–516.
14. Niu, H., Wan, L., Busygina, V., Kwon, Y., Allen, J.A., Li, X., Kunz, R.C., Kubota, K., Wang, B., Sung, P., Shokat, K.M., Gygi, S.P., and Hollingsworth, N.M. (2009) Regulation of meiotic recombination via Mek1-mediated Rad54 phosphorylation. *Mol Cell* **36**, 393–404.
15. Xu, L., Weiner, B.M., and Kleckner, N. (1997) Meiotic cells monitor the status of the interhomolog recombination complex. *Genes Dev* **11**, 106–118.
16. Wan, L., Zhang, C., Shokat, K.M., and Hollingsworth, N.M. (2006) Chemical inactivation of Cdc7 kinase in budding yeast results in a reversible arrest that allows efficient cell synchronization prior to meiotic recombination. *Genetics* **174**, 1667–1774.
17. Niu, H., Li, X., Job, E., Park, C., Moazed, D., Gygi, S.P., and Hollingsworth, N.M. (2007) Mek1 kinase is regulated to suppress double-strand break repair between sister chromatids during budding yeast meiosis. *Mol Cell Biol* **27**, 5456–5467.
18. Bishop, D.K., Park, D., Xu, L., and Kleckner, N. (1992) *DMC1*: a meiosis-specific yeast homolog of *E. coli recA* required for recombination, synaptonemal complex formation and cell cycle progression. *Cell* **69**, 439–456.
19. Padmore, R., Cao, L., and Kleckner, N.R. (1991) Temporal comparison of recombination and synaptonemal complex formation during meiosis in *Saccharomyces cerevisiae*. *Cell* **66**, 1239–1256.
20. Blethrow, J.D., Zhang, C., Shokat, K.M., and Weiss, E.L. (2004) Design and use of analog-sensitive kinases. *Curr Protoc Mol Biol* **Chapter 18,** Unit 18 11.

Chapter 10

Genetic and Molecular Analysis of Mitotic Recombination in *Saccharomyces cerevisiae*

Belén Gómez-González, José F. Ruiz, and Andrés Aguilera

Abstract

Many systems have been developed for the study of mitotic homologous recombination (HR) in the yeast *Saccharomyces cerevisiae* at both genetic and molecular levels. Such systems are of great use for the analysis of different features of HR as well as of the effect of mutations, transcription, etc., on HR. Here we describe a selection of plasmid- and chromosome-borne DNA repeat assays, as well as plasmid–chromosome recombination systems, which are useful for the analysis of spontaneous and DSB-induced recombination. They can easily be used in diploid and, most importantly, in haploid yeast cells, which is a great advantage to analyze the effect of recessive mutations on HR. Such systems were designed for the analysis of a number of different HR features, which include the frequency and length of the gene conversion events, the frequency of reciprocal exchanges, the proportion of gene conversion versus reciprocal exchange, or the molecular analysis of sister chromatid exchange.

Key words: Homologous recombination, direct repeats, inverted repeats, sister chromatid exchange, gene conversion, reciprocal exchange, yeast.

1. Introduction

Homologous recombination (HR) consists of an exchange or a transfer of genetic information between homologous DNA sequences. The homology used for the recombination reaction can be found in the homologous chromosome (allelic recombination) or at any other homologous sequence located at non-allelic positions (ectopic recombination), whether or not in the same chromosome (intramolecular and intermolecular recombination, respectively). The sister chromatid can also be used as a template in mitotic cells (sister chromatid recombination, SCR), as

H. Tsubouchi (ed.), *DNA Recombination*, Methods in Molecular Biology 745,
DOI 10.1007/978-1-61779-129-1_10, © Springer Science+Business Media, LLC 2011

first demonstrated cytologically (1). Indeed, the sister chromatid appears to be the preferred template for mitotic HR when it is available, that is in S and G2 cell-cycle phases, since it promotes error-free repair. In contrast, both allelic and ectopic recombination can lead to point mutations, loss of heterozygosity, or genome rearrangements (translocations, duplications, or inversions) (2, 3).

Mitotic recombination can occur through different HR mechanisms: single-strand annealing (SSA), synthesis-dependent strand annealing (SDSA), double-strand break repair (DSBR), and break-induced replication (BIR) (reviewed in (4)). Depending on whether the transfer of information between the two recombining DNA fragments is uni- or bi-directional, the outcome of HR can be a gene conversion (GC) (unidirectional transfer of information from one molecule to the other) or a reciprocal exchange or crossover (bi-directional or reciprocal transfer of information).

Many systems have been developed for the study of spontaneous HR in the yeast *Saccharomyces cerevisiae* at both genetic and molecular levels. Most of these systems are based on two mutant heteroalleles, which reconstitute the wild-type gene by homologous recombination, so that recombinants can be selected phenotypically by a prototrophy or by drug resistance. In addition, some HR assays have been developed for the analysis of DSB-induced events. They are based on site-specific endonucleases, such as the HO or *I-Sce*I, with the cleavage site at only one heteroallele.

In diploid cells, HR systems are appropriate for the analysis of allelic recombination and constitute excellent tools for an in-depth analysis of HR. A great example is the HR system recently developed by the laboratory of Tom Petes (5). Nevertheless, HR systems in diploids are time consuming and inadequate for the study of recessive mutations or when the goal is to obtain a detailed analysis of the effect of particular mutations on different parameters of recombination, such as transcription and the type of HR mechanism preferentially used. The use of haploid cells, instead, offers a unique advantage of *Saccharomyces* for such analyses.

Direct- and inverted-repeat recombination substrates are excellent for the analysis of HR in haploids and have been largely used by a number of laboratories since the first ones were initially reported (6, 7). Recombination between direct repeats can generate two types of events, either the GC or the deletion of one repeat unit plus the intervening sequence. Recombination between inverted repeats leads to GC of one of the repeats and/or the inversion of the intervening sequence. It is widely accepted that GC occurs through DSBR or SDSA. Instead, although deletions can occur via DSBR as a result of a crossover between the repeats, the major mechanism for deletions is SSA (*see* (8), for review). Inversions can be generated by a crossover or by BIR between the inverted repeats followed by SSA (*see* (8), for review). As in either case the final product is equivalent to a reciprocal exchange, we will refer to inversions as reciprocal exchange (RE) events, regardless of the mechanism. It is also worth noting that both deletions and inversions can also occur by SCR (*see* (8) for review). Direct-repeat assays have also been useful in establishing the biological relevance of SCR as well as its genetics requirements. Despite the importance of SCR as the major HR repair mechanism, the study of SCR has been hampered by the impossibility of genetically detecting the recombination product due to the fact that both sister chromatids are identical. Nevertheless, some genetic and molecular approaches for the study of SCR have emerged over the last few years (9).

Here we describe a selection of plasmid- and chromosome-borne DNA repeat assays, as well as plasmid–chromosome recombination systems, which are useful for the analysis of spontaneous and DSB-induced recombination. They can easily be used to analyze a number of different HR features, such as frequency and length of the GC events, RE frequency, or the proportion of GC versus RE. In addition, we describe systems for the study of SCR. The goal of this chapter is to provide the tools and the methodology necessary for the analysis of different parameters of HR for any mutant of interest. First we describe the main features of the different types of HR systems selected and the rationale for their use. Then, after providing a list of materials needed, we provide a step-by-step protocol for the analysis of the different HR events in each system.

For practical reasons we have made a selection of HR systems that we use regularly in our laboratory and that are sufficiently reliable in our hands. Nevertheless, other genetic assays are available and regularly used in other laboratories that can be equally used following the protocols described here, by just using the appropriate media for the growth of the strain of interest and the media for the selection of recombinants. Such HR systems include those developed in many different laboratories, such as those of Klein (10), Petes (11), Rothstein (12), Haber (13), Fasullo (14), Jinks-Robertson (15), Roeder (16), Symington (17), Livingston (18), and Kupiec (19). Some are valid to address specific and unique questions on HR, and others can be used instead of the ones described here to address similar questions.

All systems from our laboratory described are available upon request.

2. Materials

2.1. Direct-Repeat Recombination Systems

Strains and plasmids are listed in **Table 10.1**.

2.2. Inverted-Repeat Recombination Systems

Strains and plasmids are listed in **Table 10.1**.

2.3. Systems for the Analysis of Sister Chromatid Recombination (SCR)

Strains and plasmids are listed in **Table 10.1**.

2.4. Determination of Recombination Frequencies

2.4.1. Genetic Analysis of Recombination

1. Strain carrying the appropriate recombination system (either in the appropriate background or transformed with the plasmid containing the recombination substrate).
2. Agar plates with synthetic medium (SC) lacking the appropriate amino acids either for plasmid selection when it is the case (total cells plate) or for recombinant selection (recombinant selection plate) as specified for each system (*see* **Table 10.1**).
   SC: 1.7 g/l amino acid-free yeast nitrogen base, 5 g/l ammonium sulfate, 20 g/l glucose, 20 g/l agar, and the appropriate amino acids in standard concentrations (*see* **Note 1**).
   SGal: 1.7 g/l amino acid-free yeast nitrogen base, 5 g/l ammonium sulfate, 20 g/l galactose, 20 g/l agar, and the appropriate amino acids in standard concentrations.
   In addition, the recombinant selection plates can contain FOA (US Biological, Marblehead, MA, USA). In this case, prepare two flasks: one containing 1.7 g/l amino acid-free yeast nitrogen base, 1 g/l proline, and the appropriate amino acids in standard concentrations except uracil, which is added at 10 mg/l, and the other containing 20 g/l glucose and 20 g/l agar. Both flasks must be autoclaved separately. Let the first flask cool, add 500 mg/l 5-FOA and swirl until all powder is dissolved. Mix flasks before pouring the media into petri dishes. Store at 4°C up to 1 week.
3. Sterilized toothpicks.
4. Sterile water.

2.4.2. Genetic Analysis of Recombination by Flow Cytometry

1. Strain transformed with the pGLG plasmid, containing the G-GFP recombination substrate.

Table 10.1
Recombination systems for the genetic analysis of spontaneous and DSB-induced recombination described

			Recombination event	Recombinant selection plate	Total cells plate	Strain	Ref.
Chromosome	Direct repeats	*leu2-k::ADE2-URA3::leu2-k*	Spontaneous deletion	SC+FOA	SC	A3Y3A	(10)
		leu2-112::URA3::leu2-k	Spontaneous GC Spontaneous deletion	SC-leu SC+FOA	SC	3 12-67C	(20)
		his3Δ5′::his3Δ3′	Spontaneous or DSB-induced SCR	SC-his	SC	YNN299	(29)
	Inverted repeats	*his3^P::INV*	Spontaneous GC w/ or w/o RE	SC-his	SC	AWII-2A	(26)
Plasmid	Direct repeats	**L and derivatives**	Spontaneous deletion	SC-leu-trp	SC-trp	pRS314-L	(21)
		GL and derivatives	Spontaneous deletion	SGal-leu-trp	SC-ura +dox (LT) SC-ura (HT)	pRS414-GLB	(24)
		G-GFP	Spontaneous deletion	Liquid SGal-his (direct visualization by FACS)		pGLG	(25)
	Inverted repeats	**SU**	Spontaneous RE	SC-his-trp	SC-trp	pRS314SU	(21)
		TINV	Spontaneous or DSB-induced GC and RE	SC-leu-ura	SC-ura +dox (LT) SC-ura (HT)	pRS316-TINV	(28)
Plasmid–chromosome		**chrIII::*leu2-k*** **p(G)L2-*HOr***	Spontaneous or DSB-induced GC	SC-leu (SGal-leu)	SC-trp	pCM184-L2HOr (p414-GL2HOr)	(24, 28)

SCR, sister chromatid recombination; GC, gene conversion; RE, reciprocal exchange; HT, high transcription; LT, low transcription; dox, doxycycline.

2. Total growth plate: SC-his (1.7 g/l amino acid-free yeast nitrogen base, 5 g/l ammonium sulfate, 20 g/l glucose, 20 g/l agar, and the appropriate amino acids in standard concentrations except histidine).
3. Liquid SGal-his media: 1.7 g/l amino acid-free yeast nitrogen base, 5 g/l ammonium sulfate, 20 g/l galactose, and the appropriate amino acids in standard concentrations except histidine.

2.5. Physical Analysis of Equal or Unequal SCE

2.5.1. Time-Course Experiment

1. pL2-*HOr* or TINV transformant of the strain of interest with a *MAT*a-*inc ade3::GAL-HO leu2Δ::SFA* background.
2. Liquid SC-ura media: 1.7 g/l amino acid-free yeast nitrogen base, 5 g/l ammonium sulfate, 20 g/l glucose, 20 g/l agar, and the appropriate amino acids in standard concentrations except uracil. Add 30 ml of sterile 100% glycerol per litre after auto-claving.
3. Liquid SGL-ura media: 1.7 g/l amino acid-free yeast nitrogen base without ammonium sulfate, 5 g/l ammonium sulfate, 25.6 ml Na lactate 60% (v/v) with the appropriate amino acids in standard concentrations except uracil.
4. 20% galactose stock solution.
5. Doxycycline stock (5 mg/ml) (Sigma-Aldrich Química, Madrid, Spain), store in the dark at 4°C up to 1 month.

2.5.2. DNA Extraction Protocol

1. 1 M spermidine stock solution (Sigma-Aldrich), store at –20°C.
2. 0.5 M spermine stock solution (Sigma-Aldrich), store at 4°C.
3. Nuclei-isolating buffer (NIB) pH 7.2, store at 4°C: 17% (w/v) glycerol (Sigma-Aldrich), 50 mM (3-[*N*-morpholino]propanesulfonic acid) sodium salt (MOPS) pH 7.5, 150 mM CH_3CO_2K, 2 mM $MgCl_2$, 500 μM spermidine (Sigma-Aldrich), 150 μM spermine (Sigma-Aldrich). Autoclave and store at –4°C in the dark.
4. Zymolyase 20T (15 mg/ml; USB-Affimetrix, Santa Clara, CA, USA), store at 4°C.
5. RNase A (10 mg/ml; Sigma-Aldrich). Store at –20°C.
6. 3 M sodium acetate.
7. Phenol solution (Sigma-Aldrich).
8. Chloroform.
9. Isopropanol.
10. 1× TE: Tris–HCl 10 mM, EDTA 1 mM pH 8.
11. 70% ethanol.
12. Fluorometer apparatus and appropriate cuvettes.
13. Block heater or water bath.

2.5.3. Gel Electrophoresis and Southern Hybridization

1. Ethidium bromide (10 mg/ml EtBr from Sigma-Aldrich): The stock solution is obtained by dissolving 1 g of EtBr in 100 ml H_2O. Stir on a magnetic stirrer for several hours to allow complete dissolution (*see* **Note 2**). Wrap the bottle in aluminum foil and store at room temperature.
2. Agarose (Pronadisa, Hispanlab, Spain).
3. 1× TBE: 90 mM Tris base, 90 mM boric acid, 2 mM EDTA.
4. 1 kb ladder (Invitrogen, Barcelona, Spain).
5. *Bgl*II enzyme for equal SCE and *Xho*I and *Spe*I enzymes for unequal SCE (Takara, Madrid, Spain).
6. 10× loading buffer (Takara). Store at 4°C.
7. Electrophoresis apparatus and power supply.
8. 0.25 N HCl.
9. Denaturation solution: 0.5 M NaOH, 1.5 M NaCl.
10. Neutralization solution: 1 M NH_4Ac, 0.02 M NaOH.
11. 20× SSC: 3 M NaCl, 0.3 M trisodium citrate pH 7.
12. dATG solution (Roche Farma, Madrid, Spain): Mix 0.5 mM dATP, 0.5 mM dTTP, and 0.5 mM dGTP in water and store at –20°C.
13. α^{32}P-dCTP (Perkin-Elmer Life Sciences, Waltham, MA, USA) (1 mCi [10 mCi/ml], 3,000 Ci/mmol).
14. Klenow (Sigma-Aldrich).
15. Hexanucleotide mix (Roche).
16. Sephadex G50-TE: Dissolve 5 g Sephadex G50 (GE Healthcare, Barcelona, Spain) in 75 ml 1× TE, autoclave, and store at 4°C.
17. Hybridization solution: 0.5 M phosphate buffer pH 7, 7% SDS.
18. Wash solution: 0.1× SSPE, 5 mM EDTA, 0.5% SDS.
19. 20× SSPE: 3 M NaCl, 200 mM sodium phosphate, 20 mM EDTA pH 7.4
20. Block heater or water bath.
21. 254 nm UV crosslinker.

3. Methods

In this section we will first describe the different types of recombination substrates (**Sections 3.1**, **3.2**, and **3.3**) and then the genetic and molecular methods for the study of recombination (**Sections 3.4** and **3.5**).

3.1. Direct-Repeat Recombination Systems

3.1.1. Genetic Analysis of Chromosomal Deletions: The leu2-k::ADE2-URA3::leu2-k System

This chromosomal system is based on two copies of the *leu2-k* mutant allele carrying the *ADE2* and *URA3* markers in between them (**Fig. 10.1**). Whereas GC events cannot be distinguished with this system, deletion events can easily be detected and quantified in media containing 5-fluoorotic acid (FOA). In addition, deletion events can be directly visualized as red sectors in colonies if the strain used has an *ade2* non-functional allele at the *ADE2* locus, since *ade2 leu2-k* recombinants accumulate a red pigment (**Fig. 10.3a**) (10).

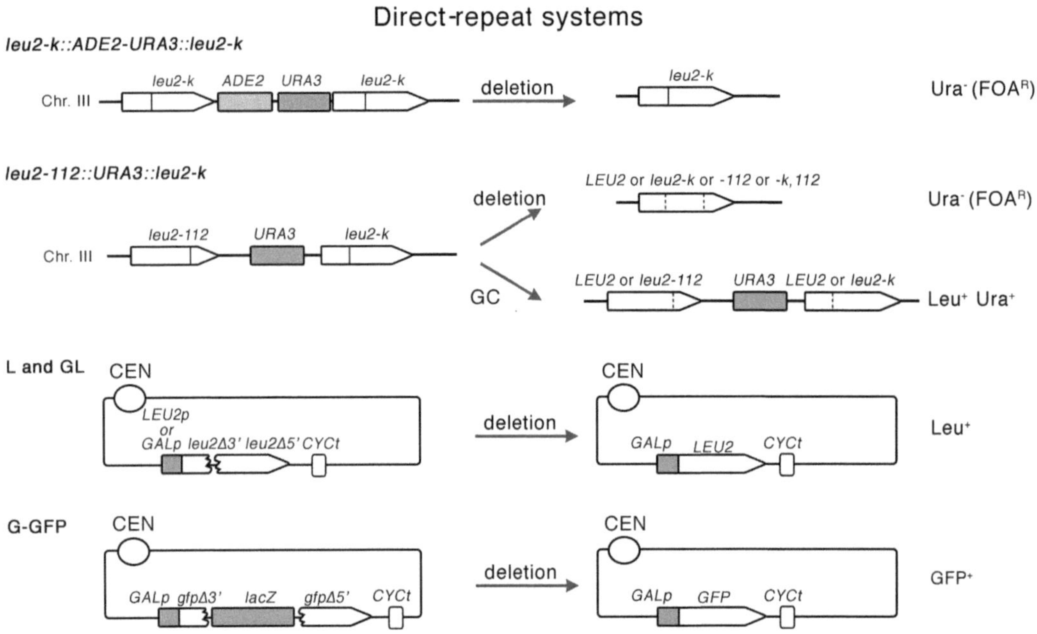

Fig. 10.1. A selection of direct-repeat recombination systems. *leu2-k::ADE2-URA::leu2-k*, *leu2-112::URA3::leu2-k*, L, GL, and G-GFP direct-repeat system diagrams and recombination products are shown. *CEN*, centromere; *GC*, gene conversion.

3.1.2. Genetic Analysis of Chromosomal Deletions and GCs: the leu2-112::URA3::leu2-k System

This chromosomal system is based on two different *leu2* mutant heteroalleles, *leu2-112* and *leu2-k*. It allows detection of GC products (**Fig. 10.1**). Leu+ colonies arise from GC or deletion, the latter being easily distinguished by the loss of the *URA3* marker and thus by the growth of recombinants in media containing FOA (20).

3.1.3. Genetic Analysis of Deletions in Plasmids: L System and Derivatives

These systems are based on two truncated repeats of the *LEU2* gene sharing 600 bp of homology and placed on a mono-copy *CEN*-based plasmid (**Fig. 10.1**). They allow the detection of deletions as Leu+. The simplest one is the L system (21). More complex systems derived from this differ in the intervening sequence located between the repeats. This is the case of the L*PHO5* system, with a 1.5 kb fragment of the yeast *PHO5*

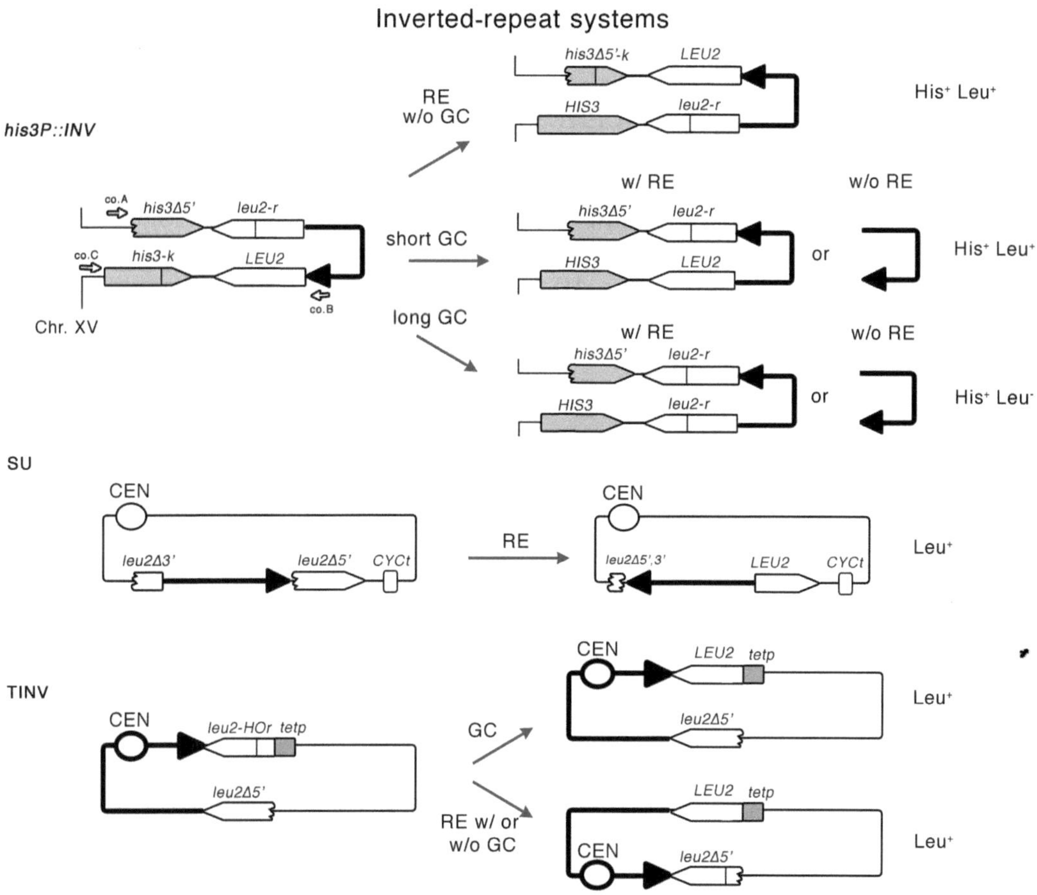

Fig. 10.2. A selection of inverted-repeat recombination systems. *his3P::INV*, SU, and TINV inverted-repeat system diagrams and recombination products are shown. *CEN*, centromere; *GC*, gene conversion; *RE*, reciprocal exchange.

gene, and the L*lacZ* system, with the 3-kb bacterial *lacZ* gene in between the repeats (22). Differences in the length of the intervening sequence can be used to determine the effect of transcription elongation in the frequency of recombination. The length of this intervening sequence is 2.2 kb in LNA, 2.5 kb in LU, 3.7 kb in LYΔNS, 5.6 kb in LY, etc. (23). One system of this series contains a *CYC1* transcriptional terminator after the first *leu2* repeat to stop transcription elongation (LNAT system) (23).

3.1.4. Genetic Analysis of Transcription-Dependent Deletions in Plasmids: GL System and Derivatives

GL systems are based on two truncated repeats of the *LEU2* gene sharing 600 bp of homology and placed on a monocopy *CEN*-based plasmid, as in the case of L systems. Nevertheless, they are under the control of the *GAL* promoter instead of the *LEU2* promoter and allow the detection of deletions as Leu+ colonies that will only grow in galactose media (**Fig. 10.1**). These systems permit the analysis of the effect of high (galactose) versus low (glucose) transcription levels on recombination. The simplest system of this series is GL with no intervening sequence in between the *leu2* repeats. Derivate versions of GL contain different intervening regions as in the case of the L systems. These are GLNA, GLNAT, GL*PHO5*, and GL*lacZ* systems (24).

3.1.5. FACS Analysis of Deletions in Plasmids: G-GFP System

This system consists of a monocopy *CEN*-based plasmid with two truncated GFP repeats sharing 200 bp of homology under the control of the *GAL* promoter, with the 3-kb *lacZ* sequence in between (**Fig. 10.1** (25)). Deletion events lead to GFP+ recombinants that can be directly scored by FACS (**Fig. 10.3b**).

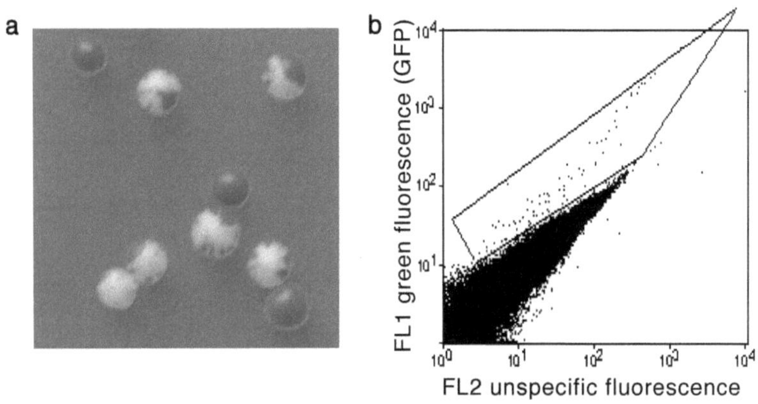

Fig. 10.3. Direct detection of recombination events. (**a**) Yeast red sectoring colonies as a result of hyper-recombination between the *leu2* direct repeats of the *leu2-k::ADE2-URA::leu2-k* and consequent deletion of the intervening *ADE2* marker. (**b**) Detection of recombination by FACS analysis. Homologous recombination between the two truncated forms of the GFP gene of the G-GFP system re-establishes a wild-type GFP copy. The recombinant population emitting green fluorescence is enclosed in a *box* for its identification in the FACS analysis.

3.2. Inverted-Repeat Recombination Systems

3.2.1. Genetic Analysis of Physical Length Variation of GCs and Their Association with REs: his3^P::INV

This chromosomal system is based on two different *his3-LEU2* mutant heteroalleles, *his3Δ5′-leu2-r* and *LEU2 his3-k* (26). His+ recombinants can arise either by RE or by GC. His+ recombinants that arise by GC can be either Leu– or Leu+, depending on whether they arise as a result of a long GC event (from 1.5 to 3 kb) covering both the *his3-k* and the *leu2-r* sites, or a short GC (shorter than 1.5 kb) not covering the *leu2-r* site, respectively (**Fig. 10.2**). The system permits the determination of the percentage of GCs associated with RE of the whole 3-kb repeat by PCR analysis of the His+ Leu– recombinants. Independent His+ Leu– recombinants must be isolated from SC-his and analyzed by multiplex PCR using three oligos at the same time (co.A, GCGTATCACGAGGCCCTTTC; co.B, TGGCAACGATAGGGACGGAG, and co.C, CGCTGCATAAACGCTGTTGG) (*see* **Fig. 10.2**). Non-RE events lead to a 4.1 kb fragment, which is PCR-amplified by oligos co.B and co.C, while RE events lead to a 3.2 kb fragment, which is PCR-amplified by oligos co.A and co.B (27).

Other systems derived from this one have been described, such as the *his3^P::VST* system, with one *his3Δ5′-k* double mutant allele in one repeat and the *his3-h* allele in the other, or the *his3^P::TER* system which contains the *CYC1* terminator sequence upstream of the *his3Δ5′* copy for more specific purposes (*see* (27)).

3.2.2. Genetic Analysis of REs in Plasmids: SU System

This system consists of two 1.36 kb truncated copies of the *LEU2* gene, placed on a monocopy *CEN*-based plasmid in an inverted orientation and separated by a 1.66 kb sequence. It allows detection of REs as Leu+ colonies (**Fig. 10.2**) (21).

3.2.3. Genetic Analysis of GCs and REs in Plasmids: TINV System

This system consists of two *leu2* inverted repeats, *leu2Δ5′* and *leu2-HOr*, sharing 1.2 kb of homology and placed on a monocopy *CEN*-based plasmid (**Fig. 10.2**). Leu+ events arise as a consequence of either GC or RE events (28). This plasmid also allows the physical measurement of SCR (see below).

3.3. Systems for the Analysis of Sister Chromatid Recombination (SCR)

3.3.1. Genetic Analysis of Unequal SCR: his3Δ5′::his3Δ3′

It is based on two truncated copies of the *HIS3* gene inserted in a direct orientation at the *TRP1* locus. Recombination can take place not only with the same repeat (equal SCR), leading to a genetically identical and thus undetectable recombinant, but also with the other repeat on the sister chromatid (unequal SCR), which leads to the formation of a triplication that results in a His+ detectable recombinant (**Fig. 10.4a**). This system exists in two variants: one without any endonuclease sites and valid for the analysis of spontaneous SCR (29); the other in which one of the repeats contains a full 117 bp HO cleavage site to directly target HO endonuclease-induced DSBs (30). Since the latter system uses the full 117 bp HO cleavage site, both sister chromatids are equally cleaved by HO endonuclease, thus impeding equal SCR. Unequal SCR can therefore be detected both spontaneously and after DSB induction. It is worthy to note that DSB repair between the repeats can also occur by intrachromatid recombination, but this event would not give rise to genetically selectable recombinants, although it could influence the overall levels of SCR detected (14).

3.3.2. Physical Analysis of Equal SCE: pL2-HOr System

The pL2-*HOr* system is a monocopy *CEN*-based plasmid with a *leu2-HOr* allele that contains a minimal 24-bp HO site (**Fig. 10.4b** (28)), in which the efficiency of cleavage is greatly reduced to < 10% with respect to the full 117-bp HO cleavage site. This allows, in most cases, only one sister chromatid to be cleaved by HO, while the other remains intact and competent to be used as a template for equal SCR (2). In this system, HO endonuclease produces mainly single-stranded DNA nicks that are converted into DSBs when they are encountered by the

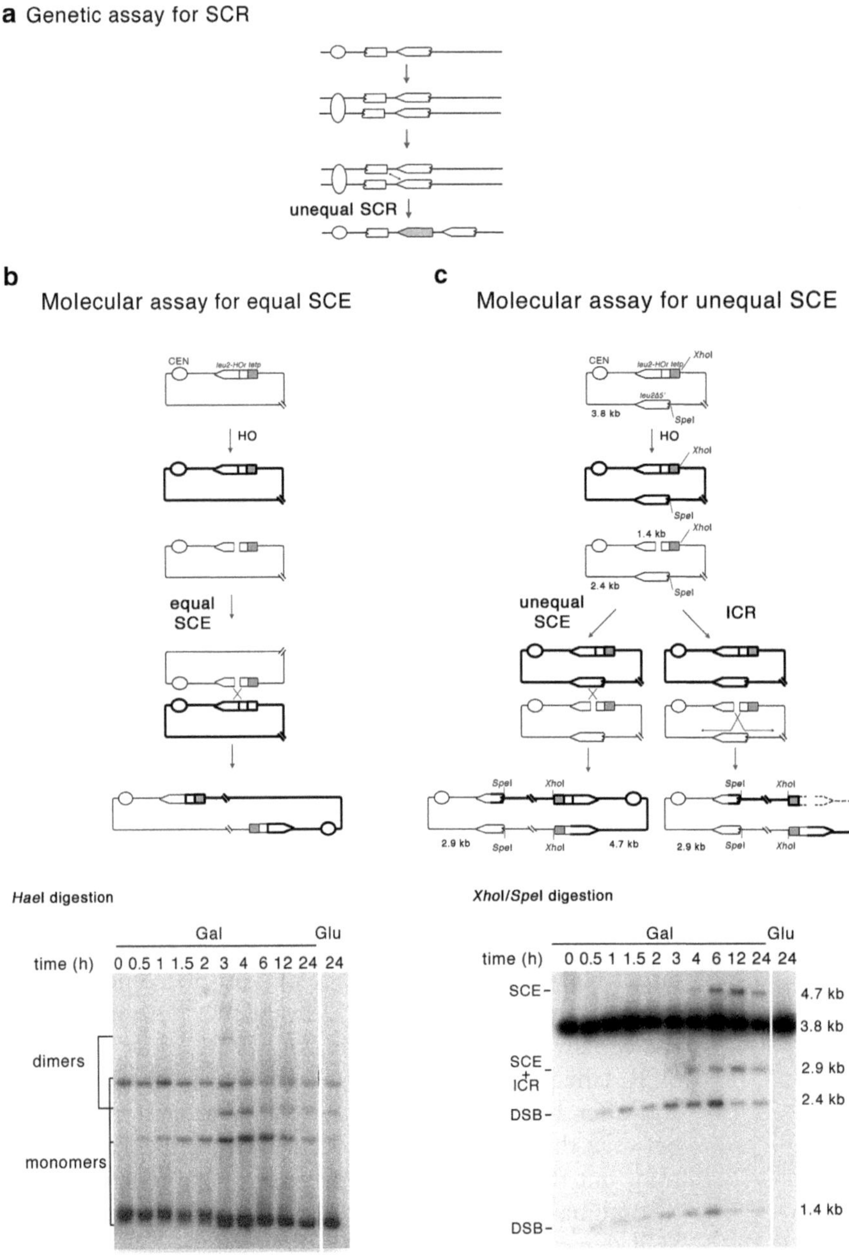

Fig. 10.4. Genetic and molecular detection of SCR. (**a**) Genetic assays of unequal SCR based on direct repeats. The repeats are displayed in an orientation in which only SCR leads to the formation of a selectable recombinant (colored in *gray*). (**b**) Molecular assay for the analysis of equal SCE. Schemes of pL2-*HOr* plasmid and the dimer produced by equal SCE are shown on *top*. Kinetics of HO-mediated DSBs and dimer formation in plasmid pL2-*HOr* after HO induction in SGal. DNA was digested with *Hae*I (*Hae*I restriction sites are not present in pL2-*HOr*). Southern blot was hybridized with the *Cla*I–*Eco*RV-specific *LEU2* probe. (**c**) Molecular assay for the analysis of unequal SCE. Schemes of plasmid pTINV and the intermediates produced by unequal SCE and ICR involving DNA synthesis by BIR are shown on *top*. Kinetics of HO-mediated DSBs and the different fragments generated by HO cleavage after *Xho*I–*Spe*I digestion. Other details as in (**b**). Unequal SCE leads to 2.9 and 4.7 kb fragments when digested with *Spe*I and *Xho*I. Equal SCE (not shown) would lead to 3.8 kb *Spe*I–*Xho*I fragments, which are not distinguished from the original TINV. (Reproduced from (2) with permission from Elsevier Inc.).

replication forks (31). Thus, this *mini*-HO site is a useful tool to mimic a natural situation in which DSBs appear as a consequence of replication failures. This system allows the direct visualization of SCR events as plasmid dimers by Southern hybridization (**Fig. 10.4b**) (2).

3.3.3. Physical Analysis of Unequal SCE: TINV System

The TINV inverted-repeat system is based on two *leu2* inverted repeats sharing 1.2 kb of homology and placed on a monocopy *CEN*-based plasmid as described in **Section 3.2.3**. One of the repeats contains the minimal 24-bp HO site (**Fig. 10.4c** (28)) and therefore it also allows equal SCE to occur. This system allows both the genetic detection of Leu+ recombinants, as described in **Section 3.2.3**, and the detection of unequal SCE and intrachromatid recombination (ICR) events in time-course experiments when the DNA is digested with *Spe*I and *Xho*I restriction enzymes (**Fig. 10.4c**) (2, 31). As can be seen in **Fig. 10.4c**, specific 2.4- and 1.4-kb *Spe*I/*Xho*I bands appear as a result of the HO-induced DSB. Approximately 30–60 min after HO induction, DSB repair leads to the formation of new 4.7- and 2.9-kb bands. While the 4.7-kb band appears exclusively as the result of unequal SCE events, the 2.9-kb band corresponds to the sum of two kinds of events: unequal SCE and ICR, which involves DNA synthesis by BIR. Therefore, ICR events can be estimated from the difference between the intensity of 2.9- and 4.7-kb bands, although the reliability of the latter measurement depends very much on the signal.

3.3.4. Plasmid–Chromosome Recombination: chrIII::leu2-k p(G)L2-HOr Systems

This plasmid–chromosome system measures both spontaneous and DSB-induced GC events (**Fig. 10.5** (28)). Recombination occurs between the *leu2-k* allele at the endogenous *LEU2* chromosomal locus and the *leu2-HOr* allele carrying the reduced

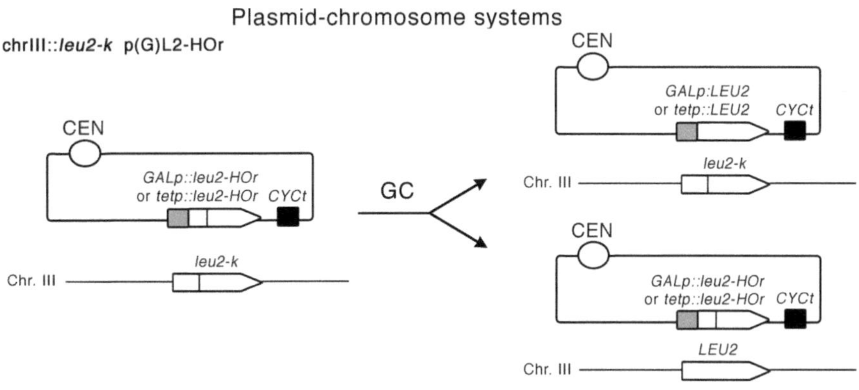

Fig. 10.5. Plasmid–chromosome recombination constructs used to study transcription-associated and DSB-induced GC. Recombination between a *CEN*-based plasmid carrying the *leu2-HOr* allele either under the *tet* or the *GAL* promoter, and chromosome III, carrying the *leu2-k* allele under its own promoter. Leu+ recombinants can arise by GC of either *leu2-HOr* or *leu2-k* alleles. *CEN*, centromere; *GC*, gene conversion.

HO recognition site, located at a monocopy *CEN*-based plasmid under the control of a regulatable promoter, such as the *tet* (24) or *GAL* promoter (28). Transcription can be switched off in these DNA templates with 5 μg/ml doxycycline or 2% glucose, respectively. In these systems, DSB-induced Leu+ colonies arise mainly as a consequence of GC of the plasmid-born *leu2-HOr* allele, which is the broken molecule acting as the receptor, although GC of the chromosomal *leu2* mutant allele can also occur (**Fig. 10.5** (28)). REs cannot be detected because they would lead to the integration of the centromeric plasmid yielding the formation of an unstable dicentric chromosome.

3.4. Determination of Recombination Frequencies

3.4.1. Genetic Analysis of Recombination

1. Streak the transformants onto growth plates (*see* **Table 10.1**) (*see* **Note 3**). This medium should contain the required drugs (*see* **Note 4**).
2. Grow the zig–zag for 3–4 days at the selected temperature (usually 30°C), until the colonies reach 2.5–3 mm diameter each (usually 3–4 days). If the colonies are too small or the expected recombination frequency is very low, inoculate each colony in liquid media (*see* **Note 5**). In the case of DSB-induced recombination, in which the HO endonuclease is under the *GAL* promoter, it is necessary to induce the HO expression by adding galactose (*see* **Note 6**).
3. Resuspend six colonies in six microtubes containing 1 ml of sterile distilled H_2O each and perform five tenfold-serial dilutions from each microtube as follows. Take 100 μl from the original microtube into a 0.9 ml distilled H_2O-containing second tube, vortex, and remove again 100 μl into a third tube. Repeat this by vortexing and taking 100 μl into the fourth 0.9 ml distilled H_2O-containing tube to finally have 1, 1/10, 1/100, and 1/1,000 dilutions.
4. Plate 25 μl of each tenfold dilution on recombinant selective plates to obtain 40, 400, 4,000, and 40,000 dilution factors for the recombinants and other additional 25 μl of the last dilution microtube (1/1,000) on growing plate to obtain the number of total cells. The plates can be divided into six sections, so that each section will be used to plate one of the dilutions of recombinants or totals from one strain or independent transformant.
5. Incubate the plates at 30°C for 2–3 days.
6. Count the number of total cells and recombinants and multiply by the appropriate dilution factor. Calculate the frequency of recombination for each of the six colonies tested by dividing the number of recombinants by the total number of cells.

7. Determine the median frequency of recombinants and use the average of at least three independent fluctuation tests as a value for the frequency of recombination (*see* **Note 7**).

3.4.2. Genetic Analysis of Recombination by Flow Cytometry

1. Streak six transformants onto SC-his plates.
2. Grow the zig–zag for 3–4 days at the selected temperature (usually 30°C), until the colonies reach 2.5–3 mm diameter each.
3. Inoculate one colony from each transformant in 5 ml of liquid SGal-his to allow GFP expression and incubate overnight at 30°C.
4. Start up CELLQuest software and optimize parameters. Create an FL1 (for GFP detection) versus FL2 (for unspecific fluorescence detection) dot plot, and adjust detector gains and voltages (instrument settings) for two-color analysis according to the software guidelines (32).
5. Dilute the samples tenfold in distilled H_2O to run them in FACSCalibur (Becton Dickinson). Acquire and store the data with the CELLQuest software. The percentage of green fluorescent cells falling above the diagonal in the FL1–FL2 dot plot represents the frequency of GFP+ recombinants for each transformant (**Fig. 10.3b**).
6. Determine the median recombination frequency of 6–12 transformants.

3.5. Physical Analysis of Equal or Unequal SCE

3.5.1. Time-Course Experiment

1. Grow a pL2-*HOr* or TINV transformant of the strain of interest (with a *MAT*a*-inc ade3::GAL-HO leu2Δ::SFA* background) overnight on a shaker at 30°C in liquid SC-ura media.
2. Dilute the culture to an OD_{600} of 0.2 in 100 ml SC-ura media and let it grow for two rounds of duplication (6–7 h) on a shaker at 30°C (until an OD_{600} of ~0.8 is reached).
3. Collect the appropriate volume of cell culture by centrifugation at 3,000×*g* for 2 min. Wash twice with fresh SGL-ura media and resuspend the pellet in 180 ml SGL-ura +doxycycline (5 μg/ml) to an OD_{600} of 0.3. Let it grow overnight on a shaker at 30°C.
4. When the culture has reached an OD_{600} of ~0.5 in SGL-ura +doxycycline, add 20 ml of 20% galactose and keep the culture on a shaker at 30°C during the entire time course.
5. Measure OD_{600} at every time point (usually 0, 0.5, 1, 1.5, 2, 3, 4, 6, and 24 h after galactose addition) and collect 50 ml of the cell culture by centrifugation at 3,000×*g* for 2 min at 4°C in 50 ml Falcon tubes, remove supernatant, and freeze

the pellet at –80°C until all time-point samples have been collected.

3.5.2. DNA Extraction

1. Wash the frozen pellet in 1 ml distilled H_2O and transfer the appropriate volume of cells as to finally have the same number of cells in every time-point sample using the OD_{600} as a reference into a 2 ml microtube.
2. Centrifuge at 3,000×*g* for 1 min and remove the supernatant.
3. Resuspend the pellet in 400 μl nucleic isolation buffer by vortexing.
4. Add 80 μl of 15 mg/ml zymolyase 20T and incubate during 20–25 min at RT.
5. Fill the microtubes with distilled H_2O to stop the zymolyase action.
6. Centrifuge at 3,000×*g* for 2 min, remove supernatant, and resuspend the pellet in 720 μl of 1× TE by vortexing.
7. Add 80 μl 10% SDS and keep 30 min on ice. Invert tubes every 10 min.
8. Add 400 μl phenol and 400 μl 24:1 chloroform/isoamylalcohol (*see* **Note 8**). Mix vigorously and separate the two phases by centrifugation at 16,000×*g* for 10 min.
9. Carefully transfer the clear upper phase into a new 2 ml microtube.
10. Repeat Steps 12 and 13 as many times as necessary to obtain a clean sample (twice or three times is usually enough, *see* **Note 9**).
11. Precipitate the DNA by adding 80 μl of 3 M sodium acetate and 800 μl isopropanol and centrifuge at 16,000×*g* for 15 min.
12. Remove the supernatant and resuspend the pellet in 500 μl 1× TE and 0.5 μl RNase A (10 mg/ml).
13. Incubate at 37°C for 30 min.
14. Add 250 μl phenol and 250 μl 24:1 chloroform/isoamylalcohol. Mix vigorously and separate the two phases by centrifugation at 16,000×*g* for 10 min.
15. Carefully transfer the clear upper phase into a new 2 ml microtube.
16. Precipitate the DNA by adding 50 μl of 3 M sodium acetate and 500 μl isopropanol and centrifuge at 16,000×*g* for 15 min.
17. Briefly wash the pellet with 1 ml 70% ethanol.

18. After centrifugation (1 min, 16,000×*g*), carefully remove the ethanol as much as possible and dissolve the DNA in 200 μl 1× TE.

After preparation of DNA samples, 1–2 μl of each DNA preparation is quantified using a DNA fluorimeter or using standard gel electrophoresis. An aliquot of the samples (usually 50 μl), corresponding to 4 μg of total DNA, is digested with *Hae*I, which does not cut the pL2-*HOr* plasmid, to see dimer formation by equal SCE (**Fig. 10.4b**) or with *Xho*I and *Spe*I restriction enzymes to distinguish between the different recombination products arising from unequal SCE and ICR (**Fig. 10.4c**). After 2 h digestion at 37°C, precipitate the DNA with sodium acetate and isopropanol, wash once with ethanol 70%, and dissolve the DNA in 20 μl of 1× TE (as in Steps 15–17).

3.5.3. Gel Electrophoresis and Southern Hybridization

1. Prepare a 0.8% agarose gel with 0.3 μg/ml ethidium bromide in fresh 1× TBE in an appropriate gel tray (we routinely use apparatus $W \times L = 20 \times 25$). After agarose polymerization, place the gel in the tray box at room temperature containing a suitable volume of 1× TBE.
2. Load the DNA samples and 5 μl of 1 kb ladder and run the gel at constant low voltage (50 V, c.a. 1 V/cm) overnight (20 h approximately).
3. Take a picture of the ethidium bromide staining with a fluorescent gel ruler.
4. Treat the gel as follows: depurinate the gel for 10 min in 0.25 N HCl, denaturate 30 min in denaturation solution, and finally neutralize for 30 min in neutralization solution.
5. Transfer the gel in standard Southern blot conditions using Hybond-N nylon (GE-Healthcare) membrane in 20× SSC and leave overnight.
6. Remove the membrane from the gel and UV cross-link the DNA to the membrane (70,000 μJ/cm^2).
7. Rinse the membranes with 2× SSC and let them dry until hybridization.

The membranes are subjected to hybridization with a radiolabeled probe consisting of the 600-bp *Cla*I–*Eco*RI fragment of *LEU2* (this is required to avoid hybridization with the endogenous *leu2ΔSFA*). Different protocols can be employed at this step; here we propose a standard rapid and efficient procedure of random prime labeling:

1. Boil 100–200 ng of purified DNA probe in 36.5 μl of H_2O for 5 min at 100°C and place it on ice.
2. Add 5μl of hexanucleotide mix, 5 μl of 0.5 mM dATG, 1 μl Klenow, and 50 μCi of α^{32}P-dCTP (*see* **Note 10**) and incubate for 1 h at 37°C.

3. Remove the non-incorporated nucleotides with a Sephadex G50-TE column. During the preparation of the radiolabeled probe, the membranes are prehybridized with 10 ml of 1× hybridization solution for at least 30 min at 65°C in a rotating tube.
4. Boil the probe 10 min at 100°C, add it to the 10 ml hybridization solution, and incubate at 65°C overnight.
5. Wash the filters twice with 50 ml wash solution during 30 min at 65°C in the rotating tube.
6. Place the hybridized membranes on a filter paper and let them air-dry briefly.

The signals are analyzed using a FUJI PhosphoImager and quantified using the Image Gauge program. SCE levels are calculated as the ratio between the 4.7 kb band and the total plasmid DNA. ICR levels are calculated by subtracting the density value of the 4.7 kb band from that of the 2.9 kb band (*see* **Note 11**).

4. Notes

1. Standard amino acid concentrations refer to adenine sulfate 20 mg/l, uracil 20 mg/l, L-tryptophan 20 mg/l, L-histidine-HCl 20 mg/l, L-arginine-HCl 20 mg/ml, L-methionine 20 mg/ml, L-tyrosine 30 mg/ml, L-leucine 30 mg/ml, L-isoleucine 30 mg/ml, L-lysine-HCl 30 mg/ml, L-phenylalanine 50 mg/ml, L-glutamic acid 100 mg/ml, L-aspartic acid 100 mg/ml, L-valine 150 mg/l, L-threonine 200 mg/l, L-serine 400 mg/ml.
2. Ethidium bromide is a powerful mutagen and is moderately toxic. Wear gloves and lab coat when handling. After use, the solutions and gels should be safely eliminated, according to your institution's safety measures for toxic waste.
3. You should always be working in a sterile environment (around a Bunsen burner), because bacteria and fungi easily contaminate yeast cultures.
4. These media should contain the required drugs to measure the effect of different chemical agents on recombination. It should also contain doxycycline to inhibit transcription in the case of systems based on the *tet* promoter or galactose as a carbon source to promote transcription in the case of recombination systems where transcription is under the *GAL* promoter.
5. If the colonies are too small or the expected recombination frequency too low, each colony can be inoculated in

liquid media and grown as long as necessary to increase the total cell number. Then, the culture is centrifuged and cells collected to perform the appropriate serial dilutions.

6. In the case of DSB-induced recombination, in which the HO endonuclease is under the *GAL* promoter (such as in the case of the pL2-*HOr*, TINV, or the plasmid–chromosome systems), it is necessary to induce the HO expression by adding galactose (as in Steps 1–4 in **Section 3.5.1**). In summary, pre-inoculate each of the six colonies in 5 ml of liquid SC medium lacking the appropriate amino acids and let them grow for two rounds of duplication (6–7 h) on a shaker at 30°C (until an OD_{600} of ~0.8 is reached). Collect the appropriate volume of cell culture by centrifugation at 3,000×*g* for 2 min, wash twice with fresh SGL media, and resuspend the pellet in SGL medium lacking the appropriate amino acids to have 5 ml with an OD_{600} of 0.3. Let it grow on a shaker at 30°C until the culture has reached an OD_{600} of ~0.5 in SGL medium (usually overnight). Take 1 ml for the dilutions of the spontaneous recombination frequencies as in Steps 3–7 in **Section 3.4.1**. Add 2% galactose to the rest of the culture and keep the culture on a shaker at 30°C for 5 h for HO induction. Finally, take 1 ml and make the appropriate dilutions for the DSB-induced recombination frequencies as in Steps 3–7 in **Section 3.4.1**.

7. The frequency of recombination can greatly vary due to its dependence on the number of divisions occurring in the culture and the cell division in which recombination takes place. To rely on median recombination frequencies it is therefore necessary that the colony size of each clone analyzed be similar, so that the number of cell divisions is the same. Otherwise, or in addition, the recombination rate can be calculated, which refers to the number of recombination events divided by the total number of cells (recombinants per cell per generation). The number of events can be estimated by several methods, the most commonly used being the *Method of Lea and Coulson*. In this method, the number of events is calculated from the median number of recombinants while the total number of cells can be averaged from all the cultures when all of the colonies are of uniform size (*see* (33), also described in (34)).

8. Wear gloves and a lab coat while working with phenol, because it is a very dangerous compound. Also, use polypropylene tubes when working with phenol. The nucleic acid will tend to separate into the organic phase if the phenol has not been adequately equilibrated to a pH of 7.8–8.0. Normally, the aqueous phase forms the upper

phase. However, if the aqueous phase is dense because of the presence of salts or sucrose, it will form the lower phase. The organic phase is easily identifiable because of the yellow color given by the hydroxyquinoline added during the phenol equilibration. You should always dispose of phenol waste in a specially sealed container and ensure that it is eliminated according to your institution's policies for dangerous wastes.

9. This step is extremely important. Phase lock gel heavy 2 ml tubes (5 PRIME GmbH, Hamburg, Deutschland) can be used to facilitate sample phenol treatment.
10. Working with radioactivity is dangerous and should be taken seriously. Always wear a lab coat, two pair of gloves, and work behind protective screens. Verify often that your hands and the materials used are not contaminated. Do this by direct verification using a hand-held Geiger counter and make a complete verification of your work space when the manipulation is complete. After use, all contaminated material must be safely stored, according to your institution's safety measures for radioactive waste.
11. To quantify the gel bands, it is essential to work with a non-saturated image, since signal saturation would sub-estimate the value obtained. Do not forget to subtract the value obtained for the specific background for each gel line from the value obtained for the gel bands. The defined area size must be the same for every specific gel band to quantify and its corresponding background.

Acknowledgments

We would like to thank H. Gaillard and M. Moriel-Carretero for reading the manuscript and D. Haun for style supervision. We apologize for not citing our colleague's work due to space limitation. Research in the A. A. laboratory was funded by research grants from the Spanish Ministry of Science and Innovation and the Junta de Andalucía.

References

1. McClintock, B. (1938) The production of homozygous deficient tissues with mutant characteristics by means of the aberrant mitotic behavior of ring-shaped chromosomes. *Genetics* **23**, 315–376.
2. Gonzalez-Barrera, S., Cortes-Ledesma, F., Wellinger, R.E., and Aguilera, A. (2003) Equal sister chromatid exchange is a major mechanism of double-strand break repair in yeast. *Mol Cell* **11**, 1661–1671.

3. Kadyk, L.C., and Hartwell, L.H. (1992) Sister chromatids are preferred over homologs as substrates for recombinational repair in *Saccharomyces cerevisiae*. *Genetics* **132**, 387–402.
4. Pardo, B., Gomez-Gonzalez, B., and Aguilera, A. (2009) DNA repair in mammalian cells: DNA double-strand break repair: how to fix a broken relationship. *Cell Mol Life Sci* **66**, 1039–1056.
5. Barbera, M.A., and Petes, T.D. (2006) Selection and analysis of spontaneous reciprocal mitotic cross-overs in *Saccharomyces cerevisiae*. *Proc Natl Acad Sci USA* **103**, 12819–12824.
6. Jackson, J.A., and Fink, G.R. (1981) Gene conversion between duplicated genetic elements in yeast. *Nature* **292**, 306–311.
7. Klein, H.L., and Petes, T.D. (1981) Intrachromosomal gene conversion in yeast. *Nature* **289**, 144–148.
8. Prado, F., Cortes-Ledesma, F., Huertas, P., and Aguilera, A. (2003) Mitotic recombination in *Saccharomyces cerevisiae*. *Curr Genet* **42**, 185–198.
9. Cortes-Ledesma, F., Prado, F., and Aguilera, A. (2007) Molecular genetics of recombination. *Top Curr Genet* **17**, 221–250.
10. Aguilera, A., and Klein, H.L. (1988) Genetic control of intrachromosomal recombination in *Saccharomyces cerevisiae*. I. Isolation and genetic characterization of hyper-recombination mutations. *Genetics* **119**, 779–790.
11. Judd, S.R., and Petes, T.D. (1988) Physical lengths of meiotic and mitotic gene conversion tracts in *Saccharomyces cerevisiae*. *Genetics* **118**, 401–410.
12. Wallis, J.W., Chrebet, G., Brodsky, G., Rolfe, M., and Rothstein, R. (1989) A hyper-recombination mutation in *S. cerevisiae* identifies a novel eukaryotic topoisomerase. *Cell* **58**, 409–419.
13. Rudin, N., Sugarman, E., and Haber, J.E. (1989) Genetic and physical analysis of double-strand break repair and recombination in *Saccharomyces cerevisiae*. *Genetics* **122**, 519–534.
14. Fasullo, M., Giallanza, P., Dong, Z., Cera, C., and Bennett, T. (2001) *Saccharomyces cerevisiae* rad51 mutants are defective in DNA damage-associated sister chromatid exchanges but exhibit increased rates of homology-directed translocations. *Genetics* **158**, 959–972.
15. Saxe, D., Datta, A., and Jinks-Robertson, S. (2000) Stimulation of mitotic recombination events by high levels of RNA polymerase II transcription in yeast. *Mol Cell Biol* **20**, 5404–5414.
16. Voelkel-Meiman, K., Keil, R.L., and Roeder, G.S. (1987) Recombination-stimulating sequences in yeast ribosomal DNA correspond to sequences regulating transcription by RNA polymerase I. *Cell* **48**, 1071–1079.
17. Rattray, A.J., and Symington, L.S. (1994) Use of a chromosomal inverted repeat to demonstrate that the RAD51 and RAD52 genes of *Saccharomyces cerevisiae* have different roles in mitotic recombination. *Genetics* **138**, 587–595.
18. Ahn, B.Y., and Livingston, D.M. (1986) Mitotic gene conversion lengths, coconversion patterns, and the incidence of reciprocal recombination in a *Saccharomyces cerevisiae* plasmid system. *Mol Cell Biol* **6**, 3685–3693.
19. Kupiec, M., and Petes, T.D. (1988) Meiotic recombination between repeated transposable elements in *Saccharomyces cerevisiae*. *Mol Cell Biol* **8**, 2942–2954.
20. Aguilera, A., and Klein, H.L. (1989) Genetic and molecular analysis of recombination events in *Saccharomyces cerevisiae* occurring in the presence of the hyper-recombination mutation hpr1. *Genetics* **122**, 503–517.
21. Prado, F., and Aguilera, A. (1995) Role of reciprocal exchange, one-ended invasion crossover and single-strand annealing on inverted and direct repeat recombination in yeast: different requirements for the RAD1, RAD10, and RAD52 genes. *Genetics* **139**, 109–123.
22. Chavez, S., and Aguilera, A. (1997) The yeast HPR1 gene has a functional role in transcriptional elongation that uncovers a novel source of genome instability. *Genes Dev* **11**, 3459–3470.
23. Prado, F., Piruat, J.I., and Aguilera, A. (1997) Recombination between DNA repeats in yeast hpr1delta cells is linked to transcription elongation. *EMBO J* **16**, 2826–2835.
24. Garcia-Rubio, M., Huertas, P., Gonzalez-Barrera, S., and Aguilera, A. (2003) Recombinogenic effects of DNA-damaging agents are synergistically increased by transcription in *Saccharomyces cerevisiae*. New insights into transcription-associated recombination. *Genetics* **165**, 457–466.
25. Gomez-Gonzalez, B., and Aguilera, A. (2007) Activation-induced cytidine deaminase action is strongly stimulated by mutations of the THO complex. *Proc Natl Acad Sci USA* **104**, 8409–8414.
26. Aguilera, A., and Klein, H.L. (1989) Yeast intrachromosomal recombination: long gene conversion tracts are preferentially associated

with reciprocal exchange and require the RAD1 and RAD3 gene products. *Genetics* **123**, 683–694.
27. Malagon, F., and Aguilera, A. (2001) Yeast spt6-140 mutation, affecting chromatin and transcription, preferentially increases recombination in which Rad51p-mediated strand exchange is dispensable. *Genetics* **158**, 597–611.
28. Gonzalez-Barrera, S., Garcia-Rubio, M., and Aguilera, A. (2002) Transcription and double-strand breaks induce similar mitotic recombination events in *Saccharomyces cerevisiae*. *Genetics* **162**, 603–614.
29. Fasullo, M.T., and Davis, R.W. (1987) Recombinational substrates designed to study recombination between unique and repetitive sequences in vivo. *Proc Natl Acad Sci USA* **84**, 6215–6219.
30. Fasullo, M., Bennett, T., and Dave, P. (1999) Expression of *Saccharomyces cerevisiae* MATa and MAT alpha enhances the HO endonuclease-stimulation of chromosomal rearrangements directed by his3 recombinational substrates. *Mutat Res* **433**, 33–44.
31. Cortes-Ledesma, F., and Aguilera, A. (2006) Double-strand breaks arising by replication through a nick are repaired by cohesin-dependent sister-chromatid exchange. *EMBO Rep* 7, 919–926.
32. Holmes, K.L., Otten, G., and Yokoyama, W.M. (2002) Flow cytometry analysis using the Becton Dickinson FACS Calibur. *Curr Protoc Immunol* **5**, 54.
33. Lea, D.E., and Coulson, C.A. (1948) The distribution of the number of mutants in bacterial population. *J Genet* **49**, 264–284.
34. Schmidt, K.H., Pennaneach, V., Putnam, C.D., and Kolodner, R.D. (2006) Analysis of gross-chromosomal rearrangements in *Saccharomyces cerevisiae*. *Meth Enzymol* **409**, 462–476.

Chapter 11

In Vivo Site-Specific Mutagenesis and Gene Collage Using the Delitto Perfetto System in Yeast *Saccharomyces cerevisiae*

Samantha Stuckey, Kuntal Mukherjee, and Francesca Storici

Abstract

Delitto perfetto is a site-specific in vivo mutagenesis system that has been developed to generate changes at will in the genome of the yeast *Saccharomyces cerevisiae*. Using this technique, it is possible to rapidly and efficiently engineer yeast strains without requiring several intermediate steps as it functions in only two steps, both of which rely on homologous recombination to drive the changes to the target DNA region. The first step involves the insertion of a cassette containing two markers at or near the locus to be altered. The second step involves complete removal of this cassette with oligonucleotides and/or other genetic material and transfer of the expected genetic modification(s) to the chosen DNA locus. Here we provide a detailed protocol of the *delitto perfetto* approach and present examples of the most common and useful applications for in vivo mutagenesis to generate base substitutions, deletions, insertions, as well as for precise in vivo assembly and integration of multiple genetic elements, or gene collage.

Key words: DNA modification, DNA oligonucleotides, site-directed mutagenesis, gene targeting, *delitto perfetto* system double-strand break, yeast *Saccharomyces cerevisiae*, gene collage.

1. Introduction

The yeast *Saccharomyces cerevisiae* is the most well-characterized eukaryotic organism as it has been long utilized for brewing and baking as well as being very easy to grow and manipulate in the laboratory (1). *Saccharomyces cerevisiae* was the first eukaryote to have the complete genome sequenced (2), and the genome sequencing project led to the discovery of many new yeast genes with unknown function (3, 4). Moreover, as an "honorary mammal," *S. cerevisiae* has a large number of genes

H. Tsubouchi (ed.), *DNA Recombination*, Methods in Molecular Biology 745,
DOI 10.1007/978-1-61779-129-1_11, © Springer Science+Business Media, LLC 2011

that are homologs of mammalian and human genes (5). Thus, functional analysis studies in the yeast model organism shed light on the roles of the corresponding genes in humans and in many other higher eukaryotes. Beyond the simplest experiments of gene disruption or gene knockout, where the original sequence of a gene is replaced with that of a genetic marker (6), site-specific mutagenesis of the genes of interest is the most powerful approach of reverse genetics to reveal what phenotypes arise as a result of the presence of particular genes and to generate novel variants of the genes. Thus, the possibility to generate specific point mutations or localized random changes at will, directly in vivo in the DNA locus of choice without leaving behind any marker or other heterologous DNA sequence, provides the opportunity to better understand and modify the role of a given genetic element, or the structure and function of a particular protein. Without leaving any trace, as in the "perfect murder," the *delitto perfetto* (Italian for perfect murder) approach to in vivo mutagenesis utilizes simple, precise, and highly efficient tools for engineering the genome of yeast cells with the desired modifications (7, 8). Exploiting the tremendous capacity of *S. cerevisiae* to perform efficient homologous recombination even when very short regions of homology are involved (30–50 bp) (6), synthetic oligonucleotides represent the most versatile and high-throughput device for genome engineering in a homology-driven manner (8). Moreover, taking advantage of the fact that a double-strand break (DSB) stimulates homologous recombination 1,000–10,000-fold, using the break-mediated *delitto perfetto* system, it is possible to simultaneously generate multiple different mutants or perform more sophisticated genetic rearrangements that would otherwise be too rare to be detected (9–11).

The first step of *delitto perfetto* involves the insertion of a COunterselectable REporter (CORE) cassette containing two markers. Prior to initiating this step, the researcher must decide which CORE cassette to use, taking into account the background of the strain (*See* **Note 1**) to be mutagenized, the markers currently present within this strain, and the kind of mutation(s) desired. Seven CORE plasmids have been created (**Fig. 11.1**), including those for a non-break system and a break system, thereby providing the researcher various choices to utilize this technique. Amplification of the chosen CORE cassette from its respective plasmid by polymerase chain reaction (PCR) is accomplished using primers which also contain 50-nucleotide (nt) tails of homology to either side of the target site (**Table 11.1** and **Fig. 11.2a**) to drive the integration of the CORE to its desired location (**Fig. 11.2b**) in the first step of *delitto perfetto*. The second step involves replacement of the entire cassette with oligonucleotides or larger pieces of DNA to yield the expected modification to the

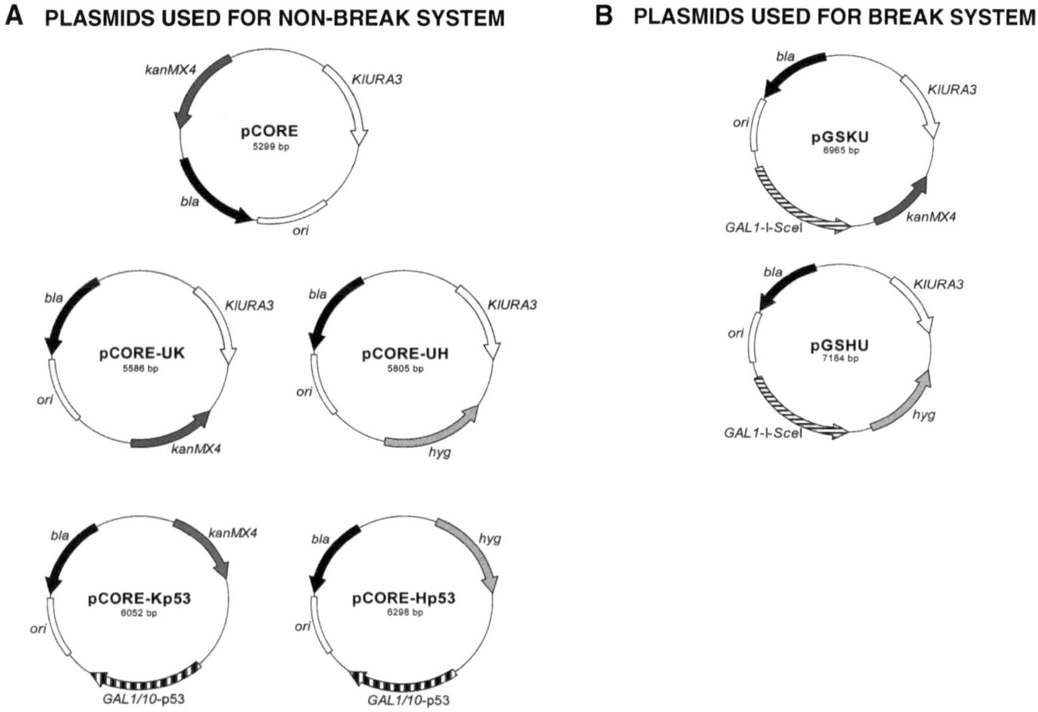

Fig. 11.1. The CORE plasmids used in the *delitto perfetto* technique. Each of the five plasmids used in the non-break system (**a**) contains a counterselectable marker, either *KlURA3* from *Kluyveromyces lactis* or a mutant form (V122A) of the human p53 cDNA, and a reporter marker, either *kanMX4* conveying resistance to Geneticin (G418) or *hyg* for resistance to the antibiotic hygromycin B. In addition to these markers, the two plasmids used in the break system (**b**) contain the inducible *GAL1* promoter and I-*Sce*I gene used to express the I-*Sce*I endonuclease and generate a DSB at the I-*Sce*I site. The origin of replication (*ori*) for all CORE plasmids is indicated as well as the *bla* marker gene, which provides resistance to the β-lactam antibiotic ampicillin and is used for selection.

original segment of chromosomal DNA (**Fig. 11.2c**). The generation of a DSB next to the CORE in the break system enhances the efficiency of targeting more than 1,000-fold (9–11), expanding the applications of the mutagenesis system. From beginning to end, *delitto perfetto* yields the final strain in less than 2 weeks and has proven to be a very useful tool in molecular biology. Examples provided in this review illustrate many changes that can be created through removal of the CORE, such as point mutations, random mutations, deletions, insertions ranging from a few nucleotides to fragments several kilobases in size, and in vivo gene collage.

2. Materials

2.1. Amplification of CORE

1. Seven CORE plasmids are available (*see* **Fig. 11.1**).

Table 11.1
Primers for CORE Cassette Amplification and Verification of CORE Cassette Insertion

Plasmid[a]	Primers to amplify CORE[b]		Cassette[c]	Markers[c]	Primers for testing cassette insertion[d]	
pCORE	P.1	5′ -... GAGCTCGTTTTCGACACTGG - 3′	CORE	*kanMX4*	**K2**	5′ - AGTCGTCACTCATGGTGATT - 3′
	P.2	5′ -... TCCTTACCATTAAGTTGATC - 3′	3.2 kb	*KlURA3*	**URA3.2**	5′ - AGACGACAAAGGCGATGCAT - 3′
pCORE-UK	P.I	5′ -... TTCGTACGCTGCAGGTCGAC - 3′	CORE-UK	*KlURA3*	**URA3.1**	5′ - TTCAATAGCTCATCAGTCGA - 3′
	P.II	5′ -... CCGCGCGTTGGCCGATTCAT - 3′	3.2 kb	*kanMX4*	**K1**	5′ - TACAATCGATAGATTGTCGCAC - 3′
pCORE-UH	P.I	5′ -... TTCGTACGCTGCAGGTCGAC - 3′	CORE-UH	*KlURA3*	**URA3.1**	5′ - TTCAATAGCTCATCAGTCGA - 3′
	P.II	5′ -... CCGCGCGTTGGCCGATTCAT - 3′	3.5 kb	*hyg*	**H1**	5′ - CCATGGCCTCCGCGACCGGCTGC - 3′
pCORE-Kp53	P.I	5′ -... TTCGTACGCTGCAGGTCGAC - 3′	CORE-Kp53	*kanMX4*	**K1**	5′ - TACAATCGATAGATTGTCGCAC - 3′
	P.II	5′ -... CCGCGCGTTGGCCGATTCAT - 3′	3.7 kb	*GAL1/10*-p53	**p53.2**	5′ - GACTGTACCACCATCCACT - 3′
pCORE-Hp53	P.I	5′ -... TTCGTACGCTGCAGGTCGAC - 3′	CORE-Hp53	*hyg*	**H1**	5′ - CCATGGCCTCCGCGACCGGCTGC - 3′
	P.II	5′ -... CCGCGCGTTGGCCGATTCAT - 3′	4.0 kb	*GAL1/10*-p53	**p53.2**	5′ - GACTGTACCACCATCCACT - 3′
pGSKU	P.I	5′ -... TTCGTACGCTGCAGGTCGAC - 3′	GSKU	*KlURA3* *kanMX4*	**URA3.1**	5′ - TTCAATAGCTCATCAGTCGA - 3′
	P.IIS	5′... **TAGGGATAACAGGGTAAT** CCGCGCGTTGGCCGATTCAT - 3′	4.6 kb	*GAL1*-I-*Sce*I	**Gal.E**	5′ - CTAAGATAATGGGGCTCTTT - 3′
pGSHU	P.I	5′ -... TTCGTACGCTGCAGGTCGAC - 3′	GSHU	*KlURA3* *hyg*	**URA3.1**	5′ - TTCAATAGCTCATCAGTCGA - 3′
	P.IIS	5′... **TAGGGATAACAGGGTAAT** CCGCGCGTTGGCCGATTCAT - 3′	4.8 kb	*GAL1*-I-*Sce*I	**Gal.E**	5′ - CTAAGATAATGGGGCTCTTT - 3′

[a]There are seven CORE plasmids available.

[b]Amplification of the plasmids by PCR can be accomplished by creating primers with the above-listed sequences that are internal to the CORE and an external region homologous to the region in which the CORE will be inserted. The primers used to amplify the cassettes from pGSKU and pGSHU require the addition of the 18-nt I-*Sce*I recognition site (**bold**) next to the *GAL1* promoter.

[c]The sizes and composition of the cassettes vary depending on the markers present.

[d]Primers and their sequences used for verification of CORE integration and replacement are provided.

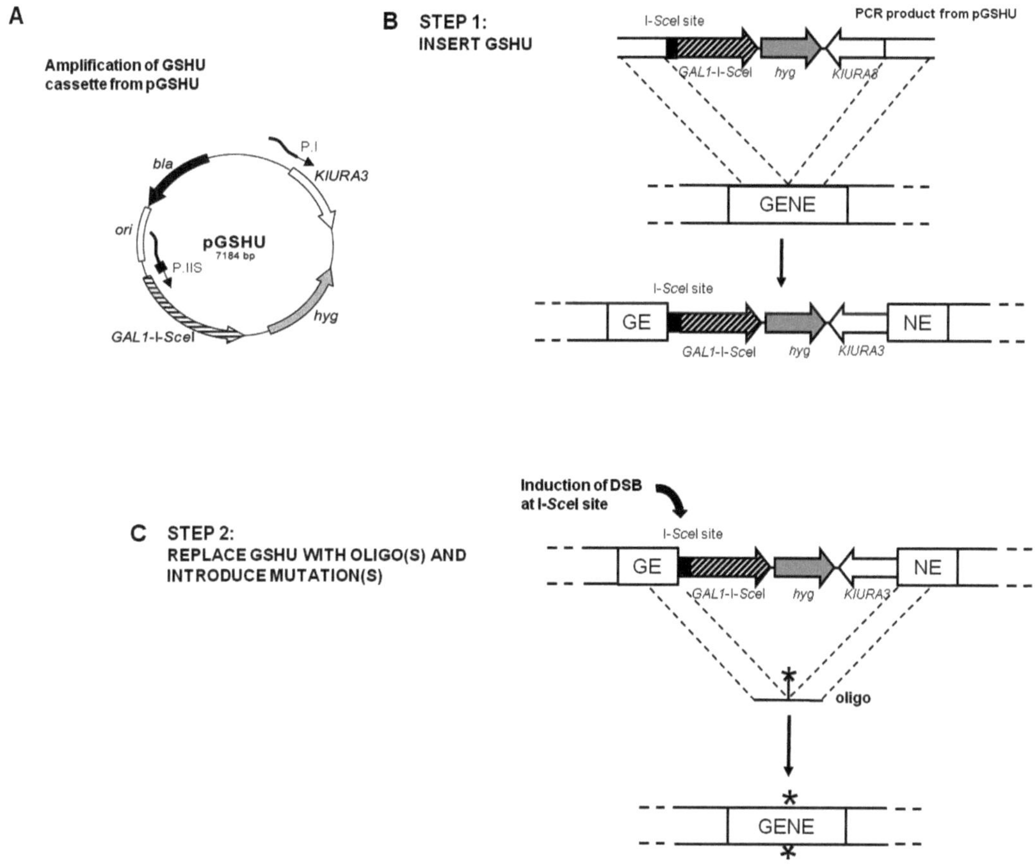

Fig. 11.2. The two-step process of *delitto perfetto*. (**a**) Step one involves the amplification of a CORE cassette by PCR (portions of primers used for amplification indicated by *thinner line* and *arrow*). (**b**) The primers create tails of homology to either side of the target region (indicated by *thicker line*) for integration into the genome using the cell's homologous recombination machinery. In this example, the use of the break system CORE cassette GSHU is illustrated. Note that the primer amplifying from the *GAL1*-I-*Sce*I side of the cassette introduces the 18-nt I-*Sce*I recognition site (*black box*). This site is utilized in the second step (**c**) when the I-*Sce*I endonuclease expression is turned on with galactose to generate a DSB prior to replacement of the CORE with an oligonucleotide sequence, which introduces the desired mutation. This example uses a single-stranded oligonucleotide to enact this change; however, a pair of complementary oligonucleotides have been shown to increase the efficiency of gene targeting.

2. DNA primers (Invitrogen, Carlsbad, CA or Alpha DNA, Montreal, Quebec, Canada), desalted and non-purified: 50 pmol/μl. Store at –20°C.
3. Ex Taq DNA polymerase, 10× buffer, 2.5 mM dNTPs (Clontech, Mountain View, CA).

2.2. Gel Electrophoresis

1. Agarose (Fisher, Pittsburgh, PA).
2. TBE running buffer (10×) (Fisher).
3. Prestained molecular weight marker (New England Biolabs, Ipswich, MA).
4. Loading dye (Fisher).

2.3. PCR Product Concentration

1. Ethanol (EtOH): 95 and 70% concentrations.
2. Sodium acetate (NaOAc; Sigma, St. Louis, MO): 3 M (pH 5.2) stock solution, filter sterilized. Store at room temperature (*see* **Note 2**).

2.4. Transformation Reagents and Media

1. YPD (per 1 l): 10 g yeast extract, 20 g soy peptone, 20 g dextrose (Difco/BD, Franklin Lakes, NJ). For solid media, add 20 g agar (Difco/BD) (*see* **Note 3**).
2. YPLac liquid (per 1 l): 10 g yeast extract, 20 g soy peptone, 27 ml lactic acid (Difco/BD), pH adjusted to 5.5 with lactic acid (Fisher).
3. Stock solution of 20% high-pure galactose (Sigma) is filter sterilized and stored at room temperature.
4. Lithium acetate (LiOAc; Sigma): Stock of 1 M concentration. Filter sterilize. Store at room temperature.
5. TE 10× stock solution: 100 mM Tris (Fisher) (pH 7.5), 10 mM ethylenediaminetetraacetic acid (EDTA; Sigma) (pH 7.5). Filter sterilize. Store at room temperature.
6. Polyethylene glycol 4000 (PEG 4000; Sigma): 50% stock solution. Store at room temperature (*see* **Note 4**).
7. Working solutions: Solution 1 (0.1 M LiOAc, TE 1×, pH 7.5) and solution 2 (0.1 M LiOAc, TE 1×, pH 7.5 in 50% PEG 4000).
8. Solution of salmon sperm DNA (SSD, Roche, Basel, Switzerland), 100 μg/ml. Store at –20°C.
9. SC-Ura (synthetic complete media lacking uracil) solid media (Fisher).
10. Glass beads, approx. 5 mm diameter (Fisher).
11. 5-Fluoroorotic acid (5-FOA; per 1 l): Solution of 5-FOA is prepared by dissolving 1 g 5-FOA (US Biological, Swampscott, MA) in 300 ml of water prior to filter sterilization. 700 ml SD-complete (synthetic dextrose-complete) agar media is autoclaved, then cooled to 55–60°C, and the filtered solution of 5-FOA is then mixed with media prior to pouring.
12. G418 (per 1 l): YPD agar media is autoclaved, then cooled to 55–60°C, and G418 solution (200 μg/ml; US Biological) is then mixed with media prior to pouring. Stock solution is prepared in 50 mg/ml filter-sterilized aliquots and stored at 4°C.
13. Hygromycin B (Hygro; per 1 l): YPD agar media is autoclaved, then cooled to 55–60°C, and Hygro solution (300 μg/ml; Invitrogen) is then mixed with media prior to pouring.

14. YPG (per 1 l): 10 g yeast extract, 20 g soy peptone, 30 ml glycerol (Difco/BD), 20 g agar.
15. Sterile velveteens (Fisher).

2.5. Genotypic Testing of Transformants

1. Lyticase (Sigma) is dissolved at 2,000 U/ml and stored in 1 ml aliquots at –20°C.
2. Taq DNA polymerase, 10× buffer, 10 mM dNTPs (Roche).

2.6. Design of DNA Oligonucleotides for Removal of CORE and Generation of Mutations

1. DNA oligonucleotides (Invitrogen or Alpha DNA): 50–100mers, desalted and non-purified (50 pmol/μl). Store at –20°C.

3. Methods

Despite the efficiency of recombination when a DSB is induced, induction of a DSB may not be required depending on the strain being mutagenized and the type of modification. The DSB system is preferred when multiple mutations are desired simultaneously; when the modification involves gross deletions, insertions, gene fusions, or other genomic rearrangements (11); and when the strain is deficient in homologous recombination functions (10).

Several combinations of two markers can be used for the *delitto perfetto* technique and are contained within the various CORE cassettes on plasmids (**Fig. 11.1**). The two CORE markers are used for selection purposes and consist of the following: an antibiotic resistance marker (REporter) – which confers resistance to the antibiotics hygromycin B or Geneticin (G418) – and a COunterselectable marker, either the *KlURA3* gene (a *URA3* homolog from *Kluyveromyces lactis*), which can be selected against using 5-FOA, or a marker coding for the human p53 mutant V122A, which is toxic to yeast when overexpressed and can be selected against using a galactose-containing media. In addition, the break system cassettes include the inducible *GAL1* promoter and the gene, used to induce the DSB at the 18-nt I-*Sce*I break site.

In the first step of *delitto perfetto*, the CORE is amplified through PCR to attach the tails of homology to the desired chromosomal locus (**Fig. 11.2a**) and its PCR product is inserted into the cells by transformation (**Fig. 11.2b**). The CORE cassette will then integrate at the desired genomic locus in approx. $1/10^6$ yeast cells via homologous recombination. Following transformation, the transformant colonies are isolated to observe for insertion of the CORE through phenotypic and genotypic testing. The second step of this technique is a transformation using oligonucleotides or other DNA to remove the entire CORE cassette and introduce the desired mutation(s) (**11.2c**). *See* **Section 3.6** for details on oligonucleotide design to remove the CORE.

3.1. Amplification of CORE from Plasmid

1. DNA primers will first be used to amplify the CORE from the chosen plasmid. These primers range from 70 to 100 nt in length with an overlap of at least 50 nt with the genomic targeting region and an overlap of 20 nt with the CORE cassette sequence (**Table 11.1**). Additionally, in the break-induced system, the 18-nt recognition sequence for the I-*Sce*I endonuclease is included in one of the two primers (**Table 11.1**).
2. PCR conditions: Amplification of the CORE cassette from circular plasmid (about 50 ng) using 50 pmol/μl of each primer is performed with high yield in a final volume of 40 μl using Ex Taq DNA polymerase with a 2 min cycle at 94°C; 32 cycles of 30 s at 94°C, 30 s at 57°C, and 4 min at 72°C (or 5 min at 72°C for cassettes over 4 kb in size); a final extension time of 7 min at 72°C; and samples are held at 4°C. Ex Taq DNA polymerase consistently produces a higher yield of CORE cassette amplification than does Taq DNA polymerase. dNTPs (10 mM) are used for this reaction. An extension time of 1 min/kb is assumed for this reaction.
3. Following PCR, the samples are ready for gel electrophoresis and PCR product concentration.

3.2. Gel Electrophoresis

1. We use a dilution of 0.5× TBE running buffer, which is obtained from 10× TBE by mixing 50 ml of 10× TBE buffer with 950 ml deionized water prior to use.
2. A small aliquot (about 2 μl) of PCR product is run on a 0.8% agarose gel to observe anticipated band.

3.3. PCR Product Concentration

1. The product of six reactions of PCR is combined for precipitation with a 2.5× volume of 95% EtOH and 1/10 3 M NaOAc (pH 5.2) in a microcentrifuge tube. Centrifugation is carried out at maximum speed for 10 min. A small pellet should be visible on the bottom of the tube.
2. The supernatant is discarded and the pellet is washed with 100 μl of 70% EtOH, paying attention not to detach the pellet. If the pellet is detached, it is necessary to spin again for 5 min and then discard the supernatant. Then, as much EtOH as possible is removed without detaching the pellet.
3. The pellet is then dried in a speed vac for about 15 min and resuspended in 50 μl of water. Five to 10 μl are used for each transformation.

3.4. Step 1: Transformation to Insert the CORE

The following transformation protocol is used to first insert the CORE PCR product into the strain of choice and then to drive replacement of the CORE with DNA oligonucleotides or other segments of DNA. This transformation procedure has been modified from the lithium acetate protocol described by Wach et al. (6). During the transformation, the LiOAc acts to make the cell wall permeable. The presence of PEG 4000 is used to adhere the DNA to the cells such that the proximity allows for entry into the cells. When transforming to insert the CORE PCR product, SSD is used as carrier DNA and serves as a buffer between the targeting DNA from the PCR and any DNA degradation factors present within the cell. In the second transformation using oligonucleotides to remove the CORE, the use of SSD is unnecessary as the oligonucleotides at the concentration of 1 nmol/20 μl act as carrier DNA themselves:

1. Inoculate 5 ml of YPD liquid medium with chosen strain and shake at 30°C overnight (O/N) (*see* **Note 5**).
2. Inoculate 50 ml of YPD liquid medium with 1.5 ml of the O/N culture in a 250-ml glass flask and shake vigorously at 30°C for 3 h.
3. Solutions 1 and 2 are prepared immediately prior to transformation.
4. Transfer culture to a 50-ml conical tube and spin at 1,562×*g* for 2 min.
5. Remove the supernatant and wash cells with 50 ml of sterile water and spin as stated previously.
6. Remove the supernatant and resuspend cells in 5 ml of solution 1 and spin as stated previously.
7. Remove the supernatant and resuspend cells in 250 μl of solution 1. This amount of cells is sufficient for approx. 7–8 transformations.
8. Aliquot 50 μl of the cell suspension in microcentrifuge tubes and add 5–10 μl of concentrated CORE PCR product and 5 μl of SSD (heat-denatured for 5 min at 100°C prior to use and immediately kept on ice), then gently mix by tapping the tube.
9. Add 300 μl of solution 2 for each transformation reaction. Mix briefly by vortexing.
10. Incubate transformation reactions at 30°C for 30 min with shaking.
11. Heat shock at 42°C for 15 min to drive the DNA into the cells.
12. Collect cells by centrifugation at 2,236×*g* for 4 min.
13. Remove the supernatant and resuspend cells well in 100 μl of water.

14. Plate all cells from each transformation tube on one SC-Ura plate using approx. 8–12 sterile glass beads and incubate at 30°C for 2–3 days (*see* **Note 6**).
15. Using sterile velveteen, replica-plate from SC-Ura to G418- or Hygro-containing media (depending on the CORE used) and incubate at 30°C O/N.
16. Once transformants are observed (typically 5–30 colonies per plate), streak for single-colony isolates on YPD solid media. Incubate at 30°C for 2 days.
17. Make patches of the single colonies on new YPD solid media, along with the original strain, and incubate at 30°C O/N.
18. Replica-plate the grown patches to YPD, SC-Ura, G418, Hygro, YPG to select against petite cells, and any other various selective media depending on the background of your strain, and incubate at 30°C O/N.
19. Following observation of correct phenotype, the samples are ready for genotypic testing.

3.5. Colony PCR of Transformants

1. Resuspend cells (approx. 1 mm^3) in 50 μl water containing 1 U of lyticase. Incubate at room temperature for 10 min, followed by incubation in a heat block at 100°C for 5 min.
2. PCR conditions: Colony PCR of the transformant patches presenting the expected phenotypes using 10 μl of the cell resuspension solution is accomplished with 50 pmol of each primer, with an expected amplification region between 300 bp and 1.2 kb (*see* **Fig. 11.3**). dNTPs (10 mM) are

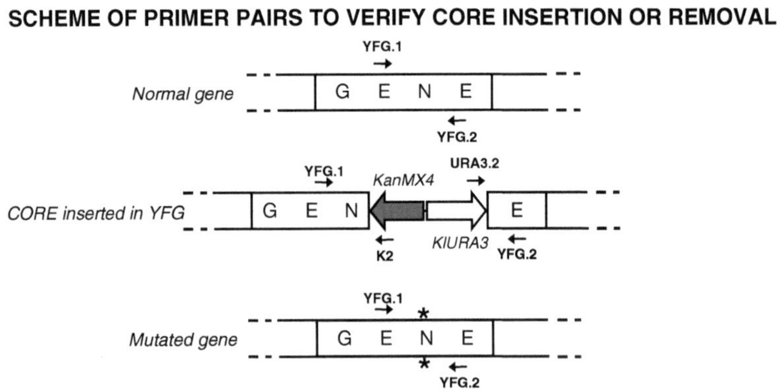

Fig. 11.3. Scheme of primer pairs used for colony PCR. Primers should be designed to allow for amplification of the target region in addition to being paired with primers internal to the CORE. The sizes of colony PCR products should range between 300 bp and 1.2 kb. Using this approach, verification of the CORE's integration as well as its replacement can be made. *See* **Table 11.1** for a list of primers and their sequences used to verify the integration of the various CORE cassettes.

used for this reaction. PCR is performed in a final volume of 50 μl using Taq DNA polymerase (Roche) with a 2 min cycle at 95°C; 32 cycles of 30 s at 95°C, 30 s at 55°C, and 1 min at 72°C; a final extension time of 7 min at 72°C; and samples are held at 4°C. An extension time of 1 min/kb is assumed for this reaction (*see* **Note 7**).

3. Following PCR, samples are run on a 1% agarose gel (*See* **Section 3.2**) for observation of PCR product.
4. Strains are now ready for step 2 to remove the CORE.

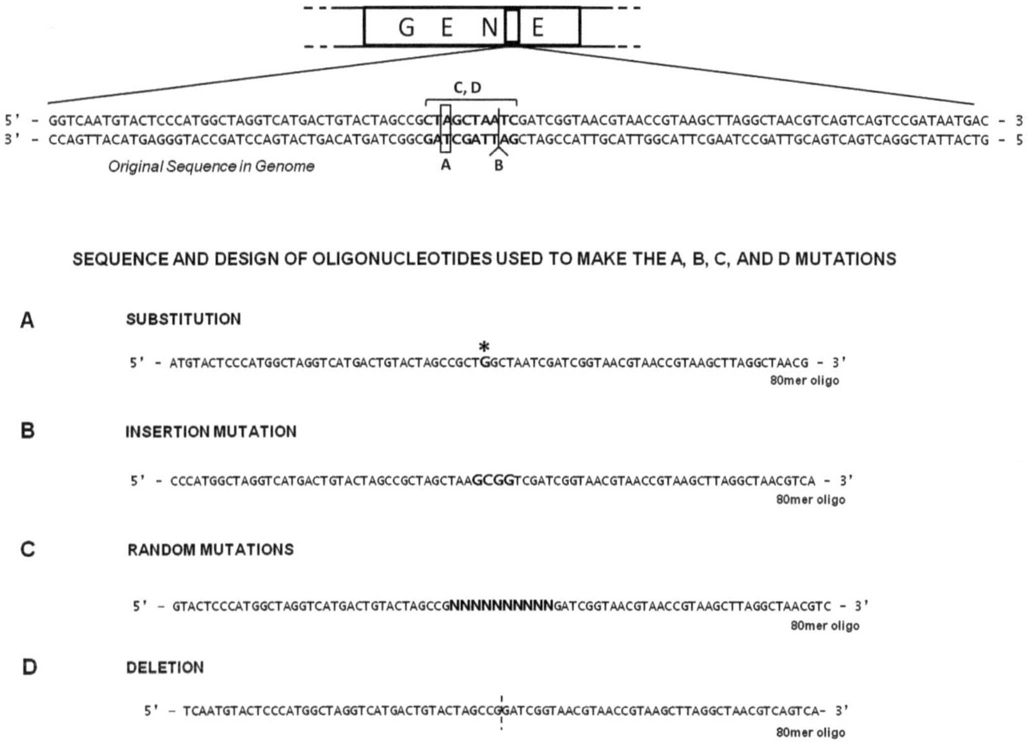

Fig. 11.4. Examples of single oligonucleotide-driven mutations generated using the *delitto perfetto* technique. When a substitution or an insertion mutation is desired (*A*, *B*), the CORE should be placed next to the target region prior to replacement with a single or complementary oligonucleotide(s). In this example, the original sequence in the genome is provided as a reference at the top of the figure. (*A*) A substitution of a guanine, marked by an *asterisk* above the bolded G on the oligonucleotide, is made in place of the adenine residue on the top strand of the reference sequence (*boxed*). (*B*) An insertion mutation in the original sequence is created through the use of an oligonucleotide containing additional nucleotides (GCGG, marked in *bold*) which are inserted between the adenine and thymine indicated in the reference. When random mutations or small deletions (<5 kb) are desired (*C, D*) in a specific region, it is preferred to delete the region of interest along with the CORE insertion, as the successive targeting event with the oligonucleotides or other DNA will then eliminate the CORE and introduce the desired changes. The region of mutagenesis in the original sequence is bolded and indicated by the *bracket*. (*C*) For the generation of random mutations, the oligonucleotide sequence contains a stretch of 10 bolded Ns, which indicate that any of the four DNA bases can be used when the oligonucleotides are synthesized. The exact degree of this randomness is determined by the investigator. (*D*) The segment of the reference sequence indicated by the *bracket* can be removed through the use of oligonucleotides missing this fragment. In the example, the location of the deleted nucleotides is indicated by the *dashed line*.

3.6. Design of DNA Oligonucleotides for Removal of CORE and Generation of Mutations

Numerous mutations can be accomplished through the use of the *delitto perfetto* technique. These include substitutions, insertions, the generation of random mutations through the use of degenerate oligonucleotides, and deletions. **Figure 11.4** illustrates the sequence of oligonucleotides (A–D) needed to produce all of these mutations at the genomic locus indicated in the figure. When substitutions or insertion mutations are desired, the location of the CORE insertion should be next to the region of modification. Conversely, the CORE should replace the entire targeted region when a small deletion or a random mutation is desired. If a large deletion is desired, a CORE with the break system is inserted within the region to be removed (11).

To remove the CORE and generate the mutation with DNA oligonucleotides, the following considerations should be made. The use of a single oligonucleotide is sufficient; however, a pair of complementary oligonucleotides increases the frequency of integration 5–10-fold (11). Additionally, while shorter oligonucleotides ($\approx$40 nt) can be used to effectively transform the strain, longer oligonucleotides approaching lengths of 80 nt are more favorable as they increase the efficiency of targeting as well as the window of mutagenesis (11). The external 30–40 nt of the oligonucleotide or oligonucleotide pair are used for efficient homologous recombination to introduce the desired mutation and allow for loss of the CORE. It is of note that once the CORE cassette has been integrated in a specific chromosomal locus, many gene variants can be generated by transforming the cells with oligonucleotides designed to produce different alterations.

3.7. Step 2: Transformation Using DNA Oligonucleotides in Non-break System

1. Inoculate 5 ml of YPD liquid medium with chosen strain and shake at 30°C (O/N).
2. Inoculate 50 ml of YPD liquid medium with 1.5 ml of the O/N culture in a 250-ml glass flask and shake vigorously at 30°C for 3 h.
3. Solutions 1 and 2 are prepared immediately prior to transformation.
4. Transfer culture to a 50-ml conical tube and spin at 1,562×*g* for 2 min.
5. Remove the supernatant and wash cells with 50 ml of sterile water and spin as stated previously.
6. Remove the supernatant and resuspend cells in 5 ml of solution 1 and spin as stated previously.
7. Remove the supernatant and resuspend cells in 250 μl of solution 1. This amount of cells is sufficient for approx. 7–8 transformations.
8. Aliquot 50 μl of the cell suspension in microcentrifuge tubes and add 1 nmole of DNA oligonucleotides

(heat-denatured for 2 min at 100°C, then immediately kept on ice prior to use). When using a single oligonucleotide, a 20 μl volume (at 50 pmol/μl) is used or when using a complementary DNA oligonucleotide pair, 10 μl of each is used. Gently mix the tube by tapping.

9. Add 300 μl of solution 2 for each transformation reaction. Mix briefly by vortexing.
10. Incubate transformation reactions at 30°C for 30 min with shaking.
11. Heat shock at 42°C for 15 min to drive the DNA into the cells.
12. Collect cells by centrifugation at 2,236×*g* for 4 min.
13. Remove the supernatant and resuspend cells well in 100 μl of water.
14. Plate cells from each transformation tube on one YPD solid plate using approx. 8–12 sterile glass beads and incubate at 30°C O/N.
15. Using sterile velveteen, replica-plate from YPD to 5-FOA and incubate at 30°C for 2 days. If necessary, replica-plate again on 5-FOA media to allow for growth of Ura^- colonies clearly distinct from the background (*see* **Note 8**).
16. Using sterile velveteen, replica-plate from 5-FOA to YPD and G418- or Hygro-containing media (depending on the CORE used) and incubate at 30°C O/N.
17. Mark G418-sensitive or Hygro-sensitive colonies on the YPD media and streak for single colonies on new YPD solid media. Incubate at 30°C for 2 days.
18. Make patches of the single colonies on new YPD solid media, along with the original strain, and incubate at 30°C O/N.
19. Replica-plate patches to YPD; SC-Ura; G418; Hygro; YPG, which selects against cells with defective mtDNA; and any other various selective media depending on the background of your strain and incubate at 30°C O/N.
20. Following observation of correct phenotype, the samples are ready for genotypic testing (*see* **Section 3.5**).
21. PCR samples containing the mutagenized region are now ready for DNA purification and sequencing analysis (*see* **Note 9**).

3.8. Step 2: Transformation Using DNA Oligonucleotides in Break System

1. Inoculate 50 ml of YPLac liquid medium with chosen strain in a 250-ml glass flask and shake at 30°C (O/N) (*see* **Note 10**).

2. Add 5 ml of galactose from a 20% solution into the O/N culture to obtain a 2% galactose solution and shake vigorously at 30°C for 3–6 h (*see* **Note 11**).
3. Solutions 1 and 2 are prepared immediately prior to transformation.
4. Transfer culture to a 50-ml conical tube and spin at 1,562 × *g* for 2 min.
5. Remove the supernatant and wash cells with 50 ml of sterile water and spin as stated previously.
6. Remove the supernatant and resuspend cells in 5 ml of solution 1 and spin as stated previously.
7. Remove the supernatant and resuspend cells in 250 μl of solution 1. This amount of cells is sufficient for approx. 7–8 transformations.
8. Aliquot 50 μl of the cell suspension in microcentrifuge tubes and add 1 nmole of DNA oligonucleotides (heat-denatured for 2 min at 100°C, then immediately kept on ice prior to use). When using a single oligonucleotide, a 20 μl volume (at 50 pmol/μl) is used or when using a complementary DNA oligonucleotide pair, 10 μl of each is used. Gently mix the tube by tapping.
9. Add 300 μl of solution 2 for each transformation reaction. Mix briefly by vortexing.
10. Incubate transformation reactions at 30°C for 30 min with shaking.
11. Heat shock at 42°C for 15 min to drive the DNA into the cells.
12. Collect cells by centrifugation at 2,236 × *g* for 4 min.
13. Remove the supernatant and resuspend cells well in 100 μl of water.
14. Plate cells from each transformation tube on one YPD solid plate using approx. 8–12 sterile glass beads and incubate at 30°C O/N. Dilutions may be necessary prior to plating due to the efficiency of oligonucleotide recombination following DSB induction.
15. Using sterile velveteen, replica-plate from YPD to 5-FOA and incubate at 30°C for 2 days. If necessary, replica-plate again on 5-FOA media to allow for growth of Ura^- colonies clearly distinct from the background (*see* **Note 8**).
16. Using sterile velveteen, replica-plate from 5-FOA to YPD and G418- or Hygro-containing media (depending on the CORE used) and incubate at 30°C O/N.

17. Mark G418-sensitive or Hygro-sensitive colonies on the YPD media and streak for single colonies on new YPD solid media. Incubate at 30°C for 2 days.
18. Make patches of the single colonies on new YPD solid media, along with the original strain, and incubate at 30°C O/N.
19. Replica-plate patches to YPD; SC-Ura; G418; Hygro; YPG, which selects against cells with defective mtDNA; and any other various selective media depending on the background of your strain and incubate at 30°C O/N.
20. Following observation of correct phenotype, the samples are ready for genotypic testing (*see* **Section 3.5**).

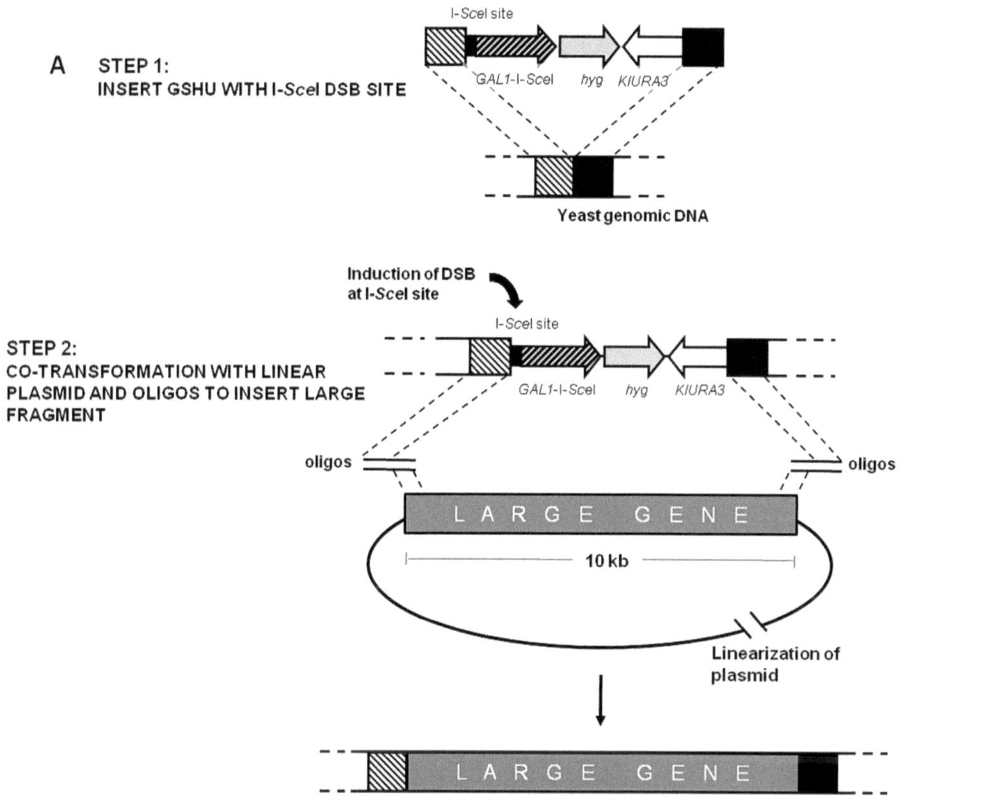

Fig. 11.5. Insertion of a large segment of DNA from a plasmid. In this example, the break system is used to drive the integration of a 10-kb fragment from a plasmid into the target region. (**a**) First, the GSHU CORE cassette and the 18-nt I-*SceI* break site are inserted into the target region. (**b**) Next, the plasmid carrying the sequence of interest to be integrated within the genome is linearized by restriction digestion outside of the fragment to be inserted. Complementary pairs of oligonucleotides have regions of homology to both the upstream and downstream portions of the sequence of interest to be integrated, as well as to either side of the target region. The linearized plasmid and oligonucleotides are co-transformed into yeast cells following DSB induction at the I-*SceI* site. By homologous recombination, the large sequence of interest is integrated into the genomic DNA at the specific site without the need for PCR amplification, which otherwise increases the likelihood of unwanted mutations during the polymerization process.

21. PCR samples containing the mutagenized region are now ready for DNA purification and sequencing analysis (*see* **Note 9**).

3.9. The Delitto Perfetto Approach to Insert a Large DNA Fragment

In **Fig. 11.5**, the two-step process shown illustrates the insertion of a large segment of DNA, 10 kb in size. Generally, an insert of these proportions is obtained through amplification of the sequence through PCR, which, although possible, greatly increases the risk of introducing several mutations through the extension process. In our system, the large DNA of interest is carried on a plasmid which is linearized prior to transformation. Linearization of the plasmid is required to generate free DNA ends and stimulate homologous recombination. The large segment of plasmid DNA is integrated into genomic DNA at a chosen location by in vivo recombination following co-transformation of the linearized plasmid carrying the fragment and two pairs of complementary oligonucleotides. Each pair contains regions of

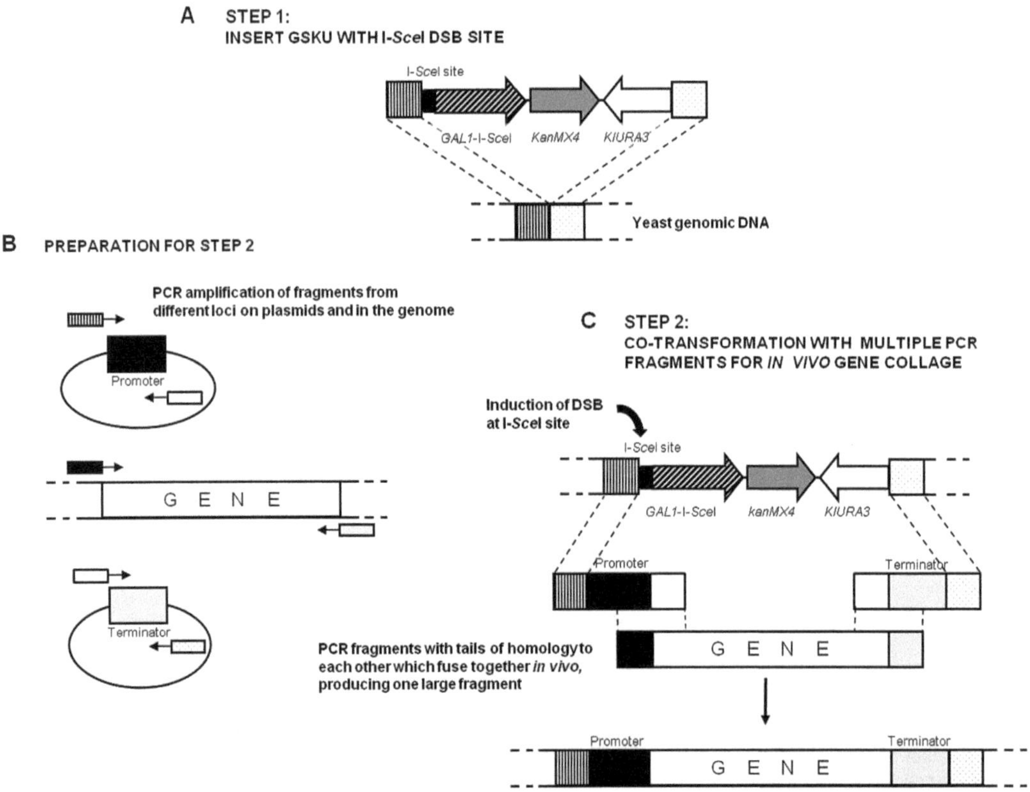

Fig. 11.6. Mechanism of in vivo gene collage by the *delitto perfetto* approach. (**a**) In step one, the GSKU CORE cassette is amplified through PCR, with the 18-nt I-*SceI* break site included within the sequence of one primer. This product is inserted into the target locus. (**b**) Prior to replacement of the cassette, a preparation step to generate PCR fragments is performed. For this example, a gene of interest will be attached to a chosen promoter and terminator sequences and all components will be inserted at a chosen locus. (**c**) In step two, the multiple PCR fragments assemble together in vivo by recombination to form a large fragment, which then replaces the GSKU cassette as it integrates into its specific region of the chromosome.

homology on either side of the target site in addition to homology with the 10-kb fragment, thereby directly driving it into its desired locus. This way, sequencing analysis is not required following integration of the large fragment.

3.10. The Delitto Perfetto Approach to Insert Multiple Sequences for Gene Collage

It is also possible to insert two or more sequences or genes next to each other simultaneously using *delitto perfetto*, as seen in **Fig. 11.6**. To accomplish this, the genes or segments of interest are amplified in such a way that the primers of each PCR fragment have tails of homology to the sequence of the contiguous segment and the most external primers contain homology to the target site. Through co-transformation with these multiple PCR products, the individual pieces recombine in vivo as a form of gene collage, while the outlying primers drive integration into the genome at the desired locus.

4. Notes

1. This review focuses on the generation of engineered haploid strains of yeast. For the use of *delitto perfetto* in diploid cells, refer to (11) for a detailed explanation of modifications to the protocol.
2. For media preparation, deionized water is used. All other uses of the term "water" in this chapter, however, refer to deionized water that was sterilized by filtration or autoclaving.
3. Unless otherwise noted, all solid media are to be stored at 4°C. Exceptions include YPD liquid and agar, which we store at room temperature.
4. PEG 4000 solution will be extremely viscous, so filter sterilizing can take up to 1 h depending on the volume. Autoclaving is an alternative means of sterilization for this solution.
5. Using 50-ml conical tubes for O/N growth is preferred to using 15-ml tubes, as the larger size allows for greater dispersion of the nutrients in the broth to each of the cells. Additionally, *S. cerevisiae* is an aerobic species, so lids should not be capped tightly but instead loosely cover the tube and secured with tape.
6. When inserting CORE, if growth on SC-Ura media is not observed after 3 days (*see* **Note 8** below), the transformation can be performed by plating onto YPD and incubating at 30°C O/N followed by replica-plating to G418- or Hygro-containing media, depending on the CORE used, and incubating at 30°C for 2–3 days until large colonies

appear. This would then be followed by replica-plating to SC-Ura and incubating at 30°C O/N.

7. As little as 5 μl of the cell resuspension solution can be used per reaction; however, this is not optimal when sequencing is necessary and a volume of 10 μl is suggested for this process.
8. The following is specific to using CORE-UK, CORE-UH, GSKU, and GSHU, as this does not apply to the other cassettes. When *KlURA3* is inserted in the same orientation as the targeted gene, interference from that gene's promoter during transcription may lead to delayed growth on SC-Ura in the first step of *delitto perfetto* (*see* **Note 6** above) and may increase the number of background cells on 5-FOA in the second step. Depending on insertion orientation of the CORE, a second round of replica-plating to 5-FOA may be needed. Therefore, it is optimal to insert the cassette in such a way that *KlURA3* is oriented opposite to the gene being targeted.
9. Upon successful colony PCR of transformants containing the newly introduced CORE sequence, sequencing analysis is not required. The resulting antibiotic resistance and Ura^+ phenotype of the strain in addition to the results of the colony PCR are sufficient to provide evidence for successful incorporation of the CORE into the targeted site. Sequencing is, however, necessary to verify the correct insertion of the desired mutation(s). Since the oligonucleotides used are non-purified, the expected additional mutations are in the range of 10–20%. Therefore, it is always better to obtain 3–5 clones for sequencing.
10. YPLac is used to provide a neutral carbon source for the cells prior to addition of galactose. However, cells grow much slower in this medium. It is, therefore, optimal to inoculate cells into YPLac at least 18–20 h prior to the transformation.
11. Addition of galactose activates the inducible *GAL1* promoter which regulates the I-*Sce*I gene. Experience has shown that longer induction (5–6 h) produces greater efficiency.

Acknowledgments

We thank the members of our lab for their contributions to the editing and revision of this work, notably Rekha Pai, Patrick Ruff, and Ying Shen. We also thank Lee Katz for assistance in proofreading and revision. This work was funded in part by the Georgia Cancer Coalition grant R9028 and the NIH R21EB9228.

References

1. Sherman, F. (2002) Getting started with yeast. *Methods Enzymol* **350**, 3–41.
2. Dujon, B. (1996) The yeast genome project: what did we learn? *Trends Genet* **12**, 263–270.
3. Oliver, S.G. (1996) From DNA sequence to biological function. *Nature* **379**, 653–654.
4. Winzeler, E.A., and Davis, R.W. (1997) Functional analysis of the yeast genome. *Curr Opin Genet Dev* 7, 771–776.
5. Resnick, M.A., and Cox, B.S. (2000) Yeast as an honorary mammal. *Mutat Res* **451**, 1–11.
6. Wach, A., Brachat, A., Pohlmann, R., and Philippsen, P. (1994) New heterologous modules for classical or PCR-based gene disruptions in *Saccharomyces cerevisiae*. *Yeast* **10**, 1793–1808.
7. Storici, F., Lewis, L.K., and Resnick, M.A. (2001) In vivo site-directed mutagenesis using oligonucleotides. *Nat Biotechnol* **19**, 773–776.
8. Storici, F., and Resnick, M.A. (2003) *Delitto perfetto* targeted mutagenesis in yeast with oligonucleotides. In Genetic engineering, principle and methods, Vol. 25 J.K. Setlow, ed. (Upton, NY: Kluwer Academic/Plenum Publisher), pp. 189–207.
9. Storici, F., Durham, C., Gordenin, D., and Resnick, M. (2003) Chromosomal site-specific double-strand breaks are efficiently targeted for repair by oligonucleotides in yeast. *Proc Natl Acad Sci* **100**, 14994–14999.
10. Storici, F., Snipe, J., Chan, G., Gordenin, G., and Resnick, M. (2006) Conservative repair of a chromosomal double-strand break by single-strand DNA through two steps of annealing. *Mol Cell Biol* **26**, 7645–7657.
11. Storici, F., and Resnick, M. (2006) *The delitto perfetto* approach to in vivo site-directed mutagenesis and chromosome rearrangements with synthetic oligonucleotides in yeast. *Methods Enzymol* **409**, 329–345.

Chapter 12

Detection of RNA-Templated Double-Strand Break Repair in Yeast

Ying Shen and Francesca Storici

Abstract

The discovery of RNA-templated DNA repair has revealed a novel case where genetic information can flow directly from RNA to genomic DNA without passing through a reverse transcript intermediate. As initially demonstrated in the yeast *Saccharomyces cerevisiae* via transformation by RNA-containing oligonucleotides (oligos), RNA sequences can serve as templates for chromosomal double-strand break (DSB) repair. Synthetic oligos containing embedded RNA tracts of various sizes, or even RNA-only molecules, although with lower efficiency, can guide DNA repair synthesis at sites of broken DNA. Mechanisms and circumstances in which cells can use RNA to repair DNA damage such as a DSB are yet to be identified. Here we show the approach we utilize to detect repair of a chromosomal DSB by RNA-containing oligos in yeast cells.

Key words: RNA-containing oligonucleotides, double-strand break (DSB) repair, transformation, yeast *Saccharomyces cerevisiae*, single-strand annealing.

1. Introduction

Among the many roles ascribed to RNA in cells, recently we showed that RNA can function as a template for repair of DNA damage and directly transfer information to chromosomal DNA (1). We demonstrated that single-strand (ss) oligos containing several ribonucleotides, or molecules made of RNA only, can precisely repair a chromosomal DSB and transfer information to the genomic DNA in a homology-driven manner in the model eukaryotic organism yeast *Saccharomyces cerevisiae*. In our experiments of RNA-directed DSB repair, we utilize the highly efficient HO endonuclease to induce a site-specific DSB at a defined yeast

H. Tsubouchi (ed.), *DNA Recombination*, Methods in Molecular Biology 745,
DOI 10.1007/978-1-61779-129-1_12,

chromosomal marker gene (2). Following DSB induction and evidence of cell cycle arrest for most cells in the population due to unrepaired DSB, yeast cells are transformed with no oligos and ssDNA oligos as controls and with RNA-only oligos or RNA-containing oligos that are designed to join the broken chromosomal ends restoring the function of the disrupted marker gene and introduce a unique, in-frame short insert harboring a restriction site within the RNA bases. To accomplish DSB repair and restore a functional marker gene, the RNA-insert sequence must be used as a template for DNA repair synthesis. Repair by RNA-containing oligos can occur almost as efficiently as using DNA-only oligos when short tracts of RNA are present within the oligos. Remarkably, although with a much lower frequency, even RNA-only oligos can repair the broken marker gene sequence and generate colonies on the selective media (**Fig. 12.1**). Differently, if no oligo is added following break induction, no transformant colonies are obtained. In addition, restoration of the functional marker gene by the DNA- or the RNA-containing oligos is strongly dependent on DSB induction, even though a few colonies can also arise when no DSB is induced (1). It had never been proven before that RNA can be directly used to correct

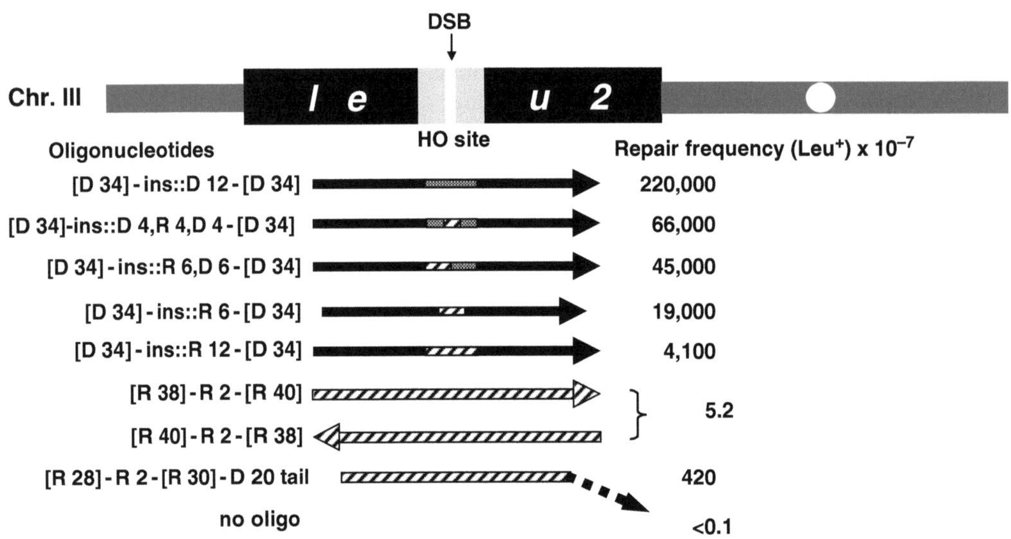

Fig. 12.1. Schematic diagram of DSB repair by RNA-containing oligos at the broken *leu2* locus. DNA-only, RNA-containing, RNA-only oligos (DNA: *black arrow*; DNA insertion: *checker board rectangle*; RNA: *upward diagonal arrow*; RNA insertion: *upward diagonal rectangle*), and no oligo used to repair the DSB in the broken *leu2* gene on chromosomal III are shown together with the corresponding frequencies of *LEU2* repair. The HO cutting site (shown in *light gray*) is in the *middle* of the *LEU2* gene. D is DNA; R is RNA. Numbers of nucleotides homologous to *LEU2* are in *square brackets*; insertions are shown as 'ins::'; *dotted tail* has no homology to *LEU2*. The repair frequency is presented as number of Leu^+ transformants per 10^7 viable cells targeted by 1 nmol of oligos.

any sort of DNA damage. In earlier reports, it was found that RNA can indirectly participate in the repair of a chromosomal DSB. RNA can be copied into DNA through reverse transcription in retroviruses, retrotransposons, and telomeres (3, 4). Reverse transcriptase (RT)-mediated events were observed in DSB repair in yeast, in which mRNA was copied into DNA (cDNA) and inserted at the break site of an HO endonuclease-induced DSB at the mating-type (MAT) locus (5, 6) or used as template for gene conversion with homologous genomic sequences (7). While RT from yeast transposon (Ty) is confined in the cytoplasm inside Ty particles (8), in human cells, retrotranscription can be primed in the nucleus by the 3′-end of a chromosomal break that captures the poly-A tail of a LINE1 retrotransposon RNA (9). However, break repair using LINE1 elements does not require RNA/DNA complementarity and is, therefore, mutagenic. We have found that RNA-containing oligos with homology to a broken chromosome can repair a DSB under conditions where the previously described cDNA-dependent RNA repair processes were inactivated (1). While there are several publications reporting RNA direct interaction with genomic DNA, none involves DNA repair. Previous studies of DSB repair in yeast with ssDNA oligos revealed a two-step single-strand annealing (SSA) repair mechanism (10). The ss repairing oligo first pairs with the ss homologous region of the exposed complementary 3′-broken strand following unwinding or 5′-strand resection and is then used as template for DNA synthesis. Successively, a second annealing interaction between the extended 3′-end and the opposite 3′-end of the break occurs, which is followed by clipping of the nonhomologous tails, gap filling synthesis, and ligation. Similar to DSB repair by ssDNA oligos, DSB repair by ssRNA oligos does not require the strand invasion function of Rad51, thus suggesting that ssRNA can repair a break via a strand annealing mechanism, forming an RNA/DNA hybrid intermediate at one break end (1). The result showing that the presence of a nonspecific DNA flap at the 3′-end of an RNA-only oligo, with no homology to the DSB ends, increases the frequency of DSB repair almost 100-fold (**Fig. 12.1**), compared to repair by the RNA-only oligo with no DNA flap, suggests that RNA-driven DNA repair may be more efficient than what estimated by using the RNA oligos. Synthetic RNA oligos, used for gene targeting, are in fact subjected to degradation by RNases in their path to the nucleus (11). Nevertheless, RNA-containing oligos represent useful starting tools to investigate the molecular mechanisms and the implications of RNA-driven genetic modifications on genome in/stability and genetic variation in cells.

2. Materials

2.1. Yeast Strains

FRO-767 (YFP17) is a derivative strain from JKM146 with the HO cutting site in *leu2(Δho Δhml::ADE1 MATa-inc Δhmr::ADE1 ade1 leu2::HOcs lys5 trp1::hisG ura3-52 ade3::GAL::HO*) (2) (*see* **Note 1**).
FRO-786 is a derivative strain from FRO-767 with a functional *LEU2* gene.

2.2. RNA-Containing Oligos

RNA-containing oligos are synthesized by Thermo Scientific Dharmacon (Lafayette, CO) and are desalted, deprotected, and non-purified.

2.3. DNA Oligos

DNA oligos are synthesized by Invitrogen (Carlsbad, CA) or Alpha DNA (Montreal, Quebec, Canada) and are desalted and non-purified.

2.4. Transformation Reagents and Media

1. YPD: For 1 l, 10 g yeast extract, 20 g soy peptone, 20 g dextrose (Difco/BD, Franklin Lakes, NJ). Add 15 g agar to make YPD solid media. Autoclave before use. Store at room temperature.
2. YPLac liquid medium: For 1 l, 12 g NaOH, 27 ml lactic acid (85% solution), 10 g yeast extract, 20 g soy peptone, pH 5.5 (Difco/BD). Autoclave before use. Store at 4°C.
3. 20% galactose (Sigma, St. Louis, MO) stock solution is filter sterilized and stored at room temperature.
4. Transformation solution 1: 0.1 M lithium acetate (LiAc, Sigma). Prepare immediately before transformation. Solution 1 is directly prepared from powder. No stock solution is made. Keep at room temperature. LiAc can increase the permeability of yeast cell wall to DNA.
5. Transformation solution 2: 0.1 M lithium acetate (LiAc, Sigma) and 50% polyethylene glycol 4000 (PEG 4000, Sigma). Solution 2 is directly prepared from powder. No stock solution is made. Keep at room temperature. PEG can deposit oligos onto yeast cell wall and facilitate the entry of oligos into yeast cells.
6. RNA-containing oligos, 50–80mers, desalted, deprotected, and non-purified. Resuspend to 250 pmol/µl. Store at –80°C.
7. DNA oligos, 50–80mers, desalted, and non-purified: 50 pmol/µl. Store at –20°C.
8. SC-Leu (synthetic complete media lacking leucine, Fisher) solid media.

9. 0.5 mm diameter glass beads.
10. RNase-off: RNase decontamination solution (Pure Biotech LLC, Middlesex, NJ).
11. 1.5 ml DNase/RNase-free centrifuge tubes.
12. 50 ml DNase/RNase-free conical tubes.
13. Sterile aerosol pipette tips with ZAP: 1–200 μl, 100–1,000 μl.

2.5. PCR Amplification of the Chromosomal Region Repaired by RNA-Containing Oligos

1. Lyticase (Sigma). Dissolve in sterile water to 2,000 U/ml. Store at –20°C.
2. DNA primers (Invitrogen; Alpha DNA) 20mers, desalted, and non-purified. Dissolve in sterile water to 50 pmol/μl. Store at –20°C.
3. Taq DNA polymerase, 10x buffer, dNTPs (Roche, Indianapolis, IN).
4. PCR tubes.

2.6. Gel Electrophoresis

1. Agarose.
2. TBE running buffer (10x).
3. Prestained molecular weight marker.

2.7. Restriction Digestion

1. Restriction enzymes, 10x buffer, BSA (New England Biolabs).

2.8. DNA Purification

1. PCR purification kit.

2.9. Alkali Treatment for the RNA-Containing Oligo

1. 1 M NaOH solution. Store at room temperature.
2. 1.2 M hydrochloric acid (HCl). Store at room temperature.
3. 1 M Tris–HCl buffer, pH 7.4 solution. Dissolve 1 M Tris (hydroxymethyl)aminomethane in water and adjust pH with HCl to 7.4. Store at room temperature.

3. Methods

3.1. Preparation of RNA-Containing Oligos

1. Wipe materials that will be used in the experiment including oligo tubes, pipettes, vortex, racks, lab gloves, and the experimental area with RNase decontamination solution to remove potential RNase contamination before everything starts. Every step in this experiment should be RNase free.
2. Resuspend RNA-containing oligos to 250 pmol/μl stock solution with RNase-free water and vortex vigorously to dissolve the pellet. Store at –80°C.

3. Before transformation thaw RNA-containing oligos on ice and dilute to 50 pmol/μl with RNase-free water in RNase-free tubes. Each transformation requires 1 nmol of RNA-containing oligos.
4. Denature chosen amount of RNA-containing oligos to eliminate secondary structures using a 100°C heat block for 2 min.
5. Then, immediately place the tube on ice to prevent re-annealing. Keep on ice till transformation.

3.2. DSB Induction and Transformation of Yeast Cells by RNA-Containing Oligos

1. Inoculate yeast cells in 50 ml of YPLac liquid medium and grow at 30°C in a shaker for 18–20 h.
2. Add 5 ml of 20% galactose to make 2% galactose-containing medium.
3. Incubate cells in the 30°C shaker for 4 h; meanwhile the HO endonuclease under the *GAL1*-inducible promoter will be overexpressed and a DSB will be induced in the middle of *leu2* gene.
4. Observe cells under the microscope after 4 h incubation with galactose. Cells should be arrested at the G2/M phase of cell cycle, showing the shape of dumbbell (*see* **Note 2**) (**Fig. 12.2**).
5. Prepare solution 1 and solution 2 immediately before transformation in RNase-free tubes.
6. Transfer cell culture to a 50 ml RNase-free tube and spin at 1,562 × *g* for 2 min. The pellet of the cell precipitation is approximately 0.5 cm^3.
7. Remove the supernatant and wash cells with 50 ml of RNase-free water and spin at 1,562 × *g* for 2 min.
8. Repeat step 7 for five times to get rid of the culture medium, galactose, and RNases that could be present in the media as much as possible.
9. Remove the supernatant and resuspend cells in 5 ml of solution 1 and spin at 1,562 × *g* for 2 min.
10. Remove supernatant and resuspend cells in 250 μl of solution 1. This amount of cells is sufficient for nine to ten transformations.
11. Aliquot 50 μl of the cell suspension in RNase-free microcentrifuge tubes, add 20 μl of 50 pmol/μl RNA-containing oligo (1 nmol) working solution and 300 μl of solution 2 for each transformation reaction (*see* **Note 3**).
12. Vortex vigorously to mix components homogenously.
13. Incubate transformation reactions at 30°C for 30 min in shaker.

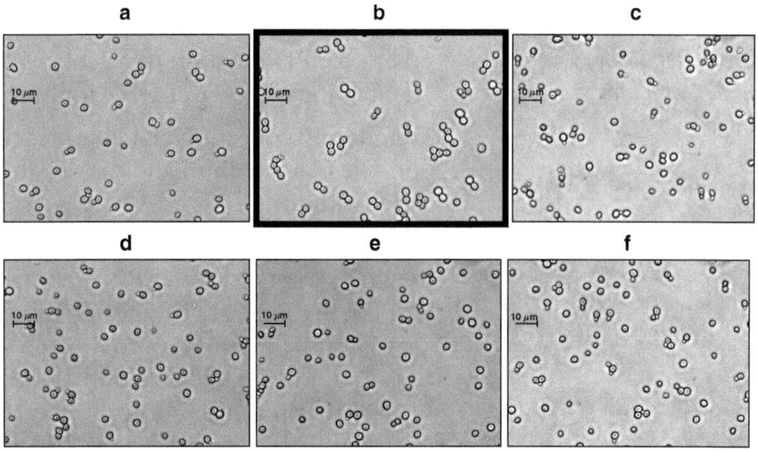

Fig. 12.2. Cell cycle arrest at G2/M phase by unrepaired HO endonuclease-induced DSB. The *leu2* mutant strain with the HO site and the *LEU2* WT strain without the HO site are treated after 19 h of growth in the YPLac neutral medium with 2% galactose to induce the DSB or with water as negative control and are incubated at 30°C for 4 h. An amount of 0.5 ml of culture is taken right before adding galactose, after 4 h in galactose, and after 4 h in water. Cells are sonicated and counted under the microscope. A total of 200 cells from each sample are analyzed under the microscope to obtain the percentage of cells that are in G1, S, and G2 phases. (**a**) *leu2* mutant strain with the HO site after 19 h growth in YPLac liquid medium. G1: 59.5%, S: 20.5%, and G2: 20%. (**b**) *leu2* mutant strain with the HO site following addition of galactose and incubation for 4 h. G1: 13.5%, S: 8%, and G2: 78.5%. The percentage of G2-arrested cells are much higher in this condition for this strain than in the other control conditions and in the same condition for the *LEU2* WT strain. (**c**) *leu2* mutant strain with the HO site following addition of water instead of galactose and incubation for 4 h. G1: 47.8%, S: 21.9%, and G2: 30.3%. (**d**) *LEU2* WT strain without the HO site incubated for 19 h in YPLac liquid medium. G1: 63.4%, S: 16.3%, and G2: 20.3%. (**e**) *LEU2* WT strain without the HO site following addition of galactose and incubation for 4 h. G1: 45%, S: 20.5%, and G2: 38.5%. (**f**) *LEU2* WT strain without the HO site following addition of water instead of galactose and incubation for 4 h. G1: 52.5%, S: 19%, and G2: 28.5%. A 10 μm bar is shown in each picture.

14. Heat shock at 42°C for 15 min.
15. Spin down cells at 2,236×*g* for 4 min.
16. Remove supernatant and resuspend cells in 100 μl of RNase-free water.
17. Dilute cell resuspension with sterile water by 10- to 100-fold and plate cells on one SC-Leu plate using approximately 15 sterile glass beads. Take out glass beads and incubate at 30°C for 3–4 days.
18. Dilute cell resuspension with sterile water by 100,000-fold and plate cells on one YPD plate using approximately 15 sterile glass beads. Take out glass beads and incubate plates at 30°C for 2 days.

3.3. Analysis of DNA Break Repair by RNA-Containing Oligos

1. Count the number of colonies on selective medium as well as that on YPD medium to calculate repair frequency by RNA-containing oligos and DNA-only oligos (*see* **Note 4**).
2. Randomly collect several colonies of transformants growing on selective medium and make patches onto the same selective medium.
3. Design a pair of primers to PCR amplify the chromosomal region targeted by the RNA-containing oligos. Procedures for colony PCR are as follows, modified from (12).
   (a) Resuspend cells (approximately 1 mm^3) in 50 μl of water and add 0.5 μl of 2,000 U/ml lyticase solution. Incubate at room temperature for 10 min, followed by incubation in a heat block at 100°C for 5 min to break the cell wall and release genomic DNA to the solution.
   (b) PCR conditions: The PCR system includes 10 μl of the cell resuspension solution, 1 μl of 50 pmol forward and reverse primer, 1 μl of 10 mM dNTPs, 0.2 μl of 5 U/μl Taq polymerase, 5 μl of 10x buffer and is adjusted with sterile water to a final volume of 50 μl. The PCR program is 3 min at 95°C; 30 cycles of 30 s at 95°C, 30 s at 55°C, and 1 min at 72°C; a final extension time of 7 min at 72°C; and samples are held at 4°C. An extension time of 1 min/kb is assumed for this reaction.
   (c) Following PCR, samples are run on a 1% agarose gel for observation of PCR products.
4. If the genetic information transferred by the RNA-containing oligo generates a new restriction site in the repair region, it is possible to verify the correct transfer of information by digesting the PCR product with the appropriate restriction enzyme. If no restriction site is generated by the RNA-containing oligo, go to step 6. Digest PCR products using a specific restriction enzyme. The digestion reaction includes 6 μl of PCR product, buffer, BSA (may not be needed for some enzymes, see instruction for the enzyme used), 0.2 μl of restriction enzyme, and sterile water to 15 μl. Samples are incubated for 1 h at the temperature specific for the enzyme used.
5. Run an undigested sample together with the digested samples on the same row of a 1.5% agarose gel to observe the genetic modification transferred by the RNA tract of the RNA-containing oligo (**Fig. 12.3**).
6. Purify the PCR products by using a PCR purification kit and prepare them for DNA sequencing. Submit samples for sequencing with the same primers used to amplify the PCR products.

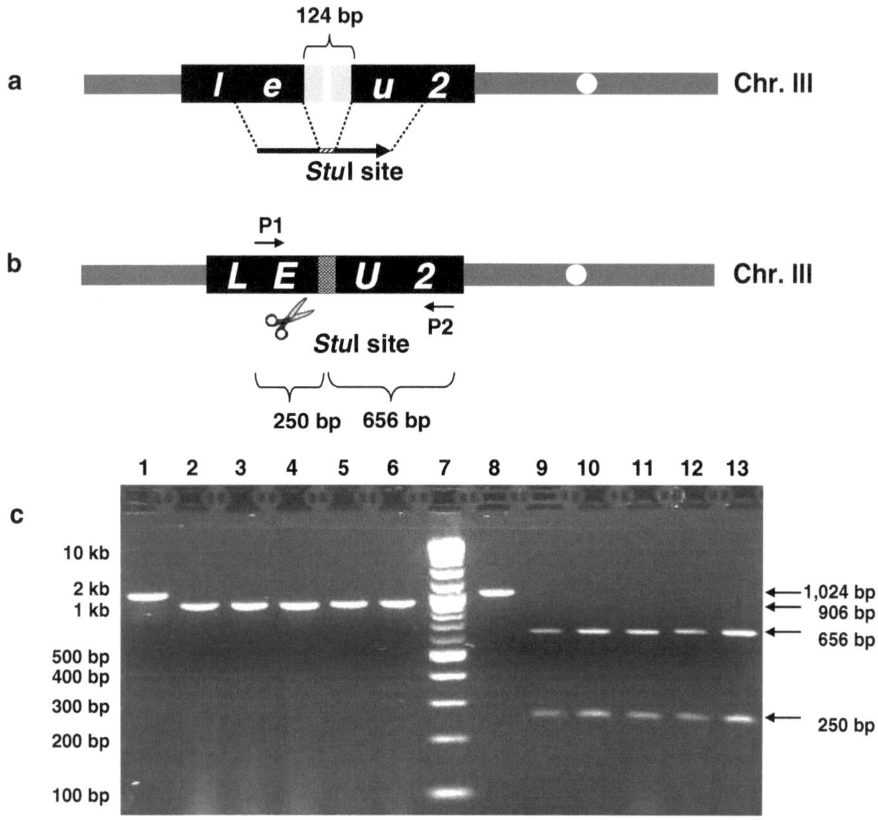

Fig. 12.3. DSB repair by a 6-base RNA-containing oligo. (**a**) Sketch of broken chromosomal *leu2* gene. The HO cutting site (124 bp) in *leu2* is cut by HO endonuclease, resulting in a DSB in the *middle* of the *leu2* gene. Yeast cells are transformed with the RNA-containing oligo containing the *StuI* restriction site insert to repair the DSB. (**b**) After the *LEU2* gene is repaired by the RNA-containing oligo, the *StuI* site is incorporated into the *LEU2* gene. A DNA fragment including only one *StuI* restriction site in the *LEU2* gene is PCR amplified by a pair of primers, P1 and P2, from the *leu2* mutant strain with the HO site before the oligo transformation and from Leu$^+$ colonies repaired with the DNA-only oligo or RNA-containing oligo. The *StuI* restriction enzyme (scissors) is utilized to digest the PCR products. (**c**) PCR products and their digestion products from **b**. *Lane 1*, PCR product of the *leu2* locus amplified from the genomic DNA of the *leu2* mutant strain with the 124 bp HO site (1,024 bp shown by the *arrow* on the *right*); *lane 2*, PCR product amplified from genomic DNA obtained from one Leu$^+$ colony targeted by the DNA-only oligo (906 bp shown by the *arrow* on the *right*); *lanes 3–6*, PCR products amplified from genomic DNA obtained from four Leu$^+$ colonies targeted by the RNA-containing oligo (906 bp); *lane 7*, DNA ladder with sizes of 100 bp to 10 kb; some band sizes are shown on the *left*; *lanes 8–13*, *StuI* restriction digestion of the PCR products from *lanes 1* to *6*. The digestion products of *StuI* have the sizes of 250 and 656 bp (shown by the *arrow* on the *right*).

7. Analyze DNA sequencing results using software that allows alignment of multiple sequences with the appropriate reference sequence.

3.4. Alkali Treatment of the RNA-Containing Oligo (See Note 5)

1. For each reaction, transfer 1 nmol (4 μl of 250 pmol/μl stock solution) of the RNA-containing oligo or DNA-containing oligo to a 1.5 ml centrifuge tube.

2. Add 4 μl of 1 M NaOH for hydrolysis, or alternatively add 4 μl H_2O as negative control, and incubate at 65°C in a water bath for 1 h. Then move from the water bath to ice.
3. Neutralize with 2 μl of 1.2 M HCl, 4 μl of 1 M Tris–HCl, and 4 μl of H_2O or alternatively 6 μl of H_2O and 4 μl of 1 M Tris–HCl for negative control. Keep on ice till transformation.

3.5. Example

The yeast strain used in this example contains one HO site in the middle of the *LEU2* gene on yeast chromosome III. Following induction of the DSB at the HO site, we introduce the chosen RNA-containing oligo or the corresponding DNA-containing oligo into the yeast cells to repair the break. In this example we present an RNA-containing oligo that is a 74mer with 6 bases of RNA embedded in DNA. The 6-base RNA insertion carries the sequence of the *Stu*I restriction site, which is not present in the *LEU2* locus or the *leu2* disrupted by the HO site. The sequence of the oligo is as follows: 5′-TGTTAGGTGCTGTGGGTGGTCCTAAATGGGGTAC-rAr GrGrCrCrT-CGGTAGTGTTAGACCTGAACAAGGTTTACTA AAA-3′ (*see* **Note 6**). In order to repair the DSB, cells must use the RNA tract as template for DNA synthesis. Yeast cells having the *leu2* gene repaired by the RNA-containing oligo can grow on leucine-lacking medium and form colonies. To confirm that the RNA tract of the RNA-containing oligo serves as a template for the DNA synthesis during DSB repair, the repaired region of the *LEU2* gene from several Leu^+ colonies is PCR amplified and digested with the *Stu*I restriction endonuclease (**Fig. 12.3**).

4. Notes

1. The endogenous HO gene and the silent mating cassettes *HML* and *HMR* are deleted in the yeast strain to prevent mating-type switching. The gene encoding the HO endonuclease has been reintroduced to replace the *ADE3* gene and is controlled under the *GAL1* promoter, which can be induced by galactose. The HO site, which can be cut by the HO endonuclease, is inserted in the *LEU2* marker gene, disrupting its normal function, thus preventing yeast growth on media without leucine. Following the generation of a DSB at the HO site, the RNA-containing oligo serves as a template for DSB repair and restores the normal function of the *LEU2* gene. Cells with the repaired *LEU2* gene can grow on SC medium lacking leucine.

2. A single DSB induced by HO endonuclease is sufficient to activate the DNA damage checkpoint and cause yeast cells to arrest at G2/M phase as dumbbell shape, thus allowing cells to repair DNA damage before entering mitosis (13).
3. Salmon sperm DNA (SSD) normally facilitates DNA uptake during yeast cell transformation. SSD does not, however, facilitate the uptake of the ssDNA or RNA-containing oligos (used at 1 nmol amount). Therefore, there is no need to add SSD in this transformation.
4. A negative control without adding RNA-containing oligos is required in the transformation. Spontaneous reversion to Leu^+ phenotype is less than 10^{-9} in the FRO-767 strain, thus no colonies are expected to grow when cells are transformed with no oligos.
5. Alkali treatment is to confirm that the DSB repair and gene modification are specifically due to the RNA-containing oligo and not to a contamination with DNA-only oligo. Since RNA is not stable at high pH, the RNA part of the RNA-containing oligo is degraded by alkaline hydrolysis. On the contrary, DNA is stable at high pH and is not degraded. Therefore, if the RNA-containing oligo is not contaminated with the DNA-only oligo, the frequency of DSB repair by the RNA-containing oligo following treatment with NaOH should drop dramatically, whereas the frequency of DSB repair by the DNA-only oligo following treatment with NaOH should remain the same as that obtained with a corresponding non-treated DNA-only oligo.
6. RNA-containing oligos are designed with homology to the broken region of the *leu2* gene. The 74mer oligo shown in this example consists of 34 bases of DNA on both ends with homology with the *leu2* gene and 6 bases of nonhomologous RNA in the middle containing a *Stu*I site. The insertion of 6 bases does not alter the function of the *LEU2* gene.

References

1. Storici, F., Bebenek, K., Kunkel, T.A., Gordenin, D.A., and Resnick, M.A. (2007) RNA-templated DNA repair. *Nature* **447**, 338–341.
2. Paques, F., Leung, W.Y., and Haber, J.E. (1998) Expansions and contractions in a tandem repeat induced by double-strand break repair. *Mol Cell Biol* **18**, 2045–2054.
3. Baltimore, D. (1985) Retroviruses and retrotransposons: the role of reverse transcription in shaping the eukaryotic genome. *Cell* **40**, 481–482.
4. Autexier, C., and Lue, N.F. (2006) The structure and function of telomerase reverse transcriptase. *Annu Rev Biochem* **75**, 493–517.
5. Moore, J.K., and Haber, J.E. (1996) Capture of retrotransposon DNA at the sites of chromosomal double-strand breaks. *Nature* **383**, 644–646.

6. Teng, S.C., Kim, B., and Gabriel, A. (1996) Retrotransposon reverse-transcriptase-mediated repair of chromosomal breaks. *Nature* **383**, 641–644.
7. Derr, L.K., and Strathern, J.N. (1993) A role for reverse transcripts in gene conversion. *Nature* **361**, 170–173.
8. Lesage, P., and Todeschini, A.L. (2005) Happy together: the life and times of Ty retrotransposons and their hosts. *Cytogenet Genome Res* **110**, 70–90.
9. Morrish, T.A., Gilbert, N., Myers, J.S., Vincent, B.J., Stamato, T.D., Taccioli, G.E., Batzer, M.A., and Moran, J.V. (2002) DNA repair mediated by endonuclease-independent LINE-1 retrotransposition. *Nat Genet* **31**, 159–165.
10. Storici, F., Snipe, J.R., Chan, G.K., Gordenin, D.A., and Resnick, M.A. (2006) Conservative repair of a chromosomal double-strand break by single-strand DNA through two steps of annealing. *Mol Cell Biol* **26**, 7645–7657.
11. Houseley, J., LaCava, J., and Tollervey, D. (2006) RNA-quality control by the exosome. *Nat Rev Mol Cell Biol* 7, 529–539.
12. Storici, F., and Resnick, M.A. (2006) The delitto perfetto approach to in vivo site-directed mutagenesis and chromosome rearrangements with synthetic oligonucleotides in yeast. *Methods Enzymol* **409**, 329–345.
13. Harrison, J.C., and Haber, J.E. (2006) Surviving the breakup: the DNA damage checkpoint. *Annu Rev Genet* **40**, 209–235.

Section II

Genetic and Molecular Biological Approaches with Higher Eukaryotes

Chapter 13

SNP-Based Mapping of Crossover Recombination in *Caenorhabditis elegans*

Grace C. Bazan and Kenneth J. Hillers

Abstract

Caenorhabditis elegans is an important experimental organism for the study of recombination during meiosis. Here, we provide methods for the use of single-nucleotide polymorphisms (SNPs) for the study of crossing over in *C. elegans*.

Key words: Crossing over, recombination, PCR, snip-SNP.

1. Introduction

Crossing over is a key event during meiosis in *Caenorhabditis elegans* and many other eukaryotes. Crossovers, in conjunction with sister chromatid cohesion, form the basis of physical connections between homologous chromosomes. These connections play an integral role in helping ensure proper chromosome segregation at meiosis I anaphase. In addition, crossovers result in recombination, the exchange of genetic information between homologous chromosomes. Mapping crossover locations involves detection of these recombination events and necessitates the use of homologous chromosomes that are distinguishable in some way. Here, we summarize approaches for mapping crossovers through the use of single-nucleotide polymorphisms (SNPs) that exist between two laboratory strains of *C. elegans*.

Traditional approaches to mapping crossovers in *C. elegans* have relied on use of animals heterozygous for morphological markers. The chief limitation of this approach is that studies are

H. Tsubouchi (ed.), *DNA Recombination*, Methods in Molecular Biology 745,
DOI 10.1007/978-1-61779-129-1_13, © Springer Science+Business Media, LLC 2011

limited in most cases to two markers (due to the relative paucity of morphological phenotypes in *C. elegans*). As a result, each experiment typically measures crossover frequency within a single interval, which prevents detection of chromosomes with multiple crossovers and complicates determination of crossover distribution along chromosomes. In addition, some morphological markers can have effects on the viability of homozygotes. An alternative approach, first pioneered by Wicks et al. (1) for gene mapping, involves the use of mapped sequence differences between two laboratory strains of *C. elegans*.

The wild-type *C. elegans* strain CB4856 (the Hawaiian strain) differs from wild-type N2 Bristol at approximately 0.1% of bases. These differences are broadly dispersed throughout the genome and provide a dense array of potential genetic markers for use in measurement of recombination. These markers have the advantage of being phenotypically neutral (in general) and codominant, thus avoiding potential complications due to viability and simplifying scoring. In addition, multiple markers can be followed in a single cross (limited only by the number of PCRs one can carry out on the DNA sample obtained). A subset of these polymorphisms alter (create or destroy) cleavage sites for restriction endonucleases. Such polymorphisms, referred to as snip-SNPs, have been exploited for use in a PCR-based approach for mapping genes and measuring meiotic crossing over (1–3). The basic approaches are similar to those used in traditional recombination studies; however, analysis of marker segregation involves molecular approaches, rather than examination of morphological characters. For more detailed background information and additional technical notes, *see* (4) and references therein.

A major advantage of this approach is that multiple intervals can be simultaneously assayed for crossing over, allowing determination of the distribution of crossover events along chromosomes and also allowing detection of chromosomes that have enjoyed multiple crossovers. Thus, use of SNP markers has now largely supplanted the use of morphological markers for analysis of crossover distribution in *C. elegans* (3, 5–12). Looking forward, we envision that the use of multiplex approaches for SNP genotyping may supplement current PCR-based approaches for mapping crossovers; an example of such an approach is the Illumina GoldenGate Assay (12). Another recent example involves high-throughput SNP genotyping using SNP-specific primers and qPCR (6). However, these high-throughput approaches tend to be expensive and complicated, requiring specialized equipment and/or reagents. The PCR-based approach described here has the advantage of being both simple and inexpensive; thus, this approach is likely to remain an important method for detecting crossover recombination in *C. elegans* in the future.

2. Materials

1. 1 M Potassium phosphate buffer, pH 6.0: 108.3 g KH_2PO_4, 35.6 g K_2HPO_4, H_2O to 1 l; autoclave.
2. 5 mg/ml Cholesterol in 95% ethanol (do not autoclave).
3. NGM plates: Combine and autoclave: 3 g NaCl, 17 g agar, 2.5 g peptone, 975 ml H_2O. Cool to 55°C. Add and mix well: 1 ml of 1 M $CaCl_2$, 1 ml of 5 mg/ml cholesterol in 95% ethanol (see above), 1 ml of 1 M $MgSO_4$, 25 ml of 1 M potassium phosphate buffer (see above). Dispense into 60-mm Petri dishes, using sterile technique.
4. *Escherichia coli* OP-50.
5. 10 mM Tris–HCl, pH 8.0.
6. 10 mg/ml Proteinase K in H_2O.
7. 2× Single-worm lysis buffer: 100 mM KCl, 20 mM Tris–HCl pH 8.3, 5.0 mM $MgCl_2$, 0.9% NP-40, 0.9% Tween-20, 0.02% gelatin. Immediately before use, add proteinase K to 120 μg/ml (using 10 mg/ml stock).
8. Reagents for polymerase chain reaction: Taq DNA polymerase and PCR buffer (any supplier); dNTPs (any supplier).
9. Primers: A large and growing number of snip-SNPs have been identified, mostly through the efforts of the Genome Sequencing Center at Washington University at Saint Louis and of Exelixis. Both data sets are available on the Web: http://genome.wustl.edu/genome/celegans/celegans_snp.cgi (Washington University) and http://www.exelixis.com/discovery_acad_c_ele.shtml (Exelixis). For further information and suggestions on primer design, *see* (4) and references therein. *See* also **Table 13.1**.
10. Restriction digestion master mix: The restriction master mix contains the appropriate restriction enzyme (specific for each snip-SNP marker) and 10× buffer, plus H_2O. To each 15 μl PCR reaction to be digested, add 5 μl of a solution containing 2 μl of the appropriate 10× restriction buffer, 3–5 U of restriction enzyme, and water to make 5 μl.

3. Methods

Section 3.1 gives an overview of the basic approaches used when measuring crossing over using snip-SNP markers, as well as providing information about snip-SNP markers that have been

Table 13.1
snip-SNP allele sets for assaying crossovers along each of the six *C. elegans* chromosomes

SNP	Cosmid	Map position	Primer sequence (5′–3′)	Restriction enzyme	N2 restriction fragments (bp)	CB4856 restriction fragments (bp)
Chromosome I[a]						
I A	ZC123	–18.6	F: CCTACAACAGGCAAAGAAGC R: AATTCCTACCAAAGCTCCGC	*Ssp*I	643	324, 319
I B*	Y71G12	–12.3	F: GACAATGACCAATAAGACG R: GATCCGTGAAATTGTTCCG	*Bsr*I	440, 125	364, 125, 76
I B	F32B5	–7.7	F: TAATGTACCACCTCACGTGACG R: CTTTCACCAGAACCCTCTATTC	*Sfu*I	348	188, 160
I C	K04F10	0.9	F: ATCATTCTCCAGGCCACGTTAC R: CTGAACTAGTCGAACAAACCCC	*Nde*I	594	300, 294
I D	T07D10	13.6	F: CTTGGTGTGGGGAGAGTATAGG R: TTTGTCCGGATTGACTCTGC	*Sau*3AI	303, 63	207, 96, 63
I E	ZK909	28.8	F: CACAAGTGGTTTGGAAGTACCG R: CAACAAAGGGATAGATCACGGG	*Hin*dIII	450	236, 214
Chromosome II[a]						
II A	T25D3	–17.9	F: CGGAGATAGTCTCGTGGTACTG R: CAGTCATGCTCCAAACATTCTC	*Dra*I	336,93	288, 93, 48
II B	R52	–14.5	F: TCCATCTTCGCAATCAGATTTC R: AACGTACTGCTTCCCATGCTC	*Ahu*I	368	203, 165
II C	M03A1	–4	F: TCATCTGTCGAGTGCTTTTG R: CGATCGCTCAAATGGTTG	*Taq*I	291, 81, 80	210, 81, 80, 70
II D	F37HB	3.3	F: TTCTCACAACTTCTTTTCCAAG R: TTCACTATTTCCCTCGCTGG	*Taq*I	572, 112, 15	382, 190, 112, 15
II E	Y38F1	13.6	F: TAGGAAAGTTGTGTCCACCTGG R: TGATGACTCCTTCTTCAGCTGC	*Hin*fI	449	288, 160
II F	Y51HI	20.9	F: GATTCGGAATGGGTGTTG R: TCTTGAATGCGTGGTGTG	*Taq*I	482	340, 142

Table 13.1 (continued)

SNP	Cosmid	Map position	Primer sequence (5′–3′)	Restriction enzyme	N2 restriction fragments (bp)	CB4856 restriction fragments (bp)
Chromosome III [b]						
III A	T17H7	−26	F: CTGCTTATAGTCTTCCTGTCG R: GCAACCCCACCTTCAATGAC	*Ssp*I	910	580, 330
III B	H0614	−10	F: AAACCACCTGAAACTGGAGC R: CTCGAGATTCTGCGTGAAAC	*Spe*I	438	268, 170
III C	F10E9	0.5	F: AGCAGATGAAAGTTCCGACG R: CCCCGCTGTGGTTATTATAC	*Acc*I	598, 225	854
III D	T28D6	8.5	F: TTTCGTGTACGAACGTCTCC R: CATTTCTCCCACTCTTGCTG	*Dra*I	500	283, 217
III E	F54F12	20	F: TTGACTCTTCTGGAGTAGCTGC R: GGATTCCAGGGATTGAAGAG	*Rsa*I	385, 76, 11	207, 178, 76, 11
Chromosome IV [c]						
IV A	Y38C1AA	−2.7	F: AAATAACAGGCACCTACCGC R: CTTTGAAGGAGGACTAACGG	*Xba*I	882	481, 397
IV B	F52C12	−14.9	F: ACATTTAGTCACGCGTAGGG R: GCCCGAATCTAGCACATAAG	*Hpa*II	191, 137, 22	328, 22
IV C	C09G12	−3.7	F: TGTCTACCGTATACCTGGAC R: ATCCAGCTCAAAAGTGTGCG	*Rsa*I	163, 131	294
IV D	B0273	1.8	F: AATACAGCAGTCGTTCCGTTC R: TGAACTTCATGAACCAGCTTG	*Dra*I	288, 144	432
IV E	D2096	3.8	F: ACGAAAAATCACAGAGCGGG R: AATCAACAACGGACGACGAG	*Eco*RI	648, 326	852
IV F	K10D11	6.7	F: GATTATTTCAGAGGAGCAGAGC R: CATAGCACGTGGAATAACCAC	*Hin*dIII	420	245, 175
IV G	T02D1	16.8	F: TGCTTAAAGTCATCGTGTCCAC R: TGTAAACCGTATCGAATCCGAC	*Ear*I	174, 235	408

(continued)

Table 13.1 (continued)

SNP	Cosmid	Map position	Primer sequence (5′–3′)	Restriction enzyme	N2 restriction fragments (bp)	CB4856 restriction fragments (bp)
Chromosome V[d]						
V A	Y38C9B	−20.0	F: TGTAGGGCGAGTAACCAAGC R: CCGCACTTCCTTCAGAAATG	*Bam*HI	318	268, 50
V B	H10D18	−7.9	F: ATTGATCCCATGATCTCGG R: AATCGCTACTTCCGATAACTTC	*Ssp*I	436	263, 173
V C	F57F5	3.6	F: ATCAATCACATGATGCCGT R: TTTCAGCTAGACCTCCCATG	*Hpy*188III	578	326, 252
V D	F57G8	10.0	F: GGCGGAAAGCAATTTCTATC R: AGCTGCAACCAACACTGCTC	*Dra*I	528	272, 256
V E	F48F5.2	25.00	F: GCTTTGGAGACATTGAGCCGTG R: ATGCTCTTCACATTTTCCTGG	*Hpy*188III	439	258, 181
Chromosome X[a]						
X A	F28C10	−19	F: GGTATACCGATCCCTTCAACAAG R: TGGCAAAACACATCCCTGTG	*Bsp*HI	208, 156	364
X B	EGAP7	−15.5	F: AGAATCTGGGAGGTAAATGG R: CCCATTGAAACTACTCCACCTG	*Sfc*I	700, 246	577, 369
X C	F11D5	−11.1	F: TCGTGGCACCATAACGATGTGG R: GATTCAGATCAAACAGAGGTGG	*Dra*I	243	128,115
X D	F45E1	−0.76	F: GGTTCCTGGACGATAACGATGTGG R: CGACCTGAAAGATGTGAGGTTCCTTATC	*Eco*RI	540, 228	768
X E	C05E7	10.1	F: GGCTCTGAGAAACCAACAAG R: TGTTTGCGATGACGTGCAG	*Sau*3AI	318, 149	467
X F	C33AII	20.8	F: CGAGCAGAGATGCAGAGTTCTCAACTG R: CGACCTGAAAGATGTGAGGTTCCTTATC	*Hae*III	280, 300	580

[a]From (10).
[b]From (7).
[c]Henzel, Turner, and Hillers, unpublished.
[d]From (5).

used in previous studies of recombination. **Section 3.2** describes a method for measuring crossing over during both oogenesis and spermatogenesis in hermaphrodites using snip-SNP markers. The major advantage of this approach is its simplicity – recombination is assayed by determining the genotype of self-progeny of heterozygous individuals (11). The chief disadvantage of this approach is that crossing over can occur during both sperm and egg production; thus, only a subset of double crossover chromosomes can be unambiguously detected (11).

As an alternative, crossing over can be assayed during meiosis in a single germline; in this case, all double crossover chromosomes can be detected. **Section 3.3** describes a method for measuring crossing over during oogenesis in hermaphrodites. This approach has the advantage that each progeny worm assayed represents the product of a single meiosis from the heterozygous hermaphrodite parent; this allows unambiguous detection of all multiply recombinant chromosomes. In addition, the codominant nature of snip-SNP markers means that crossovers can be detected without the additional complication of progeny testing (which is necessary to assay recombination during oogenesis using recessive markers). Therefore, use of snip-SNP markers to assay recombination during oogenesis is preferable to use of traditional recessive morphological markers. **Section 3.4** describes a method for measuring crossing over during spermatogenesis in males.

When measuring crossing over in meiotic mutants, it is often necessary to assay crossover formation in many individuals heterozygous for linked genetic markers. This is because mutations affecting meiosis and gametogenesis typically reduce the number of progeny produced, often drastically. Thus, when measuring recombination in meiotic mutants, the following protocols should be modified to involve increased numbers of heterozygous parents.

3.1. Using Snip-SNP Markers to Map Crossovers in *Caenorhabditis elegans*: Basic Approach

Mapping crossovers relies upon detectable differences between homologous chromosomes. The approach described herein uses single-nucleotide polymorphisms that create or destroy restriction endonuclease recognition sites (referred to as snip-SNPs) as markers for determining the location of crossover events. A large number of SNPs have been identified in the Hawaiian *C. elegans* strain CB4856; these represent potential markers for use in crossover mapping in animals heterozygous for CB4856- and N2-derived chromosomes. Several online databases exist which summarize identified SNPs (*see* **Section 2** (4)). Davis et al. (13) identified a set of snip-SNPs spanning all chromosomes that can all be analyzed under similar conditions; these represent convenient choices for use as markers to map crossovers.

A number of studies have used snip-SNPs as markers for crossover detection during meiosis (3, 5, 7–11). Use of the same markers in future experiments facilitates comparisons between studies. **Table 13.1** provides a set of snip-SNP markers on each of the six *C. elegans* chromosomes, as well as primer sequences and digestion information. These markers have been used in previous studies to map meiotic crossovers (*see* references in **Table 13.1**); researchers designing new experiments involving snip-SNP mapping of crossovers could do worse than to use these same markers.

snip-SNPs represent sequence differences between chromosomes that typically are not associated with phenotypic differences; thus, analyzing segregation of snip-SNP markers requires physical detection of the alleles. The basic approach for doing so detailed herein involves amplification of the DNA region containing the snip-SNP through PCR; once amplified, the DNA is digested with a restriction endonuclease whose recognition site is affected by the snip-SNP. Digested DNA is then analyzed through agarose gel electrophoresis. N2- and CB4856-derived DNA can be distinguished by whether or not the restriction endonuclease cleaves the DNA sample (**Fig. 13.1**).

Using snip-SNP markers to assay meiotic recombination involves production of animals heterozygous for N2- and CB4856-derived chromosomes. Doing so in an otherwise wild-type background is simple, requiring only a cross between N2 and CB4856. Use of snip-SNP markers to assay recombination in mutant backgrounds, however, requires introgression of

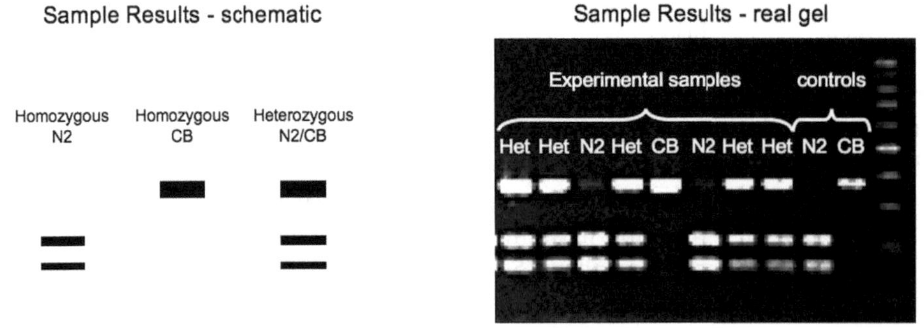

Fig. 13.1. Basic principle of snip-SNP genotyping. snip-SNPs are sequence differences that result in altered sensitivity to a restriction endonuclease (*SspI*, in this example). The DNA region containing the snip-SNP is amplified through PCR, using primers that flank the snip-SNP and recognize both N2 and CB4856 DNA. Following amplification, DNA is digested with restriction endonuclease and analyzed through agarose gel electrophoresis. Analysis of bands seen in each lane allows determination of the genotype of the individual tested. *See* **Note 8**.

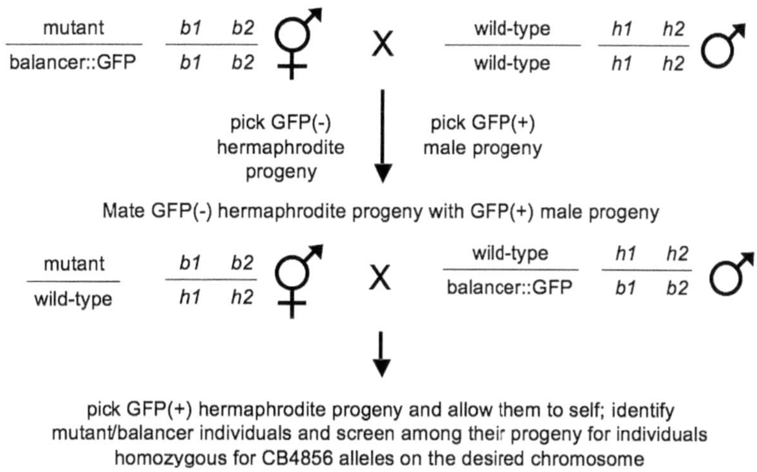

Fig. 13.2. Scheme for introgression of CB4856-derived chromosome into mutant background. This scheme assumes that the mutation of interest is balanced by a balancer chromosome that expresses GFP. *b1* and *b2* are N2-derived snip-SNP alleles; *h1* and *h2* are CB4856-derived alleles. Note, only two snip-SNP alleles are shown on each chromosome for clarity; SNP-based recombination mapping typically involves 5–6 markers per chromosome.

CB4856-derived chromosomes into the mutant strain through repeated backcrossing. This can be particularly challenging in situations where the mutation has a substantial effect upon fertility or viability. One approach for introgression of CB4856-derived chromosomes into a mutant strain is given in **Fig. 13.2**.

Once CB4856-derived chromosomes have been introgressed into a meiotic mutant background, the next step is production of animals homozygous for the mutation of interest and heterozygous for N2- and CB4856-derived chromosomes. This is accomplished through crossing, as in **Fig. 13.3**.

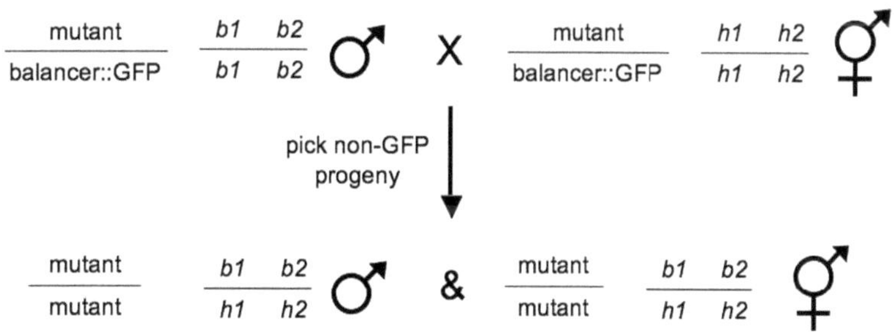

Fig. 13.3. Scheme for production of animals that are both homozygous for a meiotic mutation of interest and heterozygous for snip-SNP markers. Males heterozygous for the mutation of interest ("mutant") and a balancer chromosome marked by a gene insertion which leads to GFP expression ("balancer::GFP") are mated to hermaphrodite partners heterozygous for the mutation of interest (balanced by the GFP-marked balancer chromosome) and homozygous for a chromosome derived from CB4856 (unlinked to the mutation of interest). Male and hermaphrodite progeny from this cross that do not express GFP will be homozygous for the meiotic mutation of interest and heterozygous for the linked phenotypic markers.

Meiotic crossing over can be directly assayed among the self-progeny of N2/CB4856 heterozygous hermaphrodites (**Section 3.2**). Alternatively, recombination occurring during oogenesis in hermaphrodites or spermatogenesis in males can be assayed among the outcross progeny of N2/CB4856 heterozygous hermaphrodites or males (**Sections 3.3** and **3.4,** respectively).

3.2. Measuring the Incidence of Crossing Over During Both Spermatogenesis and Oogenesis in Hermaphrodites Through the Use of snip-SNP Markers

1. Generation of heterozygous hermaphrodites: On a small (60 mm) NGM plate seeded with *E. coli*, mate Bristol N2-derived hermaphrodites homozygous for a selected morphological marker to homozygous Hawaiian CB4856 males. After 48 h, remove both male and hermaphrodite parents from the plate and allow progeny to develop (*see* **Notes 1** and **2**).
2. Pick heterozygous (phenotypically wild type) F1 hermaphrodites (as L4 or younger) individually to small seeded NGM plates.
3. Move F1 hermaphrodites to new plates every 12–24 h until they cease producing progeny (*see* **Note 3**).
4. Scoring markers transmitted to self-progeny: As F2 progeny reach adulthood, pick individually into 0.2-ml, thin-walled tubes containing 10 μl of 10 mM Tris–HCl, pH 8.0 (*see* **Notes 4**, **5**, and **6**).
5. To each tube, add 10 μl of 2× single-worm lysis buffer and mix well.
6. Lyse worms: Freeze at –80°C, incubate at 65°C for 1 h and 95°C for 15 min (*see* **Note 5**).
7. PCR analysis: Each snip-SNP marker is amplified using a specific primer pair. Thus, PCR conditions should be empirically optimized for each marker to be analyzed. However, the following general conditions have worked well in our hands: use 0.5 μl of worm lysate in each 15 μl reaction. PCR cycling: 94°C for 2 min; 35 cycles of {94°C for 20 s; 60°C for 30 s; 72°C for 40 s}; 72°C for 10 min (*see* **Note** 7).
8. Restriction digestion: Add an appropriate volume of restriction enzyme master mix to each PCR reaction and digest for 4 h overnight.
9. Agarose gel analysis: Restriction enzyme-digested PCR products can be analyzed through agarose gel electrophoresis. As expected, DNA fragments are often small (<300 bp), we use 2.5% agarose gels in 0.5× TBE.
10. After electrophoresis, score each sample for the presence or the absence of the N2- and CB4856-specific band(s).

In cases of ambiguity, PCR analysis and restriction enzyme digestion should be repeated. *See* **Note 8**.

11. Identifiable recombinant progeny will fall into two types: (a) those in which crossing over between the assayed markers occurred during production of either sperm or egg but not both. This case results in progeny heterozygous for one marker and homozygous for the other (e.g., [*b1 h2/b1 b2*], where *b1* and *b2* represent N2-derived alleles at loci 1 and 2, respectively, and *h1* and *h2* represent the CB4856 alleles) and (b) those in which crossing over between the assayed markers occurred during production of both sperm and eggs. Detectable recombinants in this case will be homozygous for recombinant chromosomes (e.g., [*b1 h2/b1 h2*]). Note that an equal number of progeny resulting from this case will be heterozygous for both alleles (e.g., [*b1 h2/h1 b2*]) and thus indistinguishable from non-recombinants.
12. The recombination frequency (p) is calculated using the following equation: $p = 1 - (1 - R)^{1/2}$, where $R =$ ((number of animals heterozygous for one marker and homozygous for the other) + 2 × (number of animals homozygous for recombinant chromosomes))/total number of animals scored (14).

3.3. Measuring the Incidence of Crossing Over During Oogenesis in Hermaphrodites Through the Use of snip-SNP Markers

1. Generation of heterozygous hermaphrodites: On a small (60 mm) NGM plate seeded with *E. coli*, mate Bristol N2-derived hermaphrodites homozygous for a selected phenotypic marker to homozygous Hawaiian CB4856 males. After 48 h, remove both male and hermaphrodite parents from the plate and allow progeny to develop (*see* **Notes 1** and **2**).
2. Pick heterozygous (phenotypically wild type) F1 hermaphrodites (as L4) individually to small seeded NGM plates along with 5–8 males of N2 background. To aid in identification of outcross progeny, it is often convenient to use GFP-expressing males (*see* **Note 9**).
3. After 24 h, each heterozygous hermaphrodite should have mated with the N2 males present on the plate. Thus, progeny produced after 24 h of mating are likely to be outcross progeny (allowing measurement of crossing over that occurred solely during oogenesis). Move heterozygous hermaphrodites to new plates. Each 24 h thereafter for several days (or until they cease producing outcross progeny), move individually to fresh plates (*see* **Note 3**).
4. Scoring markers transmitted to progeny: As the outcross progeny of the heterozygous hermaphrodite

reach adulthood, pick individually into 0.2-ml, thin-walled tubes containing 10 μl of 10 mM Tris–HCl, pH 8.0 (*see* **Notes 4**, **5**, **6**, and **9**).

5. To each tube, add 10 μl of 2× single-worm lysis buffer and mix well.
6. Carry out worm lysis, PCR, restriction analysis, electrophoresis, and scoring as in **Section 3.2**, steps 6–10.
7. For each interval assayed, outcross progeny will fall into four classes: homozygous N2 (nonrecombinant; *b1 b2/b1 b2*), heterozygous N2/CB4856 (nonrecombinant; *b1 b2/h1 h2*), heterozygous for marker 1 (recombinant; *b1 b2/h1 b2*), and heterozygous for marker 2 (recombinant; *b1 b2/b1 h2*). (*b1* and *b2* represent N2-derived alleles and *h1* and *h2* represent CB4856-derived alleles.)
8. The recombination frequency $p = R$, where R is the fraction of progeny with recombinant genotypes.

3.4. Measuring Crossing Over in Males Using snip-SNP Markers

1. Generation of heterozygous males: On a small (60 mm) NGM plate seeded with *E. coli*, mate Bristol N2-derived hermaphrodites to homozygous Hawaiian CB4856 males (or vice versa). After 24 h of mating, remove all the male parents from the plate, which will facilitate detection of progeny males in step 2 (*see* **Note 2**).
2. Pick heterozygous F1 males individually to small seeded NGM plates with several N2-derived late L4 stage hermaphrodites homozygous for some phenotypic mutation (e.g., *unc-3*).
3. After 24 h of mating, transfer the mated hermaphrodite partners (but not the heterozygous males) individually to fresh plates. Each of these animals should have mated with the heterozygous males and will thus produce outcross progeny. Transfer these mated hermaphrodites to fresh plates every 24 h for several days (or until they cease production of outcross progeny) (*see* **Note 3**).
4. Scoring markers transmitted to progeny: Outcross progeny from mated hermaphrodites will consist of phenotypically wild-type hermaphrodites and males (if the hermaphrodite partners are homozygous for an X-linked marker such as *unc-3*, outcross males will be mutant (and thus distinguishable from their phenotypically WT fathers)). As outcross progeny reach adulthood, pick individually into 0.2-ml, thin-walled tubes containing 10 μl of 10 mM Tris–HCl, pH 8.0 (*see* **Notes 4**, **5**, and **6**).
5. To each tube, add 10 μl of 2× single-worm lysis buffer and mix well.

6. Carry out worm lysis, PCR, restriction analysis, electrophoresis, and scoring as in **Section 3.2**, steps 6–10.
7. For each interval assayed, outcross progeny will fall into four classes: homozygous N2 (nonrecombinant; *b1 b2/b1 b2*), heterozygous N2/CB4856 (nonrecombinant; *b1 b2/h1 h2*), heterozygous for marker 1 (recombinant; *b1 b2/h1 b2*), and heterozygous for marker 2 (recombinant; *b1 b2/b1 h2*). (*b1* and *b2* represent N2-derived alleles and *h1* and *h2* represent CB4856-derived alleles.)
8. The recombination frequency $p = R$, where R is the fraction of progeny with recombinant genotypes.

4. Notes

1. The N2-derived parent in this cross is homozygous for a recessive morphological marker to facilitate identification of outcross progeny, which will be wild type; self-progeny will be a homozygous mutant and thus morphologically distinguishable. This is not necessary but simplifies identification of outcross progeny. Alternative approaches for identification of outcross progeny are detailed in **Note 9**.
2. Measurement of recombination in animals homozygous for mutations affecting meiosis requires construction of worms homozygous for the meiotic mutation under study and heterozygous for linked genetic markers. However, many meiotic mutants become aneuploid only after a few generations (due to the chromosome missegregation induced by many mutations affecting meiosis); this can greatly complicate both genetic and physical measures of recombination. Thus, it is vitally important to assay recombination in the germlines of euploid mutant animals derived from parents that were heterozygous for the meiotic mutation in question. The simplest approach for doing so involves use of balancer chromosomes marked with a GFP insertion. One way to do so is shown in **Fig. 13.3**. Note that animals heterozygous for balancer chromosomes should not be used as "wild-type" controls for experiments measuring crossing over in meiotic mutant backgrounds. In balancer chromosome heterozygotes, nonhomologous chromosome synapsis occurs, with subsequent effects on meiotic recombination (e.g., (15, 16)). For more information about balancer chromosomes in *C. elegans*, *see* (17). In cases where a suitable balancer chromosome is not available, worms of the appropriate genotype should be derived as in (18).

3. A single hermaphrodite produces 250–300 progeny over a 3- to 4-day period. For measurement of recombination frequencies, it is important to assay all progeny produced by the animal under study during a given time period. By moving hermaphrodites every 24 h, "broods" of roughly 100 progeny are collected. As all of these animals hatched from eggs produced during a single 24-h period, they will all reach adulthood within a relatively narrow time window (but *see* **Note 4**); this greatly simplifies subsequent analyses.
4. As different genotypes may have different growth rates, it is important to score all progeny produced during a given time period; failure to do so may result in undercounting the number of individuals in certain genotypic class(es) and thus reduce the accuracy of the map distance measurement. Thus, each plate of progeny (each "brood"; *see* **Note 3**) should be checked for progeny multiple times over a span of several days; this will increase the likelihood that all progeny will be scored.
5. At this point, samples can be stored at –80°C until ready for further analysis.
6. Analysis can also be carried out in 96-well plates.
7. Always amplify N2 and CB4856 controls for amplification and digestion.
8. Incomplete digestion by the restriction endonuclease can give spurious uncut bands, which can complicate analysis of results. Thus, it is important to always include N2 and CB4856 controls for amplification and digestion on each gel. True heterozygotes will have N2 and CB alleles in equal abundance. Thus, the uncut band (which is larger and binds more ethidium bromide) will be brighter than the cut bands; for example, *see* lanes 1 and 2 (from L) in **Fig. 13.1**. Incomplete digestion can commonly be distinguished from heterozygosity because the smaller bands will be brighter than the larger band, as in lanes 3 and 6 (from L) in **Fig. 13.1**.
9. To measure the frequency of recombination in the oocyte germline, it is important to only score outcross progeny from the heterozygous hermaphrodite. In crosses of this sort, outcross progeny can be identified in a number of ways:
   - Only score hermaphrodite progeny picked from plates with roughly equal numbers of males and hermaphrodites; these should represent outcross offspring. However, if the animals being assayed are mutant for meiotic function, then self-progeny may also have a high proportion of male offspring (the Him phenotype); in that case, use one of the following approaches.

- Generate outcross progeny using males homozygous for a third, dominant, marker. One example that has been successfully used is the transgene insertion *ccIs4251*, which expresses GFP under control of the *myo-3* promoter (19). In this case, outcross progeny can be distinguished due to GFP expression.
- In experiments measuring recombination in animals homozygous for a deletion allele of a gene of interest (such as a gene involved in meiosis), outcross progeny will be heterozygous for the deletion allele, while self-progeny will be homozygous for the deletion. These genotypes can be assayed by PCR; this allows the researcher a molecular assay to confirm that each progeny animal assayed is truly outcross.

Acknowledgments

Anne Villeneuve helped with the preparation of a previous version of this manuscript. K.J.H. was supported by Award Number R15HD059093 from the Eunice Kennedy Shriver National Institute of Child Health and Human Development.

References

1. Wicks, S.R., Yeh, R.T., Gish, W.R., Waterston, R.H., and Plasterk, R.H. (2001) Rapid gene mapping in *Caenorhabditis elegans* using a high density polymorphism map. *Nat Genet* **28**, 160–164.
2. Hillers, K.J., and Villeneuve, A.M. (2009) Analysis of meiotic recombination in *Caenorhabditis elegans. Methods Mol Biol* **557**, 77–97.
3. Hillers, K.J., and Villeneuve, A.M. (2003) Chromosome-wide control of meiotic crossing over in *C. elegans. Curr Biol* **13**, 1641–1647.
4. Davis, M.W., and Hammarlund, M. (2006) Single-nucleotide polymorphism mapping. *Methods Mol Biol* **351**, 75–92.
5. Carlton, P.M., Farruggio, A.P., and Dernburg, A.F. (2006) A link between meiotic progression and crossover control. *PLoS Genet* **2**, e12.
6. Lim, J.G., Stine, R.R., and Yanowitz, J.L. (2008) Domain-specific regulation of recombination in *Caenorhabditis elegans* in response to temperature, age and sex. *Genetics* **180**, 715–726.
7. Mets, D.G., and Meyer, B.J. (2009) Condensins regulate meiotic DNA break distribution, thus crossover frequency, by controlling chromosome structure. *Cell* **139**, 73–86.
8. Nabeshima, K., Villeneuve, A.M., and Hillers, K.J. (2004) Chromosome-wide regulation of meiotic crossover formation in *Caenorhabditis elegans* requires properly assembled chromosome axes. *Genetics* **168**, 1275–1292.
9. Saito, T.T., Youds, J.L., Boulton, S.J., and Colaiacovo, M.P. (2009) *Caenorhabditis elegans* HIM-18/SLX-4 interacts with SLX-1 and XPF-1 and maintains genomic integrity in the germline by processing recombination intermediates. *PLoS Genet* **5**, e1000735.
10. Tsai, C.J., Mets, D.G., Albrecht, M.R., Nix, P., Chan, A., and Meyer, B.J. (2008) Meiotic crossover number and distribution are regulated by a dosage compensation protein that resembles a condensin subunit. *Genes Dev* **22**, 194–211.
11. Hammarlund, M., Davis, M.W., Nguyen, H., Dayton, D., and Jorgensen, E.M. (2005)

Heterozygous insertions alter crossover distribution but allow crossover interference in *Caenorhabditis elegans*. *Genetics* **171**, 1047–1056.
12. Rockman, M.V., and Kruglyak, L. (2009) Recombinational landscape and population genomics of *Caenorhabditis elegans*. *PLoS Genet* **5**, e1000419.
13. Davis, M.W., Hammarlund, M., Harrach, T., Hullett, P., Olsen, S., and Jorgensen, E.M. (2005) Rapid single nucleotide polymorphism mapping in *C. elegans*. *BMC Genomics* **6**, 118.
14. Brenner, S. (1974) The genetics of *Caenorhabditis elegans*. *Genetics* 77, 71–94.
15. McKim, K.S., Howell, A.M., and Rose, A.M. (1988) The effects of translocations on recombination frequency in *Caenorhabditis elegans*. *Genetics* **120**, 987–1001.
16. MacQueen, A.J., Phillips, C.M., Bhalla, N., Weiser, P., Villeneuve, A.M., and Dernburg, A.F. (2005) Chromosome sites play dual roles to establish homologous synapsis during meiosis in *C. elegans*. *Cell* **123**, 1037–1050.
17. Edgley, M.L., Baillie, D.L., Riddle, D.L., and Rose, A.M. (April 6, 2006) Genetic balancers In WormBook, The C. elegans Research Community, WormBook, ed. doi/10.1895/wormbook.1.89.1, http://www.wormbook.org.
18. Kelly, K.O., Dernburg, A.F., Stanfield, G.M., and Villeneuve, A.M. (2000) *Caenorhabditis elegans msh-5* is required for both normal and radiation-induced meiotic crossing over but not for completion of meiosis. *Genetics* **156**, 617–630.
19. Fire, A., Xu, S., Montgomery, M.K., Kostas, S.A., Driver, S.E., and Mello, C.C. (1998) Potent and specific genetic interference by double-stranded RNA in *Caenorhabditis elegans*. *Nature* **391**, 806–811.

Chapter 14

Characterization of Meiotic Crossovers in Pollen from *Arabidopsis thaliana*

Jan Drouaud and Christine Mézard

Abstract

Homologous recombination processes, which occur during the prophase of the first meiotic division, while generating new allelic combinations, are mechanistically important for the regular segregation of homologous chromosomes. They generate either crossovers, which are reciprocal exchanges between chromosome segments, or gene conversions. Both kinds of events occur in narrow regions (less than 10 kb) called hotspots, which are distributed along chromosomes. Classical genetic methods for CO characterization, which rely on the building of large populations and require appropriately located markers, are not well suited to the study of meiotic recombination hotspots. Here, we present a method based on allele-specific PCR amplification of single molecules from pollen genomic DNA. It allows detection, quantification and characterization of CO events arising at low frequencies in recombination hotspots.

Key words: Meiosis, crossover, pollen DNA, allele-specific PCR.

1. Introduction

During meiosis, the ploidy level is halved through a series of two cell divisions following a single round of DNA replication. The first division segregates homologous chromosomes, while the second division segregates sister chromatids, similar to mitosis. Hence, meiosis yields four cells each having half the number of chromosomes of the progenitor cell.

The creation of physical connections between homologues is an absolute requirement for their proper segregation at the first division. In most species, this is achieved through the formation of crossovers (COs), which are reciprocal exchange of segments between homologous non-sister chromatids.

H. Tsubouchi (ed.), *DNA Recombination*, Methods in Molecular Biology 745,
DOI 10.1007/978-1-61779-129-1_14, © Springer Science+Business Media, LLC 2011

COs result from the homologous repair of double-strand breaks (DSBs) formed early during the meiotic prophase. Other outcomes of that repair process are non-crossovers (NCOs, local non-reciprocal exchange, sometimes called gene conversions) and sister chromatid exchanges (SCEs) (1, 2).

Besides their crucial mechanical role in the segregation of homologues, COs generate new allelic combinations. This provides both a substrate for natural selection and the basis of genetic maps.

It has been noticed from the very beginning of genetics and cytogenetics that neither the rate of COs nor their distribution along chromosomes is uniform. These two features are controlled at several levels. First, meiotic DSBs cluster in small regions (few kilobases) called hotspots, which are not homogenously distributed across chromosomes. Second, the odds of a DSB to yield a CO or a NCO or a SCE are likely intrinsically specific to each hotspot. Third, the phenomenon of interference prevents COs from occurring close to each other (3).

The distributions of COs at whole chromosome and genome levels have been extensively studied by analysing the segregation of genetic markers in the offspring of hybrid parents.

On the other hand, cytological methods describe the distribution along chromosomes at the pachytene stage of meiotic prophase of structures, either electron-dense nodules or immunostained foci that correspond to COs (4).

Recently, a completely new type of approach has arisen, which relies on the analysis of linkage disequilibrium (LD) among populations. This allows building haplotype maps of whole genomes, whose boundaries are thought to represent the preferential location of meiotic COs over evolutionary times. In human, 50,000 such 'historical' hotspots have thus been described (5–8).

Such a global analysis of currently active meiotic hotspots is by no way feasible using the available tools of molecular biology. Nevertheless, some have been extensively studied, either in baker's yeast or in mammals (mouse and human) (for review, see (9, 10)).

Among the tens of hotspots analysed so far in higher eukaryotes, the reported frequency of COs does not exceed 1% of meioses. So it is clear that classical genetic analysis on siblings would need huge populations to get enough COs to characterize a hotspot accurately. On the other hand, cytological methods are not spatially accurate enough to describe hotspots. LD analysis methods cannot distinguish between actual and historical hotspots and they cannot detect evolutionary uprising hotspots, which have not yet significantly been involved in the reshuffling of haplotypes.

Jeffreys and collaborators (11) set up a technique called 'sperm typing' that allows the recovery of CO molecules from

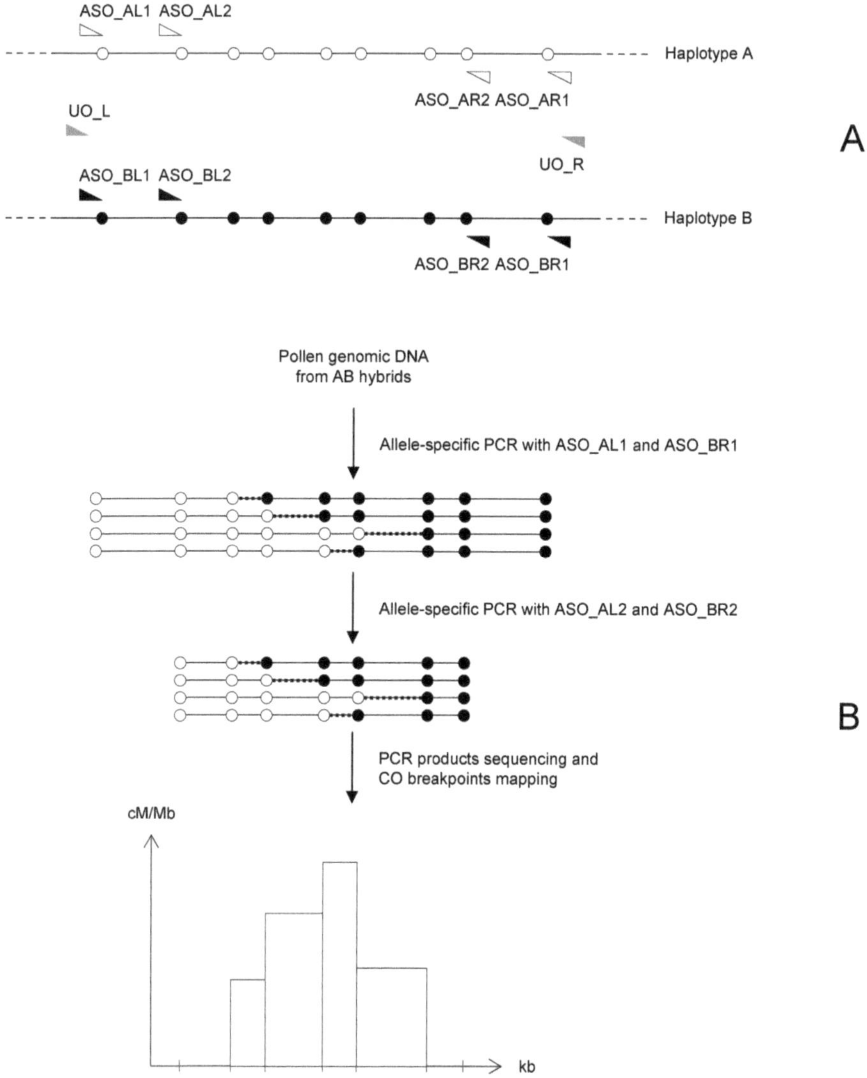

Fig. 14.1. Outline of the pollen typing method. (**a**) Polymorphisms in parental A (*white*) and B (*black*) sequences are represented by *circles*. A- and B-type allele-specific oligonucleotides (ASOs) are depicted as *white* and *black triangles*, respectively. Universal oligonucleotides (UOs) are displayed as *grey triangles*. (**b**) Procedure for PCR amplification and mapping of CO events.

sperm DNA. The procedure is based on a series of nested PCRs for isolating single recombinant molecules from pools of parental non-recombinant molecules. Several reviews have extensively described the use of this method in humans or mice (12, 13). Here, we present an adaptation of this technique to the study of meiotic hotspots in DNA extracted from pollen grains, which are haploid structures producing sperm cells in higher plants. We focus on the specificities of the design of allele-specific oligonucleotides that allow performing long PCR (up to 12 kb) on single molecules. We will also give protocols to perform such long PCR and to extract DNA from pollen grains. Using this 'pollen typing' strategy, virtually any hotspot can be studied. Moreover, it can be adapted to the characterization of any kind of rare DNA variation in a population.

The principle of this method is outlined in **Fig. 14.1**.

Briefly, single molecules, either parental or COs, are detected in genomic DNA (gDNA) extracted from hybrid pollen grains, with two rounds of allele-specific PCR. This allows measuring the overall CO frequency in the studied region. Next, PCR-amplified CO molecules are sequenced to map breakpoints, allowing the analysis of CO frequencies across the hotspot.

2. Materials

2.1. DNA Extraction

2.1.1. Extraction of DNA from Pollen

1. 10% saccharose.
2. Lysis buffer: 100 mM NaCl, 50 mM Tris–HCl (pH 8), 1 mM EDTA, 1% SDS. Add dithiothreitol (DTT) to 1 mM just prior to use.
3. Proteinase K (20 mg/ml).
4. Liquid phenol, saturated with 1 M Tris–HCl (pH 8).
5. Chloroform/isoamyl alcohol (IAA) (24/1, v/v).
6. 3 M sodium acetate (pH 5.2).
7. Isopropanol.
8. 70% ethanol.
9. RNase A (10 mg/ml), DNase free.
10. TE buffer: 10 mM Tris–HCl (pH 8), 1 mM EDTA.

2.1.2. Extraction of DNA from Leaves

1. Polyvinylpolypyrrolidone (PVPP).
2. Lysis buffer: 2% cetyltrimethylammonium (CTAB), 100 mM Tris–HCl (pH 8), 1.4 M NaCl, 20 mM EDTA. Add 2-mercaptoethanol to 10 mM just prior to use.
3. Chloroform/isoamyl alcohol (IAA) mix (24/1).
4. 3 M sodium acetate (pH 5.2).
5. Isopropanol.
6. 70% ethanol.
7. TE buffer: 10 mM Tris–HCl (pH 8), 1 mM EDTA.

2.1.3. DNA Purification

1. DNase-free RNase A (10 mg/ml).
2. Denaturation/binding buffer: 5 M guanidine isothiocyanate, 50 mM Tris–HCl (pH 8).

3. DNeasy Plant Mini Kit (Qiagen).
4. TE buffer: 10 mM Tris–HCl (pH 8), 1 mM EDTA.

2.2. Allele-Specific Long PCR

1. 10× PCR buffer: 450 mM Tris–HCl (pH 8.8), 110 mM $(NH_4)2SO_4$, 45 mM $MgCl_2$, 67 mM 2-mercaptoethanol, 44 μM EDTA, 8 mM dATP, 8 mM dCTP, 8 mM dGTP, 8 mM dTTP, 1.13 mg/ml BSA. Store in 500 μl aliquots at −20°C. Buffer quality is of utmost importance for the outcome of all PCR experiments described in this chapter. Only high-quality reagents can be used, and each buffer batch must be tested prior to any routine use (*see* **Note 1**).
2. Mix of Taq and Pfu DNA polymerases (*see* **Note 2**).
3. Desalted oligonucleotides. Stock solutions: 100 μM in 5 mM Tris (pH 8.8). 10× working solutions: 4 μM in 5 mM Tris–HCl (pH 8.8).

3. Methods

3.1. Genomic DNA Preparation

3.1.1. Extraction of Genomic DNA from Pollen

In the course of this procedure, pollen is first isolated from inflorescences:

1. Harvest *Arabidopsis thaliana* whole inflorescences in ice-cold 10% saccharose.
2. Store at −20°C or proceed directly to step 3.
3. Grind inflorescences in a minimal volume of 10% saccharose, using a 'Waring blender'.

   In most plant species, including *A. thaliana*, pollen wall is much more resistant to mechanical disruption than are other tissues. This treatment bruises floral organs so that intact pollen grains are released from anther locules.
4. Filter the homogenate through a 80-μm mesh (nylon or steel).
5. Centrifuge the filtrate at 350×*g* for 10 min at 4°C.
6. Discard the supernatant.
7. Wash the pellet with ice-cold 10% saccharose.
8. Centrifuge the cell suspension at 100×*g* for 10 min at 4°C.

   After this step, pollen grains are pelleted, while small cell and tissue fragments remain in the supernatant.
9. Discard the supernatant.

10. Repeat steps 7–9.
11. Store pollen at –20°C or proceed directly to step 12.
12. Resuspend the cell pellet in four volumes of lysis buffer.
13. Add proteinase K to 20 μg/ml.
14. Incubate for 4 h at 65°C with occasional gentle homogenisation.
15. Add five to ten 2 mm diameter glass beads.
16. Vortex at full speed for 30 s.
17. Add 1 μl of the suspension into 10 μl of water onto a microscope glass slide. Confirm the disruption of the cells with a microscope (*see* **Note 3**).
18. Proceed again to steps 16–17 until ~90% of pollen grains are disrupted.
19. In a chemical hood, add 1 volume of phenol saturated with 1 M Tris–HCl (pH 8).
20. Mix on a rocking wheel for 30 min.
21. Centrifuge at 15,000×*g* for 10 min.
22. Transfer the supernatant to a new tube and avoid pipetting any solid material.
23. In a chemical hood, add an equal volume of chloroform/IAA. Homogenate by gentle shaking.
24. Centrifuge at 15,000×*g* for 10 min.
25. Transfer the supernatant to a new tube.
26. Add 0.7 volume of isopropanol.
27. Centrifuge at 15,000×*g* for 10 min at 4°C. Discard the supernatant.
28. Wash the pellet with 1 ml of 70% ethanol.
29. Centrifuge at 15,000×*g* for 2 min at 4°C. Discard the supernatant. Drain residual ethanol.
30. Let the pellet dry for 15 min at room temperature.
31. Dissolve the pellet in 100 μl of TE buffer per gram of fresh material.

3.1.2. Extraction of Genomic DNA from Leaves

Genomic DNA extracts from parents are used for testing the specificity of allele-specific oligonucleotides (ASOs) (**Section 3.2.4**), while an extract from an F1 hybrid is used for testing its efficiency (**Section 3.2.7**) and performing control reactions (**Section 3.2.8**):

1. Pre-incubate 100 ml of lysis buffer at 65°C.
2. Weigh an empty 50-ml Falcon-type tube. Keep it on ice.

3. Harvest *A. thaliana* young rosette leaves in the tube.
4. Weigh the tube again. Deduce the weight of fresh material.
5. Freeze leaves with liquid nitrogen in a mortar.
6. Add an equal amount of PVPP.
7. Grind with a pestle until getting a fine powder, taking care that the mortar keeps cold (add liquid nitrogen if necessary).
8. Grind again to get an even finer powder.
9. Transfer the powder to a 30-ml centrifuge tube and allow it to thaw to room temperature.
10. Add 5 ml of hot (65°C) lysis buffer for each gram of fresh material and homogenate thoroughly.
11. Incubate for 30 min at 65°C, with occasional gentle shaking.
12. Let the lysate cool to room temperature.
13. In a chemical hood, add an equal volume of chloroform/IAA. Homogenate by vigorous shaking.
14. Centrifuge at 15,000×*g* for 10 min.
15. Transfer the supernatant to a new tube and avoid pipetting any solid material.
16. If needed, centrifuge again and transfer the supernatant to a new tube.
17. Add 0.7 volume of isopropanol.
18. Centrifuge at 15,000×*g* for 10 min at 4°C. Discard the supernatant.
19. Wash the pellet with 2 ml of 70% ethanol.
20. Centrifuge at 15,000×*g* for 2 min at 4°C. Discard the supernatant. Drain residual ethanol.
21. Let the pellet dry for 15 min at room temperature.
22. Dissolve the pellet in 40 μl of TE buffer per gram of fresh material.
23. Run 1 μl of solution on a 0.8% agarose gel (containing 0.2 μg/ml ethidium bromide) in 1× TBE. Photograph the gel under UV, including a high exposure time, in order to see faint bands (*see* **Note 4**).
24. Check DNA integrity. A noticeable smear is indicative of extensive DNA shearing. In such a case, DNA is expected to have a low amplifiability (*see* **Section 3.1.4**) and should not be used for setting up pollen typing experiments. Then proceed again to step 1.

3.1.3. Purification of Genomic DNA

1. Add 1/100 volume of 10 mg/ml RNase A to DNA samples to be purified. Incubate for 10 min at room temperature.
2. Add 4 volumes of binding buffer.
3. Pipet up to 650 μl of mixture into the DNeasy mini spin column.
4. Proceed to step 14 of the Qiagen procedure (Mini protocol).
5. Adjust the volume of DNA solution to 40 μl per gram of fresh material with TE buffer.

3.1.4. Quantification of Genomic DNA

During the processes of genomic DNA extraction and purification, a variable number of breaks intervene along chromosomes. Consequently, the proportion of molecules that can be actually PCR amplified (amplifiability) drops. Typically, it ranges from 20 to 50% for amplicons longer than 8 kb.

If high amounts of DNA are yielded, its mass concentration can be readily measured by spectrophotometry, or gel electrophoresis and ethidium bromide staining. In addition, the latter method allows monitoring the integrity of DNA, which is roughly indicative of its amplifiability:

1. Run 1 μl of solution along with 100, 200, 400 and 800 ng of phage lambda DNA on a 0.8% agarose gel (containing 0.2 μg/ml ethidium bromide) in 1× TBE.
2. Photograph the gel under UV, including a high exposure time, in order to see faint bands.

Mass concentration is always an overestimate of the concentration of amplifiable molecules. Nevertheless, it can first be considered for carrying out the design procedure of oligonucleotides, as long as only one batch of gDNA is used.

Ultimately, the concentration of amplifiable molecules will be determined by nested PCR amplification of single molecules, as described in **Section 3.3**. Only this value should be considered for subsequent experiments (*see* **Note 5**).

3.2. Designing and Testing Oligonucleotides

Two kinds of oligonucleotides are used for PCR experiments described here. Those which anneal to a site which is not specific to any haplotype (i.e. nonpolymorphic) will be referred to as 'universal oligonucleotides' (UOs). On the other hand, some are intentionally positioned at polymorphic sites so that they are intended to anneal specifically to DNA from one haplotype. The latter are coined 'allele-specific oligonucleotides' (ASOs) (*see* **Fig. 14.1a**).

3.2.1. Length of Amplicons

The outcome of a PCR reaction performed on single molecules of template DNA depends primarily on the size of amplicons. We routinely amplify DNA fragments whose size reaches 10 kb from *A. thaliana* genomic DNA. Longer fragments can be obtained (up to 15 kb), but the consistency of results is expected to drop with amplicon size.

Since the distribution pattern of COs across the hotspot is generally unknown, the size of the region over which PCR reactions will be performed should not be purposely restricted. Ideally, PCR fragments should also encompass the hotspot flanking regions, which are devoid of COs (*see* **Section 3.3.2**).

3.2.2. Designing and Testing UOs

UOs are primarily intended to be used for testing ASOs. Two UOs are needed, which must be designed in the vicinity of the most outer ASOs (*see* **Fig. 14.1a**). UO–ASO and ASO–ASO pairs will generate fragments of similar size and thus are expected to perform similarly. This allows the indirect assessment of the quality of ASOs used for nested PCR amplification of long single molecule. UO_L is used for testing ASO_AR1, ASO_BR1, ASO_AR2 and ASO_BR2. UO_R is used for testing ASO_AL1, ASO_BL1, ASO_AL2 and ASO_BL2 (*see* **Fig. 14.1**).

UOs must anneal with gDNA at high temperatures (68°C or above) so that ASOs with a lower T_m are the only limiting factor with respect to annealing with the DNA template. UOs must also be highly efficient, i.e. yield consistently high amounts of DNA. These two conditions should be assessed first using UO_L and UO_R together as described in **Section 3.2.3**.

Subsequently, UOs will be used in combination with candidate ASOs for determining their T_{opt}, which is intended to be around 60°C (*see* **Section 3.2.4**), and for evaluating their efficiency (*see* **Section 3.2.7**).

The efficiency of UOs can be assessed by performing a series of reactions with a decreasing amount of template gDNA:

1. Prepare a reaction pre-mix for 14 reactions as follows:

   28 μl of 4 μM UO_L;

   28 μl of 4 μM UO_R;

   28 μl of 10× PCR buffer;

   14 μl of 0.5 U/μl Taq:Pfu mix;

   196 μl of H_2O.
2. Combine 42 μl of pre-mix and 2 μl of 1.5 ng/μl F1 leaf gDNA in PCR tube/well '1'.
3. Add 12 μl of H_2O to the remaining pre-mix. Then aliquot 22 μl into PCR tubes/wells '2'–'12'. *See* **Fig. 14.2** for the rationale of serial dilution.
4. Transfer 22 μl from tube/well '1' to tube/well '2'. Mix.
5. Transfer 22 μl from tube/well '2' to tube/well '3'. Mix. Continue the process until tube/well '12'. gDNA is serially diluted at 1/2 from tube '1' to tube '12', starting from 1.5 ng.

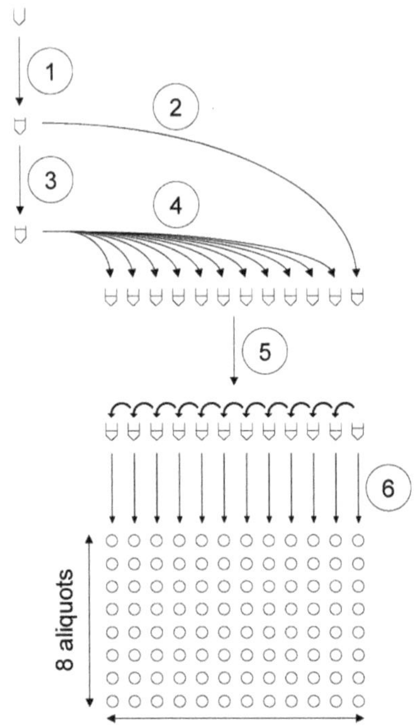

Fig. 14.2. A generic process for testing serial dilutions of genomic DNA. (1) Prepare a mix for (8+1) (12+1)+1 = 118 reactions without template DNA. (2) Add 2×(8+1) = 18 reactions into tube 1. Add DNA. Add water to final volume. (3) Add water to final volume into the remaining mix, which contains 118–18 = 100 reactions. (4) Aliquot 8+1 = 9 reactions into tubes 2–12. (5) Transfer 8+1 = 9 reactions from tube 1 to tube 2. Mix. Transfer 8+1 = 9 reactions from tube 2 to tube 3. Mix. Continue the process until tube 12. (6) Aliquot tube 1 into column 1. Aliquot tube 2 into column 2. Continue the process until tube 12.

6. Proceed to thermal cycling as follows:

$$(92^\circ\mathrm{C}; 2\ \mathrm{min})\{(92^\circ\mathrm{C}; 20\ \mathrm{s})(68^\circ\mathrm{C}; 30\ \mathrm{s} + 45\ \mathrm{s/kb})\}\times 30(68^\circ\mathrm{C}; 90\,\mathrm{s/kb})(4^\circ\mathrm{C}; \infty).$$

7. Add 5 μl of DNA loading dye and run 10 μl on a 0.8% agarose gel (containing 0.2 μg/ml ethidium bromide) in 1× TBE. Photograph the gel under UV, including a high exposure time, in order to see faint bands.

High-efficiency UOs allow synthesizing an amount of DNA which can be visualized starting from as few as 6 pg (40 genomes, *see* **Section 3.1.4**) of *A. thaliana*, which corresponds to the ninth dilution.

If UOs fail to produce this amount of DNA starting from 8 or 16 times more initial gDNA, a new combination must be tested until performing well.

3.2.3. Designing ASOs

The specificity of ASOs is the key to successful detection of CO molecules. Hence, while sometimes tedious, this step requires extreme care.

ASOs must be highly discriminative of parental polymorphic templates when annealing during PCR. It means that at a given temperature, a type A primer must anneal efficiently to type A genomic DNA, but not to type B. Therefore, genomic DNA regions where parental sequences are most dissimilar are favourite candidates for positioning ASOs.

At candidate polymorphic sites, ASOs must be designed so that their 3′-ends diverge as much as possible (*see* **Fig. 14.3a**).

Large insertions/deletions (INDELs) are the most favourable situation, provided they are not repeated. In the latter case, oligonucleotides encompassing the deletion should be absolutely avoided, because they always will anneal along their 3′-end to DNA of the other type (*see* **Fig. 14.3b**).

Nonetheless, even if insertions are not repetitions, some caution must be taken when positioning ASOs across deletions, because looping of genomic DNA can cause non-specific annealing to occur. To avoid this, the 3′-end of the ASOs located beyond the site of insertion must be short so that the hybridization of the 3′-end of the ASO to genomic DNA of the other type will be destabilized by the adjacent loop (*see* **Fig. 14.3c**). We currently limit the length of this 3′-end to 6°C equivalent (*see* **Note 6**).

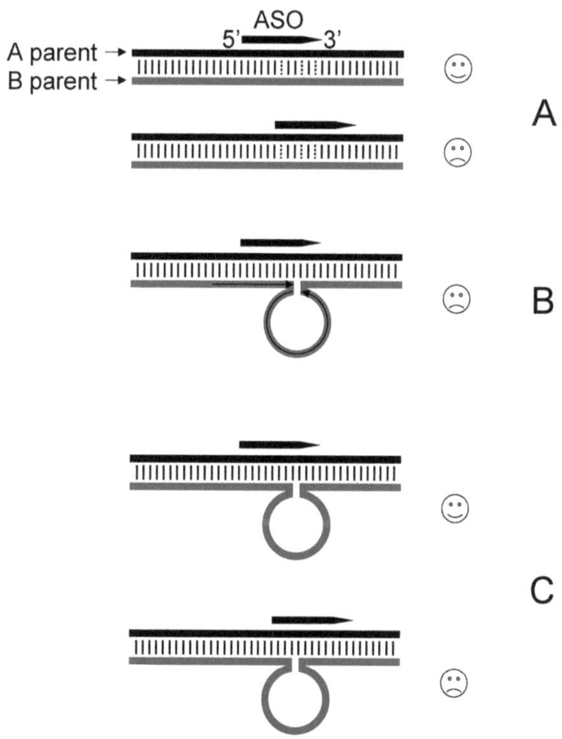

Fig. 14.3. Sample cases for designing ASOs. Aligned A and B parental sequences are represented by *grey* and *black horizontal lines*, respectively. Identities are represented by *vertical plain lines* and SNPs by *vertical dotted lines*. A-specific candidate ASOs are represented by *arrowed lines*. Insertions in B parental sequence with respect to A parental sequence are represented as a *loop*. Direct repeats in B parental sequence are indicated as *thin arrowed lines*.

It happens most of the time that only SNPs are available in the region of interest. Multiple SNPs close to each other are preferred over isolated ones. Only groups of SNPs less than 10 bases apart from each other should be considered as more interesting than isolated ones.

Nevertheless, isolated SNPs can sometimes prove to be sufficient for highly discriminative ASOs. Only preliminary set-up experiments can provide such evidence.

3.2.4. Assessing the Optimal Annealing Temperature of ASOs

The length of the ASOs has to be chosen so that they anneal to their target site at a convenient temperature. Given that the annealing behaviour of an oligonucleotide depends heavily on PCR conditions, it cannot be predicted by dedicated software, which provide starting point temperatures only. Instead it should be empirically characterized by gradient PCR, considering the following rationale:

- Gradient PCR is informative about the less stable oligonucleotide used in the reaction: this is the ASO to be studied, while the other one is an UO purposely chosen to be very stable (annealing above 68°C).
- Two informative temperatures can be determined from the amplification pattern along a gradient. T_{opt} is the highest temperature for which the reaction yield reaches the maximum amount of product. T_{max} is the highest temperature at which a product can be detected by gel electrophoresis.
- For every ASO, gradient PCR experiments must be performed in parallel with each type of parental gDNA, in order to define the range of temperatures over which it anneals efficiently with its specific template only, i.e. between non-specific T_{max} and specific T_{opt} (ΔT, *see* **Fig. 14.4**).
- Ideally, ASOs will anneal to its specific template only, even at the lower end of the gradient. Nevertheless, ASOs with ΔT higher than 6°C are also good candidates.
- Optimal results have been obtained in our laboratory for ASOs with specific T_{opt} around 60°C.

1. Prepare a reaction pre-mix for 28 reactions as follows:

   56 μl of 4 μM UO;

   56 μl of 4 μM ASO;

   56 μl of 10× PCR buffer;

   28 μl of 0.5 U/μl Taq:Pfu mix;

   378 μl of H_2O.
2. For each parent, combine 260 μl of reaction pre-mix and 26 μl of 1.5 ng/μl leaf gDNA.
3. Aliquot 22 μl of each mix into the wells of one plate row.

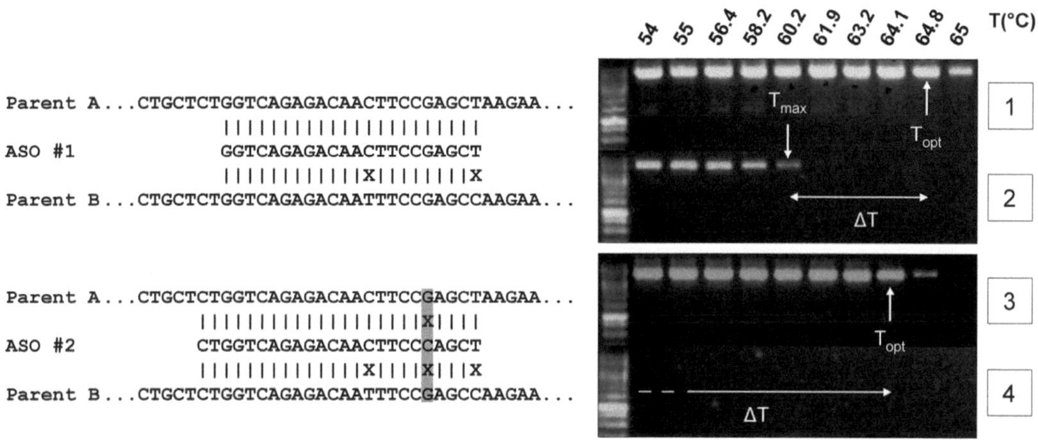

Fig. 14.4. Sample cases of ASO testing by gradient PCR. (*Left*) ASO #1 and #2 sequences aligned to parental sequences. Mismatched nucleotide in ASO #2 is highlighted. (*Right*) *Panels* 1 and 3: PCR is performed on gDNA from parent A. *Panels* 2 and 4: PCR is performed on gDNA from parent B. *Panels* 1 and 2: ASO #1. The difference between specific T_{opt} and non-specific T_{max} (ΔT) is 4.6°C. The ASO is poorly specific. *Panels* 3 and 4: ASO #2. Extra mismatch in ASO sequence increases ΔT to more than 10°C. This version of the ASO is highly specific.

4. Proceed to thermal cycling as follows:

$$(92°\text{C}; 2\ \text{min})\{(92°\text{C}; 20\ \text{s})(\text{gradient from 56 to }68°\text{C}; 30\ \text{s})(68°\text{C}; 45\ \text{s/kb})\} \times 30(68°\text{C}; 90\ \text{s/kb})(4°\text{C}; \infty).$$

5. Add 5 μl of DNA loading dye and run 10 μl on a 0.8% agarose gel (containing 0.2 μg/ml ethidium bromide) in 1× TBE. Photograph the gel under UV, including a high exposure time, in order to see faint bands.

3.2.5. Maximizing the Discrimination Between Polymorphic Genomic Targets of ASOs

Whenever only moderately discriminative ASOs are found, their specificity can be improved by introducing additional mismatches close to their 3′-end. This aims to decrease the stability of ASO/genomic DNA duplexes, but much more for the non-specific target than for the specific one. Such mismatches must be chosen carefully, neither too close to the 3′-end, in order to keep annealing of the ASO to cognate template, nor too far, in order to decrease enough annealing of the ASO to non-cognate template. **Figure 14.4** provides an example of successfully designing such an 'extra-mismatch' ASO, whereas the 'non-mismatch' version was not discriminative enough.

It must be noted that these mismatches decrease the annealing temperature of ASOs to specific gDNA sites. Consequently, nucleotides must be added at the 5′-end of ASOs to compensate this lowering and keep the melting temperature around the optimum (*see* **Note 6**).

3.2.6. Harmonizing the T_{opt} of ASOs

The use of ASOs with different T_{opt} in the same reaction should be avoided. More generally it seems desirable to harmonize T_{opt} for all ASOs used in the analysis of one particular hotspot, as it does not preclude their use whatever be their association in future experiments, either planned or not yet planned. This harmonization step involves a 'trial-and-error' process:

1. Choose an ASO, taking the T_m predicted by your favourite primer design software as an estimation of its T_{opt}.
2. Measure T_{opt} by gradient PCR.
3. If necessary, add or remove nucleotides at the 5′-end, in order to increase or decrease the T_m, respectively (*see* **Note 6**).
4. Proceed again to step 2.

3.2.7. Testing the Efficiency of ASOs

The efficiency requirement for ASOs is not as decisive as for UOs. Indeed, ASOs are used in nested PCR experiments. The yield of the second PCR is generally high enough for sensitive detection purposes, even if the efficiency of oligonucleotides is not tremendous. Moreover raising the number of PCR cycles can generally compensate for a moderate efficiency of ASOs.

However, ASOs sometimes happen to perform very poorly so that their use should be avoided. Hence, the efficiency of ASOs should be evaluated, each in combination with a high-efficiency UO (UOL with ASO_AR1, ASO_BR1, ASO_AR2 or ASO_BR2; UOR with ASO_AL1, ASO_BL1, ASO_AL2 or ASO_BL2) using the procedure described in **Section 3.2.3**.

If necessary, the number of cycles can be adjusted in successive testing experiments.

If an ASO fails to produce any detectable amount of DNA, starting from 50 pg (320 genomes, *see* **Section 3.1.4**) of template or less, then its use for nested PCR amplification of single molecules is not recommended.

3.2.8. Testing the Specificity of ASOs: Control Reactions with Somatic DNA

When amplifying single CO molecules by nested PCR, some undetectable non-specific products arise during the first reaction, as a consequence of misannealing of ASOs, most likely with DNA molecules of the other parental type at the homologous site, but also possibly at other genomic locations. They can nonetheless be abundant in terms of number of molecules. It may happen that these products are in turn amplified non-specifically during the second reaction, yielding this time a detectable product.

In the course of the first PCR round, it may also happen that short fragments, either broken molecules initially present in the gDNA extract or truncated PCR products (resulting from the untimely termination of DNA synthesis, because of a break in template DNA, for example), anneal with longer complementary fragments, next priming DNA synthesis. Whenever template and priming molecules are allelic, chimeric molecules will arise, which cannot be distinguished from true CO molecules.

The occurrence of both these kinds of events must be evaluated by carrying out nested PCR control experiments using high amounts of leaf gDNA from F1 (A×B) hybrids, which is not intended to contain any CO molecule.

All potential pairs of ASOs designed for the amplification of CO molecules should be tested: ASO_AL1/ASO_BR1 then ASO_AL2/ASO_BR2, and ASO_BL1/ASO_AR1 then ASO_BL2/ASO_AR2.

Given that artefactual products are likely to arise infrequently during the first PCR, stochastically, multiple (e.g. 6) identical and independent reactions are to be run in parallel:

1. Prepare a reaction pre-mix for 64 reactions as follows:
   128 μl of 4 μM ASO_AL1;

   128 μl of 4 μM ASO_AR2;

   128 μl of 10× PCR buffer;

   64 μl of 0.5 U/μl Taq:Pfu mix;

   755.2 μl of H_2O.
2. Combine 338.4 μl of pre-mix and 57.6 μl of 1.5 ng/μl F1 leaf gDNA in tube '1'. This is a mix for 18 reactions each containing 32,000 genomes.
3. Add 147.2 μl of H_2O to the remaining pre-mix. Aliquot 198 μl into tubes '2'–'6'. Each of these mixes contains nine reactions without gDNA.
4. Transfer 198 μl from tube '1' to tube '2'. Mix.
5. Continue the process until tube '6'. gDNA is serially diluted at 1/2 from tube '1' to tube '6'.
6. Aliquot 22 μl of tube '6' into each well of column '6' of PCR plate 1.
7. Aliquot 22 μl of tube '5' into column '5' of PCR plate 1.
8. Continue the process until tube '1'.
9. Proceed to thermal cycling as follows:

$$(92^\circ\text{C}; 2\ \text{min})\{(92^\circ\text{C}; 20\ \text{s})(T_{\text{opt}}; 30\ \text{s})(68^\circ\text{C}; 45\ \text{s/kb})\}\times 30(68^\circ\text{C}; 90\ \text{s/kb})(4^\circ\text{C}; \infty).$$

10. Dilute 1 μl of PCR products in 50 μl of 5 mM Tris–HCl (pH 8) and 0.01% Triton X-100.
11. Prepare a reaction pre-mix for 56 reactions as follows:
    112 μl of 4 μM ASO_AL2;

    112 μl of 4 μM ASO_AR2;

112 μl of 10× PCR buffer;

56 μl of 0.5 U/μl Taq:Pfu mix;

784 μl of H_2O.

12. Aliquot 21 μl of the pre-mix into the wells of PCR plate 2.
13. Transfer 1 μl of diluted products from PCR plate 1 into PCR plate 2.
14. Proceed to thermal cycling as follows:

$$(92°\text{C}; 2\ \text{min})\{(92°\text{C}; 20\ \text{s})(T_{\text{opt}}; 30\ \text{s})(68°\text{C}; 45\ \text{s/kb})\} \times 30(68°\text{C}; 90\ \text{s/kb})(4°\text{C}; \infty).$$

15. Add 5 μl of DNA loading dye and run 10 μl on a 0.8% agarose gel (containing 0.2 μg/ml ethidium bromide) in 1× TBE. Photograph the gel under UV, including a high exposure time, in order to see faint bands.

The occurrence of faint bands for moderate to high amounts of DNA is not uncommon in the first two dilutions (16,000 amplifiable genomes or more). This is 30 times the amount of pollen gDNA which is typically needed for the amplification of single CO molecules. Hence, it is not problematic as long as the actual CO rate is not exceedingly low.

Of course, this is all the less worrying when the size of those faint bands is obviously different from that of specific products.

Conversely, if products are detected for moderate to low amounts of DNA (i.e. at concentrations which are used for amplifying single CO molecules from F1 pollen gDNA), then one or several ASOs can be suspected to lack specificity, despite the outcome of gradient PCR analysis.

In such cases, the exchange points between haplotypes are all expected to be located either before the second polymorphism or after the penultimate one, the priming sites of left and right ASOs being defined as the first and the last ones, respectively. This can be assessed readily by sequencing PCR products. If it turns out that ASOs are indeed not specific enough, they should not be used for pollen typing experiments.

3.3. Amplification of Single Molecules

The characterization and quantification of CO events in a gDNA extract rely on the specific amplification of single (or quasi-single, see below) molecules, of either parental or recombinant (CO) type. In this way, two rounds of PCR, using nested sets of ASOs, are performed (*see* **Fig. 14.1a**).

Nested PCR serves two goals:

- A single round of PCR is not sufficient to get a detectable amount of product. Usually, two rounds yield a strong and consistent amplification. This allows detecting unambiguously all target molecules initially present in the reactions.

- The use of nested ASOs enhances the specificity of PCR amplification, because non-specific products from the first round of PCR, which are not expected to be abundant, are in turn unlikely to give rise to any detectable product during the second round of PCR.

Single target molecules are isolated by dilution. When their average number per sample in a series of aliquots drops below 1, at least one reaction is expected to contain no template, and therefore to yield no PCR product. Actually, the proportion of these negative reactions can be estimated using the Poisson law. Given m, the mean number of molecules per reaction, it provides the probability of a reaction to contain exactly k molecules:

$$p(k) = \frac{m^k \times e^{-m}}{k!}$$

So

$$p(0) = \frac{m_0 \times e^{-m}}{0!} = e^{-m}$$

For example, if the mean number of molecules is 0.2, $p(0) = e^{-0.2} \approx 82\%$ of reactions contain no molecule.

Hence, the mean number of molecules per well, m, can be readily estimated from the proportion of negative wells P_0 (which itself approximates $p(0)$):

$$m \approx -\ln(P_0)$$

In turn, m provides an estimate of the actual concentration of amplifiable molecules in the gDNA stock solution. *See* **Fig. 14.5** for an illustrated example.

The variance of m can be calculated as described by (12).

3.3.1. Quantification of gDNA by Single-Molecule PCR

Given an uncharacterized gDNA extract, the concentration of amplifiable molecules 'C' is approximated by successive measurements, each performed on a narrower range of dilutions but with more aliquot reactions than the previous one, in order to increase the accuracy of the estimation of C. Each estimate of C is used as a starting point for the following step. Eventually, a large series of reactions is carried out for a single suitable dilution only, providing a definitive estimate of C.

It is first necessary to perform nested PCR on a broad series of dilutions (e.g. 12), in order to get a rough, preliminary estimate of C, called thereafter 'C_1'. For each dilution, a small number (e.g. 8) of aliquot reactions are carried out. For dilutions in which negative wells appear, Poisson formula allows calculating the concentration of molecules in the gDNA extract:

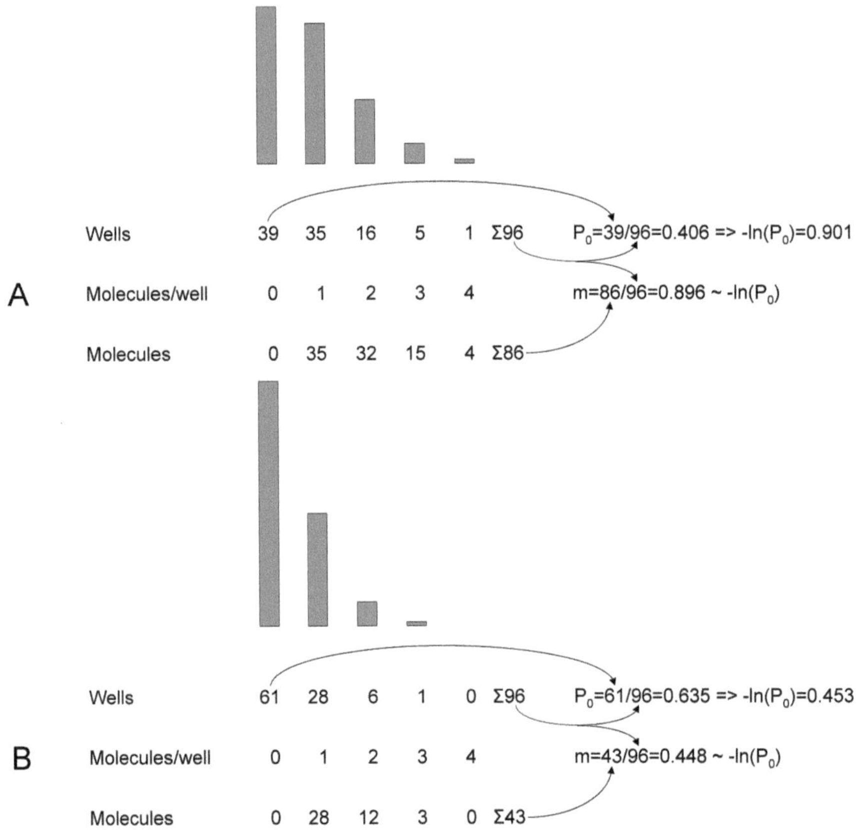

Fig. 14.5. Quantification of Poisson distributed molecules. A genomic DNA extract is diluted 1/10,000 (**a**) or 1/20,000 (**b**). Next 96 reactions are performed, each with 1 μl of diluted DNA. The number of wells containing either 0, 1, 2, 3 or 4 molecules follows a Poisson distribution. They are indicated in the first line of each panel and displayed above as *grey bars*. The corresponding number of molecules is shown in the last line. The real value of average molecule number per well m is very close to its theoretical value which is calculated as $-\ln(P_0)$. In that case, the estimated concentration of amplifiable DNA molecules is $0.453 \times 20{,}000 \sim 0.901 \times 10{,}000 \sim 9{,}039$ molecules/μl, which corresponds to 1.36 ng/μl.

1. Prepare a reaction pre-mix for 115 reactions as follows:
   230 μl of 4 μM ASO_AL1;
   230 μl of 4 μM ASO_AR1;
   230 μl of 10× PCR buffer;
   115 μl of 10 ng/μl carrier DNA (*see* **Note 7**);
   115 μl of 0.5 U/μl Taq:Pfu mix;
   1,598.5 μl of H_2O.
2. Combine 262.8 μl of pre-mix and 1.2 μl of gDNA in tube '1'. This is a mix for 12 reactions each containing 0.1 μl of gDNA.
3. Add 10.3 μl of H_2O to the remaining pre-mix (2,255.7 μl). Aliquot 198 μl into tubes '2'–'12'.

4. Transfer 66 μl from tube '1' to tube '2'. Mix.
5. Transfer 66 μl from tube '2' to tube '3'. Mix.
6. Continue the process until tube '12'. Amplifiable gDNA is serially diluted at 1/4 from tube '1' to tube '12'.
7. Aliquot 22 μl of tube '12' into the eight wells of column '12'.
8. Aliquot 22 μl of tube '11' into the eight wells of column '11'.
9. Continue the process until tube '1'.
10. Proceed to thermal cycling as follows:

$$(92°\text{C}; 2\ \text{min})\{(92°\text{C}; 20\ \text{s})(T_{\text{opt}}; 30\ \text{s})(68°\text{C}; 45\ \text{s/kb})\}\times 30(68°\text{C}; 90\ \text{s/kb})(4°\text{C}; \infty).$$

11. Dilute 1 μl of PCR products in 50 μl of 5 mM Tris–HCl (pH 8) and 0.01% Triton X-100.
12. Prepare a reaction pre-mix for 98 reactions as follows:
    196 μl of 4 μM AL2;
    196 μl of 4 μM AR2;
    196 μl of 10× PCR buffer;
    98 μl of 0.5 U/μl Taq:Pfu mix;
    1,372 μl of H_2O.
13. Aliquot 21 μl of the pre-mix into the 96 wells of PCR plate 2.
14. Transfer 1 μl of diluted products from PCR plate 1 into PCR plate 2.
15. Proceed to thermal cycling as follows:

$$(92°\text{C}; 2\ \text{min})\{(92°\text{C}; 20\ \text{s})(T_{\text{opt}}; 30\ \text{s})(68°\text{C}; 45\ \text{s/kb})\}\times 30(68°\text{C}; 90\ \text{s/kb})(4°\text{C}; \infty).$$

16. Add 5 μl of DNA loading dye and run 10 μl on a 0.8% agarose gel (containing 0.2 μg/ml ethidium bromide) in 1× TBE. Photograph the gel under UV, including a high exposure time, in order to see faint bands.
17. Assuming that the proportion of negative wells P_0 in column i is the closest to 0.5, approximate C as follows:

$$C_1 = -\ln(P_0) \times \frac{1}{0.1} \times 4^{(i-1)}$$

–ln(P_0) is divided by the volume of DNA used in each reaction (0.1 μl), then multiplied by a correction factor accounting for dilution in the ith column ($4^{(i-1)}$).

Next, carry out PCR upon a narrower range of 1/2 serial dilutions (e.g. 4), each with a higher number of aliquot reactions (e.g. 24), in order to get a more accurate estimate 'C_2' of C.

18. Prepare a reaction pre-mix for 128 reactions as follows:
    256 μl of 4 μM ASO_AL1;
    256 μl of 4 μM ASO_AR1;
    256 μl of 10× PCR buffer;
    128 μl of 0.5 U/μl Taq:Pfu mix;
    1,536 μl of H_2O.
19. Combine in tube '1' 1,000 μl of pre-mix and 100 μl of gDNA diluted at $1/C_1$ in 5 ng/μl carrier DNA. This is a mix for 50 reactions each expected to contain two amplifiable molecules.
20. Add 156 μl of 5 ng/μl carrier DNA to the remaining pre-mix. Aliquot 550 μl into tubes '2'–'4'. Each of these mixes contains 25 reactions without amplifiable gDNA.
21. Transfer 550 μl from tube '1' to tube '2'. Mix. Continue the process until tube '4'.
22. Amplifiable gDNA is serially diluted at 1/2 from tube '1' to tube '4'.
23. Aliquot 22 μl of tube '4' into the 24 wells of rows 'G' and 'H'. Continue the process until tube '1'.
24. Proceed to thermal cycling as follows:

$$(92^\circ\mathrm{C}; 2\ \mathrm{min})\{(92^\circ\mathrm{C}; 20\ \mathrm{s})(T_{\mathrm{opt}}; 30\ \mathrm{s})(68^\circ\mathrm{C}; 45\ \mathrm{s/kb})\}\times 30(68^\circ\mathrm{C}; 90\ \mathrm{s/kb})(4^\circ\mathrm{C}; \infty).$$

25. Dilute 1 μl of PCR products in 50 μl of 5 mM Tris–HCl (pH 8) and 0.01% Triton X-100.
26. Prepare a reaction pre-mix for 98 reactions as follows:
    196 μl of 4 μM AL2;
    196 μl of 4 μM AR2;
    196 μl of 10× PCR buffer;
    98 μl of 0.5 U/μl Taq:Pfu mix;
    1,372 μl of H_2O.
27. Aliquot 21 μl of the pre-mix into the 96 wells of PCR plate 2.
28. Transfer 1 μl of diluted products from PCR plate 1 into PCR plate 2.

29. Proceed to thermal cycling as follows:

$$(92^{\circ}\mathrm{C}; 2\ \mathrm{min})\{(92^{\circ}\mathrm{C}; 20\ \mathrm{s})(T_{\mathrm{opt}}; 30\ \mathrm{s})(68^{\circ}\mathrm{C}; 45\ \mathrm{s/kb})\} \times 30(68^{\circ}\mathrm{C}; 90\ \mathrm{s/kb})(4^{\circ}\mathrm{C}; \infty).$$

30. Add 5 μl of DNA loading dye and run 10 μl on a 0.8% agarose gel (containing 0.2 μg/ml ethidium bromide) in 1× TBE. Photograph the gel under UV, including a high exposure time, in order to see faint bands.
31. Assuming that the proportion of negative wells P_0 in dilution '*j*' is the closest to 0.5, a finer estimate C_2 of C is calculated as follows:

$$C_2 = -\ln(P_0) \times \frac{1}{2} \times 2^{(j-1)} \times C_1$$

–ln(P_0) is divided by the volume of DNA used in each reaction (2 μl), then multiplied by a correction factor accounting for dilution in the *j*th column: $2^{(j-1)} \times C_1$.

At last, 96 reactions are performed for one dilution only, providing a definitive estimate 'C_3' of target gDNA concentration in the stock solution.
32. Prepare a reaction mix for 98 reactions as follows:
196 μl of 4 μM ASO_AL1;
196 μl of 4 μM ASO_AR1;
196 μl of 10× PCR buffer;
98 μl of 0.5 U/μl Taq:Pfu mix;
69.3 μl of gDNA diluted at 1/C_2 in 5 ng/μl carrier DNA;
126.7 μl of 5 ng/μl carrier DNA;
1,274 μl of H_2O.
33. Aliquot 22 μl of the mix into the 96 wells of PCR plate 1.
34. Proceed to thermal cycling as follows:

$$(92^{\circ}\mathrm{C}; 2\ \mathrm{min})\{(92^{\circ}\mathrm{C}; 20\ \mathrm{s})(T_{\mathrm{opt}}; 30\ \mathrm{s})(68^{\circ}\mathrm{C}; 45\ \mathrm{s/kb})\} \times 30(68^{\circ}\mathrm{C}; 90\ \mathrm{s/kb})(4^{\circ}\mathrm{C}; \infty).$$

35. Dilute 1 μl of PCR products in 50 μl of 5 mM Tris–HCl (pH 8) and 0.01% Triton X-100.
36. Prepare a reaction pre-mix for 98 reactions as follows:
196 μl of 4 μM AL2;
196 μl of 4 μM AR2;
196 μl of 10× PCR buffer;
98 μl of 0.5 U/μl Taq:Pfu mix;
1,372 μl of H_2O.

37. Aliquot 21 μl of the pre-mix into the 96 wells of PCR plate 2.
38. Transfer 1 μl of diluted products from PCR plate 1 into PCR plate 2.
39. Proceed to thermal cycling as follows:

$$(92^\circ\mathrm{C}; 2\ \mathrm{min})\{(92^\circ\mathrm{C}; 20\ \mathrm{s})(T_{\mathrm{opt}}; 30\ \mathrm{s})(68^\circ\mathrm{C}; 45\ \mathrm{s/kb})\}\times 30(68^\circ\mathrm{C}; 90\ \mathrm{s/kb})(4^\circ\mathrm{C}; \infty).$$

40. Add 5 μl of DNA loading dye and run 10 μl on a 0.8% agarose gel (containing 0.2 μg/ml ethidium bromide) in 1× TBE. Photograph the gel under UV, including a high exposure time, in order to see faint bands.
41. The proportion of negative wells is expected to be $e^{-0.693} = 0.5$. Given the real value P_0, calculate the concentration of molecules in the gDNA extract as follows:

$$C \approx C_3 = \frac{\ln(p_0)}{\ln(0.5)} \times C_2$$

–ln(P_0) is divided by the volume of DNA used in each reaction (–ln(0.5) = 0.693 μl), then multiplied by a correction factor accounting for dilution (C_2).

Provided that very efficient ASOs have been used (*see* **Section 3.2.7**), high PCR yields should be achieved. If it is not the case, the number of PCR cycles can be increased to 35, either for the first reaction, or for the second one, or for both.

The last steps of this procedure (33–41) should then be carried out using sets of ASOs designed for detecting type ‘B’ parental molecules: BL1/BR1 and BL2/BR2 for the first and second PCRs, respectively. Of course, results are expected to be the same as for type ‘A’ parental molecules. The two values must then be summed, providing the total concentration of parental molecules in the hybrid.

3.3.2. Amplification and Characterization of Single CO Molecules

Single CO molecules are intended to be amplified by two rounds of PCR, using a procedure similar to that of **Section 3.3.1**.

Assuming that control reactions performed with F1 somatic DNA (*see* **Section 3.2.8**) allow a priori ruling out the possibility that recombined molecules could be amplified non-specifically, i.e. from parental-type molecules, the outcome of single CO amplification experiments can be confidently envisioned.

Nevertheless, during the first cycles of the first PCR round, parental molecules are present in very large excess over a single CO molecule to be amplified. Hence, they can possibly compete for annealing with ASOs. Consequently, the amplification efficiency of single CO molecules might be sub-optimal, yielding low amounts of DNA at the end of second PCR. When arising, this problem can possibly be solved by increasing the number of first PCR cycles.

The method for quantifying CO molecules is identical to that for parental molecules, described in **Section 3.3.1**, with the following modifications:

- ASO_AL1/ASO_BR1 and then ASO_AL2/ASO_BR2 are used for detecting COs from A to B, and ASO_BL1/ASO_AR1 and then ASO_BL2/ASO_AR2 are used for detecting COs from B to A (*see* **Fig. 14.1a**).
- Carrier DNA is not required and can be replaced by H_2O, since CO molecules are to be detected among a large excess of parental molecules.

Whenever positive reactions can be easily distinguished from negative ones at some DNA dilution, they can be considered to arise most probably from CO molecules.

Figure 14.6 displays a typical result of single CO molecule quantification.

Since the crossing over process always generates reciprocal exchanges, COs from A to B haplotype are expected to be as frequent as those from B to A. Then, once CO concentration has been determined for one orientation (e.g. A to B), only the last steps of the procedure (i.e. one dilution only, steps 33–41 in **Section 3.3.1**) are to be repeated for the other orientation (B to A).

Once the concentration of CO molecules has been determined, it is divided by the concentration of parental molecules to get the CO frequency R. Note that since a crossing over always

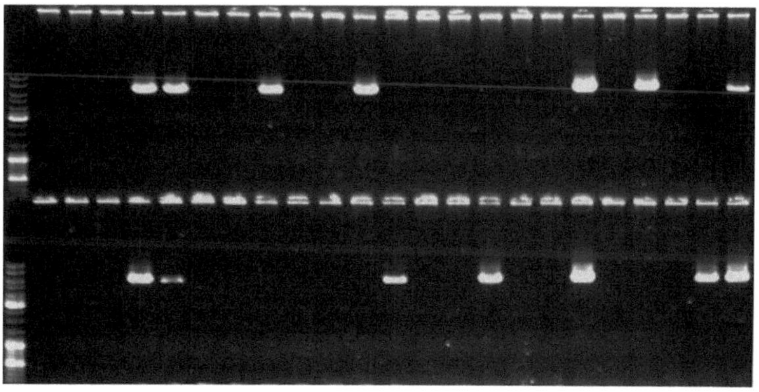

Fig. 14.6. Sample PCR amplification of single CO molecules. Each reaction has been performed using 1 μl of pollen gDNA, diluted at 1/64. Stock solution contains 32,550 parental molecules/μl. Among these 46 reactions, 14 are positive and 32 are negative. The estimated mean number of CO molecules per reaction is $-\ln(32/46) = 0.363$. The estimated total number of CO molecules is $0.363 \times 46 = 16.7$. The concentration of CO molecules in the gDNA extract is $0.363 \times 64 = 23.2$ CO/μl. CO rate is 23.2/32,550 = 1/350.

generates two symmetrical molecules, CO rate is not the sum of A to B and B to A CO rates, but the average.

Once CO rate has been measured, single CO events can be amplified in order to map CO breakpoints along the hotspot. This is most readily performed by sequencing PCR products. As discussed in **Section 3.3**, positive wells among a series of aliquot reactions can result from the amplification of single or multiple molecules. The proportion of positive reactions issued from a single CO molecule can be expressed as a function of the proportion of negative wells:

$$P_0 \times \frac{\ln(P_0)}{P_0 - 1}$$

For example, if $P_0 = 80\%$ of reactions are negative, $\frac{0.8 \times \ln(0.8)}{0.8-1} = 89\%$of positive reactions proceed from a single molecule.

Hence, those PCR amplification products that are yielded by two CO molecules will generate two overlapping sequences over the region extending between the two CO breakpoints. As long as no INDEL-type polymorphism is encountered, the mixed sequence will be readable, in particular at SNP positions where two overlapping peaks should be detected. In theory, the two CO breakpoints can thus be mapped. In practice though, the analysis of mixed products often turns out unworkable, because of the occurrence of INDELs and/or the differential amplification of parental molecules.

It should be noted that such mixed sequences cannot arise from the amplification of heteroduplex pollen DNA. Indeed, unlike spermatozoids in animals, pollen development includes mitotic divisions following meiosis (*see* **Note 3**).

If ASOs are located far enough from the hotspot, no CO breakpoint should be detected in the most outer intervals of PCR products, that is, between the first and second polymorphisms or between the penultimate and last ones (the first and last polymorphisms are priming sites for left and right inner ASOs, respectively). Whenever some events of this kind are observed, their status depends on the occurrence of COs in inner intervals:

- If the distribution is actually truncated, say on one side, then the origin of CO breakpoints in the corresponding outer interval cannot be ascertained. ASOs that are more external are required for amplifying the whole hotspot.
- If most of the outer CO breakpoints in inner intervals are located far away from the ASOs, then CO breakpoints in outer intervals should be suspected to arise from misannealing of ASOs, as discussed in **Section 3.2.8**, and consequently discarded from the analysis.

Given R, the CO rate over the whole hotspot; N, the total number of mapped CO breakpoints; $n_1, n_2, \ldots, n_k$, the number of CO breakpoints mapped in intervals 1, 2,. . ., k, respectively; and $l_1, l_2, \ldots, l_k$, the size (in pb) of intervals 1, 2,. . ., k, respectively, the CO rates (in cM/Mb) in intervals 1, 2,. . ., k are, respectively,

$$\frac{(n_1/N) \times R \times 10^2}{10^{-6} \times l_1}, \frac{(n_2/N) \times R \times 10^2}{10^{-6} \times l_2}, \ldots, \frac{(n_k/N) \times R \times 10^2}{10^{-6} \times l_k}$$

Results can be plotted as in **Fig. 14.1b**.

4. Notes

1. Only ultra-pure dNTPs must be used (e.g. Roche 03622614001). Indeed, depending on the supplier, the yield of PCR reactions can vary over several orders of magnitude, especially for long amplicons. This is most likely due to a poisoning effect of Pfu by dUTP, which is generated at high temperature by dCTP deamination, but is also present as a trace contaminant in all commercial batches of dNTPs.

   BSA should be of molecular biology grade (e.g. MP Biochemicals, #BSAS2001). Depending on the supplier, slight variations in PCR yield can occur.

   A white precipitate sometimes appears in the 10× PCR buffer when preparing a new batch or when thawing an aliquot. It must not be discarded. Instead, buffer should be carefully homogenized before aliquoting or using.

   Prior to any routine use, the performances of each new batch of buffer should be evaluated. For this purpose, control PCR assays are carried out using serial dilutions of a gDNA template, and their yields are compared to those obtained with the previous batch of buffer in the same conditions. *See* **Section 3.2.3** for setting up such an assay. If one gets only a moderate shift (no more than twofold) in gDNA amount required for getting a fixed PCR yield, the buffer is assumed to be satisfactory.

2. Commercial blends of Taq and Pfu (or another *Pyrococcus* species) DNA polymerases that perform task very well are available (e.g. 'long PCR enzyme' mix from Fermentas). Alternatively, homemade mixes of enzymes can be used, but this requires setting up thorough procedures for quality control.

   Whatever be the origin of polymerases, the amount to be used for getting a high yield of clean PCR product has to be determined for each batch. Carry out a series of PCR reactions with an increasing amount of polymerase mix, using universal oligonucleotides designed for setting up pollen typing experiments. Usually, a smear is observed for a given amount of enzyme and above. The optimum should be fixed at half of this quantity.

3. This harsh treatment is required for disrupting cells efficiently. At the same time, it breaks chromosomal DNA, thus lowering its amplifiability. It should then be used with much care. This method is highly efficient for breaking the wall of pollen grain, but is rather ineffective for immature gametophytes (microspores and young bicellular pollen). Hence gDNA is extracted mostly from tricellular pollen grains.
4. Since the extract contains mostly RNA, photometric quantification of DNA is impossible. In order to avoid trapping of ethidium bromide by RNA, add loading buffer supplemented with RNase A (500 μg/ml) to loaded samples.
5. One hundred and twenty-five megabase is probably a gross underestimate of *A. thaliana* Col-0 genome size. Considering 147 Mb as a more plausible value, the weight of one haploid genome is 0.15 pg (14).
6. The contribution of individual nucleotides upon the T_m of an oligonucleotide cannot be accurately predicted, because it depends on many factors, including their sequence context and their position along the oligonucleotide. As a first approach, A or T nucleotides are considered to contribute 2°C, and G or C 4°C. These values are only indicative. In particular, nucleotides added at the 5′-end of an oligonucleotide are expected to have a lesser effect upon its T_m.
7. At low DNA concentrations, a significant proportion of molecules are thought to adsorb onto plastic surfaces, where their availability as templates for polymerization becomes questionable. This concern is readily settled by adding carrier DNA (from calf thymus or salmon sperm) to the reaction.

Acknowledgements

We are grateful to Wayne Crismani, Mathilde Grelon, Anouchka Guyon, Arnaud Ronceret and Nathalie Vrielynck for critical reading of the manuscript and helpful comments.

This work was supported by grants from INRA and ANR (COMEREC1 and COPATH).

References

1. Hunter, N. (2007) Meiotic recombination. In Molecular Genetics of Recombination. A. Aguilera, R. Rothstein, eds. (Berlin: Springer), pp. 381–442.
2. Whitby, M.C. (2005) Making crossovers during meiosis. *Biochem Soc Trans* **33**, 1451–1455.
3. Mezard, C., Vignard, J., Drouaud, J., and Mercier, R. (2007) The road to crossovers: plants have their say. *Trends Genet* **23**, 91–99.
4. Anderson, L.K., and Stack, S.M. (2005) Recombination nodules in plants. *Cytogenet Genome Res* **109**, 198–204.
5. Gabriel, S.B., Schaffner, S.F., Nguyen, H., Moore, J.M., Roy, J., Blumenstiel, B., Higgins, J., DeFelice, M., Lochner, A., Faggart, M., Liu-Cordero, S.N., Rotimi, C., Adeyemo, A., Cooper, R., Ward, R., Lander, E.S., Daly, M.J., and Altshuler, D. (2002) The structure of haplotype blocks in the human genome. *Science* **296**, 2225–2229.
6. McVean, G.A., Myers, S.R., Hunt, S., Deloukas, P., Bentley, D.R., and Donnelly, P. (2004) The fine-scale structure of recombination rate variation in the human genome. *Science* **304**, 581–584.
7. HapMap (2005) A haplotype map of the human genome. *Nature* **437**, 1299–1320.
8. Myers, S., Bottolo, L., Freeman, C., McVean, G., and Donnelly, P. (2005) A fine-scale map of recombination rates and hotspots across the human genome. *Science* **310**, 321–324.
9. Petes, T.D. (2001) Meiotic recombination hot spots and cold spots. *Nat Rev Genet* **2**, 360–369.
10. Kauppi, L., Jeffreys, A.J., and Keeney, S. (2004) Where the crossovers are: recombination distributions in mammals. *Nat Rev Genet* **5**, 413–424.
11. Jeffreys, A.J., Murray, J., and Neumann, R. (1998) High-resolution mapping of crossovers in human sperm defines a minisatellite-associated recombination hotspot. *Mol Cell* **2**, 267–273.
12. Baudat, F., and de Massy, B. (2009) Parallel detection of crossovers and noncrossovers in mouse germ cells. *Methods Mol Biol* **557**, 305–322.
13. Kauppi, L., May, C.A., and Jeffreys, A.J. (2009) Analysis of meiotic recombination products from human sperm. *Methods Mol Biol* **557**, 323–355.
14. Bennett, M.D., Leitch, I.J., Price, H.J., and Johnston, J.S. (2003) Comparisons with *Caenorhabditis* (approximately 100 Mb) and *Drosophila* (approximately 175 Mb) using flow cytometry show genome size in *Arabidopsis* to be approximately 157 Mb and thus approximately 25% larger than the *Arabidopsis* genome initiative estimate of approximately 125 Mb. *Ann Bot* **91**, 547–557.

Chapter 15

Isolation of Meiotic Recombinants from Mouse Sperm

Francesca Cole and Maria Jasin

Abstract

Homologous recombination during meiosis is critical for the formation of gametes. Recombination is initiated by programmed DNA double-strand breaks which preferentially occur at hotspots dispersed throughout the genome. These double-strand breaks are repaired from the homolog, resulting in either a crossover or noncrossover product. Multiple noncrossover events are required for homolog pairing, and at least one crossover is critical for proper chromosome segregation at the first meiotic division. Consequently, homologous recombination in meiosis occurs at high frequencies. This chapter describes how to characterize crossovers and noncrossovers at a hotspot in mice using allele-specific PCR. Amplification of recombinant products directly from sperm DNA is a powerful approach to determine recombination frequencies and map recombination breakpoints, providing insight into homologous recombination mechanisms.

Key words: Meiotic recombination, sperm, crossover, noncrossover, hotspot, allele-specific PCR, F1 hybrid mice, gene conversion, homolog.

1. Introduction

Meiotic recombination does not occur evenly throughout the genome but instead clusters at "hotspots" which are estimated to occur every 25–100 kb in the human genome (1–4). Hotspots are presumed sites for programmed DNA double-strand breaks (DSBs) which initiate meiotic recombination between homologs (5, 6). Crossover (CO) recombination results in reciprocal exchange of flanking sequences (**Fig. 15.1**) and is essential for proper chromosome segregation during meiosis. Each chromosome requires at least one CO, termed the obligate CO, for proper alignment on the metaphase plate and to avoid meiotic non-disjunction which generates aneuploid gametes. However,

H. Tsubouchi (ed.), *DNA Recombination*, Methods in Molecular Biology 745,
DOI 10.1007/978-1-61779-129-1_15, © Springer Science+Business Media, LLC 2011

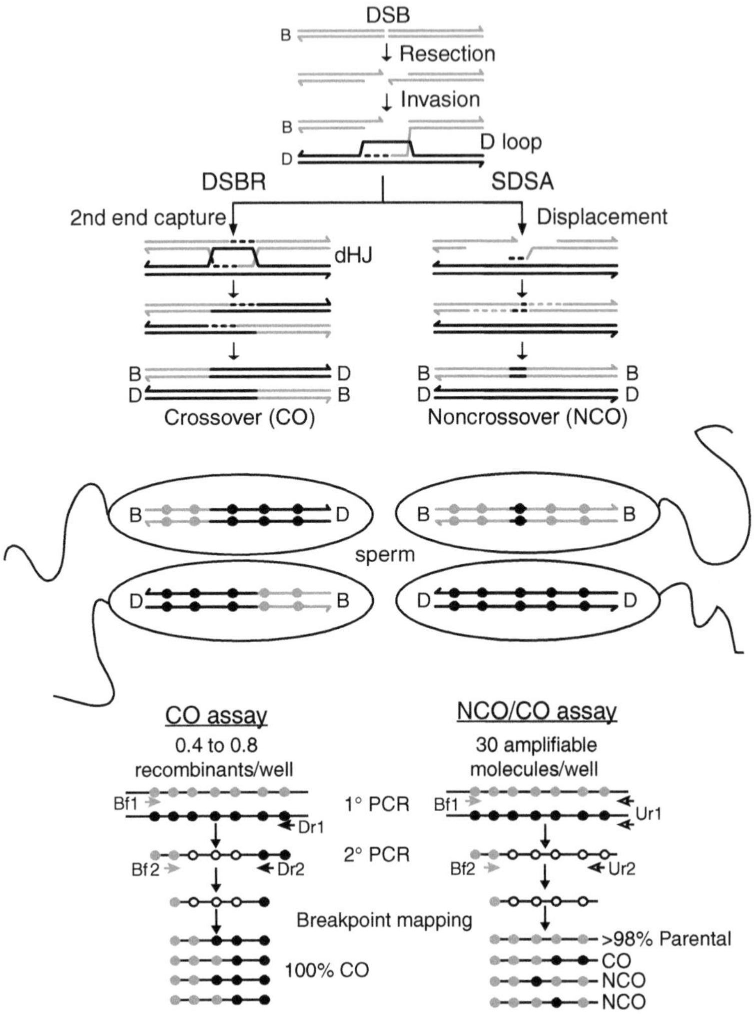

Fig. 15.1. Homologous recombination pathways and assays to amplify COs and NCOs. *Top*: A double-strand break (DSB) is generated on the *gray* chromosome (e.g., B, C57BL/6 J) and the *black* chromosome (e.g., D, DBA/2 J) serves as the homologous donor for repair. After the DSB is generated, 5′ ends are resected and one 3′ single-stranded tail invades the homolog to generate a displacement loop (D-loop). The invading strand serves as a primer to initiate DNA repair synthesis (*dotted lines*). At this point the two major pathways diverge. Most crossovers (CO) are generated by the canonical DSB repair (DSBR) pathway and most noncrossovers (NCO) are generated by synthesis-dependent strand annealing (SDSA). In DSBR, the second 3′ end of the DSB is "captured" to generate a double Holliday junction (dHJ) intermediate which can be resolved to form COs. In SDSA, the invading strand is displaced and anneals to the other 3′ end of the DSB and subsequently repaired to form NCOs. Only one chromatid from each homolog is shown for simplicity, but importantly, sister chromatids are present throughout. *Middle*: After meiosis, recombining chromatids segregate, and after spermiogenesis sperm are formed. *Circles* represent polymorphisms between the B and D genotypes. *Bottom*: Assays to isolate COs and NCOs by PCR in microtiter plates. In the CO assay, nested, allele-specific PCR is performed on small pools of sperm DNA, using primers that flank the hotspot. Only COs are amplified in this assay. In the NCO/CO assay, nested PCR is also performed but only one set of primers is allele-specific, whereas the other set is

COs are only one outcome of recombination. Estimates of DSBs indicate that COs are a fraction of recombinants (7), suggesting that noncrossovers (NCOs) account for a substantial proportion of inter-homolog DSB repair events. NCOs are the result of a patch-like repair of DSBs without exchange of flanking sequences (**Fig. 15.1**).

Experiments in yeast have shown that COs and NCOs derive from the same initiation events which diverge into distinct molecular pathways (8–11). The majority of COs are generated by the canonical DSB repair (DSBR) pathway via a double Holliday junction (dHJ) intermediate (12), while NCOs are thought to be primarily generated by the synthesis-dependent strand annealing (SDSA) pathway (**Fig. 15.1**). After DSB formation and 5′–3′ resection, a 3′ single-stranded tail invades the unbroken homologous template forming a D-loop. In SDSA, the invading strand is extended by a DNA polymerase and is then displaced to re-anneal with the other 3′ end. COs are thought to derive from polymerizing a more stable single-end invasion intermediate which then "captures" the other 3′ end (second-end capture), generating a dHJ. Studies in yeast have shown that resolution of the dHJ primarily generates COs (13).

Meiotic recombination is an excellent system for gaining insight into DSB repair mechanisms, with a detailed understanding of these mechanisms requiring the isolation of both CO and NCO products. The powerful technique of sperm typing pioneered by the laboratories of Norman Arnheim and Alec Jeffreys enables characterization of recombination products (14–16). PCR with allele-specific forward and reverse primers is used to selectively amplify CO recombination products from isolated sperm DNA (CO assay, **Fig. 15.1**). PCR with one allele-specific primer and one universal primer (which can recognize both alleles equally well), followed by genotyping of products, can identify NCO and CO recombination products in the same analysis (NCO/CO assay, **Fig. 15.1**) (17). Analyses of human and mouse hotspots have shown that CO recombination activity can vary substantially between hotspots, from as low as 0.0004% to as high as 2% (18, 19). CO recombination breakpoints at mammalian hotspots typically span 1–2 kb with the peak of CO and NCO recombination breakpoints clustering in the center (17, 20), consistent with both COs and NCOs deriving from the same initiation events, presumably DSBs. The ratio of NCOs to COs at mammalian hotspots ranges between <1:12 and 9:1 (21,

Fig. 15.1. (continued) "universal" (recognizing both parents). In this assay, the majority of products are from the parental genotype, but NCOs and COs are also amplified. In both assays, *white circles* represent polymorphisms derived from one or the other parent.

22). Although it is likely that different hotspots have different propensities for CO or NCO formation, this broad range likely also reflects the ability to identify NCOs in the tested hotspots. NCO gene conversion tracts are very short (17), estimated to be ~100 bp in mouse (22). Thus, the ability to score NCOs at any hotspot is dependent upon the polymorphism density throughout the hotspot but especially close to the most frequent position(s) of recombination initiation.

Mouse is an ideal model in which to study meiotic recombination because of the evolutionary conservation of recombination mechanisms, the ability to generate and utilize mutants, and the availability of a large number of inbred strains, which provide genetic variation across meiotic hotspots. In this chapter, we describe methods to characterize COs and NCOs from mouse meiotic hotspots. For additional information with a focus on human hotspots, see (23).

2. Materials

To prevent contamination, keep all reagents and materials used for PCR in a separate area. Additionally, designated micropipettors are highly recommended.

2.1. PCR Buffers and Reagents

1. 10X Jeffreys' buffer (24): 450 mM Tris-HCl pH 8.8, 110 mM $(NH_4)_2SO_4$, 45 mM $MgCl_2$, 67 mM β-mercaptoethanol, 44 μM EDTA, 10 mM each dATP, dTTP, dGTP, and dCTP, and 1.13 mg/ml non-acetylated bovine serum albumin (BSA). Store in aliquots at –20°C (*see* **Note 1**).
2. 2 M Tris base (2-amino-2-(hydroxymethyl)-1,3-propanediol).
3. Taq DNA polymerase (Abgene AB-0192 http://www.abgene.com/).
4. Cloned Pfu DNA Polymerase.
5. Mouse inbred genomic DNA can be obtained from the Jackson Laboratory (http://www.jax.org/dnares/index.html).
6. S1 nuclease diluted to 10 U/μl in 20 mM Tris-HCl pH 7.5, 50 mM NaCl, 0.1 mM $(CH_3CO_2)_2Zn$ (Zinc acetate), and 50% (v/v) glycerol. Can be stored at –20°C for up to 9 months.

7. 10X S1 buffer: 0.2 M CH_3COONa (Sodium acetate) pH 4.9, 10 mM Zn acetate, 1 M NaCl. Store in aliquots at –20°C.
8. S1 mix: 1X S1 buffer supplemented with 4.8 ng/μl sonicated salmon sperm DNA and 0.7 U/μl S1 nuclease. Freshly prepared.
9. Dilution mix: 10 mM Tris-HCl pH 7.5 and 5 μg/ml sonicated salmon sperm DNA (25). Freshly prepared.
10. 10X Tris-borate-EDTA (TBE) buffer.
11. Loading dye: 0.5X TBE, 30% (v/v) glycerol, supplemented with bromophenol blue.
12. 10 mg/ml ethidium bromide

2.2. Genomic DNA Isolation

1. F1 hybrid mice (e.g., C57BL/6 J × DBA/2 J).
2. 20X Sodium chloride–sodium citrate (SSC) buffer: 3 M NaCl and 0.3 M $HOC(COONa)(CH_2COONa)_2{\cdot}2H_2O$ (citric acid trisodium salt dihydrate), pH 7.0. Prepare 2X, adjust pH to 7.0 prior to use and dilute to 1X and 0.2X SSC from this stock.
3. 80 μm metal mesh (Sigma-Aldrich S3770).
4. β-Mercaptoethanol.
5. 10% (w/v) $CH_3(CH_2)_{11}OSO_3Na$ (SDS).
6. 20 mg/ml proteinase K.
7. Phenol/chloroform/isoamyl alcohol 25:24:1 (v/v/v); saturated with 100 mM Tris pH 8.0.
8. Ethanol, 100% and 70%.
9. 3 M Na acetate, pH 5.2.
10. 5 mM Tris-HCl, pH 7.5.

2.3. Quantification and Quality Assessment of Genomic DNA

1. SYBR green included in a qPCR master mix (e.g., Brilliant® II SYBR® Green QPCR Master Mix, Stratagene).
2. ROX reference dye (1:500).
3. Stratagene Mx3005 real-time PCR instrument or equivalent.

2.4. Allele-Specific Oligonucleotide (ASO) Hybridization

1. 96-Well dot-blot manifold.
2. Denaturation buffer: 0.5 M NaOH, 2 M NaCl, 25 mM EDTA.
3. Whatman filter paper, grade 3.
4. Nylon hybridization membranes (e.g., Hybond™-XL).
5. Multichannel 30–300 μl pipettor.
6. Stratalinker or equivalent.

7. 10 U/μl T4 polynucleotide kinase.
8. 10X kinase mix: 700 mM Tris-HCl pH 7.5, 100 mM $MgCl_2$, 50 mM spermidine trichloride (Sigma S2501), and 20 mM dithiothreitol. Aliquot and store at –20°C.
9. Kinase Stop Solution: 25 mM EDTA, 0.1% SDS, 10 μM ATP. Aliquot and store at –20°C.
10. 10 mCi/ml (γ-^{32}P)ATP.
11. 50X Denhardt's solution: 1% (w/v) Ficoll 400, 1% (w/v) polyvinylpyrrolidone, 1% (w/v) BSA (Fraction V). Filter sterilize and store at 4°C.
12. Tetramethylammonium chloride (TMAC) hybridization solution: 3 M TMAC, 0.6% SDS, 10 mM $NaPO_4$ pH 6.8, 1 mM EDTA, 4 μg/ml yeast RNA in 5X Denhardt's solution. Store at 4°C.
13. TMAC wash solution: 3 M TMAC, 0.6% SDS, 10 mM $NaPO_4$ pH 6.8, 1 mM EDTA. Store at 4°C.
14. Hybridization mesh.
15. 3 mg/ml sonicated salmon sperm DNA.
16. Rotisserie hybridization oven and bottles.
17. Phosphorimager and screen.
18. 0.1% SDS for stripping membranes.
19. 2X and 3X SSC.

2.5. Cloning and Confirmation of NCOs

1. TOPO® TA cloning kit (Invitrogen).
2. Competent cells (e.g., TOP10® chemically competent, Invitrogen).
3. LB agar plates supplemented with 50 μg/ml of ampicillin.
4. 40 mg/ml 5-bromo-4-chloro-3-indolyl-$^{\beta}$-D-galactopyranoside (X-gal).
5. 82 mm nylon hybridization membranes (e.g., Hybond™-XL).
6. India ink.
7. Cloning denaturation buffer: 0.5 M NaCl, 0.5 M NaOH.
8. Cloning neutralization buffer: 1.5 M NaCl, 0.5 M Tris-HCl pH 7.5.
9. 2X and 3X SSC.

3. Methods

Several mouse recombination hotspots have been characterized, some of which have been shown to be differentially active in different strain combinations (e.g., (26–30)). These characterized hotspots represent a tiny fraction of hotspots within the mouse genome, however, such that the approximate position of many additional hotspots can be inferred from crossover maps from a number of sources, for example, recombinant inbred strains (27) or recombinants directly from F1 hybrids (30). Once a known hotspot has been chosen for further analysis or a candidate hotspot region has been narrowed down for investigation, polymorphisms between the two mouse strains used in the crosses must be identified. A goal of the Mouse Genome Project is a haplotype map at 20 kb resolution of 48 mouse strains, with in-depth sequencing of 17 mouse strains (http://www.sanger.ac.uk/resources/mouse/genomes/). However, as yet, DNA sequencing may be necessary to identify/confirm polymorphisms between strains in the region of interest. Once polymorphisms are identified, the methods below detail how to design primers, extract DNA, amplify recombinants from sperm from F1 hybrids, and map recombination breakpoints.

The following nomenclature will be used throughout (*see* **Fig. 15.1**), with the example of C57BL/6J (the "B" strain) and DBA/2J (the "D" strain) as the strain combination to generate F1 hybrids (B × D). Reciprocal recombinants are referred to as B to D and D to B, which can be amplified using forward (Bf) and reverse (Dr) primers for the B to D orientation, with 1° (Bf1 and Dr1) or 2° (Bf2 and Dr2) PCR primer sets. Universal primers that recognize both strains equally well are designated as "U."

3.1. Designing and Assessing Allele-Specific and Universal Primers

Allele-specific PCR primers must be designed and tested to be both efficient in amplifying a single recombinant from a complex genomic milieu and highly specific for the recombinant. In both CO and NCO assays, two serial rounds of amplification are used to ensure efficiency and specificity. Nested sets of allele-specific PCR primers that encompass the hotspot are required. Most mammalian hotspots span 1–2 kb, but the allele-specific primer binding sites may not immediately flank the hotspot. Long-range PCR is reliably efficient and accurate up to 10 kb, although with high-quality DNA and highly efficient allele-specific PCR primers up to 12 kb products can be isolated. The first external round of PCR (1° PCR) requires the primers be both specific and highly efficient. Primers for the second, nested PCR (2° PCR) can be less efficient as the input DNA is not as complex and recombinant molecules should represent a substantial proportion; however, these primers should still have high specificity.

3.1.1. General Guidelines for Primer Design

1. Analyze the sequence of interest for repeat regions and areas of low complexity. Programs for this purpose can be found in sequence analysis software and on the web, for example, http://www.repeatmasker.org/. Avoid repetitive regions for design of universal primers and, if possible, for allele-specific primers.

2. Universal primers are targeted to regions with no polymorphisms. Design universal primers with 50% G/C content and a length of 20–22 nucleotides. Universal primers should also be tested as detailed below to ensure that they efficiently and equivalently amplify both strains and do not compromise the efficiency of the allele-specific reactions.
3. For allele-specific primers, the locations of polymorphisms dictate the sequence. Some sites readily allow for efficient and specific amplification, while others do not; the characteristics that define "good" sites vs. "bad" are not simple and so sites must be tested empirically. Allele-specific primers can range between 16 (high G/C content) and 24 (low G/C content) nucleotides, but most are in the 18–20 nucleotide range. Importantly, the polymorphic nucleotide (or nucleotides in the case of an insertion/deletion polymorphism) is located at the 3′ end of the primer. Four allele-specific primer sets are shown in **Table 15.1**.
4. To increase specificity of a mildly non-specific primer, the primer can be shortened by one or two nucleotides. To increase efficiency, the primer length can be increased (e.g., compare primers to POLY4020 and POLY5465 in **Table 15.1**) or, especially for primers with low G/C content, additional non-templated G or C nucleotides (e.g., GGGG) can be added to the 5′ end of the primer (19). The non-templated nucleotides serve to increase the efficiency of the PCR only after the initial target has been amplified. If amplification of a particular polymorphic site is critical and allele specificity cannot be achieved with these modifications, an intentional mismatch can be included one to two nucleotides 5′ to the polymorphic site (*see* **Note 2**). In this

Table 15.1
Allele-specific primers

Polymorphism	PCR	Primer	Sequence	Length	%GC
POLY665	1°	Bf3	ATAAGCACGTATTTGAGGCC	20	45
		Df3-1	AAGCACGTGTTTGAGGCG	18	56
POLY697	2°	Bf4-1	CAGCAGCTGAGTTAAAACT	19	42
		Df4-1	CAGCAGCTGAGTTAAAACA	19	42
POLY4020	2°	Br4020	TCTCCAACAGTGGGGGAT	18	56
		Dr4020	TCTCCAACAGTGGGGGAC	18	61
POLY5465	1°	Br5465	GTGTCACATTTCAGTTGATGT	21	38
		Dr5465	GTGTCACATTTCAGTTGATGC	21	43

case, optimization of $MgCl_2$ concentration and increasing the length of the primer may be required to get efficient amplification.

3.1.2. Testing Specificity and Efficiency of Primers

1. Set up PCR optimization reactions in a total volume of 10 μl with 1X Jeffreys' PCR buffer, supplemented with an additional 12.5 mM Tris base, 0.03 U/μl Taq polymerase, 0.006 U/μl Pfu polymerase, 0.2 μM of each primer, and 12.5 ng of genomic DNA. *See* **Table 15.2** for an example optimization experiment that tests two forward, two reverse, and one universal primer sets at four different annealing temperatures.
2. Each reaction should be carried out with genomic DNA from each inbred strain (in this example, B and D) used to generate the F1 hybrid to be examined (B × D). Test several annealing temperatures ranging from 56 to 67°C.
3. Use universal primers such that the amplified products approach the size predicted for your CO or NCO assay. Avoid amplifying regions larger than your intended CO or NCO products to avoid potential contamination issues. Conditions for the optimization of PCR are denaturation (1 min at 96°C) followed by 30 cycles of amplification (20 s at 96°C, 30 s at the annealing temperature, and 60 s/kb at 65°C for extension) (*see* **Note 3**).
4. Analyze PCR products by agarose gel electrophoresis. To aid visualization of weak bands, include additional ethidium bromide in the loading buffer (50 ng/ml) and/or transfer the gel for Southern blotting and take a long exposure of the membrane. Examples of good and bad allele-specific primers are shown in Fig. **15.2**, including a bad primer that can be modified.

3.2. Genomic DNA Isolation

To prevent contamination, all samples are preferably prepared in a PCR hood with laminar flow or alternatively in a separate area from what will be used for analysis of PCR products. Preparation of somatic DNA should always be performed separately and with cleaned (*see* **Note 4**) or designated instruments. All steps are performed at room temperature except where noted.

3.2.1. Extracting DNA from Sperm and Somatic Tissue

1. Somatic and sperm DNA should always be prepared from the same mice. Dissect the somatic tissue of choice (e.g., spleen, brain, or liver) first and then dissect the cauda epididymides (**Fig. 15.3**) from 6- to 8-week-old male mice (*see* **Notes 5** and **6**). Place the tissue to be extracted in a clean Petri dish. With a fresh razor blade, finely chop the sample until homogenized. Add 1 ml of 1X SSC and continue to homogenize with the razor blade.

Table 15.2
Primer optimization example

Master mix	1X	132X	Remaining components per reaction (see below)		
10X Jeffreys' PCR buffer	1.0	132			
2 M Tris base	0.0625	8.25	10 μM f primer	0.2	
Taq polymerase (5 U/μl)	0.06	7.92	10 μM r primer	0.2	
Pfu polymerase (2.5 U/μl)	0.024	3.17	ddH_2O	7.49	
Total	1.109	151.34	12.5 μg/ml DNA	1.0	
Forward mix	**55X (μl)**	**Reverse mix**	**55X (μl)**	**Positive mix**	**10X**
Master mix, 132X	61	Master mix, 132X	61	Master mix, 132X	11.09
10 μM universal r primer	11	10 μM universal f primer	11	10 μM universal f primer	2 μl
ddH_2O	412	ddH_2O	412	10 μM universal r primer	2 μl
				ddH_2O	74.9 μl
88 μl each into 5 tubes		88 μl each into 5 tubes		9 μl each into 9 tubes (see below)	
Forward primer mixes			**Reverse primer mixes**		
Rx	**Primer**	**Amt (μl)**	**Rx**	**Primer**	**Amt (μl)**
A	10 μM Bf1	2	F	10 μM Br1	2
B	10 μM Df1	2	G	10 μM Dr1	2
C	10 μM Bf2	2	H	10 μM Br2	2
D	10 μM Df2	2	I	10 μM Dr2	2
E	10 μM Uf1	2	J	10 μM Ur1	2
9 μl each into 9 tubes/wells			9 μl each into 9 tubes/wells		

(continued)

Table 15.2 (continued)

PCRs (in tube or well)			Annealing temperature			
Rx	DNA	Amt (μl)	56°C	59°C	62°C	65°C
1	12.5 μg/ml C57BL/6 J	1.0	Pos1	Pos1	Pos1	Pos1
2	12.5 μg/ml DBA/2 J	1.0	Pos2	Pos2	Pos2	Pos2
–	No DNA		Pos-			
1	12.5 μg/ml C57BL/6 J	1.0	A1	A1	A1	A1
2	12.5 μg/ml DBA/2 J	1.0	A2	A2	A2	A2
–	No DNA		A-			
1	12.5 μg/ml C57BL/6 J	1.0	B1	B1	B1	B1
2	12.5 μg/ml DBA/2 J	1.0	B2	B2	B2	B2
–	No DNA		B-			
			Continue for all samples. . .			

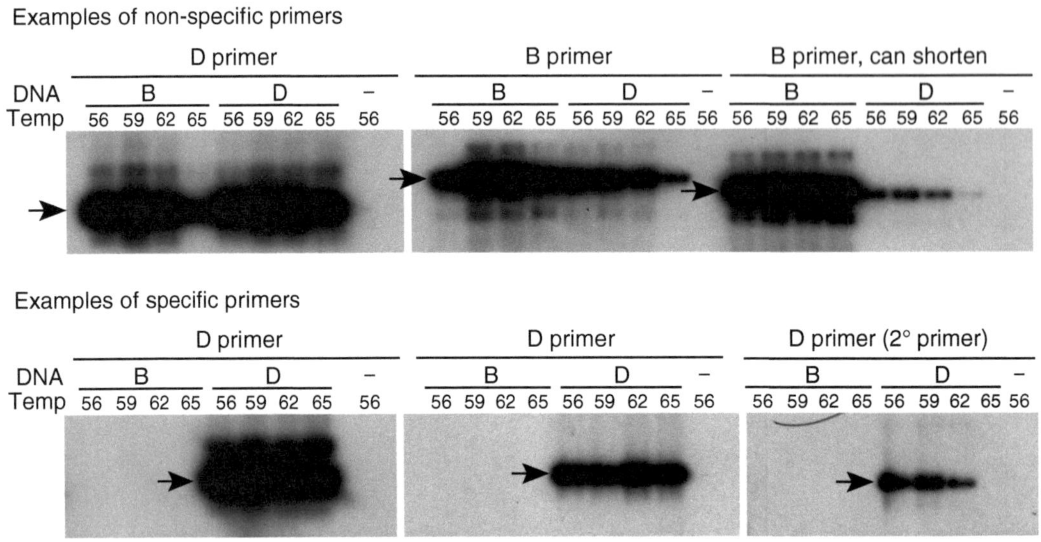

Fig. 15.2. Assessing allele-specific primers. Southern blots of six allele-specific primer optimizations are shown, three of which are assessed to be non-specific and three of which are specific. Each PCR comprises one allele-specific primer, as indicated, and one universal primer. The input DNA (B, D, or no DNA) and annealing temperatures of the PCRs are also indicated. Arrows indicate the expected size of the amplified DNA. Note that the Southern blots are overexposed to detect non-specific amplification. *Top*: Three non-specific primers. The primer on the far *right* shows highly efficient amplification of B DNA, but also amplifies D DNA at a significantly lower level; this primer may have improved performance if shortened by one or two nucleotides. *Bottom*: Three allele-specific primers. The primer on the far *right* is highly specific, but is not as efficient as the other two. This primer can be used successfully for 2° PCRs.

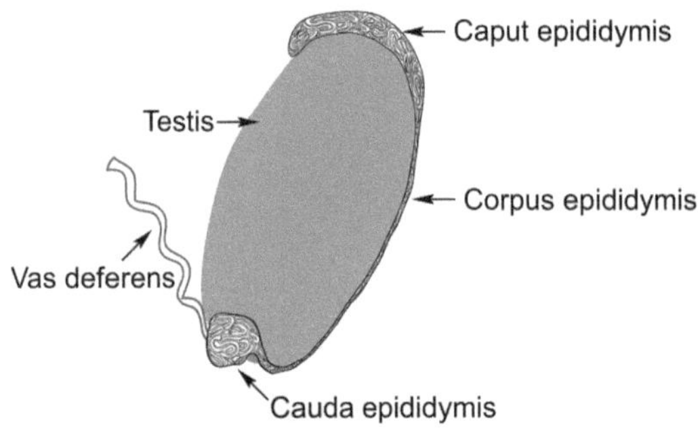

Fig. 15.3. Diagram of the *left* male mouse testis. Mature sperm are isolated from the cauda epididymis.

2. Repeatedly pipet up and down with a transfer pipet to resuspend and retrieve the sample to filter through an 80 μm mesh into a 1.5 ml screw cap tube. Rinse the plate with an additional 0.5 ml of 1X SSC and add through the filter into the same tube.
3. Pellet the cells in a microcentrifuge at 2,700 rcf (~5,000 rpm) for 1 min (*see* **Note 7**). Carefully remove

the supernatant with a 1 ml pipet tip and resuspend the pellet in 1 ml of 1X SSC by vortexing. Repeat one time.

4. For epididymides: Remove 5 μl and assess the cells under a microscope. Sperm should represent >95% of the cells in the mix. If so, go on to Step 5. If somatic cells represent a more significant portion, *see* **Note 8**.
5. Briefly spin the cells, and remove the residual 1X SSC and resuspend in 960 μl of 0.2X SSC. Pellet the cells as in Step 3. Repeat one time and resuspend in 960 μl of 0.2X SSC.
6. Add 120 μl of β-mercaptoethanol (*see* **Note 9**), 20 μl of freshly prepared 20 mg/ml proteinase K, and 100 μl of 10% SDS. Invert to mix and incubate at 55°C for 2 h, inverting occasionally.
7. Split the resulting slurry into four tubes of ~320 μl each. Add an equivalent volume of phenol:chloroform:isoamyl alcohol (25:24:1) to extract the protein. Mix well and spin at 15,000 rcf for 5 min.
8. For the subsequent phenol extractions: Use 1 ml pipet tips with the ends removed at an angle with a clean razor blade. Transfer the top aqueous layer into a clean screw cap 1.5 ml tube. Leave a significant portion of the aqueous phase at the interphase to prevent protein contamination.
9. Re-extract the phenol layer from Step 8 by adding 200 μl of 1X SSC and 20 μl of 10% SDS. Mix well by inverting vigorously and spin as in Step 7. Combine the aqueous phase from the re-extraction to the previous sample.
10. Add 500 μl of phenol:chloroform:isoamyl alcohol (25:24:1) to the combined aqueous phases and spin as in Step 7. Remove the aqueous phase to a fresh tube. Precipitate the DNA by adding 1 ml of –20°C 100% ethanol. Mix well and spin at 15,000 rcf for 5 min. Decant supernatant and wash the pellet with 70% ethanol. Repeat the centrifugation step and decant the supernatant. Briefly spin and remove the last of the supernatant with a pipet tip.
11. Immediately resuspend the pellet in 75 μl of ddH_2O and pool the four aliquots together. Add 1/10 volume of 3 M Na acetate (pH 5.2) and three volumes of –20°C 100% ethanol. Mix well and spin as in Step 10. Decant supernatant and wash the pellet with 70% ethanol. Air dry the pellet (*see* **Note 10**). Resuspend in 100 μl of 5 mM Tris pH 7.5. Incubate at 55°C for 1 h with occasional mixing, transfer to 4°C overnight. Store indefinitely at –20°C.

3.2.2. Quantification and Quality Assessment of Genomic DNA

1. Test several amounts of your genomic DNA (e.g., 1, 2, and 3 μl into 1 ml each) and quantify the concentration by measuring the absorbance at 260 nm with a spectrophotometer (*see* **Note 11**). Make a working stock concentration of DNA of 20 ng/μl in 5 mM Tris pH 7.5 and retest the absorbance at 260 nm. Adjust the concentration as necessary.
2. Run 1, 5, and 10 μl samples of the working stock concentration of genomic DNA preparations on an agarose gel to verify that the DNA is of high molecular weight and to confirm the relative concentrations. Make sure to include sperm and somatic DNA on the same gel for comparison.
3. If comparison of absolute recombination frequency between samples is essential to your study, the concentration of different DNA samples can be equalized after assessment by quantitative PCR (qPCR). Design universal primers that generate a 100–150 bp amplified product of the same sequence regardless of the parental strain (i.e., containing no polymorphisms) and completely located within your experimental primary (1°) PCR product for CO and NCO assays (see below). Ensure that these primers equivalently amplify both inbred strains used to generate the F1 hybrid of interest, by testing them with serial dilutions of known quantities of pure inbred genomic DNA. Compare samples with at least three replicates containing 0.2 ng of genomic DNA with SYBR green for qPCR in a final volume of 20 μl. qPCR conditions in a Stratagene Mx3005 or similar are denaturation (10 min at 95°C) followed by 40 cycles of amplification (30 s at 95°C, 1 min at 60°C, and 30 s at 72°C) and finishing with 1 min at 95°C, 30 s at 55°C, and 30 s at 95°C.

3.2.3. Assessing the Amplification Efficiency of Genomic DNA

Even with careful genomic DNA extraction, shearing will occur and proteinase K digestion can be incomplete, both of which can result in non-amplifiable genomic DNA. In order to accurately quantify recombination activity at your locus, the amplification efficiency across your hotspot for each genomic DNA sample used in your study must be calculated.

1. For each genomic sample to be assayed, perform a series of PCRs using allele-specific primers directed against one side of the hotspot and universal primers on the other (**Fig. 15.4**). Because each strain and allele-specific primer must be checked separately, four separate sets of reactions are generated for all samples.
2. Each PCR is seeded with 12 pg of genomic DNA from a dilution generated from your working stock. As the haploid mouse genome is ~3 pg, this corresponds to two copies of the region of interest from each parental strain. Multiple

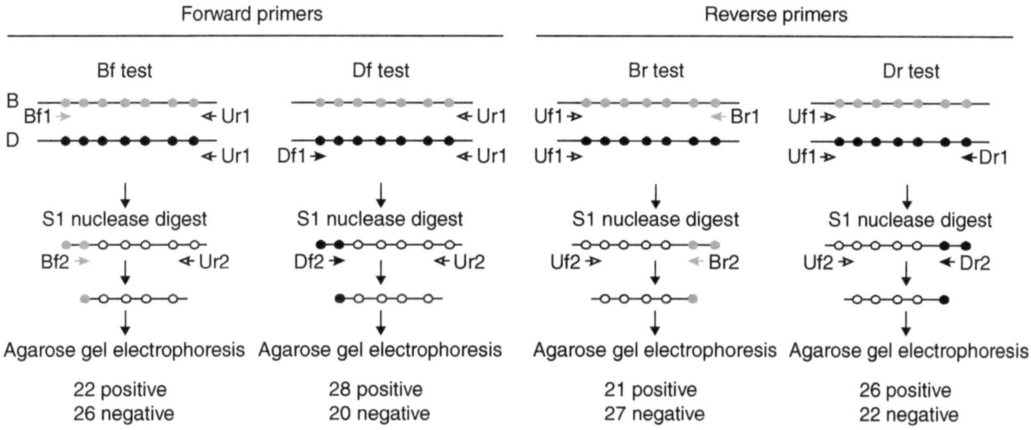

Fig. 15.4. Strategy for determining amplification efficiency. For each DNA preparation, four separate PCRs are required to determine amplification efficiency. Each PCR contains 12 pg of DNA per well, and 48 wells are typically assayed. At the *bottom* of each experimental design, sample results are shown.

sets of reactions are required to get an accurate assessment of amplification efficiency, so we routinely perform 24–48 PCRs per set, such that each DNA sample is tested in 1–2 96-well plates.

3. Perform the primary (1°) PCRs in a total volume of 8 μl, as described in **Section 3.1.2**. Conditions for the PCR are denaturation (1 min at 96°C) followed by 26 cycles of amplification (20 s at 96°C, 30 s at the optimized annealing temp, and 60 s/kb at 65°C for extension).
4. Immediately upon completion of the 1° PCR, S1 nuclease digest the reactions to eliminate single-stranded DNA by seeding 0.5 μl of each 1° PCR into 5 μl of S1 Mix per reaction and incubating for 20 min at room temperature. Add 45 μl of dilution buffer to each well.
5. Use 1.6 μl of the S1-digested samples to seed a secondary (2°) 8 μl PCR with nested allele-specific and universal primers. There is no need to inactivate the S1 nuclease prior to addition to the 2° reaction. Conditions for the 2° PCR are denaturation (1 min at 96°C) followed by 27 cycles of amplification (20 s at 96°C, 30 s at the optimized annealing temp, and 60 s/kb at 65°C for extension).
6. The samples are then analyzed by agarose gel electrophoresis and the number of positive wells (N_{pos}), negative wells (N_{neg}), and total wells (N_{tot}) are determined. To calculate the amplification efficiency that corresponds to a particular combination of allele-specific primers in an experiment (e.g., a CO assay using Bf1 to Dr1 followed by Bf2 to Dr2), use the well-count numbers from the allele combinations that had the smaller N_{pos}. In the example shown in **Fig. 15.4**,

the smaller N_{pos} corresponds to the Bf test ($N_{pos} = 22$ for Bf test vs. $N_{pos} = 26$ for the Dr test).

7. Use the Poisson approximation $p(n) = \frac{\mu^n e^{-u}}{n!}$ to estimate the amplification efficiency. Determine μ_{amp}, the average number of molecules amplified/well. By setting $n = 0$, μ_{amp} can be calculated because $p(0) = \frac{N_{neg}}{N_{tot}}$ and therefore, $\mu_{amp} = -\ln\left(\frac{N_{neg}}{N_{tot}}\right)$. The amplification adjustment factor is the average number of amplifiable molecules per seeded molecule and is calculated as $\frac{\mu_{amp}}{m}$, where m is the number of molecules seeded in each reaction ($m = 2$). In later calculations, the variance for $\mu_{amp}(\sigma^2_{\mu\ amp})$ will need to be estimated, which is $\sigma_\mu{}^2 = \frac{1}{N_{neg}} - \frac{1}{N_{tot}}$. For the example depicted in **Fig. 15.4**, $\mu_{amp} = 0.6131$, the amplification adjustment factor is 0.3065, and $\sigma^2_{\mu_{amp}} = 0.0177$. The amplification adjustment factor is typically 0.3–0.8.

3.3. Amplification of Recombinants from Sperm DNA

For amplification of COs, two rounds of nested allele-specific PCR are performed (e.g., Bf1 to Dr1 and Bf2 to Dr2, **Fig. 15.1**). For amplification of NCOs, two rounds of nested PCR are performed, using allele-specific primers to only one side of the hotspot with universal primers to the other side (e.g., Bf1 to Ur1 and Bf2 to Ur2, **Fig. 15.1**). Note that the latter primer sets amplify parental, non-recombinant DNA (the major amplification product), while also amplifying NCOs and COs. An advantage of this method is that the relationship between NCOs and COs at the hotspot can be discerned; moreover, NCOs are amplified nonselectively. Other approaches have been developed that selectively isolate NCOs based on either hybridization or PCR (23, 31).

3.3.1. Amplification of COs

As the recombination activity of mammalian hotspots is quite variable, a series of input DNA concentrations (hereafter referred to as "pools") should be tested to determine the ideal range before performing large-scale experiments. As the haploid sperm DNA content is ~3 pg, 6 pg of mouse sperm DNA contains one copy of the allele being amplified. Calculate the amplification efficiency as outlined in **Section 3.2.3** to determine the number of amplifiable molecules in the input pools by dividing the desired number of molecules by the amplification adjustment factor. In the example shown in **Section 3.2.3**, one amplifiable molecule would be found in 19.58 pg of DNA (6 pg/0.3065). To control for contamination and any PCR-derived artifacts, include negative controls with no DNA (mock) and somatic DNA.

1. Set up a series of PCR amplifications (*see* **Fig. 15.1**, CO assay) with multiple small pools containing varying concentrations of sperm DNA. For example, in a 96-well plate,

set up 12 reactions of 100, 200, 500, 1,000, and 3,000 amplifiable molecules per well. Include 12 wells with no DNA for contamination controls. Also, set up 24 wells with 2,400 amplifiable molecules of somatic DNA, such that the total amount of amplifiable molecules tested for somatic and sperm DNA is equivalent (57,600 molecules, in this example).

2. Set up the PCRs in a total volume of 8 μl with 1X Jeffreys' PCR buffer, supplemented with an additional 12.5 mM Tris base, 0.03 U/μl Taq polymerase, 0.006 U/μl Pfu polymerase, 0.2 μM of each primer, and the desired amount of genomic DNA. Conditions for the 1° PCR are denaturation (1 min at 96°C) followed by 26 cycles of amplification (20 s at 96°C, 30 s at the optimized annealing temp, and 60 s/kb at 65°C for extension). *See* **Note 12**.

3. Immediately upon completion of the 1° PCR, S1 nuclease digest the reactions as outlined in **Section 3.2.3**, Step 4. Add 45 μl of dilution buffer and use 1.6 μl to seed the 2° PCRs.

4. Set up and perform the 2° PCRs in a total volume of 8 μl, as outlined above with nested allele-specific primers. Thirty cycles should be sufficient to see positive reactions by 2° PCR.

5. Add 2.5 μl of loading dye and run 3 μl of each sample on an agarose gel. The somatic and mock controls should show no positive reactions (*see* **Note 12**). It should be apparent that increasing concentrations of input DNA result in a larger fraction of positive wells.

6. Use the calculations outlined in **Section 3.2.3**, Step 7 to estimate μ, the average number of recombinants per well in each pool size. Select a range of input DNA that corresponds to 0.4–0.8 recombinants per pool and perform a number of CO amplifications. A low (0.4) and high (0.8) input in 1–2 96-well plates per orientation (Bf to Dr vs. Df to Br) is ideal to map CO breakpoints. This range gives an ample number of CO events but not so many as to make accurately estimating the frequency by Poisson correction problematic. Make sure to include mock and somatic DNA controls in each experiment. For accurate assessment of the distribution of CO breakpoints, aim to isolate ~100 COs per orientation. *See* **Table 15.3** for an example of the PCR set-up for a CO assay.

7. To later estimate the recombination frequency in total and per interval, note the number of positive, negative, and total wells/experiment. Keep a record of these data for each DNA sample and each input amount.

Table 15.3
CO assay example

Master mix	1X	232X		
10X Jeffreys' PCR buffer	0.8	185.6		
2 M Tris base	0.05	11.6		
Bfl	0.16	37.1		
Drl	0.16	37.1		
Taq polymerase (5 U/μl)	0.048	11.1		
Pfu polymerase (2.5 U/μl)	0.0216	5.0		
Total	1.24	287.5		
Wells	**18 mock (no DNA)** (μl)	**36 somatic DNA (2,025 mol/well)** (μl)	**80 sperm DNA (300 mol/well)** (μl)	**80 sperm DNA (600 mol/well)** (μl)
Master mix, 232X	22.3	44.6	99.2	99.2
DNA (20 ng/μl)[a]	0	71.4	23.5	47
ddH_2O	121.7	171.96	517.3	493.8

Plate 8 μl per well in 2X 96-well plates with 8 mock, 16 somatic controls, 36 sperm DNA at 300 amplifiable molecules/well and 36 sperm DNA at 600 amplifiable molecules/well.

[a]Note that the amplification adjustment factor was previously calculated to be 0.3065 for the sperm DNA preparation. We are simplifying here by using the same number for somatic DNA.

3.3.2. Amplification of NCOs and COs: the NCO/CO Assay

As the large majority of DNA amplified in this assay will be of the parental genotype, the input pool sizes are much smaller than in the CO assay. With smaller pools, NCOs and COs can be detected within the context of a large excess of non-recombinant, parental DNAs. Because NCO gene conversion tracts are short and frequently encompass only a single polymorphism, the ability to score NCOs at any hotspot is highly dependent upon the polymorphism density.

1. Set up 8 μl PCRs as detailed in **Section 3.3.1** using allele-specific forward primers against universal reverse primers (e.g., Bfl to Url) or vice versa. Perform the assay in bulk in 96-well plates with 30 amplifiable molecules per well (*see* **Note 13**). In a separate PCR machine, set up eight positive hybridization control reactions (*see* **Note 14**) with the alternate allele-specific primer (e.g., Dfl to Url).
2. Immediately upon completion of the 1° PCRs, add 35 μl of dilution buffer to each well (*see* **Note 15**).
3. Use 0.6 μl of the diluted 1° PCRs to seed 15 μl 2° PCRs containing the nested allele-specific and universal primers for both the experimental plate(s) and the positive hybridization controls.

4. Conditions for the 2° PCR are denaturation (1 min at 96°C) followed by 36 cycles of amplification (20 s at 96°C, 30 s at the optimized annealing temperature, and 60 s/kb at 65°C for extension).
5. To later estimate the recombination frequency in total and per interval, note the number of positive, negative, and total wells/experiment. Frequencies will be determined after allele-specific oligonucleotide (ASO) mapping detailed in the next section. Keep a record of these data for each DNA sample and each input amount.
6. Once the NCO frequency has been determined (**Section 3.4.3**), a suitable number of PCRs should be performed with somatic DNA, as a negative control, to assess the frequency of PCR mis-incorporations at particular alleles and to confirm that the NCOs observed are meiosis-specific.

3.4. Genotyping Recombinants by Allele-Specific Oligonucleotide (ASO) Hybridization

For CO breakpoint mapping, most of the reactions should contain a single recombinant. Amplified DNA from positive 2° PCRs can be purified using standard techniques and cloned or sequenced. If higher yields are needed from the 2° PCRs, a tertiary (3°) round of PCR can be performed using either nested allele-specific or universal primers. Alternatively, positive 2° reactions (and some negative reactions for controls) can be subjected to 3° PCR to generate enough DNA for dot-blotting and allele-specific oligonucleotide (ASO) hybridization. For breakpoint mapping in the NCO/CO assay, the ASO approach is the preferred method and is outlined here.

3.4.1. Generation of Replica Dot-Blots for ASO Hybridization

1. For the CO assay, perform 3° PCRs with either nested allele-specific or universal primers in a total volume of 30 μl seeded with 0.75 μl of the 2° PCRs as outlined in **Section 3.3.1**, Step 2. Use all positive 2° PCRs and include some somatic controls. Also, for a positive hybridization control, PCR amplify genomic DNA from each parental strain that generated the F1 hybrid. Conditions for the 3° PCR are denaturation (1 min at 96°C) followed by 21 cycles of amplification (20 s at 96°C, 30 s at the optimized annealing temperature, and 60 s/kb at 65°C for extension). Add 7.5 μl of loading dye to each sample.
2. For the NCO/CO assay, directly use the 2° PCRs. We routinely perform a duplicate 2° PCR to generate a larger amount of amplified DNA for dot-blotting (*see* **Note 16**). Combine the two 2° PCRs and add 7.5 μl of loading dye to each sample. The eight positive control tubes from the first round of the 2° PCR should be combined and 60 μl of loading dye added.

3. Prior to generating the dot-blots, load 1.0 μl of randomly chosen samples (always include the positive hybridization controls) on an agarose gel to check the quality and quantity of the PCRs.

4. For the NCO/CO assay, replace one well of the 96-well plate (*see* **Note 17**) with 30 μl of positive hybridization control (**Fig. 15.5**). Additionally add positive hybridization control DNA to 5 wells of the 96-well plate at ratios of 1:10, 1:30, 1:100, 1:300, and 1:1,000 (**Fig. 15.5**). In the ideal scenario, the positive signal from an NCO or CO product should be close to or exceed the signal observed in the 1:30 dilution.

5. Following manufacturer's instructions for the 96-well dot-blot manifold, cut Whatman filter paper (two to three pieces) and nylon membranes (10 replicates) to the appropriate sizes. Wet the filter paper and a membrane with ddH_2O and assemble the manifold, applying a vacuum.

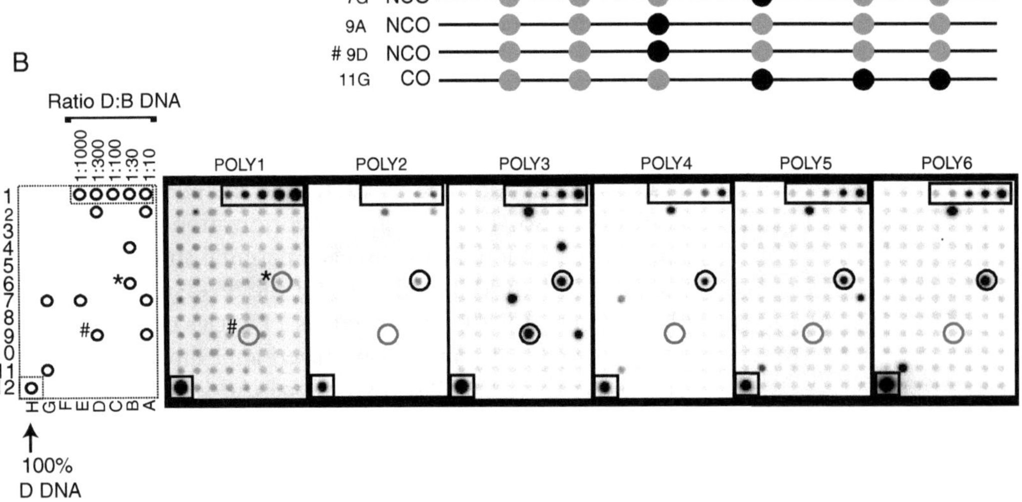

Fig. 15.5. Mapping NCOs and COs using ASO hybridization. (**a**) PCR strategy for the NCO/CO assay. In this example, B parental DNA and recombinants are amplified, such that dot-blots are probed with ASOs that recognize D polymorphisms. (**b**) DNA is amplified in a 96-well plate and then replica dot-blots are generated using a 96-well manifold for ASO hybridization. Six separate polymorphisms (POLY1-6) are tested in this example. All recombinants identified in this example are indicated with *circles* in the *left* diagram, with two recombinants highlighted on each blot (*, CO in well 6B; #, NCO in well 9D). Positive controls (*boxes*) are amplified D DNA located at the 12H location and a dilution series at the indicated ratios of D into B DNA located at 1A through 1E. (**c**) Maps of all NCOs and COs identified in B.

6. Add 280 μl of denaturation buffer to each PCR and pipet up and down to mix (*see* **Note 18**). Generate replica dot-blots by loading 30 μl/well onto the assembled dot-blot manifold. Rinse each well with 150 μl of 2X SSC. Remove the membrane and repeat until all replicas are generated. UV crosslink the membranes with a Stratalinker or similar apparatus and proceed to hybridization.

3.4.2. Probe Preparation and ASO Hybridization Conditions

For CO assays, for each polymorphism prepare probes to each parental genotype. Perform one round of hybridization with one parental genotype, followed by stripping the ASO probe and reprobing with the other genotype. For NCO/CO assays, hybridize only with the probes for the parental genotype that was not amplified. For example, if the B genotype was amplified (**Fig. 15.1**), probe only with D genotype ASOs (**Fig. 15.5**).

1. ASOs are 18-mers that contain the polymorphism in the middle of the oligonucleotide, typically the 8th or 12th nucleotide from the 5′ end in the case of a single nucleotide polymorphism. Frequently short insertion/deletion polymorphisms are found at hotspots and can be used to generate excellent ASOs (and allele-specific primers); design the ASO with the insertion/deletion polymorphism in the center as well. ASOs are stored in ddH_2O at 0.8 mg/ml. Prior to use, dilute in ddH_2O to 8 μg/ml (1:100). For each ASO hybridization, ASOs specific to both parental DNAs are required – one to hybridize and the other as a competitor. For example in **Fig. 15.5**, ASOs to D were labeled and ASOs to B served as the competitor.
2. In screw cap microcentrifuge tubes, assemble the kinase reaction in a final volume of 10 μl containing 1X kinase mix, 0.35 μl T4 polynucleotide kinase (10 U/μl), 0.2 μl (γ-^{32}P)ATP (10 mCi/ml), and 1 μl ASO (8 μg/ml). Incubate at 37°C for 45 min. Add 20 μl of Kinase Stop Solution, mix and centrifuge the sample. Add 20 μl unlabeled competitor ASO (8 μg/ml) (*see* **Note 19**).
3. Wet the dot-blot membranes with 3X SSC and place in a small hybridization bottle (DNA facing inside). Multiple dot-blots from separate experiments can be hybridized with the same probe by separating the membranes with hybridization mesh and increasing the volume of solutions accordingly.
4. Pre-hybridize a single membrane in a rotisserie hybridization oven with 3 ml of pre-warmed TMAC hybridization buffer (*see* **Note 20**) at 56°C for 10–15 min. Pour off this buffer and add 2.5 ml of fresh TMAC hybridization buffer supplemented with 7 μl of 3 mg/ml sonicated salmon sperm DNA (freshly boiled for 5 min and stored on ice until use). Reduce

the hybridization oven temperature to 53°C and rotate for an additional 10 min.

5. Add the ASO probe directly into the hybridization buffer at the bottom of the bottle, swirl the bottle to distribute, and continue to incubate at 53°C with rotation for 45 min.
6. Wash the membranes three times with 2.5 ml of pre-warmed TMAC wash buffer over 20 min with rotation at 56°C. Change solution to 4 ml of pre-warmed TMAC wash buffer for a final wash of 15 min.
7. Rinse the membrane in the bottle two times with 3X SSC and transfer to a tray containing 3X SSC. Blot off excess liquid, wrap in plastic, and expose for 45 min to 1 h on a phosphorimager screen (*see* **Note 21**).
8. For CO breakpoint mapping, strip the membranes and reprobe the same membrane with the labeled ASO of the opposite genotype. It is important to use the same membrane to account for dot-blot to dot-blot variation. Strip probes from membranes by repeatedly washing with boiled 0.1% SDS until the signal is sufficiently reduced when scanning with a Geiger counter.

3.4.3. Scoring COs and NCOs

Once the ASO hybridization has been performed, the information can be assembled to generate CO and NCO maps. It is important that polymorphisms that flank the hotspot are included in the ASO hybridization to assess the genotype of recombinants and to avoid including parental DNA that was non-specifically amplified in the quantification.

3.4.4. Scoring COs from the CO Assay

1. In the CO assays, the majority of amplification-positive reactions should contain only a single recombinant (*see* **Fig. 15.6**). The genotype of these recombinants and the location of the CO breakpoints should be readily apparent, as a positive signal will be seen for only one ASO at every site. However, particularly in larger pool sizes, two or more recombinants can be amplified in the same well, and if they do not share the same breakpoints, polymorphisms will score positively for both ASOs at some sites (in **Fig. 15.6c**, wells 3 and 9). In these cases, the two exchanges from the unique to mixed genotypes are scored as two CO breakpoints. A small number of pools may contain two independent recombinants with identical breakpoints (i.e., invisible doubles); the problem of multiples is kept to a minimum by keeping the fraction of positive pools within the suggested range (0.4–0.8).
2. Positive signals from somatic controls will likely reflect non-specific amplification of parental DNA and can help identify problematic allele-specific primers. If such non-specific

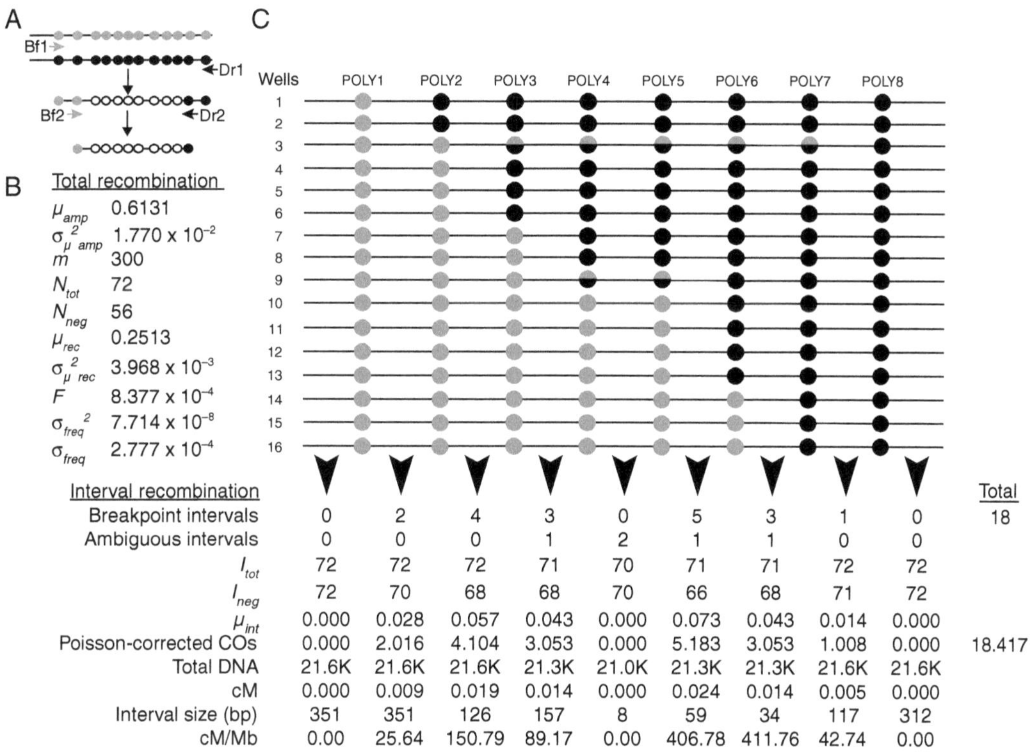

Fig. 15.6. Calculating recombination frequency. (**a**) PCR strategy for the CO assay in which COs are isolated in the B to D orientation. (**b**) Calculations to determine the total CO recombination frequency across the hotspot. (**c**) Breakpoint maps for COs amplified from the 16 positive wells. Polymorphisms that hybridized to both parental genotypes in wells 3 and 9 are indicated with *half gray/half black circles*. CO frequency calculations for each interval are indicated below.

amplification is observed at a significant frequency, any breakpoints that appear to occur between the outermost genotyped polymorphisms and the internal allele-specific primers are also likely to be from non-specific amplification of parental DNA and should be discounted.

3. Occasionally a CO will flip between genotypes multiple times across the hotspot (*see* **Note 22**). These could be bona fide discontinuous gene conversion tracts; however, as they complicate the mapping of CO breakpoints, they may be counted in the overall CO frequency but eliminated from the breakpoint map. Such events are rare in our experience, so omitting them has little impact on the shape of the breakpoint map.

3.4.5. Scoring NCOs and COs from the NCO/CO Assay

1. Examples of NCO/CO assay dot-blots are shown in **Fig. 15.5**. COs show no hybridization with the ASO probe(s) for polymorphisms on one side of the hotspot and then flip to showing hybridization with the remaining polymorphisms, whereas the NCO conversion tracts often encompass only a single polymorphism. As the pool size is 30

amplifiable molecules, the intensity of the positive hybridization signals for both COs and NCOs should fall between those observed from the 1:100 and the 1:10 dilutions of the positive control. The CO frequency should concur with what was observed in the CO assay and can be used as a gauge for the success of the NCO/CO assay.

2. Very faint hybridization signals should be evaluated carefully, for example, by comparing with somatic control plates. A signal fainter than the 1:100 dilution control should be discounted as potential PCR mis-incorporation or a hybridization artifact.

3. Due to the fact that the pool size is small, multiple recombinants per well are rare, but they do occur. *See* **Section 3.5** for statistical analysis which takes this into account and *see* **Section 3.6** for a method for cloning recombinants derived from the NCO/CO assay.

4. If a hotspot shows preferential initiation of meiotic recombination on one parental chromosome over the other (e.g., B >> D, **Fig. 15.1**), the frequency of NCOs can be quite low for the non-initiating parent (D). Nonetheless, both orientations of COs (e.g., B to D and D to B; **Fig. 15.1**) are equivalent in frequency and serve as a critical control in this instance.

3.5. Calculation of CO and NCO Frequencies

CO and NCO frequencies are estimated using similar approaches. To account for multiple events per well (some of which may have identical breakpoints), overall numbers of COs and NCOs are Poisson-corrected to estimate the true frequency. **Figure 15.6** shows a model CO experiment with 16 polymorphisms typed across the hotspot. In this experiment, 300 amplifiable molecules per well were assayed in 72 wells, with 56 negative wells and 16 positive wells, 2 of which contained at least two COs. Total activity across the hotspot and activity for each interval on **Fig. 15.6** were calculated as described below. For further information on recombination statistics, see (31, 32).

1. For total CO activity across the hotspot (**Fig. 15.6b**), calculate the mean number of recombinants per well (μ_{rec}) and the variance of μ_{rec}($\sigma_{\mu}{}^{2}$rec) as shown above in **Section 3.2.3**, Step 7 for different DNAs (i.e., biological replicates) and pool sizes separately. The frequency of recombination (F) equals μ_{rec} divided by the number of amplifiable molecules per well (m). The variance of the frequency ($\sigma_{freq}{}^{2}$) takes into account the variance from both the estimation of μ_{rec} and the μ_{amp} and equals $F^2\left(\frac{\sigma_{\mu}{}^{2}{}_{rec}}{\mu_{rec}{}^{2}}+\frac{\sigma_{\mu}{}^{2}{}_{amp}}{\mu_{amp}{}^{2}}\right)$. The standard deviation (σ_{freq}) is the square root of the variance.

2. For CO activity for each interval, determine the number of Poisson-corrected COs per interval. Quantify different DNAs and input pool sizes separately. Order the COs in a tabular format and count the number of breakpoints within each interval (**Fig. 15.6c**). Include the intervals between the nested allele-specific primers and the outermost tested polymorphisms. In the event of wells with more than one CO (e.g., wells 3 and 9), the first and last breakpoints are considered unambiguous, whereas the ones in between are considered ambiguous (as they may contain hidden CO breakpoints). On a per interval basis, calculate the number of wells with ambiguous breakpoints and subtract that from the total number (72 in this example), to obtain the total number of unambiguous intervals (I_{tot}). Calculate the negative intervals (I_{neg}) by subtracting the number of breakpoint intervals from I_{tot}. Note that the number of positive wells in the experiment shown in **Fig. 15.6c** is 16, but there are 18 breakpoint intervals because of two wells with multiple events. Similar to **Section 3.2.3**, Step 7, calculate the average number of breakpoints per interval, μ_{int}. The Poisson-corrected number of COs per interval is $\mu_{int} \times I_{tot}$. To calculate centiMorgans (cM), divide the Poisson-corrected number of COs per interval by the total number of corrected input molecules per interval (Total DNA = $m \times I_{tot}$), and multiply by 100. For CO activity in cM/Mb, divide cM by the number of bp in the interval and multiply by 10^6. See the intervals in the example in **Fig. 15.6c** for calculations.
3. For NCO frequency across the entire hotspot or at a particular polymorphism, calculate as in Steps 1 and 2, respectively.

3.6. Cloning and Confirmation of NCOs

ASO hybridization is a powerful method to detect NCOs, but further information can be gleaned by cloning. For example, while most wells containing NCOs will display conversion of only a single polymorphism, a significant fraction may encompass two or more polymorphisms. These latter wells may arise from a single NCO by co-conversion of adjacent polymorphisms or from two separate NCOs found in the same well by chance, which can be distinguished by cloning. NCOs derived from the NCO/CO assay represent a small fraction of the amplified DNA, with the non-recombinant parental genotype in large excess (i.e., ~1 in 30). Here is a straightforward method to clone and confirm the identity of NCOs from a sea of parental DNA.

1. After identifying wells with NCOs of interest, use 0.6 μl from the 1° PCRs to seed a 2° PCR in a total volume of 15 μl. These PCRs are identical to the 2° PCRs performed for the NCO/CO assay in **Section 3.3.2**, Steps 3 and 4, with two exceptions: they do not include Pfu polymerase and

the concentration of Taq polymerase is increased to 0.0037 U/μl. Also, after 36 PCR cycles, an extended extension step at 65°C for 7 min is added.

2. Run 1–2 μl of these 2° PCRs on an agarose gel to confirm the quality and quantity of the amplified DNA.
3. Perform a TOPO® TA (Invitrogen) ligation reaction with 1 μl of the 2° PCR (*see* **Note 23**), following the manufacturer's instructions.
4. Transform 1–2 μl of the ligation reaction into competent cells (e.g., TOP10® chemically competent) by standard protocols. Prepare LB agar plates supplemented with 50 μg/ml ampicillin (or other antibiotics depending upon the vector used) and spread with 40 μl of 40 mg/ml X-gal. Plate the transformants at several concentrations to obtain an adequate number of colonies to ensure the presence of colonies with the NCO, but not too dense to impair picking these colonies later. A good range to plate is 5, 20, and 75% of the transformation mix. Incubate overnight at 37°C.
5. Let plates come to room temperature. Place dry 82 mm disc nylon membranes (e.g., Hybond™-XL) on the plates and leave for 30 s. At this time, puncture the membrane at three spots around the perimeter of the disc with an 18-gauge needle soaked in India ink to orient the membranes (*see* **Note 24**).
6. Place the membrane colony side up on Whatman filter paper soaked with cloning denaturation buffer for 2–5 min, followed by two subsequent incubations on Whatman filter paper soaked with cloning neutralization buffer for 3 min.
7. Wash in 2X SSC, dry the membranes, and crosslink in a Stratalinker or equivalent.
8. Store the agar plates wrapped at 4°C.
9. Perform ASO hybridization as outlined in **Section 3.4.2** to identify NCOs. For example, if these were amplifications with Bf and Ur primers, probe with an ASO to detect the D polymorphism in bacterial colonies containing NCOs (e.g., black polymorphism 4 in **Fig. 15.7a**). Next, strip and re-probe (*see* **Note 25**) with the ASO to detect the B polymorphism at this position (e.g., gray polymorphism 4). The B probe will hybridize to the vast majority of the bacterial colonies, but will not hybridize to a bona fide NCO at this polymorphism. For potential co-conversions, strip and re-probe for the other converted D polymorphism (e.g., black polymorphism 5 in **Fig. 15.7a**). As the same colonies hybridize to both D-specific ASOs, this NCO is surmised to be a co-conversion of both polymorphisms.

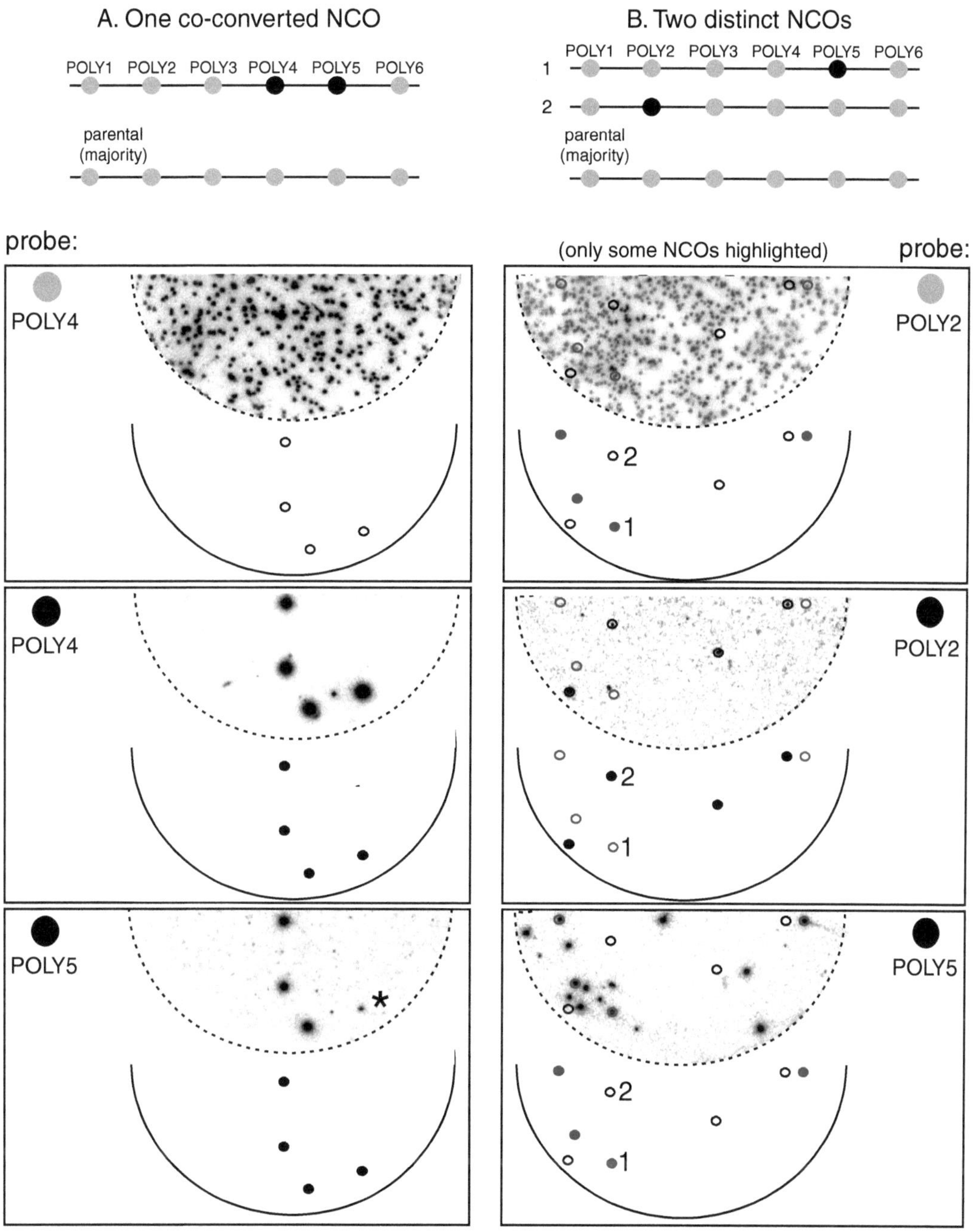

Fig. 15.7. Cloning NCOs to determine whether NCOs have undergone co-conversion. NCOs are amplified and cloned with a large excess of parental DNA and identified by ASO hybridization. The phosphorimager scans of a membrane hybridized with three ASOs that recognize the indicated polymorphisms are shown for each case. (**a**) A co-converted NCO with the indicated genotype. Hybridization patterns for four colonies containing NCOs are depicted below each scan. Note that the hybridization signal from one of the colonies was reduced after multiple re-probings (*asterisk*). (**b**) Two distinct NCOs with the indicated genotypes. Hybridization patterns for eight colonies containing NCOs are depicted below each scan. The colonies hybridize to different probes, indicating that they are derived from distinct NCOs (four colonies are type 1 and four colonies are type 2). *Open circles*, hybridization negative; *filled circles*, hybridization positive.

Confirmation of this interpretation can be obtained by probing with additional ASOs, for example, to rule out D parental DNA contamination. **Figure 15.7b** shows the expected hybridization patterns in the case of two separate NCOs.

10. Print the phosphorimager scans onto transparencies and align with the agar plates. Pick colonies of interest to isolate plasmid DNA to confirm the identities of NCOs surmised from ASO hybridization, using restriction mapping and/or DNA sequencing.

4. Notes

1. The original Jeffreys' buffer was formulated at 11.1X strength (24), but the 10X formulation described here is equally effective. Avoid repeated freezing and thawing. It is essential that the BSA be of high quality and non-acetylated (e.g., Ambion Ultrapure BSA, AM2616 http://www.ambion.com/) (33). Different preparations of 10X Jeffreys' buffer can affect the efficiency and specificity of PCRs. Make a large-scale batch (e.g., 10 ml) to allow progression through an entire experiment, but if a new preparation is needed, re-assess the primers, as in **Section 3.1.2**.
2. The addition of an intentional mismatch to an allele-specific primer (Amplification-Refractory Mutation System, or ARMS) has been used successfully in multiple contexts (34). The mismatch can be placed between 1 and 5 nucleotides 5′ to the polymorphic site.
3. The 65°C extension temperature for long-range PCR is lower than the commonly utilized 72°C, but it increases efficiency of the amplification and may reduce PCR artifacts caused by annealing of incompletely synthesized products (23).
4. We routinely expose our dissecting instruments, pipettors, and tube racks to ultraviolet light from a Stratalinker or equivalent instrument to prevent DNA crosscontamination (follow the instructions provided for your instrument). Also, aerosol-resistant tips are used for all pre-PCR steps.
5. Some mice have delayed development of the epididymides and should be dissected at 8 weeks of age (e.g., F1 hybrid A/J x DBA/2 J mice).
6. Dissected epididymides and control tissues can be stored at –80°C until use.

7. This pellet will be flocculent: Do not overspin the samples as this will result in a white precipitate that will be resistant to resuspension.
8. If somatic cells are a significant proportion of the sperm preparation, specifically lyse somatic cells by adding SDS to a final concentration of 0.15%. Somatic cells will lyse, but sperm are resistant to this lower concentration of SDS (35). Mix briefly by inverting and spin at 10,000 rpm for 3 min. Remove the supernatant with a 1 ml pipet tip and discard.
9. β-Mercaptoethanol reduces disulfide bonds that inter- and intramolecularly connect protamines which highly compact sperm DNA. Protamines are cysteine- and arginine-rich proteins that replace histones in sperm.
10. Do not overdry the pellet as this will hamper resuspension. It is best to air dry in a PCR laminar flow hood to prevent potential contamination.
11. If the genomic DNA is viscous, abrogating accurate pipetting, a portion can be digested with a restriction enzyme(s) to reduce viscosity. Choose an enzyme that does not cleave within any regions being amplified. After digesting the DNA overnight, the sample can be phenol:choloroform extracted and ethanol precipitated. Proceed with quantification and assessment of DNA quality.
12. The number of amplification cycles in 1°, 2°, and 3° PCRs can be modified as needed to reduce potential low-grade, non-specific amplification of non-recombinant DNA or to enhance less efficient reactions.
13. If NCOs are hard to distinguish by ASO hybridization (described in **Section 3.4**), reduce the concentration of amplifiable molecules in the 1° PCRs. We routinely use between 10 and 40 molecules per reaction.
14. Generate a master mix without the allele-specific primer and remove an aliquot for the positive control PCRs. Then add the allele-specific primers to each reaction to a final concentration of 0.2 μM. Generate the positive control reactions in a separate PCR machine and handle with extreme care throughout the experiment. Routinely perform all steps with experimental samples prior to removing the positive hybridization controls from the PCR machine.
15. S1 nuclease digestion is not routinely used for these assays but can be added for increased stringency if necessary as outlined in **Section 3.2.3**, Step 4.
16. A single 15 μl 2° PCR is sufficient to generate five replica blots for ASO hybridization. If your assays require more replicas, a duplicate 2° PCR can be performed and com-

bined with the previous to generate 10 replica blots. Alternatively, a 20 μl reaction can be seeded with 0.8 μl of the 1° PCR to generate seven to eight replica blots. Note: the 2° PCR for the eight positive hybridization controls provides DNA for up to 20 replica blots.

17. Remove the amplification reaction from one well and rinse the well several times with water. For example, use the lower right-hand well, 12H, to avoid potential crosscontamination of wells during the 2X SSC rinses of the dot-blots.
18. The loading dye marks the membrane, assisting in dot-blot preparation, but fades in the denaturation buffer, so prepare dot-blots immediately.
19. A single ASO kinase reaction can be used for hybridizing two dot-blots. A master kinase mix can be prepared and used to label multiple ASOs. Labeled ASOs are stable at 4°C or –20°C for several days.
20. TMAC binds to A/T-rich sequences and eliminates the hybridization bias for G/C base pairing and allows different ASO hybridizations to be performed at the same temperature regardless of sequence content (36). TMAC is highly toxic and must be handled and disposed of following institutional guidelines.
21. Do not allow the membranes to dry at any point as they will be repeatedly stripped and re-probed. If a phosphorimager is not available, expose the membranes to autoradiography film with an intensifying screen at –80°C for ~3 h to overnight. If the ASO probe is somewhat non-specific, rewash the membranes with TMAC wash buffer at increasing temperatures up to ~62°C.
22. The frequency of discontinuous gene conversion tracts and other more complex events has been reported to be ~5% in some mouse hotspots (27, 28).
23. TOPO® TA cloning is highly efficient, but blunt-end cloning can also be used. Blunt-end PCR products with T4 DNA polymerase and ligate into a blue-white selection plasmid. In this case, Pfu polymerase can be included in the 2° PCRs.
24. Cutting one or two notches in the membrane also helps to orient the membrane to the film.
25. It is important to determine the potential genotypes of NCOs by ASO hybridizations of dot-blots first, and based on that decide which ASOs to probe membranes from colony lifts. Colony lifts are more sensitive to stripping than dot-blots, so some colonies can be lost after multiple reprobing (for example, *see* asterisk in **Fig. 15.7a**).

Acknowledgments

We thank members of the Jasin Laboratory, especially Erika Brunet, and Scott Keeney and members of his laboratory, especially Liisa Kauppi and Esther de Boer, for comments on the manuscript and suggestions on the techniques described in this chapter. This work was supported by a Ruth L. Kirschstein National Research Service Award F32HD51392 (F.C.) and National Institutes of Health Grant RO1HD40916 (M.J.).

References

1. Coop, G., Wen, X., Ober, C., Pritchard, J.K., and Przeworski, M. (2008) High-resolution mapping of crossovers reveals extensive variation in fine-scale recombination patterns among humans. *Science* **319**, 1395–1398.
2. Fearnhead, P., and Smith, N.G. (2005) A novel method with improved power to detect recombination hotspots from polymorphism data reveals multiple hotspots in human genes. *Am J Hum Genet* 77, 781–794.
3. Khil, P.P., and Camerini-Otero, R.D. (2010) Genetic crossovers are predicted accurately by the computed human recombination map. *PLoS Genet* **6**, e1000831.
4. Myers, S., Bottolo, L., Freeman, C., McVean, G., and Donnelly, P. (2005) A fine-scale map of recombination rates and hotspots across the human genome. *Science* **310**, 321–324.
5. Cole, F., Keeney, S., and Jasin, M. (2010) Evolutionary conservation of meiotic DSB proteins: more than just Spo11. *Genes Dev* **24**, 1201–1207.
6. Kauppi, L., Jeffreys, A.J., and Keeney, S. (2004) Where the crossovers are: recombination distributions in mammals. *Nat Rev Genet* **5**, 413–424.
7. Baudat, F., and de Massy, B. (2007) Regulating double-stranded DNA break repair towards crossover or non-crossover during mammalian meiosis. *Chromosome Res* **15**, 565–577.
8. Allers, T., and Lichten, M. (2001) Intermediates of yeast meiotic recombination contain heteroduplex DNA. *Mol Cell* **8**, 225–231.
9. Hunter, N., and Kleckner, N. (2001) The single-end invasion: an asymmetric intermediate at the double-strand break to double-Holliday junction transition of meiotic recombination. *Cell* **106**, 59–70.
10. McMahill, M.S., Sham, C.W., and Bishop, D.K. (2007) Synthesis-dependent strand annealing in meiosis. *PLoS Biol* **5**, e299.
11. Borner, G.V., Kleckner, N., and Hunter, N. (2004) Crossover/noncrossover differentiation, synaptonemal complex formation, and regulatory surveillance at the leptotene/zygotene transition of meiosis. *Cell* **117**, 29–45.
12. Szostak, J.W., Orr-Weaver, T.L., Rothstein, R.J., and Stahl, F.W. (1983) The double-strand-break repair model for recombination. *Cell* **33**, 25–35.
13. Bishop, D.K., and Zickler, D. (2004) Early decision; meiotic crossover interference prior to stable strand exchange and synapsis. *Cell* **117**, 9–15.
14. Hubert, R., MacDonald, M., Gusella, J., and Arnheim, N. (1994) High resolution localization of recombination hot spots using sperm typing. *Nat Genet* 7, 420–424.
15. Jeffreys, A.J., Murray, J., and Neumann, R. (1998) High-resolution mapping of crossovers in human sperm defines a minisatellite-associated recombination hotspot. *Mol Cell* **2**, 267–273.
16. Jeffreys, A.J., Tamaki, K., MacLeod, A., Monckton, D.G., Neil, D.L., and Armour, J.A. (1994) Complex gene conversion events in germline mutation at human minisatellites. *Nat Genet* **6**, 136–145.
17. Jeffreys, A.J., and May, C.A. (2004) Intense and highly localized gene conversion activity in human meiotic crossover hot spots. *Nat Genet* **36**, 151–156.
18. Guillon, H., and de Massy, B. (2002) An initiation site for meiotic crossing-over and gene conversion in the mouse. *Nat Genet* **32**, 296–299.
19. Jeffreys, A.J., Kauppi, L., and Neumann, R. (2001) Intensely punctate meiotic recombination in the class II region of the major histocompatibility complex. *Nat Genet* **29**, 217–222.

20. Guillon, H., Baudat, F., Grey, C., Liskay, R.M., and de Massy, B. (2005) Crossover and noncrossover pathways in mouse meiosis. *Mol Cell* **20**, 563–573.
21. Holloway, K., Lawson, V.E., and Jeffreys, A.J. (2006) Allelic recombination and de novo deletions in sperm in the human beta-globin gene region. *Hum Mol Genet* **15**, 1099–1111.
22. Cole, F., Keeney, S., and Jasin, M. (2010) Comprehensive, fine-scale dissection of homologous recombination outcomes at a hot spot in mouse meiosis. *Mol Cell* **10**, 700–710.
23. Kauppi, L., May, C.A., and Jeffreys, A.J. (2009) Analysis of meiotic recombination products from human sperm. *Methods Mol Biol* **557**, 323–355.
24. Jeffreys, A.J., Neumann, R., and Wilson, V. (1990) Repeat unit sequence variation in minisatellites: a novel source of DNA polymorphism for studying variation and mutation by single molecule analysis. *Cell* **60**, 473–485.
25. Jeffreys, A.J., and Neumann, R. (1997) Somatic mutation processes at a human minisatellite. *Hum Mol Genet* **6**, 129–132; 134–126.
26. Baudat, F., and de Massy, B. (2007) Cis- and trans-acting elements regulate the mouse Psmb9 meiotic recombination hotspot. *PLoS Genet* **3**, e100.
27. Bois, P.R. (2007) A highly polymorphic meiotic recombination mouse hot spot exhibits incomplete repair. *Mol Cell Biol* **27**, 7053–7062.
28. Kauppi, L., Jasin, M., and Keeney, S. (2007) Meiotic crossover hotspots contained in haplotype block boundaries of the mouse genome. *Proc Natl Acad Sci USA* **104**, 13396–13401.
29. Kelmenson, P.M., Petkov, P., Wang, X., Higgins, D.C., Paigen, B.J., and Paigen, K. (2005) A torrid zone on mouse chromosome 1 containing a cluster of recombinational hotspots. *Genetics* **169**, 833–841.
30. Paigen, K., Szatkiewicz, J.P., Sawyer, K., Leahy, N., Parvanov, E.D., Ng, S.H., Graber, J.H., Broman, K.W., and Petkov, P.M. (2008) The recombinational anatomy of a mouse chromosome. *PLoS Genet* **4**, e1000119.
31. Baudat, F., and de Massy, B. (2009) Parallel detection of crossovers and noncrossovers in mouse germ cells. *Methods Mol Biol* **557**, 305–322.
32. Zheng, N., Monckton, D.G., Wilson, G., Hagemeister, F., Chakraborty, R., Connor, T.H., Siciliano, M.J., and Meistrich, M.L. (2000) Frequency of minisatellite repeat number changes at the MS205 locus in human sperm before and after cancer chemotherapy. *Environ Mol Mutagen* **36**, 134–145.
33. Kreader, C.A. (1996) Relief of amplification inhibition in PCR with bovine serum albumin or T4 gene 32 protein. *Appl Environ Microbiol* **62**, 1102–1106.
34. Newton, C.R., Graham, A., Heptinstall, L.E., Powell, S.J., Summers, C., Kalsheker, N., Smith, J.C., and Markham, A.F. (1989) Analysis of any point mutation in DNA. The amplification refractory mutation system (ARMS). *Nucleic Acids Res* **17**, 2503–2516.
35. Meistrich, M.L., Reid, B.O., and Barcellona, W.J. (1976) Changes sperm culei during sperimogensis and epidymal maturation. *Exp Cell Res* **99**, 72–78.
36. Wood, W.I., Gitschier, J., Lasky, L.A., and Lawn, R.M. (1985) Base composition-independent hybridization in tetramethylammonium chloride: a method for oligonucleotide screening of highly complex gene libraries. *Proc Natl Acad Sci USA* **82**, 1585–1588.

Chapter 16

Homologous Recombination Assay for Interstrand Cross-Link Repair

Koji Nakanishi, Francesca Cavallo, Erika Brunet, and Maria Jasin

Abstract

DNA interstrand cross-links (ICLs) covalently link both strands of the DNA duplex, impeding cellular processes like DNA replication. Homologous recombination (HR) is considered to be a major pathway for the repair of ICLs in mammalian cells as mutants for HR components are highly sensitive to DNA-damaging agents that cause ICLs. This chapter describes GFP assays to measure HR following site-specific ICL formation with psoralen through DNA triplex technology. This approach can be used to determine the genetic requirements for ICL-induced HR in relation to those involved in HR repair of other DNA lesions such as double-strand breaks.

Key words: Homologous recombination, interstrand cross-link repair, triplex-forming oligonucleotide, GFP reporters.

1. Introduction

DNA interstrand cross-links (ICLs) are toxic to dividing cells because they impede DNA replication and other cellular processes, and as a result, agents that cause ICLs such as cisplatin are frequently used in cancer chemotherapy (1). In addition to causing lethal damage, ICLs can induce mutations and gross chromosomal rearrangements. Multiple pathways have been implicated in ICL repair, including nucleotide excision repair, translesion synthesis, and homologous recombination (HR) (2). In mammalian cells, a role for HR in ICL repair is postulated based on the extreme sensitivity of cells deficient in HR components, such as BRCA1 (3, 4) and BRCA2 (5, 6), to various agents that cause ICLs. Cells deficient in Fanconi anemia pathway components are

H. Tsubouchi (ed.), *DNA Recombination*, Methods in Molecular Biology 745,
DOI 10.1007/978-1-61779-129-1_16, © Springer Science+Business Media, LLC 2011

also sensitive to ICL agents (7, 8) and show defects in HR, especially HR coupled to DNA replication (9, 10).

Given the many open questions about the relationship between ICL repair and HR, we developed an approach to assay ICL-induced HR in mammalian cells based on the DR-GFP reporter which has previously been developed to detect HR induced by another type of lesion, a DNA double-strand break (DSB) (**Fig. 16.1a**) (11). DR-GFP is composed of two differentially mutated green fluorescent protein (*GFP*) genes oriented as direct repeats (hence, "DR"): the upstream repeat contains the recognition site for the rare-cutting I-*Sce*I endonuclease and the downstream repeat is a 5′ and 3′ truncated *GFP* fragment. Transient expression of I-*Sce*I leads to a DSB in the upstream *GFP* gene; HR to repair the DSB results in GFP+ cells which are quantified by flow cytometry (12). This assay has been widely used to identify proteins required for HR repair, such as BRCA1 and BRCA2, and to determine which pathways suppress HR repair, using both candidate gene approaches (13) and whole genome screens (14). While developed for chromosomal DSB repair assays, the DR-GFP reporter can also be used to assay repair in plasmids.

To elucidate the role of HR in the repair of ICLs, we modified DR-GFP to contain a specific site for ICL formation, creating the TR-GFP reporter (TR, triplex, and repeats of *GFP*; **Fig. 16.1b, c**) (10). The modification was accomplished by replacing the I-*Sce*I site with a sequence that can bind a triplex-forming oligonucleotide (TFO) conjugated with psoralen at its 5′-end (pso-TFO) (15). Following triplex formation between pso-TFO and TR-GFP (pso-TFO:TR-GFP), intercalation of the psoralen into duplex DNA, and exposure to 365 nm ultraviolet light (UVA), a site-specific ICL forms in TR-GFP (**Fig. 16.1d**). Although typically not as high as with DSB-induced HR, GFP+ cells are obtained (i.e., several percent for DR-GFP compared with a few percent or less for TR-GFP), indicating ICL-induced HR. HR is dependent on known HR factors such as BRCA1 and BRCA2 (10). Such an approach has previously been used with a *supF* gene reporter (16). Several modifications of this approach are possible. For example, the TR-GFP reporter has been modified to contain an origin of replication (OriP) from Epstein–Barr virus (EBV) for replication in human cells expressing the EBV nuclear antigen, EBNA1 (10, 17). This modification allows the examination of HR coupled to DNA replication. Further, the TFO "tail" can be removed, by using a pso-TFO with a disulfide bond between the psoralen moiety and the TFO (pso-SS-TFO) (10, 15).

In this chapter, we provide a protocol to quantify ICL-induced HR in which the site-specific ICL is formed in cells

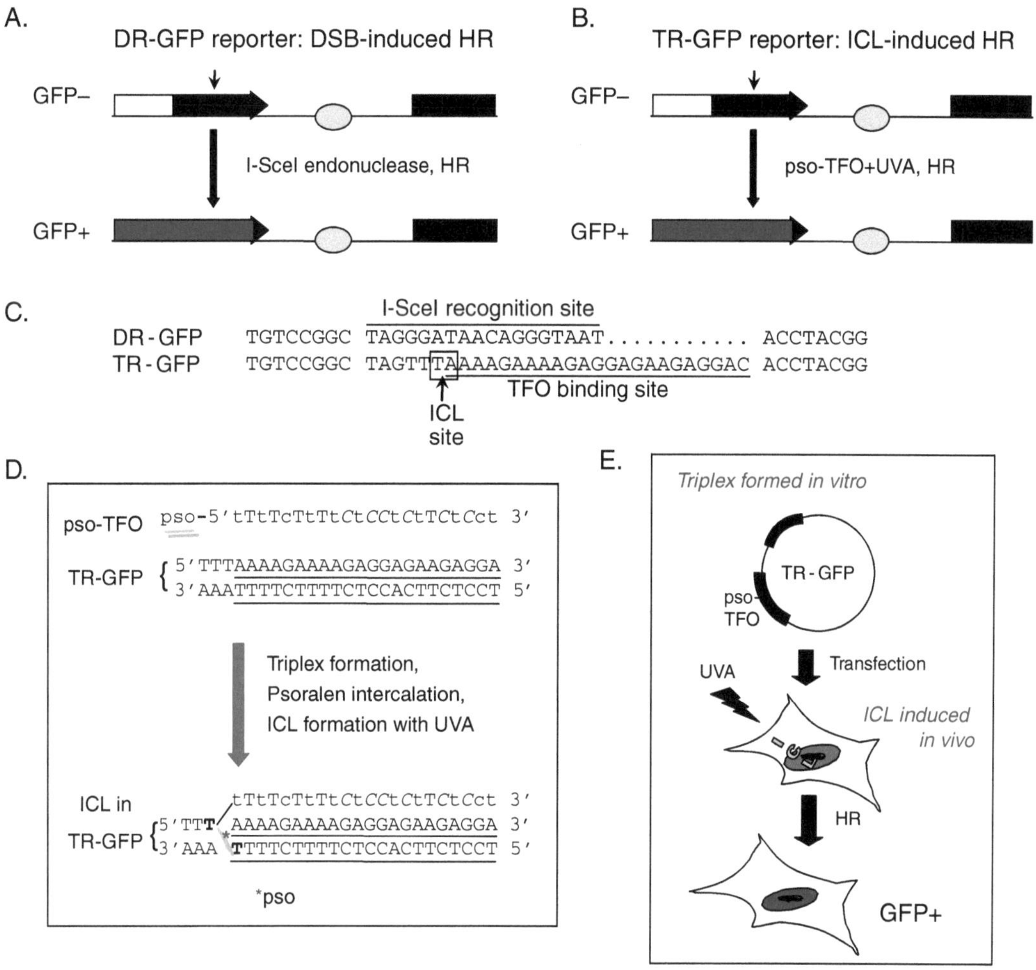

Fig. 16.1. Design of HR assays. **a** DR-GFP reporter measures DSB-induced HR (11, 12, 18). This reporter consists of two defective *GFP* genes, the first of which contains an I-*Sce*I endonuclease site (*arrow*) such that cells are GFP–. Cellular expression of I-*Sce*I leads to a DSB which can be repaired by HR using the downstream wild-type *GFP* sequence as a template, resulting in GFP+ cells. Two different DR-GFP plasmids have been created: pDR-OriP-GFP has an EBV OriP placed between the *GFP* repeats (*light gray circle*) to allow the plasmid to replicate in the presence of EBNA1; in pDR-GFP-hprt, the reporter is flanked by *Hprt* genomic sequences for targeting the reporter to the mouse *Hprt* locus (not shown). *Short black bar box*, *GFP* repeat; *white box* and *horizontal arrow* head, non-repetitive parts of the *GFP* gene; *long black bar*, GFP+ gene. **b** TR-GFP reporter measures ICL-induced HR (10). A TFO binding site (*arrow*) replaces the I-*Sce*I site of the DR-GFP reporter. After triplex formation at the TFO binding site with a psoralen-conjugated TFO (pso-TFO:TR-GFP), followed by UVA irradiation (ICL formation), ICL-induced HR repair restores an intact *GFP* gene, giving rise to GFP+ cells. Similar to DR-GFP, two TR-GFP plasmids have been constructed, pTR-GFP-hprt and TR-OriP-GFP. **c** Sequences of the I-*Sce*I recognition site in DR-GFP and the TFO binding site in TR-GFP. The ICL site – a TA/AT sequence suitable for psoralen intercalation and ICL formation – is *boxed.* The TAG of the I-*Sce*I site and the same triplet in TR-GFP are stop codons, truncating the GFP protein. Eight base pairs of flanking sequences in DR/TR-GFP are shown on both sides of the damage sites. **d** Details of ICL formation. After triplex formation between the pso-TFO and the TR-GFP, the 5′-psoralen moiety of pso-TFO intercalates at the TA/AT site. UVA irradiation cross-links psoralen (ICL, *gray line*) to the two Ts (*bold*) of the double-stranded DNA of TR-GFP. See the legend for **Fig. 16.2a** for more details about the pso-TFO. **e** Scheme for the TR-GFP assay. The pso-TFO:TR-GFP triplex is formed in vitro and transfected into cells which are then treated with UVA to form the ICL. Cells are incubated for 48 h and GFP+ cells arising by HR are quantified by flow cytometry.

after transfection of the pso-TFO:TR-GFP triplex and exposure to UVA (**Fig. 16.1e**). A procedure to verify in vivo ICL formation is also provided.

2. Materials

2.1. Cell Culture

1. Human osteosarcoma cell line U2OS (ATCC HTB-96) which grows in D-MEM high glucose (GIBCO 31053-036) and 15% fetal calf serum, supplemented with 1× penicillin–streptomycin–L-glutamine (GIBCO 10378-016).
2. U2OS-CEP cells, which are U2OS cells stably transfected with the pCEP4 plasmid (Invitrogen V044-50), selected with 0.4 mg/ml hygromycin B (Roche 10843555001) supplemented in the medium.
3. Phosphate buffered saline (PBS).
4. 37°C, 5% CO_2 incubator.

2.2. Plasmids

Prepare with standard protocols. Unless plasmids are prepared to high purity with cesium chloride ultracentrifugation, avoid repeated freeze thawing:

1. DSB recombination reporter pDR-GFP-hprt (18) (*see* **Note 1**).
2. I-*Sce*I endonuclease expression vector pCBASce (19).
3. The empty expression vector pCAGGS (20), used as a negative control.
4. ICL recombination reporter pTR-GFP-hprt (10).
5. Replicating DSB and ICL recombination reporters DR-OriP-GFP and TR-OriP-GFP, respectively (10) (*see* **Note 2**).

2.3. Transfection

1. Electroporation system (Bio-Rad Gene Pulser II).
2. 0.4-cm-Gap cuvettes (Bio-Rad).
3. Opti-MEM media (Invitrogen).
4. 10-cm Tissue culture plates.

2.4. ICL Formation

TFO-containing oligonucleotides are from Eurogentec. Sequences are presented in **Fig. 16.2a**. Nucleotide modifications to enhance the stability of the triplex are locked nucleic acids (LNA, 2′O-4′C methylene bridge) (lower case) and 5′-methylated cytosines (italics) (21) (*see* **Note 3**). The 10× stock solution is 100 μM in H_2O for each:

1. As a negative control for ICL formation, the TFO without psoralen.

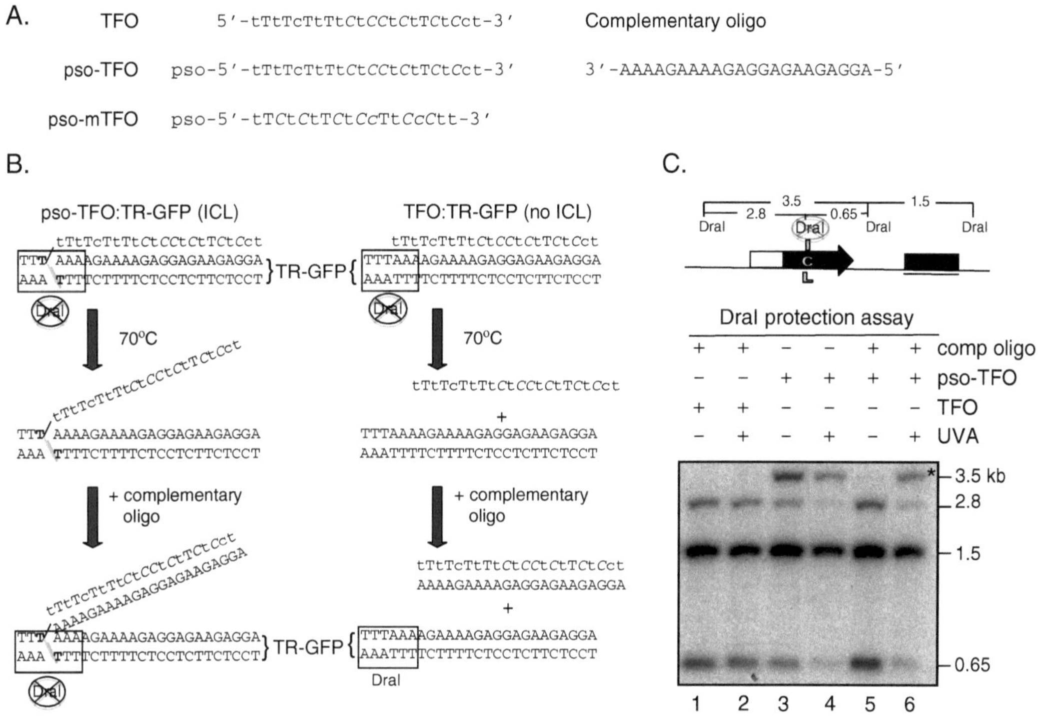

Fig. 16.2. *DraI* protection assay. **a** Oligonucleotide sequences of TFOs and the oligonucleotide complementary to the TFO and pso-TFO used in this study. Several nucleotides within the TFO are modified to enhance the stability of the triplex. Lower case nucleotides, locked nucleic acids (LNA); cytosines in italics, methylated at the 5′ position. Psoralen (pso) is attached to the 5′ nucleotide through a $(CH_2)_6$ linkage to the phosphate. **b** *DraI* protection assay to distinguish ICL formation from triplex formation. After UVA, the triplex is heated to dissociate the TFO from the duplex; with cooling, the free TFO anneals to the complementary oligonucleotide preventing it from reannealing to the duplex, permitting *DraI* cleavage (*right panel*). By contrast for cross-linked TR-GFP, pso-TFO does not dissociate from TR-GFP because of the covalent linkage of psoralen with TR-GFP (*left panel*). *DraI* is unable to cleave in this case. However, heating and cooling steps with the complementary oligonucleotide traps pso-TFO that is not cross-linked and prevents non-covalent triplex formation during analysis (*not shown*). **c** *DraI* protection assay results. After extraction of DNA from the transfected cells and digestion by *DraI*, samples were analyzed by Southern blotting. The *GFP* probe is underlined. The fragment protected by the TFO and/or the ICL is 3.5 kb; if unprotected, the *DraI*-cleaved fragments are 2.8 and 0.65 kb. For the unconjugated TFO samples, no 3.5-kb fragment was detected without or with UVA irradiation when the complementary oligonucleotide was added prior to digestion with *DraI* (*lanes 1* and *2*, respectively) (**Fig. 16.2b**, *right panel*). In the absence of the complementary oligonucleotide, both unirradiated and UVA-irradiated pso-TFO samples showed the 3.5-kb *DraI*-resistant fragment (*lanes 3* and *4*, respectively). In contrast, in presence of the complementary oligonucleotide (*lanes 5* and *6*), only the pso-TFO treated sample which was UVA irradiated showed the 3.5-kb fragment corresponding to ICL formation (*asterisk*) (**Fig. 16.2b**, *left panel*).

2. For ICL formation, the pso-TFO, which is the TFO conjugated with psoralen at the 5′-nucleotide through a $(CH_2)_6$ linkage to the phosphate.
3. As a negative control for triplex formation, pso-mTFO, which cannot form a triplex with the target sequence in the TR-GFP plasmids.
4. 10× TFO buffer: 500 mM HEPES (pH 7.2), 500 mM NaCl, 100 mM $MgCl_2$, 5 mM spermine.

5. UVA irradiator: A UVA lamp with a sensor to accurately measure the dose (J/cm^2). Alternatively, we are using a UV Stratalinker 2400 (Stratagene) with UVA bulbs (365 nm).

2.5. DraI Protection Assay

1. Complementary oligonucleotide (**Fig. 16.2a**), 10× stock solution at 500 μM.
2. *Dra*I restriction enzyme (40 U/μl; Roche).
3. *Hin*dIII restriction enzyme (New England Biolabs).
4. Materials for DNA extraction from mammalian cells. We use the QIAamp DNA Blood Mini Kit (Qiagen).
5. Standard materials for agarose gel electrophoresis and Southern blotting (12).

2.6. Flow Cytometry

We use a FACScan (BD), although any flow cytometry analyzer will suffice. Cells are gated by forward and side scatter, and fluorescence is analyzed on the FL1 and FL2 channels (12). GFP+ cells are determined from the FL1 shift from the majority of negative population. Consult a flow cytometry facility if you are uncertain about using a flow cytometer or do not have one accessible for your own use.

3. Methods

The assays involve transient transfection of the HR reporter plasmids into mammalian cells followed by flow cytometry 48 h later to quantify GFP+ cells. Although the DR-GFP assay was originally developed with the reporter integrated into the chromosome (11), the non-integrated reporter assays can speed the analysis tremendously. However, transfections must give reproducible efficiencies, which can be evaluated using a second marker that fluoresces in a different channel from GFP.

3.1. DR/TR-GFP Recombination Assays

We typically perform DR-GFP and TR-GFP assays in parallel to compare DSB and ICL-induced HR.

3.1.1. DR-GFP Assay

1. The day before transfection, plate U2OS cells at ~50% confluence such that on the day of transfection, they are still subconfluent. Using confluent cells reduces transfection efficiency and HR levels.
2. For transfection, trypsinize cells, pellet, and rinse once. Each transfection uses 5×10^6 cells/800 μl in Opti-MEM.
3. Add cells to cuvette.
4. Add 20 μg pDR-GFP-hprt and 20 μg pCBASce or pCAGGS to cuvette, mix cells and DNA well, but gently, and electroporate immediately at 950 μF/250 V.

5. Plate cells in 10-cm plates and incubate for 48 h.
6. Perform flow cytometry analysis. We typically get up to 10% GFP+ cells under these conditions.

3.1.2. TR-GFP Assay

The overall scheme is presented in **Fig. 16.1e** (*see* **Note 4**):

1. For triplex formation, mix 20 μg pTR-GFP-hprt with 4 μl TFO (TFO:TR-GFP), pso-TFO (pso-TFO:TR-GFP), or pso-mTFO (pso-mTFO/TR-GFP), each at a final concentration of 10 μM, and 4 μl of 10× TFO buffer, for a final volume of 40 μl.
2. Incubate at room temperature for 30 min to allow the triplex to form. The efficiency of triplex formation is checked using the *Dra*I protection assay as described in **Section 3.2**.
3. Add 40 μl triplex mix to cells in cuvette and electroporate, as described in steps 1–4 of **Section 3.1.1**.
4. Plate cells in 10-cm plates and incubate for 1 h at 37°C.
5. Aspirate media completely, as residual phenol red in the media may absorb UVA, and rinse cells once with PBS. Be careful not to dislodge newly attached cells.
6. Add 1 ml PBS.
7. Place cells in Stratalinker and irradiate at 0.15 J/cm^2 UVA. Avoid drying the cells. For unirradiated control samples, skip this step.
8. Add medium and incubate for 48 h at 37°C.
9. Perform flow cytometry analysis. We typically get up to a few percent GFP+ cells under these conditions.

3.2. *Dra*I Protection Assay

The ICL is formed within a *Dra*I restriction site such that the efficiency of ICL formation can be tested by resistance to *Dra*I cleavage (**Fig. 16.2b**). As TFO binding also protects from *Dra*I cleavage, the unconjugated TFO is removed by heating and then trapped by the addition of a complementary oligonucleotide (22):

1. After UVA irradiation (step 7 of **Section 3.1.2**), extract DNA from cells (QIAamp DNA Blood Mini kit). Measure the DNA concentration.
2. Prepare duplicates of 1 μg of each DNA preparation in 20 μl of 1X *Dra*I buffer. In one set, have a 50 μM final concentration of complementary oligonucleotide.
3. Incubate at 70°C for 10 min to dissociate the TFO from the plasmid DNA. Slowly cool to room temperature. At this step, the complementary oligonucleotide binds the dissociated TFO, preventing it from reannealing to the plasmid. The excess complementary oligonucleotide captures all of the dissociated TFO.
4. Add 1 μl *Dra*I to each sample and incubate for 1 h at 37°C.
5. Heat inactivate at 65°C for 20 min.

6. Run samples on a 0.8% agarose gel and perform Southern blotting.
7. Probe with the 800-bp *Hin*dIII fragment from pDR-GFP-hprt.
8. With complete ICL formation, the 2.8- and 0.65-kb *Dra*I fragments are converted to a 3.45-kb fragment. The example shows substantial but not complete ICL formation (**Fig. 16.2c**).

3.3. TR-OriP-GFP Assay

As ICL repair may be coupled with DNA replication (23), the TR-GFP assay was modified so that the reporter could replicate in human cells, by adding OriP to the plasmid, forming TR-OriP-GFP (**Fig. 16.1b**), and expressing EBNA1 in U2OS cells (10):

1. The assays are identical to those described in **Section 3.1.2**, except that the plasmid (TR-OriP-GFP) and cells (U2OS-CEP) are different. U2OS cells can also be used as a negative control, since the TR-OriP-GFP plasmid will not replicate in those cells.

4. Notes

1. This plasmid is based on pDR-GFP but additionally contains *Hprt* genomic sequences which can be used for gene targeting the reporter in mouse cells (18).
2. These plasmids contain an EBV origin of replication cloned between the GFP repeats of pDR-GFP and its derivative pTR-GFP.
3. These nucleotide modifications decrease the dissociation rate constant for triplex formation and confer an entropic gain (21).
4. The ICL can also be formed in vitro prior to transfection of pso-TFO:TR-GFP triplexes, although we usually obtain lower HR levels than with ICL formation in cells (10).

Acknowledgments

This work was supported by the Byrne Fund and National Institutes for Health grants P01CA94060 (M.J.) and R01GM54668 (M.J.).

References

1. Guainazzi, A., and Schärer, O.D. (2010) Using synthetic DNA interstrand crosslinks to elucidate repair pathways and identify new therapeutic targets for cancer chemotherapy. *Cell Mol Life Sci* **67**, 3683–3697.
2. Hinz, J.M. (2010) Role of homologous recombination in DNA interstrand crosslink repair. *Environ Mol Mutagen* **51**, 582–603.
3. Moynahan, M.E., Chiu, J.W., Koller, B.H., and Jasin, M. (1999) Brca1 controls homology-directed DNA repair. *Mol Cell* **4**, 511–518.
4. Moynahan, M.E., Cui, T.Y., and Jasin, M. (2001) Homology-directed DNA repair, mitomycin-c resistance, and chromosome stability is restored with correction of a Brca1 mutation. *Cancer Res* **61**, 4842–4850.
5. Moynahan, M.E., Pierce, A.J., and Jasin, M. (2001) BRCA2 is required for homology-directed repair of chromosomal breaks. *Mol Cell* 7, 263–272.
6. Kraakman-van der Zwet, M., et al. (2002) Brca2 (XRCC11) deficiency results in radioresistant DNA synthesis and a higher frequency of spontaneous deletions. *Mol Cell Biol* **22**, 669–679.
7. Wang, W. (2007) Emergence of a DNA-damage response network consisting of Fanconi anaemia and BRCA proteins. *Nat Rev Genet* **8**, 735–748.
8. Auerbach, A.D. (2009) Fanconi anemia and its diagnosis. *Mutat Res* **668**, 4–10.
9. Nakanishi, K., et al. (2005) Human Fanconi anemia monoubiquitination pathway promotes homologous DNA repair. *Proc Natl Acad Sci USA* **102**, 1110–1115.
10. Nakanishi, K., Cavallo, F., Perrouault, L., Giovannangeli, C., Moynahan, M.E., Barchi, M., Brunet, E., and Jasin, M. (2011) Homology-directed Fanconi anemia pathway cross-link repair is dependent on DNA replication. *Nat Struct Mol Biol* doi:10.1038/nsmb.2029.
11. Pierce, A.J., Johnson, R.D., Thompson, L.H., and Jasin, M. (1999) XRCC3 promotes homology-directed repair of DNA damage in mammalian cells. *Genes Dev* **13**, 2633–2638.
12. Pierce, A.J., and Jasin, M. (2005) Measuring recombination proficiency in mouse embryonic stem cells. *Methods Mol Biol* **291**, 373–384.
13. Moynahan, M.E., and Jasin, M. (2010) Mitotic homologous recombination maintains genomic stability and suppresses tumorigenesis. *Nat Rev Mol Cell Biol* **11**, 196–207.
14. Slabicki, M., et al. (2010) A genome-scale DNA repair RNAi screen identifies SPG48 as a novel gene associated with hereditary spastic paraplegia. *PLoS Biol* **8**, e1000408.
15. Chin, J.Y., and Glazer, P.M. (2009) Repair of DNA lesions associated with triplex-forming oligonucleotides. *Mol Carcinog* **48**, 389–399.
16. Raha, M., Wang, G., Seidman, M.M., and Glazer, P.M. (1996) Mutagenesis by third-strand-directed psoralen adducts in repair-deficient human cells: high frequency and altered spectrum in a xeroderma pigmentosum variant. *Proc Natl Acad Sci USA* **93**, 2941–2946.
17. Reisman, D., Yates, J., and Sugden, B. (1985) A putative origin of replication of plasmids derived from Epstein-Barr virus is composed of two cis-acting components. *Mol Cell Biol* **5**, 1822–1832.
18. Pierce, A.J., Hu, P., Han, M., Ellis, N., and Jasin, M. (2001) Ku DNA end-binding protein modulates homologous repair of double-strand breaks in mammalian cells. *Genes Dev* **15**, 3237–3242.
19. Richardson, C., Moynahan, M.E., and Jasin, M. (1998) Double-strand break repair by interchromosomal recombination: suppression of chromosomal translocations. *Genes Dev* **12**, 3831–3842.
20. Niwa, H., Yamamura, K., and Miyazaki, J. (1991) Efficient selection for high-expression transfectants with a novel eukaryotic vector. *Gene* **108**, 193–199.
21. Brunet, E., et al. (2005) Exploring cellular activity of locked nucleic acid-modified triplex-forming oligonucleotides and defining its molecular basis. *J Biol Chem* **280**, 20076–20085.
22. Brunet, E., Corgnali, M., Cannata, F., Perrouault, L., and Giovannangeli, C. (2006) Targeting chromosomal sites with locked nucleic acid-modified triplex-forming oligonucleotides: study of efficiency dependence on DNA nuclear environment. *Nucleic Acids Res* **34**, 4546–4553.
23. Raschle, M., et al. (2008) Mechanism of replication-coupled DNA interstrand crosslink repair. *Cell* **134**, 969–980.

Chapter 17

Evaluation of Homologous Recombinational Repair in Chicken B Lymphoma Cell Line, DT40

Hiroyuki Kitao, Seiki Hirano, and Minoru Takata

Abstract

Homologous recombination (HR) is a mode of double-strand break (DSB) repair required for cell viability in vertebrate cells. Targeted integration of homologous DNA fragment by HR is usually a very rare event in vertebrate cells; however, in chicken B lymphoma cell line DT40, the ratio of targeted to random integration is extremely high. Although the underlying mechanism of this phenotype is not fully understood, DT40 has been utilized as a model cell line for a number of genetic analyses. Here we describe three assays for evaluating homologous recombinational repair (HRR) using DT40 as a model system, measuring gene-targeting frequency, analyzing HRR process of single DSB induced by yeast homing endonuclease I-*Sce*I, and measuring sister chromatid exchange frequency. Combined with generation of gene-disrupted DT40 mutant cell line, these assays have been highly useful to investigate the mechanisms in HRR. Using these techniques, a role of HRR of not only Rad52 epistasis group genes but also genes whose mutation cause hereditary cancer syndrome, such as Fanconi anemia, has been established.

Key words: Homologous recombinational repair (HRR), double-strand break (DSB), homing endonuclease I-*Sce*I, chicken B lymphoma cell line DT40, sister chromatid exchange.

1. Introduction

A DNA double-strand break (DSB) is the most serious damage to the cells. Even one DSB prevents cell proliferation and/or induces cell death if left unrepaired. DSB is introduced on a chromosome not only when cells are exposed to ionizing radiation or chemotherapeutic drugs but also during normal cellular metabolism, such as DNA replication. Homologous recombinational repair (HRR) and nonhomologous end joining (NHEJ) are the two major cellular systems to repair DNA DSBs. In

H. Tsubouchi (ed.), *DNA Recombination*, Methods in Molecular Biology 745,
DOI 10.1007/978-1-61779-129-1_17, © Springer Science+Business Media, LLC 2011

general, HRR plays a dominant role in DSB repair during late S to G2 phase of the cell cycle, while NHEJ during G1 to early S phase (1).

In HRR process, Rad51 is the center player. Rad51 accumulates at single-strand DNA around DSB after the processing of DSB ends (resection) and forms a structure called "nucleoprotein filaments." This enables the DNA end to carry out homology search and strand invasion into the homologous sister chromatid that serves as a template sequence, resulting in D-loop formation. Then using the invaded DNA end as a primer, DNA synthesis occurs. If the first end with some extension is released from the template strand and is captured by the second end, two ends are somehow ligated. This mechanism is something like copy–paste of genetic information and is called gene conversion. On the other hand, if the D-loop is converted to the two four-way DNA junctions (termed Holliday junction), the crossing over event could be the result, depending on the mode of resolution. The mechanisms of HRR are highly complicated and are still poorly understood; however, it is clear that Rad51 requires a plethora of co-factors such as Rad52 epistasis group genes, *BRCA* genes, and genes involved in end resection for its full function. In addition, recent studies have identified genes involved at the late-stage HRR such as Holliday junction resolvases (2).

It had been long believed that NHEJ is the dominant repair process of DSBs in vertebrate cells. This was partly because, compared with bacteria and yeast, it was very difficult to integrate DNA element at specific site with sequence homology in vertebrate genome. However, chicken B cell line DT40 was the exception. DT40 cell line was originally established from avian leukemia virus (ALV)-infected chicken B cell lymphoma in the bursa of Fabricius (3). It was discovered by Buerstedde and Takeda in 1991 that extremely high frequency of targeted integration into the homologous gene loci can be achieved in DT40 (3). Although the underlying mechanism of this phenomenon is still enigmatic, DT40 has been utilized as a model cell line for genetic analysis of essential cellular metabolisms, such as DNA repair, chromosome segregation, and cell cycle checkpoint. Numerous gene-disrupted or gene-manipulated DT40 cells have been established so far. It is now possible to create a knockout cell line lacking a gene of interest in DT40 and to evaluate HRR efficiency in those clones by measuring the ratio of targeted to random integration events.

The 18-bp recognition sequence of yeast homing endonuclease I-*Sce*I probably does not exist in vertebrate genome. Thus, expression of I-*Sce*I to cells carrying an artificial chromosomal reporter construct, which contain a single I-*Sce*I cutting site, introduces only one DSB in vertebrate genome (4). The reporter contains a tandem repeat of defective *neo* gene in which I-*Sce*I recognition site is located in one of them. HRR of induced DSB would reconstitute functional neo gene using the other neo gene as a template. Thus we can measure efficiency of HRR by counting number of neo-resistant colonies. In addition, the structure of the repair products can be clarified by PCR and direct sequencing. Dr Maria Jasin originally developed this system in 1994, and there have been many variations so far developed. For example, another version of the reporter utilizes *GFP* gene, and in this case HR efficiency is represented by percentage of GFP-positive cells as measured by flow cytometry (5).

This Jasin's assay has revealed the important role of HRR in DSB repair in mammalian cells (6), although this had been already apparent in bacteria and yeast. For example, hereditary breast cancer-susceptible genes *BRCA1* (7) and *BRCA2* (8), and genes involved in other cancer-prone genetic disease, such as Fanconi anemia (9), have been shown to be involved in HRR process. This elegant assay system has been introduced in DT40 (10), and the critical involvement of *BRCA2* (11) and *FancD2* (12) in HRR process has also been shown. In addition, it was established that *Nbs1* has an essential role in HRR by using this assay system in DT40 (13).

Another useful assay for HRR in DT40 (and in mammalian cells) is the measurement of levels of sister chromatid exchange (SCE). In mammalian cells cultured for two cell cycles in the presence of nucleotide analog bromodeoxyuridine, two sister chromatids can be differentially stained and discriminated using Hoechst dye and Giemsa staining on the metaphase chromosome (14). This classical method gives darker staining on sister chromatid with BrdU incorporated in only one DNA strand than on the other sister chromatid with BrdU in both strands. Thus SCEs can be recognized as abrupt discontinuities in the staining patterns of the two chromatids of a metaphase chromosome with reciprocal switching from one chromatid to the other (15). This technique was applied to DT40, and the dependency of SCE levels on *Rad51* and other *HRR* genes has been established. For example, SCEs were significantly reduced for chicken DT40 cells lacking the key HR genes *Rad51* and *Rad54* but not for nonhomologous DNA end joining (NHEJ)-defective $KU70^{-/-}$ cells (16). Thus SCE is now considered as the product of HRR during replication which is accompanied by crossing-over between sister chromatids. It is well known that SCE frequency was extremely high in lymphocytes isolated from the patient of Bloom syndrome (mutated in a gene encoding BLM helicase) (17), indicating that the crossing-over events during mitotic HRR process are under strict control by mechanisms involving BLM helicase (18). We have shown increased levels of spontaneous SCE events in *FancD2*- (12) or *FancC*-deficient DT40 cells (19), with only marginal increase following MMC treatment. Moreover, FancC is in an epistatic relationship with BLM helicase, and consistently, FancC and FancD2 regulate localization of BLM helicase in response to MMC (19).

2. Materials

2.1. Cell Culture

1. A well-characterized line of DT40 cells (e.g., CL.18, DT40Cre1).
2. DT40 cell culture medium: RPMI1640 (Invitrogen, Carlsbad, CA) supplemented with 10% fetal bovine serum (FBS; HyClone, Ogden, UT), 1% chicken serum (Sigma, St. Louis, MO), 20 mM L-glutamine, 1% penicillin–streptomycin, and 55 μM β-mercaptoethanol.
3. Tissue culture incubator.
4. Laminar flow tissue culture hood/safety cabinet.
5. 96-Well and 24-well tissue culture plates, and 10-cm dishes.

2.2. Gene Targeting in DT40

1. Gene Pulser Xcell electroporation system (Bio-Rad, Hercules, CA).
2. Selection marker gene cassettes and final concentrations of the corresponding drugs: G418 (2 mg/ml; Nakalai), hygromycin B (2.5 mg/ml; Calbiochem), blastcisin-S (25 μg/ml; Calbiochem), L-histidinol (1 mg/ml; Sigma), puromycin (0.5 μg/ml; Sigma), zeocin (100 μg/ml; Invitrogen), and mycophenolic acid (15 μg/ml; Sigma). G418, hygromycin B, and zeocin are obtained as solutions. All others are dissolved in distilled water.
3. Targeting vector plasmids: *FANCC* KO-bsd
4. 0.4-cm Cuvette (Bio-Rad).

2.3. I-SceI Assay

1. Plasmid: pBluescript-KS (+), pCAGGS-I-*Sce*I (I-*Sce*I expression vector, a gift of Dr Maria Jasin, Memorial Sloan-Kettering Cancer Center, NY), OVA-SCneo/puro^{R} (a gift of Dr Shunichi Takeda, Kyoto University, Kyoto).
2. G418 (Gibco-BRL, Grand Island, NY).
3. FACSCalibur (Becton-Dickinson, Franklin Lakes, NJ).

2.4. Southern Blotting and Hybridization

1. Lysing solution: 200 mM NaCl, 20 mM EDTA, 40 mM Tris–HCl (pH 8.0), 0.5% SDS, 71.5 mM β-mercaptoethanol (should be added before use), 200 μg/ml proteinase K (should be added before use).
2. 6 M NaCl.
3. Agarose.

4. 0.5× TBE: 45 mM Tris–borate, 1 mM EDTA.
5. Gel electrophoresis equipment.
6. 3 MM chromatography paper (Whatman, Maidstone, UK).
7. Biodyne® B 0.45 μm (Pall, Pensacola, FL)/Hybond N+ (GE Healthcare Biosciences, Pittsburgh, PA).
8. Denaturing solution: 0.5 N NaOH, 1.5 M NaCl.
9. Neutralizing solution: 0.5 M Tris–HCl (pH 7.5), 1.5 M NaCl.
10. 20× SSC: 3 M NaCl, 0.3 M sodium citrate.
11. 20× SSPE: 3 M NaCl, 0.2 M NaH_2PO_4, 20 mM EDTA.
12. 50× Denhardt's solution: 5 g Ficoll (type 400; Pharmacia), 5 g polyvinylpyrrolidone, 5 g bovine serum albumin (fraction V; Sigma), and H_2O to 500 ml.
13. Sonicated salmon sperm DNA (10 mg/ml) (ssDNA) (Sigma).
14. Hybridization oven.
15. Rediprime II DNA labeling kit (GE Healthcare Biosciences).
16. [α-^{32}P]dCTP (Pharmacia).
17. Hybridization buffer: 50% formamide, 5× Denhardt's solution, 5× SSPE, 1% SDS.
18. Dextran sulfate.
19. Washing solution 1: 2× SSC, 0.1% SDS.
20. Washing solution 2: 0.1× SSC, 0.1% SDS.
21. BAS2500 plate reader (Fuji Film).
22. Fuji imaging plate/imaging cassette.

2.5. Measuring Sister Chromatid Exchanges

1. Bromodeoxyuridine.
2. KaryoMAX COLCEMID® solution (Gibco-BRL).
3. 75 mM KCl.
4. Methanol/acetic acid solution 3:1. This should be prepared fresh.
5. Clean slide glass (kept at 4°C in 50% ethanol).
6. Hot plate.
7. Hoechst 33258.
8. Phosphate buffer (pH 6.8): 2.13 g/l KH_2PO_4, 4.10 g/l $Na_2HPO_4 \bullet 12H_2O$.
9. McILvaine solution (pH 7.0): 58.9 g/l $Na_2HPO_4 \bullet 12H_2O$, 3.38 g/l citric acid$\bullet H_2O$.
10. FL20BLB (Black light; Toshiba, Japan).
11. Giemsa solution (Sigma).

3. Methods

First, we explain general methods to achieve gene targeting in DT40 cells. We show how to disrupt *FancC* gene as one example. Since targeting frequency of DT40 is extremely high, once we design and create targeting vector, we can utilize it as a tool for measuring gene-targeting frequency.

Second, we describe a method to measure the efficiency of HRR using Jasin's method in DT40 cells. Ectopic expression of I-*Sce*I can produce only one specific DSB in the chromosome carrying one copy of the *SCneo* reporter gene at specific gene locus. In DT40, *Ovalbumin* gene locus has been commonly used as the locus of reporter gene's integration.

Third, we describe how to observe sister chromatid exchanges (SCEs) and to measure SCE levels in DT40. This can be done with or without induction of DNA damage using reagents such as mitomycin C.

3.1. Gene Targeting in DT40

3.1.1. Design and Construct Targeting Vectors

1. Obtain the full cDNA sequence of your gene of interest from the NCBI database. Using the sequence as a query, get full sequence of the gene locus, including exon and intron, from UCSC Genome Bioinformatics (http://genome.ucsc.edu/). Map exon–intron structure of the gene and search recognition sites of restriction enzymes.
2. Design the targeting vector and a hybridization probe. The probe should be a short DNA fragment (300 bp–1 kb) just outside of the vector. To establish gene-disrupted cells, replace the DNA fragment containing several important exons with a selection marker cassette. Available selection cassettes give resistance to the following drugs: neomycin (neo^R), hygromycin B (hyg^R), blastocidine-S (bsd), L-histidinol (hisD), puromycin (puro^R), zeocin (zeo), and mycophenolic acid (ecogpt). Make sure that the size of the restriction fragment will change significantly and can be detected by Southern hybridization when the vector is incorporated into the expected gene locus. Here we show the map of *FancC* gene locus on the sex chromosome Z, the design of *FancC* targeting vector, and estimated size of *Sac*I-digested genome DNA fragment that can be detected with the probe by Southern hybridization (**Fig. 17.1**).

3.1.2. Transfection of Targeting Vector in the Host DT40 Cell Line

1. Digest 30 μg plasmid DNA of targeting vector with an appropriate restriction enzyme. Choose the restriction enzyme which digests only backbone vector DNA.

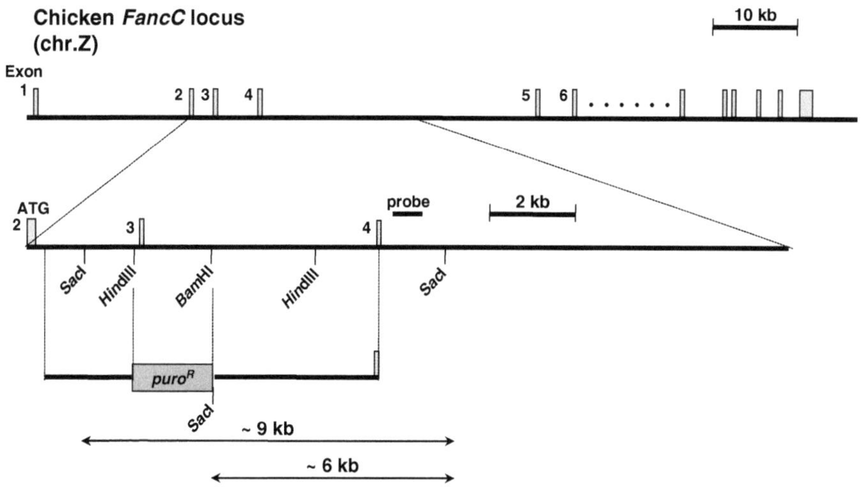

Fig. 17.1. Map of chicken *FancC* gene locus and *FancC* gene-targeting vector.

2. Purify linearized targeting vector with phenol/chloroform and ethanol precipitation. Dissolve DNA precipitate in 30 μl PBS.
3. Prepare 1×10^7 host DT40 cells and wash them with 5 ml ice-cold PBS once.
4. Resuspend the cells with 0.5 ml ice-cold PBS and transfer them into 0.4-mm cuvette.
5. Mix linearized DNA solution with the cells in the cuvette. Place the cuvette on ice for 10 min.
6. Electroporation. We usually use Gene Pulser Xcell electroporation system. The condition is 550 V voltage and 25 μF capacitance. Place the cuvette on ice for 10 min.
7. Transfer the electroporated cells to 10-cm dish with 20 ml culturing media. Incubate cells at 39.5°C in 5% CO_2 for 24 h.
8. Collect the electroporated cells and resuspend them in 80 ml culturing media containing an appropriate selection drug. Plate the cells into 96-well flat-bottom plates using a multichannel pipette.
9. Incubate the 96-well plates at 39.5°C in 5% CO_2 for 5 days–2 weeks until visible colonies emerge.

3.1.3. Extraction of the Genomic DNA from Drug-Resistant Clones

1. Pick up drug-resistant colonies. Expand those clones in 12-well or 6-well flat-bottom plates.
2. Make small-scale cell stock of each clone. Cells can be stored at –80°C or in liquid nitrogen in 50% FBS–10% DMSO stock solution.

3. Make cell pellets from ~5×10^6 cells of each clone. These can be stored at –80°C.
4. Activate cell lysis buffer by adding β-mercaptoethanol and proteinase K. Resuspend the cell pellets with 0.5 ml of this cell lysis buffer.
5. Incubate the lysates at 55°C overnight.
6. Add 250 μl of 6 M NaCl and vortex vigorously for 10 s. Put the tubes on ice for 15 min and centrifuge at 15,000 rpm for 15 min at 4°C.
7. Transfer the supernatant to a new tube.
8. Add 750 μl of 99.5% ethanol and mix by inverting 10–15 times. Genomic DNA will be visible by separation.
9. Centrifuge at 6,000 rpm for 1 min at 4°C.
10. Aspirate the supernatant. Wash the precipitate with 70% ethanol.
11. Centrifuge at 6,000 rpm for 1 min. Aspirate the supernatant and air-dry the precipitate briefly.
12. Dissolve the genomic DNA precipitate with 50 μl TE. Incubate them at 55°C for 15 min.

3.1.4. Blotting

1. Digest the purified genomic DNA with an appropriate restriction enzyme. Usually, 10–20 μg of DNA is digested in 200 μl volume with 30–40 units of restriction enzyme. Incubate them at 37°C overnight.
2. Add additional 10 units of restriction enzyme as a booster. Mix well and incubate them at 37°C for several hours.
3. Add 20 μl of 3 M sodium acetate and 600 μl of 99.5% ethanol. Mix well by inverting and incubate them on ice for 10 min.
4. Centrifuge at 15,000 rpm for 15 min.
5. Aspirate the supernatant. Wash the precipitate with 70% ethanol.
6. Centrifuge at 15,000 rpm for 5 min.
7. Aspirate the supernatant and air-dry the precipitate.
8. Dissolve the precipitate with 10 μl of 1× DNA loading dye.
9. Prepare 0.7% agarose gel in 0.5× TBE. Load the sample.
10. After electrophoresis, stain the gel with ethidium bromide and take picture of it with the scale as shown in **Fig. 17.2**.
11. Soak the gel in denaturing solution for 30 min at room temperature with gentle shaking. Repeat once.

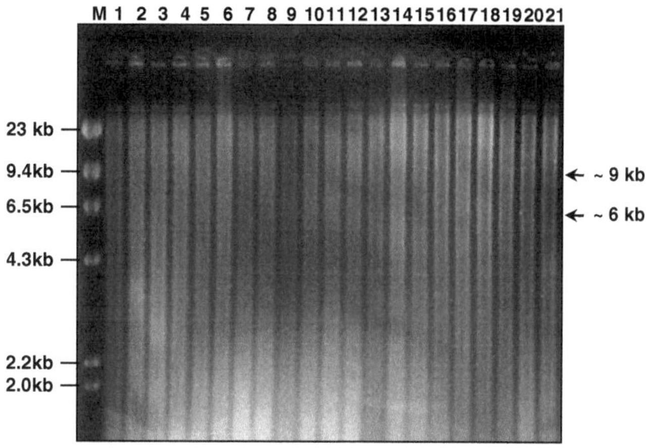

Fig. 17.2. Image of ethidium bromide-stained agarose gel.

12. Rinse the gel with distilled water. Soak the gel in neutralizing solution for 30 min at room temperature with gentle shaking. Repeat once.
13. Capillary transfer. Wet a nylon membrane of the size of the gel on surface of distilled water and then submerge. Put the membrane into 20× SSC for additional 5–10 min. Assemble Whatman 3 MM filter paper, the denatured/neutralized gel, the prewet nylon membrane, Whatman filter papers, and a stack of paper towel on a platform as shown in **Fig. 17.3**. We usually use 20× SSC as transfer buffer and Biodyne® B 0.45 μm or Hybond N$^+$ membrane. Make sure the transfer buffer does not bypass the agarose gel and nylon membrane into a stack of paper towel. Overnight transfer is recommended for genomic Southern blotting.

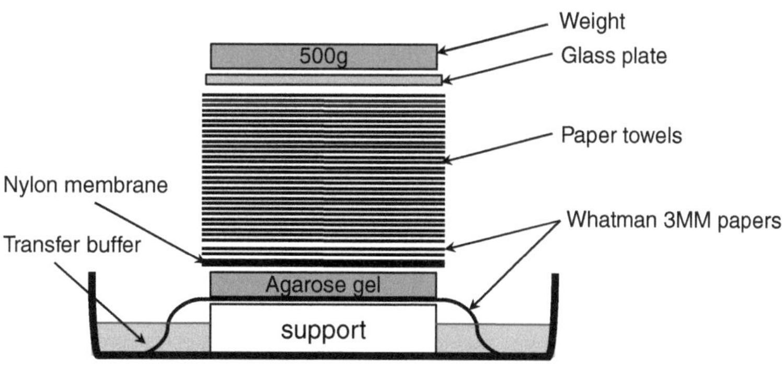

Fig. 17.3. Capillary transfer of DNA from agarose gel to nylon membrane.

14. Peel the nylon membrane off the gel. Bake the membrane at 80°C for 2 h.

3.1.5. Hybridization (Note 1)

1. Place the dry nylon membranes and hybridization buffer containing ssDNA (final 0.1 mg/ml) into the hybridization bottle.
2. Prehybridization: Rotate the bottle in hybridization oven at 42°C for more than ~4 h at 30 rpm.
3. Dissolve dextran sulfate powder in hybridization buffer at 10% (g/vol) by briefly boiling in microwave. Stir the mixture at room temperature for a while until it dissolves completely. Dextran sulfate makes the hybridization buffer viscous and it increases effective concentration of the probe.
4. Radiolabeling of probe. We use Rediprime II DNA labeling kit (GE Healthcare). Dilute 25 ng gel-purified DNA fragment with TE buffer. Denature DNA by incubating at 98°C for 5 min.
5. Immediately chill the denatured DNA on ice for 5 min.
6. Dissolve the blue pellet (contains all components for labeling reaction) with the denatured DNA solution by pipetting about 10 times.
7. Add 5 μl [α-^{32}P]dCTP (370 MBq/ml) into the DNA labeling mixture.
8. Incubate at 37°C for 15 min.
9. Stop the reaction by adding 5 μl of 0.2 M EDTA.
10. Denature the radiolabeled DNA mixture by incubating 98°C for 5 min.
11. Immediately chill the denatured/radiolabeled DNA mixture on ice for 5 min.
12. Discard the hybridization buffer used for prehybridization in the hybridization bottle. Add 10 ml hybridization buffer with 10% dextran sulfate.
13. Add the radiolabeled DNA mixture in the hybridization bottle.
14. Hybridization: Rotate the bottle in the oven at 42°C overnight at 30 rpm.
15. Carefully discard the hybridization buffer. Store this radioactive buffer for several months until the radioactive material loses the activity.
16. Wash: Add 100 ml of 2× SSC and 0.1% SDS wash buffer to the bottle.
17. Rotate the bottle at 42°C for 10 min at 30 rpm.

18. Discard the wash buffer. Store this wash buffer until the radioactive material loses the activity.
19. Repeat steps 16–18.
20. Add pre-warmed (50°C) 100 ml of 0.1× SSC and 0.1% SDS wash buffer to the bottle.
21. Rotate the bottle at 50°C for 10 min at 30 rpm.
22. Discard the wash buffer.
23. Repeat steps 20–22.
24. Take the nylon membrane from the bottle and briefly dry it on a paper towel.
25. Wrap the nylon membrane and expose it to the Fuji imaging plate. The exposure time should be from several hours to several days (depending on the signal intensity).
26. Read the signal by BAS2500 plate reader. In the experiment shown in **Fig. 17.4**, clones #1 and #16 were precisely targeted. Since there is only one *FANCC* allele in DT40 cells (it is on sex chromosome Z), single targeting is enough to disrupt *FancC* gene (**Fig. 17.4**).
27. Calculate gene-targeting frequency by the percentage of targeted clones (in this case, two clones were targeted, so the targeting frequency is 2/21 = 9.5%).

3.1.6. Evaluation of Gene-Targeting Frequency in the Mutant Cell Lines

1. Calculate gene-targeting frequency as shown above using wild-type and mutant as parent cell lines.
2. **Table 17.1** shows the relative gene-targeting efficiency in several mutants compared with that in wild type, which has already been published.

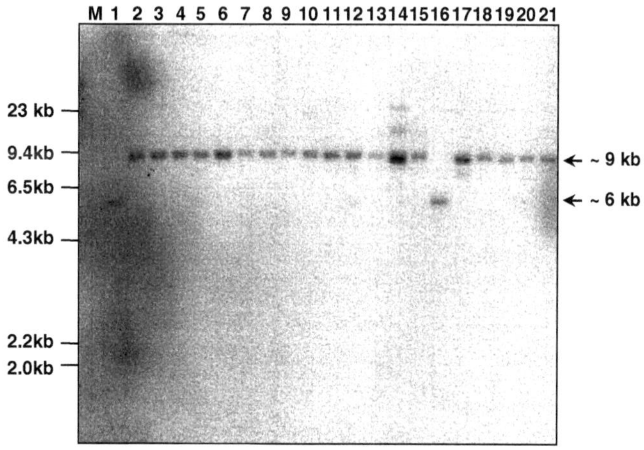

Fig. 17.4. Image of hybridized nylon membrane by FANCC probe.

Table 17.1
Relative gene-targeting efficiency in mutant cell lines

	Targeting locus							
Mutant	***Ovalbumin***	***β-Actin***	***Igλ***	***Xrcc2***	***Xrcc3***	***Ku70***	***Rad54***	**References**
rad54	0	3.1	13.2	ND	ND	ND	ND	(20)
rad52	86.7	17.0	38.4	ND	ND	ND	ND	(21)
rad51b	0	ND	ND	0	ND	0	ND	(22)
rad51c	16.8	ND	ND	6.9	ND	ND	ND	(23)
rad51d	0	ND	ND	ND	ND	ND	ND	(23)
xrcc2	ND	ND	ND	ND	ND	0	ND	(23)
xrcc3	0	ND	ND	0	ND	ND	ND	(23)
atm	66.9	ND	ND	93.9	ND	ND	ND	(24)
mre11	0	ND	ND	ND	ND	ND	ND	(25)
fancg	92.9	ND	ND	57.8	14.2	52.8	ND	(26)
fancd2	0	ND	ND	11.5	0	ND	ND	(12)
fancc	0	ND	ND	18.2	ND	36.4	ND	(19)
brca2tr	0	ND	ND	ND	ND	ND	21.1	(11)
rev3	72.9	ND	ND	ND	ND	ND	25.6	(27)
polkappa	44.5	ND	ND	ND	ND	ND	11.5	(28)
nbs1	80.0	ND	78.9	ND	ND	ND	ND	(13)

% of wild-type control; ND. not determined

3.2. Analysis of I-SceI-Induced HRR in OVA-SCneo Integrated Cells

3.2.1. Measuring the HRR Frequency

1. Prepare 5×10^6 or 1×10^7 cells of DT40 cell line, which contains *SCneo* reporter gene at the *Ovalbumin* gene locus (the map of *Ovalbumin* gene locus and *SCneo* reporter is shown in **Fig. 17.5**), for electroporation. Collect the cells and resuspend them in 0.5 ml fresh media and transfer them into 0.4-cm cuvette.
2. Prepare 30 μg DNA of I-*Sce*I expression plasmid or control plasmid (pBS-KS, for example). Plasmid DNA is ethanol precipitated and resuspended in 30 μl PBS. Mix each DNA in cell suspension in the cuvette. Incubate at room temperature for 10 min.
3. Electroporate each DNA into host DT40 cells with Gene Pulser II Xcell electroporation system. The condition is 975 V voltage and 250 μF capacitance. Incubate the cuvette at room temperature for 10 min.
4. Transfer the cell/DNA mixture to 20 ml culturing media in a 10-cm dish. Incubate at 39.5°C in 5% CO_2 for 24 h.

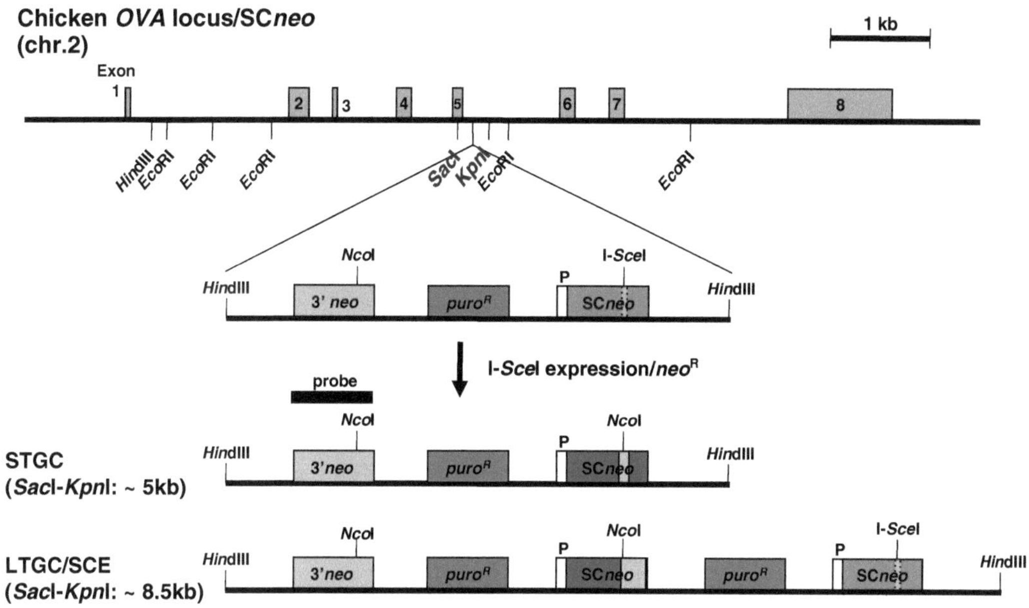

Fig. 17.5. Map of chicken *Ovalbumin* gene and *SCneo* reporter gene. Structure of neo^R SCneo reporters after I-*Sce*I-induced DNA DSB repair is also shown.

5. Dilute the electroporated cells (100 μl, 1 ml, 10 ml) in 40 ml fresh media in the presence of 2 mg/ml G418 and inoculate them into 96-well flat bottom plates.
6. Incubate for 5 days–10 days in 5% CO_2 at 39.5°C until the visible colonies emerge. Count the number of colonies. Calculate the ratio of neo^R cells by subdividing the number of colonies by the number of electroporated cells. In *FancD2*$^{-/-}$ cells, only ~300 neo^R colonies emerged when 1×10^7 cells were electroporated. In the parallel experiment, wild-type cells or *FancD2*$^{-/-}$ cells rescued by chicken *FancD2* gene, more than 1×10^4 neo^R colonies emerged. Taken into account the reduced plating efficiency of *FancD2*$^{-/-}$ cells, the decrease in HRR frequency was estimated to be ~26-fold (12).

3.2.2. Analyzing Each HRR Events

1. Select the wells with single neo^R colony originated from a single cell. Transfer the cells to 12-well flat-bottom plates with 3 ml fresh media.
2. Incubate several days until the cells grow to the confluency.
3. Make small-scale cell stock of each clone. Cells can be stored at –80°C or in liquid nitrogen in 50% FBS–10% DMSO stock solution (optional).
4. Harvest the cells.
   (The following process should be done as described in **Section 3.1**).

5. Purify the genomic DNA.
6. Digest the purified genomic DNA with *Kpn*I and *Sac*I.
7. Electrophorese them in 0.7% agarose gel in 1× TAE
8. Blot them onto nylon membranes.
9. Hybridize the nylon membranes with radiolabeled neo probe.
10. Wash the membranes and expose them to imaging plate.
11. Read the signal by BAS2500 plate reader. Short tract gene conversion (STGC) makes ~5-kb *Sac*I/*Kpn*I neo$^+$ DNA fragment, while long tract gene conversion/sister chromatid exchange (LTGC/SCE) makes ~8.5-kb fragment. For example, we did not detect significant difference in the overall frequency of STGC and LTGC/SCE between wild type and *FancD2*$^{-/-}$. However, in several neo^R clones isolated from *FancD2*$^{-/-}$ cells, the neo$^+$ restriction fragment with aberrant length was detected, which indicates the qualitative HRR defect in this mutant (12).

3.3. Sister Chromatid Exchange

1. Culture DT40 cells with 10 μM BrdU for the duration of two cell cycles (for wild-type DT40 cells, 16–18 h). For mutant cell lines whose doubling time is longer than that of wild type, labeling time should be longer. If mitomycin C is used, it should be added 8 h before harvest.
2. Pulse with 0.1 μg/ml colcemid for the last 1–2 h.
3. Harvest the cells to 15-ml centrifuge tube and centrifuge for 5 min at 1,000 rpm.
4. Add 75 mM KCl (1–2 ml) to pellet and incubate for 15–30 min at room temperature. This hypotonic treatment makes cells swollen and fragile. Treat them softly hereafter.
5. Add 5 ml freshly prepared methanol/acetic acid (3:1) solution and mix well by inverting.
6. Centrifuge for 5 min at 1,000 rpm.
7. Resuspend in 5 ml methanol/acetic acid (3:1) solution.
8. Incubate for 30 min at room temperature.
9. Centrifuge for 5 min at 1,000 rpm.
10. Resuspend the cell pellet to appropriate volume (100–200 μl) of methanol/acetic acid (3:1) solution. This sample may be stored at –20°C.
11. Prepare prewet slide glass in 50% ethanol solution.
12. Carefully drop the cell suspension (one drop at a time) onto the slide glass from ~12 in. height (**Note 2**).

13. Dry slides on the hot plate (40–42°C) covered over by a wet paper.
14. Incubate slides in Hoechst 33258 (10 μg/ml) diluted in 0.05 M phosphate buffer (pH 6.8) for 20 min at room temperature.
15. Rinse slides in McILvaine solution and directly cover the samples with coverslip (the samples should be kept wet).
16. Irradiate slides with black light (FL20BLB, Toshiba: 1 cm distance from the bulb: 0.2–0.4 $J/m^2/s$).
17. Immerse slides into McILvaine solution and remove the coverslip.
18. Immerse slides in 2× SSC solution at 62°C for 1 h.
19. Rinse slides with phosphate buffer (pH 6.8) and immerse in 3% Giemsa solution, diluted in 0.05 M phosphate buffer (pH 6.8) for 20–60 min.
20. Rinse back of the slide with water and air-dry.
21. Mount the slides in EUKITT mounting fluid.

    View the slides at 1,000× magnification under immersion oil.
22. Count the number of SCE breakpoints (shown by arrowheads in **Fig. 17.6**) on the 12 macrochromosomes (chromosomes 1 (two alleles), 2 (three alleles), 3 (two alleles), 4 (two alleles), 5 (two alleles), and Z). **Figure 17.6** shows highly elevated spontaneous SCE events in *FANCC/RAD18* double-deficient cells. We concluded that higher levels of SCE in *FANCC*- or *RAD18*-deficient cells are through different mechanisms, since *FANCC/RAD18* double-deficient cells displayed even higher SCE levels compared to either single mutant (19).

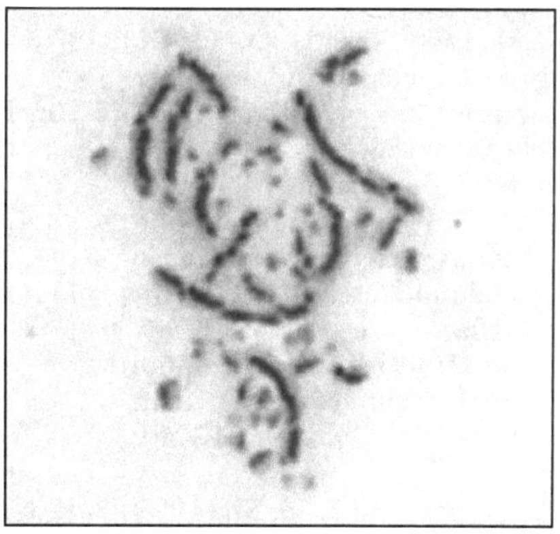

Fig. 17.6. Spontaneous SCE in *FANCC/Rad18* double-deficient cells.

4. Notes

These three assays are quite straight forward, and if you stick to general molecular/cellular biological technique, they should be simple and (relatively) easy. We just would like to draw reader's attention to only two points:

1. Detection of the gene-targeting events. Genomic Southern blotting can be difficult if you omit something from the protocol. As a result, you may then lose detection power of the procedure leading to the loss of the signal.
2. Preparation of the metaphase spread. You need a good metaphase spread of chromosome to obtain a good picture of SCEs. It is important to control the moisture (50% ethanol) on the slide glass before dropping the fixed cell suspension. Too much or too few moisture is not appropriate. Also when you drop the cell suspension, the distance between pipette and slide glass should be around 12 in. height. If the distance is too large, then chromosomes will be scattered on the glass. If the distance is too small, the chromosomes will not spread at all. Try several times until you find an appropriate distance.

References

1. Takata, M., et al. (1998) Homologous recombination and non-homologous end-joining pathways of DNA double-strand break repair have overlapping roles in the maintenance of chromosomal integrity in vertebrate cells. *EMBO J* **17**(18), 5497–5508.
2. Ip, S.C., et al. (2008) Identification of Holliday junction resolvases from humans and yeast. *Nature* **456**(7220), 357–361.
3. Buerstedde, J.M., and Takeda, S. (1991) Increased ratio of targeted to random integration after transfection of chicken B cell lines. *Cell* **67**(1), 179–188.
4. Rouet, P., Smih, F., and Jasin, M. (1994) Introduction of double-strand breaks into the genome of mouse cells by expression of a rare-cutting endonuclease. *Mol Cell Biol* **14**(12), 8096–8106.
5. Pierce, A.J., et al. (1999) XRCC3 promotes homology-directed repair of DNA damage in mammalian cells. *Genes Dev* **13**(20), 2633–2638.
6. Johnson, R.D., Liu, N., and Jasin, M. (1999) Mammalian XRCC2 promotes the repair of DNA double-strand breaks by homologous recombination. *Nature* **401**(6751), 397–399.
7. Moynahan, M.E., et al. (1999) Brca1 controls homology-directed DNA repair. *Mol Cell* **4**(4), 511–518.
8. Moynahan, M.E., Pierce, A.J., and Jasin, M. (2001) BRCA2 is required for homology-directed repair of chromosomal breaks. *Mol Cell* **7**(2), 263–272.
9. Nakanishi, K., et al. (2005) Human Fanconi anemia monoubiquitination pathway promotes homologous DNA repair. *Proc Natl Acad Sci USA* **102**(4), 1110–1115.
10. Fukushima, T., et al. (2001) Genetic analysis of the DNA-dependent protein kinase reveals an inhibitory role of Ku in late S-G2 phase DNA double-strand break repair. *J Biol Chem* **276**(48), 44413–44418.
11. Hatanaka, A., et al. (2005) Similar effects of Brca2 truncation and Rad51 paralog deficiency on immunoglobulin V gene diversification in DT40 cells support an early role for Rad51 paralogs in homologous recombination. *Mol Cell Biol* **25**(3), 1124–1134.
12. Yamamoto, K., et al. (2005) Fanconi anemia protein FANCD2 promotes immunoglobulin gene conversion and DNA repair through a mechanism related to homologous recombination. *Mol Cell Biol* **25**(1), 34–43.

13. Tauchi, H., et al. (2002) Nbsl is essential for DNA repair by homologous recombination in higher vertebrate cells. *Nature* **420**(6911), 93–98.
14. Bayani, J., and Squire, J.A. (2005) Sister chromatid exchange. *Curr Protoc Cell Biol* **Chapter 22**: p. Unit 22 7.
15. German, J., and Alhadeff, B. (2001) Analysis of sister-chromatid exchanges. *Curr Protoc Hum Genet* **Chapter 8**: p. Unit 8 6.
16. Sonoda, E., et al. (1999) Sister chromatid exchanges are mediated by homologous recombination in vertebrate cells. *Mol Cell Biol* **19**(7), 5166–5169.
17. Chaganti, R.S., Schonberg, S., and German, J. (1974) A manyfold increase in sister chromatid exchanges in Bloom's syndrome lymphocytes. *Proc Natl Acad Sci USA* **71**(11), 4508–4512.
18. Johnson, R.D., and Jasin, M. (2000) Sister chromatid gene conversion is a prominent double-strand break repair pathway in mammalian cells. *EMBO J* **19**(13), 3398–3407.
19. Hirano, S., et al. (2005) Functional relationships of FANCC to homologous recombination, translesion synthesis, and BLM. *EMBO J* **24**(2), 418–427.
20. Bezzubova, O., et al. (1997) Reduced X-ray resistance and homologous recombination frequencies in a RAD54–/– mutant of the chicken DT40 cell line. *Cell* **89**(2), 185–193.
21. Yamaguchi-Iwai, Y., et al. (1998) Homologous recombination, but not DNA repair, is reduced in vertebrate cells deficient in RAD52. *Mol Cell Biol* **18**(11), 6430–6435.
22. Takata, M., et al. (2000) The Rad51 paralog Rad51B promotes homologous recombinational repair. *Mol Cell Biol* **20**(17), 6476–6482.
23. Takata, M., et al. (2001) Chromosome instability and defective recombinational repair in knockout mutants of the five Rad51 paralogs. *Mol Cell Biol* **21**(8), 2858–2866.
24. Takao, N., et al. (1999) Disruption of ATM in p53-null cells causes multiple functional abnormalities in cellular response to ionizing radiation. *Oncogene* **18**(50), 7002–7009.
25. Yamaguchi-Iwai, Y., et al. (1999) Mre11 is essential for the maintenance of chromosomal DNA in vertebrate cells. *EMBO J* **18**(23), 6619–6629.
26. Yamamoto, K., et al. (2003) Fanconi anemia FANCG protein in mitigating radiation- and enzyme-induced DNA double-strand breaks by homologous recombination in vertebrate cells. *Mol Cell Biol* **23**(15), 5421–5430.
27. Sonoda, E., et al. (2003) Multiple roles of Rev3, the catalytic subunit of pol zeta in maintaining genome stability in vertebrates. *EMBO J* **22**(12), 3188–3197.
28. Okada, T., et al. (2002) Involvement of vertebrate polkappa in Rad18-independent postreplication repair of UV damage. *J Biol Chem* **277**(50), 48690–48695.

Chapter 18

Understanding the Immunoglobulin Locus Specificity of Hypermutation

Vera Batrak, Artem Blagodatski, and Jean-Marie Buerstedde

Abstract

The *immunoglobulin* (*Ig*) genes of B cells are diversified at high rate by point mutations whereas the *non-Ig* genes of B cells accumulate no or significantly fewer mutations. Ig hypermutations are critical for the affinity maturation of antibodies for most of jawed vertebrates and also contribute to the primary *Ig* diversity repertoire formation in some species. How the hypermutation activity is specifically targeted to the *Ig* loci is a long-standing debate. Here we describe a new experimental approach to investigate the locus specificity of *Ig* hypermutation using the chicken B-cell line DT40. One feature is the use of a *green fluorescent protein* (*GFP*) gene as a mutation reporter. Some nucleotide changes produced by somatic hypermutation can cripple the *GFP* gene which leads to a decrease or loss of the green fluorescence. Therefore such changes can be easily quantified by fluorescence-activated cell sorting (FACS). Another advantage of this approach is the targeted integration of the mutation reporter into a defined chromosomal position. This system allowed us to identify a 10 kb sequence within the *Ig light chain* (*IgL*) locus, which is both necessary and sufficient to activate hypermutation in the neighboring reporter gene. We have called this sequence *Diversification Activator* (*DIVAC*) and postulated that similar *cis*-acting sequences exist in the heavy and light chain *Ig* loci of all jawed vertebrate species. Our experimental system promises further insight into the molecular mechanism of *Ig* hypermutation. For example, it may be possible to identify smaller functional motifs within *DIVAC* and address the role of putative transacting binding factors by gene knock-outs.

Key words: Somatic hypermutation, immunoglobulin gene, AID, B cell, DT40, *DIVAC*.

1. Introduction

One of the most fascinating biological phenomena in vertebrates is the development of a vast repertoire of antibody variants to protect the organism against the menace of rapidly evolving pathogens. The genetic basis for the antibody diversity

H. Tsubouchi (ed.), *DNA Recombination*, Methods in Molecular Biology 745,
DOI 10.1007/978-1-61779-129-1_18, © Springer Science+Business Media, LLC 2011

is the assembly and diversification of the *Ig* genes during the somatic development of each individual B cell (1). In all jawed vertebrates a certain level of diversity is already produced by site-specific recombination of *V*, *D*, and *J* gene segments (*V(D)J* recombination) during *Ig* light and heavy chain gene assembly.

However, in all jawed vertebrates analyzed so far the 5′-regions of the rearranged *Ig* genes are also frequently altered by point mutations. First, evidence for the presence of a high mutation rate within the *Ig* genes came from the comparison of different amino terminal κ light chain sequences of clonal mouse antibodies which were most likely derived from the same rearranged *κ chain gene* (2). Based on the sequence diversity the mutation rate within the sequence encoding the N-terminal domain was calculated to 1 per 10^3 bases per generation. This result was subsequently confirmed by the comparison of sequences derived from the same rearranged *Ig* gene, which varied from each other by single nucleotide substitutions (3). These mutation rates for the *Ig* genes are orders of magnitude higher than the spontaneous mutation rates of vertebrate cells believed to be only about 1 per 10^9 bases per cell division.

The so-called somatic or *Ig* hypermutations occur only in B cells which express the cell type-specific activation-induced cytosine deaminase (AID) protein (4) and actively transcribe their *Ig* genes (5). The AID protein most likely acts on both strands of the DNA deaminating cytosines to uracils (6). The resulting uracils may pair with adenines during the next replication cycle leading to *C* to *T* or *G* to *A* transition mutations. Alternatively, the uracils can be excised by uracil glycosylases and error-prone repair or replication of the abasic sites which leads to transition and transversion mutations at *C/G* bases. This hypothesis of AID action is supported by the observation that AID expression in *Escherichia coli* leads to an increased rate of transition mutations at *C/G* bases, which was further increased in an uracil glycosylase-deficient background (7). Although murine and human AID-expressing B cells accumulate transition and transversion mutations at all four bases indicating more complex processing of the AID-induced uracils, hypermutating variants of the DT40 cell line mainly accumulate *Ig* hypermutations at *G/C* bases consistent with the hypothesis that AID initiates mutations by deamination of cytosines in DNA (8, 9). Furthermore, hypermutating DT40 in which uracil glycosylase is either inhibited (10) or inactivated (11) shows a remarkable change in the mutation spectrum toward transitions at *C/G* bases validating the idea that a large number of AID-induced uracils are processed by uracil glycosylases. Similar changes in the mutation spectrum are observed in B cells of uracil glycosylase knock-out mice (12).

Unregulated hypermutation represents a threat to genome integrity and has been suggested to play a role in the development of cell lymphomas when inappropriately activated (13, 14). Only *Ig* genes undergo hypermutation at a high rate, and this seems to be precisely targeted as sequence analysis of genes neighboring to the *IgL* locus in DT40 revealed no diversity (15). Sequence analysis of *non-Ig* genes in B cells revealed in most cases no mutations. Nevertheless, some non-*Ig* genes, including *BCL6*, *FAS*, *CD79A*, and *CD95B* were found to undergo somatic hypermutation in normal B cells, albeit at lower frequencies compared to the *Ig* locus (13, 16–18). In human B-cell tumors several other loci, such as *BCL6*, *MYC*, *PIM1*, *RHOH*, and *PAX5*, were demonstrated to be targeted by hypermutation (14, 18). A comprehensive sequence analysis of hundreds of genes in AID-expressing versus AID knock-out murine B cells showed that AID increases the mutation rate in many genes about two- to tenfold (19). However, the difference in the mutation rates for the *Ig* genes is orders of magnitude higher between AID-expressing versus AID knock-out B cells.

The analysis of chimeric genes consisting of *non-Ig* sequences combined with parts of the *Ig* loci in transgenic mice indicated that regions encompassing the *Ig* enhancers distinguish the *Ig* genes as hypermutation targets (20, 21). The role of *Ig* enhancers for the locus specificity of hypermutation remained however controversial, since B cells in which the *Ig* enhancer regions had been deleted still accumulated *Ig* mutations (22).

The DT40 genome can be easily modified by targeted integration. Furthermore, a *pseudo-V* gene-deleted DT40 variant hypermutates its rearranged *IgL* gene dependent on AID expression (9). Based on these experimental advantages of DT40 we developed an experimental system to address the role of *cis*-acting sequences for the locus specificity of hypermutation (23). This system promises further insight into (i) what the active sequence motifs and their arrangement within *DIVACs* are, (ii) how *DIVACs* correlate and cooperate with enhancer-like elements, and (iii) whether *DIVAC*-binding proteins recruit AID. This may allow us to understand the details how hypermutations are specifically targeted to the *Ig* transcription units.

2. Materials

Cell Line for Transfection

A straightforward option is to test *DIVAC* sequences at the chromosomal position of the deleted rearranged *IgL* locus of DT40 (**Note 1**). A cell line suitable for this is the DT40 variant ψV$^-$IgL$^-$ (**Fig. 18.2**).

Whereas the insertion of a *GFP*-based hypermutation reporter (*GFP2*) into the deleted rearranged *IgL* locus of ψV⁻IgL⁻ results in stable GFP expression presumably due to a very low rate of *GFP2* mutations, the insertion of *GFP2* combined with *cis*-regulatory *DIVAC* sequences results in unstable GFP expression due to hypermutations in the *GFP2* reporter (23). Using ψV⁻IgL⁻ the *DIVAC* activity of sequences can be easily determined and quantified after a single-step transfection and the analysis of GFP expression in transfectants having undergone targeted integration (**Note 1**).

Targeting Vector

The targeting vector designed for the insertion of the *GFP2* reporter and *DIVAC* sequences into the deleted rearranged *IgL* locus of the ψV⁻IgL⁻ cell line was named *pIgL*$^{-,GFP2}$. It consists of the *GFP2* hypermutation reporter (the *GFP* gene under the influence of the *RSV* promoter linked by an *IRES* sequence to the *blasticidin resistance* gene), an upstream and downstream target arm corresponding to the sequences flanking the *IgL* locus deletion of the ψV⁻IgL⁻ cell line as well as the *pBluescriptKS* plasmid backbone (**Fig. 18.1**). Between the *RSV* promoter and the downstream target arm are unique *Nhe*I/*Spe*I restriction sites which can be used to insert potential *DIVAC* sequences. The

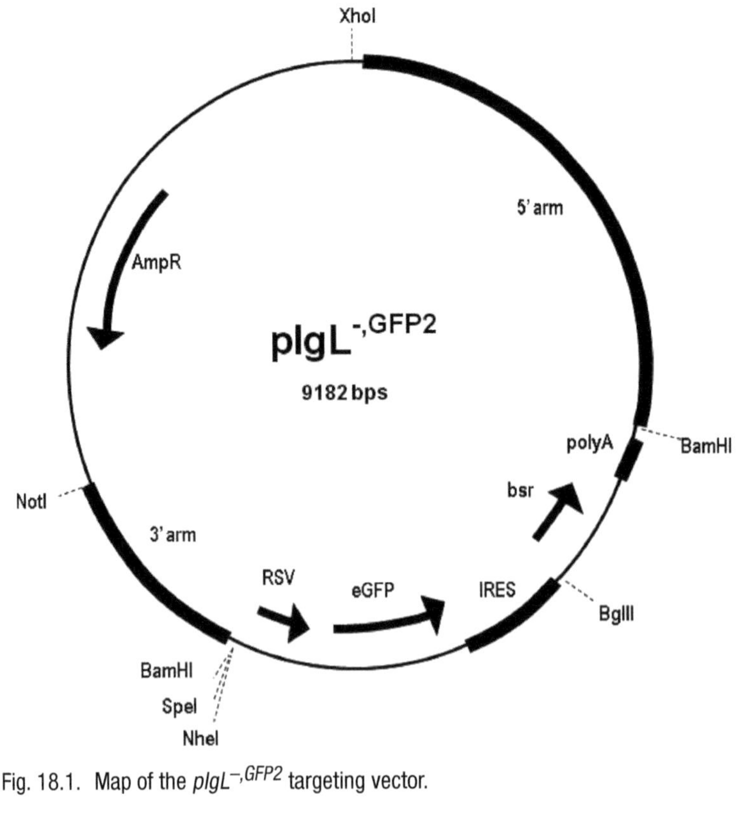

Fig. 18.1. Map of the *pIgL*$^{-,GFP2}$ targeting vector.

plasmid contains a single *Not*I site in its polylinker downstream of the 3′-target arm which can be used for its linearization before transfection, if *Not*I is not present in the *DIVAC* sequence cloned into the construct. Alternatively the unique *Xho*I site upstream of the 5′-target arm might be used. The vector and its sequence are available upon request.

Targeted integration of a $pIgL^{-,GFP2}$ derived construct into $\psi V^{-}IgL^{-}$ leads to single copy insertion of the *GFP2* reporter and potential adjacent *DIVAC* sequences at the position of the deleted rearranged *IgL* locus. A consequence of targeted integration is the removal of the puromycin resistance gene **(Fig. 18.2)**.

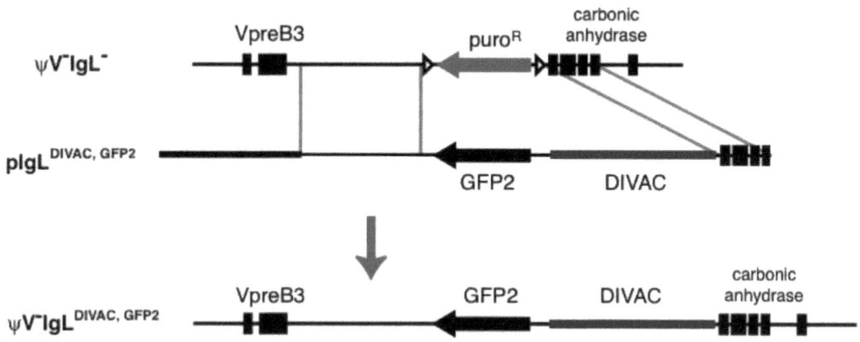

Fig. 18.2. Targeted integration of a putative $pIgL^{-,GFP2}$ derived construct containing potential *DIVAC* sequences ($pIgL^{DIVAC,GFP2}$) into the deleted *IgL* locus.

2.1. Cell Culture

1. Cell culture medium for DT40 in the following referred to as chicken medium is prepared by mixing 500 ml of Dulbecco's modified Eagle medium/F-12-Medium L-glutamine (+) (Gibco, 31330-038) referred to as DMEM/F-12-Medium, 50 ml of FBS (Biochrom, S0115), 10 ml of penicillin–streptomycin (Gibco, 15140-122), 5 ml of chicken serum (PAN-Biotech, P30-0301), and 50 μl of 1 M beta-mercaptoethanol (Sigma, M7522) (**Note 2**). The medium can be stored at 4°C.
2. Freezing medium: 70 ml of DMEM/F-12-Medium, 20 ml of FBS, and 10 ml of dimethyl sulfoxide (DMSO) (Sigma, D2650).
3. Tissue culture flasks for small-scale culture up to 10 ml (Nunc, 136196), for middle-scale culture up to 50 ml (Greiner bio-one, 690175), and for large-scale culture up to 250 ml (Greiner bio-one, 658175).
4. Laminar flow bench and CO_2 incubator.
5. Inverted microscope to check cell growth and viability.

2.2. Transfection

1. Purified plasmid DNA of the targeting construct.
2. Appropriate restriction enzyme.

3. Phenol/chloroform, chloroform, propanol, and 70% ethanol.
4. Agarose gel electrophoresis equipment.
5. Trypan blue solution, cell counter, or cell viability analyzer.
6. Electroporation System Gene Pulser Xcell (BioRad) or similar electroporator (**Note 2**).
7. Gene pulser cuvettes, 0.4 cm gap between the two electrodes (BioRad, 165-2088).
8. 96-Well flat-bottom microtiter plates (Nunc, 167008).
9. Chicken medium as described in step 1 of **Section 2.1**.
10. Blasticidin selection medium: Blasticidin S HCl (Invitrogen, R210-01) diluted by chicken medium to a concentration of 30 μg/ml, 2× the concentration used for the selection of transfectants. It can be stored for some time at 4°C.

2.3. Screening Transfectants for Puromycin Sensitivity

1. 96-Well flat-bottom microtiter plates (Nunc, 167008).
2. Chicken medium.
3. Puromycin selection medium: Puromycin (Sigma-Aldrich, P9620) is diluted by chicken medium to a concentration of 2 μg/ml, 2× the concentration used for the selection of transfectants. It can be stored for some time at 4°C.

2.4. Confirmation of Targeted Integration by PCR

1. K buffer: 1× PCR buffer of Expand Long Template System (Roche, 1681834), 0.5% Tween 20, 100 μg/ml proteinase K (Qiagen, 19131).
2. Heating block or water bath.
3. Expand Long Template PCR System (Roche, 1681842).
4. PCR cycler.
5. Agarose gel electrophoresis and gel documentation equipment.

2.5. Subcloning

1. Trypan blue solution, cell counter, or cell viability analyzer.
2. Chicken medium as described in step 1 of **Section 2.1**.
3. 96-Well flat-bottom microtiter plates (Nunc, 167008).

2.6. Analysis of Mutation Rates by FACS

1. 24-Well flat-bottom plate.
2. Chicken medium.
3. 2 ml plastic tubes for FACS analysis.
4. FACSCalibur (BD) or similar machine.

2.7. Analysis of Mutation Rates by Sequencing

1. Low concentration mycophenolic acid selection medium: Mycophenolic acid (Sigma-Aldrich, M5255) is diluted by

chicken medium to a concentration of 0.5 μg/ml, which is the concentration used for selection of *AID-IRES-gpt* expressing cells. It can be stored for some time at 4°C.

2. K buffer as described in step 1 of **Section 2.4**.
3. PfuUltra Hotstart DNA Polymerase (Stratagene, 600390).
4. BigDye Terminator v3.1 Cycle Sequencing Kit (Applied Biosystems, 4337455).
5. BigDye® Terminator v1.1/v3.1 Sequencing Buffer (5×) (1 ml) (Applied Biosystems, 4336697).

2.8. Removal of the AID Expression Cassette by Cre Recombinase Induction

1. 96-Well flat-bottom microtiter plates (Nunc, 167008).
2. Chicken medium.
3. Mycophenolic acid selection medium: Mycophenolic acid (Sigma-Aldrich, M5255) is diluted by chicken medium to a concentration of 2 μg/ml, 2× the concentration used for the selection of transfectants. It can be stored for some time at 4°C.

3. Methods

3.1. Cell Culture

1. DT40 cells and ψV⁻IgL⁻ variant cell line are not tricky to culture and propagate. However, care should be taken that the culture medium (chicken medium) and the culture conditions support vigorous growth, as otherwise problems with transfection efficiency may arise (*see* **Note 2**). The cells can be cultured in tissue culture flasks, Petri dishes, 24-well plates or 96-well microtiter plates. As long as the incubator is well humidified and the wells of the microtiter plates are filled with at least 100 μl, the medium needs to be exchanged only when it turns yellow. The optimum culture conditions for the cells are 41°C with 5% CO_2.

3.2. Transfection

1. Linearize the $pIgL^{-,GFP2}$ derived targeting construct in the plasmid polylinker or the plasmid backbone by using an appropriate restriction enzyme. At least 40 μg DNA per electroporation and an overnight digest in 500 μl total reaction volume are recommended. The rare cutter *Not*I or *Sal*I are good candidates, but which enzyme to use depends on the restriction map of the potential *DIVAC* sequences cloned next to the *GFP2* reporter.
2. Check the completeness of the digestion and the quality of the DNA by the analysis of 0.5 μg DNA on an agarose gel.

3. Purify the digested DNA once by phenol/chloroform extraction, once by chloroform extraction and precipitate with isopropanol. Rinse the precipitated DNA with 70% ethanol. Dry the pellet for 10 min inside a laminar flow bench.
4. Re-suspend the pellet in distilled water to final 1 μg/μl concentration.
5. Determine the cell density in the ψV⁻IgL⁻ culture which you would like to use for transfection. The culture should have good viability and an optimal cell density of about $0.5–1.5 \times 10^6$ cells/ml. Add a volume containing about 10 million cells to a 50 ml tube and spin down for 5 min at 1,500 rpm, 4°C.
6. Remove the supernatant and re-suspend the cell pellet in 800 μl of chicken medium.
7. Transfer the cell suspension and the re-suspended linearized DNA into an electroporation cuvette.
8. Electroporate the cuvette using 25 μF and 700 V.
9. Transfer the cell/DNA solution into a tube containing 9.5 ml of chicken medium and add 100 μl into each well of a flat-bottom microtiter plate.
10. On the following day (12–24 h after electroporation) add 100 μl blasticidin selection medium to each well.
11. Leave the plates for about 7–10 days in the incubator without changing the medium. Drug-resistant colonies should be visible at this stage to the naked eye as white spots in the wells of the microtiter plate. The number of transfectants per transfection may vary between 5 and 30. If you get no or a lower number of colonies, *see* **Note 2** for possible reasons.

3.3. Screening of ΨV⁻IgL⁻ Transfectants for Puromycin Sensitivity

Targeted integration of *pIgL⁻,GFP2* derived constructs leads to the removal of the puromycin resistance gene in ψV⁻IgL⁻ transfectants. This can be easily checked by the loss of puromycin resistance.

1. Pick up blasticidin-resistant colonies from the 96-well microtiter plate onto which the electroporated cells had been plated. This is best done by punching the tip of a 20 μl pipette into the center of a colony and pulling in 10 μl.
2. Transfer the 10 μl containing the cells of the colony into a well of a flat-bottom 96-well plate containing 200 μl of chicken medium. Make a duplicate of the colony by transferring 100 μl into a separate well of a 96-well plate.
3. Add 100 μl of puromycin selection medium to the well containing one of the duplicates, and add 100 μl of chicken medium to the well containing the other duplicate.

4. Incubate for 3 days.
5. Select clones whose duplicates have died in the presence of puromycin. These clones have almost certainly integrated the *pIgL*$^{-,GFP2}$ derived construct at the chromosomal position of the deleted *IgL* locus.
6. The ratio of puromycin sensitive to resistant clones should be approximately 1 in 8 as the targeting rate of *pIgL*$^{-,GFP2}$ derived constructs in ψV^{-}IgL^{-} cells is about 15%.

3.4. Confirmation of Targeted Integration by PCR

1. Culture duplicates of stable transfectants in wells of a flat-bottom 96-well plate using 300 μl chicken medium. Wait until the cells have grown confluent.
2. Transfer the cells into a 96-well PCR plate, spin down the cells for 5 min at 1,500 rpm, and decant the supernatant. Re-suspend the cells in 200 μl of PBS, then spin down and decant the supernatant again.
3. Re-suspend the cells in 10 μl of K buffer.
4. Incubate for 45 min at 56°C to achieve Proteinase K-mediated proteolysis.
5. Incubate at 95°C for 10 min to inactivate the Proteinase K. Use 1 μl of the resulting crude extract for PCR.
6. To confirm targeted integration of *pIgL*$^{-,GFP2}$ derived constructs we recommend using the primer PS31: TTCTGAGGGAAAAGGACGCGTGTAATTGCA from the genomic sequence upstream of the 5′-target arm of the construct together with the primer PU5: CCCACCGACTCTAGAGGATCATAATCAGCC derived from the SV40 polyA signal. In our hands the Expand Long Template PCR System (Roche) gave reliable results. PCR can be performed with the following protocol: 2-min initial incubation at 93°C, 35 cycles consisting of 93°C for 10 s, 65°C for 30 s, and 68°C for 5 min with cycle elongation of 20 s per cycle, and a final 5-min elongation step at 68°C.

3.5. Subcloning

1. Using Trypan blue staining count the viable cell density in the culture to be subcloned.
2. Dispense 10 ml chicken medium into each of three tubes adding 30 viable cells to the first tube, 100 to the second, and 300 to the third.
3. Distribute each of the three cell dilutions onto a 96-well flat-bottom microtiter plate by adding 100 μl to each well.
4. Incubate the plates for 7–10 days without changing the medium. Subclones should become visible as defined round colonies. Pick subclones starting with the plate showing the lowest number of colonies. The precision of subcloning may be further increased by not transferring all cells growing

within a well, but punching the tip of a 20 μl pipette into the center of a colony and transferring only 10 μl.

5. If the chicken medium and the culture conditions are satisfactory, the subcloning efficiency should not be (much) worse than one in five plated cells.

3.6. Analysis of GFP2 Mutation Rates by FACS

Both primary targeted transfectants and subclones there from can be analyzed. To keep the results comparable, the time from each transfection or subcloning until the FACS analysis should be kept constant. The results obtained from a limited number of primary transfectants should be considered only rough estimates of the mutation rate of a particular *GFP2-DIVAC* combination due to possible fluctuation effects.

1. Subclone primary transfectants by limiting dilution as described under **Section 3.5**.
2. Pick up 24 subclones for each transfectant. Transfer the 10 μl subclone cell suspensions into wells of 24-well flat-bottom plates containing 1 ml of chicken medium per well.
3. Whenever the chicken medium starts to turn yellow, remove the medium and part of the cells and add new chicken medium to the wells. This is necessary to avoid over-growth of the cells.
4. Two weeks after subcloning transfer the cells into tubes compatible with the FACS machine, wash twice with PBS, and re-suspend in PBS.
5. Determine the percentage of the cells possessing decreased green fluorescence by flow cytometer for each subclone (**Note 3**).

3.7. Analysis of the Mutation Rates by Sequencing

1. Culture the primary transfectants or subclones of these for 6 weeks starting from the time of transfection or subcloning. During such a prolonged culture some cells may lose the *AID* expression cassette which would stop the hypermutation activity of their progeny. To exclude this, culture the cells in low concentration mycophenolic acid selection medium. This medium will prevent the proliferation of cells having lost the *gpt* gene included in the *AID* expression cassette.
2. Prepare cell crude extract as described in steps 3, 4, and 5 of **Section 3.4**.
3. Amplify the *GFP* gene sequence of the *GFP2* reporter by PCR using PfuUltra Hotstart polymerase and the following primer combination RS11: GGGACTAGTCTGCTC-CCTGCTTGTGTGTTGGAGG, BS6: GGGCCCGGGT-TAATTTCGGGTATATTTGAGTGGA. Use 1 μl of crude extract for total 50 μl reaction volume.

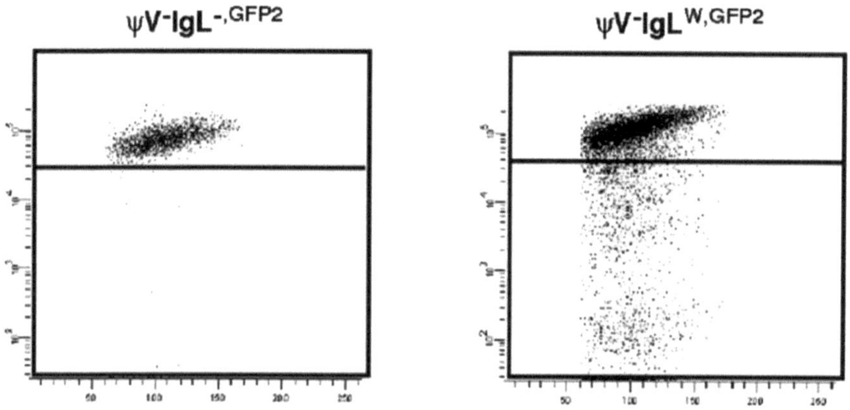

Fig. 18.3. FACS analysis of representative subclones. The $\psi V^{-} IgL^{-,GFP2}$ clone was derived from the $\psi V^{-}IgL^{-}$ cell clone by transfection of the *pIgL*$^{-,GFP2}$ vector. It contains only the GFP2 reporter inserted at the position of the deleted *IgL* locus and stably maintains GFP expression. In the $\psi V^{-} IgL^{W,GFP2}$ clone the GFP2 reporter is inserted together with *DIVAC* sequences of the *IgL* locus. The presence of cells outside the green fluorescence positive gate reflects hypermutations in the GFP reporter.

4. Purify the PCR product by ethanol precipitation, digest it by *Spe*I and *Sma*I for at least 3 h, and finally purify the 2.2 kb PCR product by gel extraction.
5. Clone the purified PCR fragment into a plasmid suitable for sequencing, for example, pUC119.
6. Prepare plasmid DNA by miniprep.
7. Perform cycle sequence reaction in the following reaction mixture: 1 μl BigDye Terminator v3.1 Cycle Sequencing Kit, 1.5 μl 5× Sequencing Buffer, 0.5 μl DMSO, 5 nM sequence primer, 500 ng plasmid DNA in total 10 μl volume. Cycle sequence reaction: 1 min initial incubation at 95°C, 40 cycles consisting of 95°C for 10 s, 50°C for 5 s, and 60°C for 2 min 30 s.
8. Sequence reactions can be analyzed by sequencer like 3730 DNA analyzer.
9. Analyze the mutation spectrum. AID-induced mutations occur almost exclusively at *C* or *G* bases and consist predominantly of transversions (**Fig. 18.4**). Mutation rates per cell division can be calculated assuming a doubling time of DT40 cells of about 10 h.

3.8. Removal of the AID Expression Cassette by Cre Recombinase Induction

1. Culture the AID-expressing cell clones in chicken medium containing 20 nM 4-hydroxytamoxifen for 3 days.
2. Subclone as described under **Section 3.5**.
3. Pick up 24 colonies by transferring 10 μl containing the cells of a subclone into a well of a flat-bottom 96-well plate containing 200 μl of chicken medium. Make a duplicate of

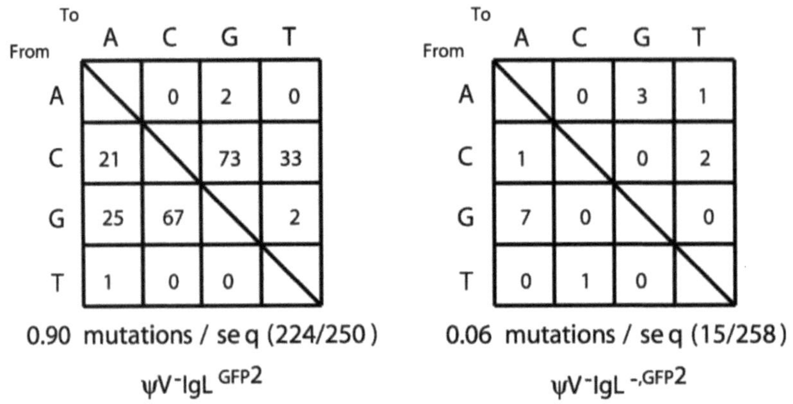

Fig. 18.4. Typical spectrum of AID- and *DIVAC*-dependent hypermutations in the *GFP* gene sequence of the *GFP2* reporter. In the hypermutating $\psi V^- IgL^{GFP2}$ clone the *GFP2* reporter is inserted into the *IgL* locus next to the *DIVAC* sequences of this locus. The $\psi V^- IgL^{-,GFP2}$ clone is a stable GFP-expressing clone whose FACS profile is shown in **Fig. 18.3** (23).

the culture by transferring 100 μl into a separate well of a 96-well plate.

4. Add 100 μl of 2 μg mycophenolic acid selection medium to the well containing one of the duplicates and add 100 μl of chicken medium to the well containing the other duplicate.
5. Incubate for 3 days.
6. Select clones whose duplicates have died in the presence of mucophenolic acid. These clones have lost the *AID-IRES-gpt* expression cassette.

4. Notes

1. Testing *DIVAC* sequences at the position of the deleted *IgL* locus and at other chromosomal positions

   The DT40 variant $\psi V^- IgL^-$ cell line proposed for the experiments is derived from the AID deleted and Cre recombinase expressing DT40 variant $AID^{-/-}$ (24). A further intermediate in the generation of $\psi V^- IgL^-$ was the $\psi V^- AID^{R1}$ variant derived from the $AID^{-/-}$ cell line by inserting a conditional AID expression cassette and deleting the *pseudo-V* gene part of the rearranged *IgL* locus (9). Finally, the remaining sequences of the functionally rearranged *IgL* locus, still present in $\psi V^- AID^{R1}$, have been deleted in $\psi V^- IgL^-$ and replaced by a puromycin resistance gene cassette (23).

   We are proposing to insert the *GFP2* reporter together with potential *DIVAC* sequences at the position of the deleted rearranged *IgL* locus. A number of reasons favor this approach. First of all this is the chromosomal context where the *IgL* transcription units in the presence of the natural *DIVAC* sequences of the *IgL* locus undergo hypermutation at high rate in the absence of pseudo-*V* gene donors (9). Furthermore, insertion of only the *GFP2* reporter at this position results in rather stable GFP expression in the presence of AID expression, whereas the insertion of the *GFP2* reporter together with the natural *DIVAC* sequences of the *IgL* locus fully reconstitutes hypermutation in AID-expressing cells (23). One other technical advantage is that transfectants having integrated the *GFP2* reporter targeted can be identified by their puromycin sensitivity.

Nevertheless, targeted integration of the *GFP2* reporter and potential *DIVACs* into other chromosomal positions might be considered too. We already showed that the targeted integration of the GFP2 reporter at various chromosomal positions leads to stable GFP expression, whereas insertion together with the *IgL-DIVAC* sequence produces instability of GFP expression at these positions presumably due to hypermutation (23). Technically it is not difficult to integrate and test *GFP2-DIVAC* combinations at various chromosomal positions in DT40.

2. Transfection efficiency

For what we have heard from other laboratories, transfection of DT40 can be difficult and may present a major stumbling block for experiments. We also encountered difficult to understand problems with transfection from time to times and would like to share our ideas on this.

Stable transfection and targeted integration of constructs in DT40 has only been reported by electroporation, although transfection using chemicals or lipid-based reagents can result in high levels of transient gene expression. The protocol used for stable transfection by electroporation has remained rather similar to the one reported for in first transfection (25). Compared to other cell lines the number of stable transfectants obtained remains modest in the order of 1–10 transfectants per 10^6 transfected cells.

For these reasons researchers should pay attention to a number of variables which may affect the transfection efficiencies. For example the viability of the cells to be transfected and the quality of the DNA as well as its linearization may be important. Given that the yield of stable transfectants is often limiting, we recommend using rather large amounts of construct DNA up to 100 μg per transfection. Another factor is the quality of the chicken medium and the culture conditions. Transfection is similar to subcloning as the single cell being transfected has to expand its growth. For this reason we recommend to check the suitability of the chicken medium and the culture conditions by determining the subcloning efficiency. Unless a good subcloning efficiency is obtained, it appears unreasonable to hope for a good transfection efficiency even if everything else is optimized.

Finally, the type of electroporator and the physical electroporation settings may be critical. We used to obtain reliable results using BioRad electroporators. A rough guide which settings are appropriate might be to achieve about 50–70% cell killing by the electroporation itself. Transient GFP expression after electroporation of a *GFP* expression construct like *pIgL*$^{-,GFP2}$ may be a more precise test for the suitability of the electroporation conditions. A satisfactory result would be at least 3–10% of the live cells transiently expressing GFP 1 day after the transfection.

3. FACS analysis

FACS analysis of GFP expression in DT40 cells should be straightforward, but to obtain reliable results a number of issues should be considered. Perhaps most important is reducing the background of false-negative events to a minimum. Setting strict gates for live cells using forward and side scatter is critical in this regard. In addition a reasonable cell viability of the cultures to be analyzed is desirable, because a large number of dead or dying cells may overwhelm the capacity of the FACS sorter. The reproducible setting of the gates for cells showing wild-type level of green fluorescence and those showing decreased green fluorescence is also important. As *DIVAC* sequences seem to overlap the *Ig* enhancers, their presence may slightly influence the transcription level of the GFP2 reporter. This may require adjustment of the green fluorescence gates. Alternatively these gates may be set in such a way that they include only green fluorescent negative cells. This would score only mutations which completely inactivate the green fluorescence of the GFP protein thereby reducing the sensitivity of the assay.

To assure the reproducibility of the results it is recommended to include a positive and negative control in each FACS analysis, such as the stable green fluorescence positive clone ψV$^-$IgL^{GFP2}AID$^{-/-}$ (23) and the green fluorescence negative clone ψV$^-$IgL$^-$.

Acknowledgments

AB was supported by the grant no. 02.740.11.5016 from the Russian Ministry of Science.

References

1. Tonegawa, S. (1983) Somatic generation of antibody diversity. *Nature* **302**, 575–581.
2. McKean, D., Huppi, K., Bell, M., Staudt, L., Gerhard, W., and Weigert, M. (1984) Generation of antibody diversity in the immune response of BALB/c mice to influenza virus hemagglutinin. *Proc Natl Acad Sci USA* **81**, 3180–3184.
3. Kocks, C., and Rajewsky, K. (1988) Stepwise intraclonal maturation of antibody affinity through somatic hypermutation. *Proc Natl Acad Sci USA* **85**, 8206–8210.
4. Muramatsu, M., Kinoshita, K., Fagarasan, S., Yamada, S., Shinkai, Y., and Honjo, T. (2000) Class switch recombination and hypermutation require activation-induced cytidine deaminase (AID), a potential RNA editing enzyme. *Cell* **102**, 553–563.
5. Peters, A., and Storb, U. (1996) Somatic hypermutation of immunoglobulin genes is linked to transcription initiation. *Immunity* **4**, 57–65.
6. Di Noia, J.M., and Neuberger, M.S. (2007) Molecular mechanisms of antibody somatic hypermutation. *Annu Rev Biochem* **76**, 1–22.
7. Petersen-Mahrt, S.K., Harris, R.S., and Neuberger, M.S. (2002) AID mutates E. coli suggesting a DNA deamination mechanism for antibody diversification. *Nature* **418**, 99–103.
8. Sale, J.E., Calandrini, D.M., Takata, M., Takeda, S., and Neuberger, M.S. (2001) Ablation of XRCC2/3 transforms immunoglobulin V gene conversion into somatic hypermutation. *Nature* **412**, 921–926.
9. Arakawa, H., Saribasak, H., and Buerstedde, J.M. (2004) Activation-induced cytidine deaminase initiates immunoglobulin gene conversion and hypermutation by a common intermediate. *PLoS Biol* **2**, E179.
10. Di Noia, J.M., and Neuberger, M.S. (2002) Altering the pathway of immunoglobulin hypermutation by inhibiting uracil-DNA glycosylase. *Nature* **419**, 43–48.
11. Saribasak, H., Saribasak, N.N., Ipek, F.M., Ellwart, J.W., Arakawa, H., and Buerstedde, J.M. (2006) Uracil DNA glycosylase disruption blocks Ig gene conversion and induces transition mutations. *J Immunol* **176**, 365–371.
12. Rada, C., Williams, G.T., Nilsen, H., Barnes, D.E., Lindahl, T., and Neuberger, M.S. (2002) Immunoglobulin isotype switching is inhibited and somatic hypermutation perturbed in UNG-deficient mice. *Curr Biol* **12**, 1748–1755.
13. Shen, H.M., Peters, A., Baron, B., Zhu, X., and Storb, U. (1998) Mutation of BCL-6 gene in normal B cells by the process of somatic hypermutation of Ig genes. *Science* **280**, 1750–1752.
14. Pasqualucci, L., Neumeister, P., Goossens, T., Nanjangud, G., Chaganti, R.S., Küppers, R., and Dalla-Favera, R. (2001) Hypermutation of multiple proto-oncogenes in B-cell diffuse large-cell lymphomas. *Nature* **412**, 341–346.
15. Gopal, A.R., and Fugmann, S.D. (2008) AID-mediated diversification within the IgL locus of chicken DT40 cells is restricted to the transcribed IgL gene. *Mol Immunol* **45**, 2062–2068.
16. Gordon, M.S., Kanegai, C.M., Doerr, J.R., and Wall, R. (2003) Somatic hypermutation of the B cell receptor genes B29 (Igbeta, CD79b) and mb1 (Igalpha, CD79a). *Proc Natl Acad Sci USA* **100**, 4126–4131.
17. Müschen, M., Re, D., Jungnickel, B., Diehl, V., Rajewsky, K., and Küppers, R. (2000) Somatic mutation of the CD95 gene in human B cells as a side-effect of the germinal center reaction. *J Exp Med* **192**, 1833–1840.
18. Pasqualucci, L., Migliazza, A., Fracchiolla, N., William, C., Neri, A., Baldini, L., et al. (1998) BCL-6 mutations in normal germinal center B cells: evidence of somatic hypermutation acting outside Ig loci. *Proc Natl Acad Sci USA* **95**, 11816–11821.
19. Liu, M., Duke, J.L., Richter, D.J., Vinuesa, C.G., Goodnow, C.C., Kleinstein, S.H., et al. (2008) Two levels of protection for the B cell genome during somatic hypermutation. *Nature* **451**, 841–845.
20. Storb, U., Peters, A., Klotz, E., Kim, N., Shen, H.M., Hackett, J., et al. (1998) Immunoglobulin transgenes as targets for somatic hypermutation. *Int J Dev Biol* **42**, 977–982.
21. Klix, N., Jolly, C.J., Davies, S.L., Brüggemann, M., Williams, G.T., and Neuberger, M.S. (1998) Multiple sequences from downstream of the J kappa cluster can combine to recruit somatic hypermutation to a heterologous, upstream mutation domain. *Eur J Immunol* **28**, 317–326.
22. Yang, S.Y., and Schatz, D.G. (2007) Targeting of AID-mediated sequence diversification by cis-acting determinants. *Adv Immunol* **94**, 109–125.
23. Blagodatski, A., Batrak, V., Schmidl, S., Schoetz, U., Caldwell, R.B., Arakawa, H.,

et al. (2009) A cis-acting diversification activator both necessary and sufficient for AID-mediated hypermutation. *PLoS Genet* **5**, e1000332.

24. Arakawa, H., Hauschild, J., and Buerstedde, J.M. (2002) Requirement of the activation-induced deaminase (AID) gene for immunoglobulin gene conversion. *Science* **295**, 1301–1306.
25. Buerstedde, J.M., and Takeda, S. (1991) Increased ratio of targeted to random integration after transfection of chicken B cell lines. *Cell* **67**, 179–188.

Section III

In Vitro Reconstitution of Homologous Recombination Reactions and Single Molecular Analysis of Recombination Proteins

Chapter 19

Quality Control of Purified Proteins Involved in Homologous Recombination

Xiao-Ping Zhang and Wolf-Dietrich Heyer

Abstract

Biochemical reconstitution using purified proteins and defined DNA substrates is a key approach to develop a mechanistic understanding of homologous recombination. The introduction of sophisticated purification tags has greatly simplified the difficult task of purifying individual proteins or protein complexes, generating a wealth of mechanistic information. Using purified proteins in reconstituted recombination assays necessitates strict quality control to eliminate the possibility that relevant protein or nucleic acid contaminations lead to misinterpretation of experimental data. Here we provide simple protocols that describe how to detect in purified protein preparations contaminating nucleic acids and relevant enzymatic activities that may interfere with in vitro recombination assays. These activities include ATPases, indicating the potential presence of helicases or translocases, endo- and exonucleases, phosphatases, and type I or type II topoisomerases.

Key words: ATPase, DNA helicase, DNA translocase, endonuclease, exonuclease, phosphatase, topoisomerase, in vitro recombination assays, protein purification.

1. Introduction

Nearly 200 genes have been identified to be involved in DNA repair processes in human cells (1). Homologs for most of these genes have also been found in other organisms, including the budding yeast *Saccharomyces cerevisiae*, underlining the value of model organisms (2). To understand the mechanisms of maintaining genomic integrity, the encoded proteins are usually overexpressed and purified to apparent homogeneity to identify their individual activities, reconstitute more complex in vitro recombination reactions, and determine their structure. Obtaining high-quality purified proteins is a prerequisite for their structural and mechanistic studies.

H. Tsubouchi (ed.), *DNA Recombination*, Methods in Molecular Biology 745,
DOI 10.1007/978-1-61779-129-1_19, © Springer Science+Business Media, LLC 2011

Reconstitution of biological processes using purified proteins and model substrates has been a spectacularly successful approach in elucidating the mechanism of DNA replication and has been canonized by the late Arthur Kornberg as the first commandment of enzymology: "Rely on enzymology to resolve and reconstitute biologic events" (3). Two additional commandments are of particular relevance for the present discussion: "Not waste clean thinking on dirty enzymes" (IV) and "Not waste clean enzymes on dirty substrates" (V). Here we provide simple protocols for quality control of purified proteins that are used in reconstituted in vitro recombination reactions. The presence of many proteins in a single reconstituted system requires that every component is scrutinized for potential contaminations. The first concern is the presence of nucleic acids (RNA/DNA) in the protein preparation, because such nucleic acids contaminate the designed substrates and compete for binding and activity by the proteins under study, leading to potential misinterpretations. During protein purification, nucleic acids are typically removed by various methods (*see* **Note 1**). However, residual nucleic acid contaminations may persist and confirming the absence of nucleic acids in purified recombination proteins is an important quality control step. The second concern is the presence of contamination by enzymatic activities that interfere with the intended assay by acting on the designed substrates, potential reaction intermediates, or reaction products (see commandment IV of (3)). Such activities include ATPases, indicating the potential presence of nucleic acid-based motor proteins such as helicases or translocases, nucleases (endo- or exonucleases), phosphatases, and topoisomerases (type I and II). Many of these activities are active at concentrations that cannot be visualized by standard techniques such as Coomassie staining of protein gels. Hence, even a protein that is apparently pure may be contaminated by relevant interfering activities. Here, we will provide simple protocols to test for such contaminating activities. It is more challenging to identify a potential contamination with the same type of activity as the intended purification target (e.g., a contaminating ATPase in a preparation of a protein that is an ATPase), but such a possibility should be taken into consideration.

2. Materials

2.1. Detection of Nucleic Acid Contaminations

2.1.1. Agarose Gel Electrophoresis

1. Agarose, LE (low electroendosmosis).
2. Agarose gel running buffer: Tris-acetate–EDTA (TAE): 40 mM Tris-acetate, final pH 8.5, 2 mM EDTA, or Tris–borate–EDTA (TBE): 89 mM Tris base, 89 mM borate, and 2 mM EDTA, final pH 8.0.

3. 10× nucleic acid agarose gel loading buffer: 20% Ficoll 400, 0.1 M EDTA, 1% SDS, 0.25% bromophenol blue, 0.25% xylene cyanol.
4. Agarose gel apparatus (Owl Easy-Cast model B1A, tray size 7×8 cm, gel volume ~60 ml or Owl Easy-Cast model B2, tray size 12×14 cm, gel volume ~100 ml).
5. Power supply.
6. Microwave oven.
7. 10 mg/ml ethidium bromide stock solution in ddH_2O or alternative dyes, for example, SYBR dyes (Molecular Probes) (*see* **Note 2**).
8. DNA size marker (EZ load 1 kb molecular ruler, Bio-Rad, #170-8355).
9. Transilluminator UV light and gel documentation system.

2.1.2. Spectrophotometry

1. NanoDrop ND-1000 (NanoDrop Technologies, Inc.).
2. A micropipette that can measure a sample of 2 μl accurately.
3. Buffer used for protein storage.

2.1.3. 5′-[^{32}P]-Labeling of Nucleic Acids

1. Equipment and safety measures for working with radioactivity (*see* **Note 3**).
2. CIA: 24:1 (v/v) chloroform/isoamyl alcohol.
3. PCIA: 1:1 (v/v) phenol/CIA, made with buffered phenol (25:24:1 (v/v/v) phenol/chloroform/isoamyl alcohol) (*see* **Note 4**).
4. 100% ethanol.
5. 70% ethanol.
6. 3 M sodium acetate (NaOAc).
7. Antarctic phosphatase (New England Biolabs [NEB] M0289S, 5,000 units/ml).
8. 10× Antarctic phosphatase buffer (NEB, pH 6.0): 500 mM Bis-Tris-propane-HCl, 10 mM $MgCl_2$, 1 mM $ZnCl_2$, pH 6.0.
9. T4 polynucleotide kinase (PNK) (NEB, M0201S, 10,000 units/ml).
10. 10× PNK buffer (NEB): 700 mM Tris-HCl, pH 7.6, 100 mM $MgCl_2$, 50 mM DTT.
11. [γ-^{32}P]-ATP (Perkin Elmer, 10 μCi/μl).
12. Proteinase K solution: 0.71% SDS, 0.357 M EDTA, and 4.2 mg/ml proteinase K (Roche #03115801001, PCR grade).

13. 0.8% agarose gel (*see* **Section 2.1.1**).
14. Phosphorimager (Storm, Molecular Dynamics) and ImageQuant software.
15. 30 and 37°C water baths, 70°C heat block.
16. Gel dryer.

2.2. Detection of ATPase Contaminations

2.2.1. Charcoal Assay

1. Equipment and safety measures for working with radioactivity (*see* **Note 3**).
2. Activated charcoal (Sigma, C3345, 100–400 mesh) solution: 5% charcoal, 0.25 N HCl, 50 mM KH_2PO_4.
3. [γ-^{32}P]-ATP (Perkin Elmer, 10 μCi/μl), dilute 100× with 10 mM Tris-HCl, pH 7.0.
4. 2× ATPase assay buffer: 66 mM Tris-HCl, pH 7.5, 26 mM $MgCl_2$, 3.6 mM DTT, 2 mM ATP, 180 μg/ml BSA.
5. Commercial RF I (form I) ΦX174 (NEB, N3021L) and virion DNA (circular ssDNA, NEB, N3023L).
6. RecA protein (NEB, M0249S) as positive control.
7. 30°C water bath.
8. 4°C bench top centrifuge.
9. Liquid scintillation vials (volume ~10 ml), liquid scintillation cocktail (EcoLume, MP Biologicals Inc., #882470), and scintillation counter (Beckman LS6500 multipurpose scintillation counter).

2.2.2. Thin-Layer Chromatography (TLC) Method

1. Equipment and safety measures for working with radioactivity (*see* **Note 3**).
2. Cellulose PEI-F (J. T. Baker, 4474-00), 5×20 cm sheets. The plates should be pre-run in 95% ethanol, air-dried, and then pre-run in ddH_2O and air-dried.
3. Glass thin-layer chromatography tank, 25 cm wide × 20 cm high.
4. Hair dryer (optional).
5. [γ-^{32}P]-ATP mixture: 10 mM Tris-HCl, pH 7.5, 9.6 mM unlabeled ATP, and 0.385 μCi/μl [γ -^{32}P]-ATP.
6. Commercial form I ΦX174 (NEB, N3021L) and virion DNA (circular ssDNA, NEB, N3023L).
7. RecA protein (NEB, M0249S) as positive control.
8. TLC running buffer: 1 M formic acid and 0.5 M LiCl.
9. Reaction stop solution: 6.7 mM ATP, 6.7 mM ADP, and 33.3 mM EDTA.
10. 30°C water bath.
11. Phosphorimager (Storm, Molecular Dynamics) and ImageQuant software.

2.3. Detection of Nuclease and Phosphatase Contaminations

2.3.1. Endonuclease

1. Commercial RF I ΦX174 (NEB, N3021L) and virion DNA (circular ssDNA, NEB, N3023L) (*see* **Note 5**).
2. DNase I (NEB, M0303S).
3. 37 and 30°C water baths, heat block.
4. 10× reaction buffer: 200 mM Tris-acetate, pH 7.9, 100 mM $MgCl_2$, 500 mM KCl, and 10 mM DTT (*see* **Note 6**).
5. BSA, 10 mg/ml in 20 mM KPO_4, pH 7.0, 50 mM NaCl, 0.1 mM EDTA, 5% glycerol.
6. 0.8% agarose gel (*see* **Section 2.1.1**).
7. Transilluminator UV light and gel documentation system.

2.3.2. Exonuclease/ Phosphatase

1. Equipment and safety measures for working with radioactivity (*see* **Note 3**).
2. Commercial RF I ΦX174 (NEB, N3021L).
3. *Xho*I (NEB, R0146S, 20,000 units/ml) and 10× buffer supplied by manufacturer.
4. T7 exonuclease (NEB, M0263S, 10,000 units/ml), a 5′ → 3′ exonuclease. Control enzyme.
5. Exonuclease III (NEB, M0206S, 100,000 units/ml), a 3′ → 5′ exonuclease. Control enzyme.
6. [γ-^{32}P]-ATP (Perkin Elmer, 10 μCi/μl).
7. T4 polynucleotide kinase (PNK) and 10× PNK buffer (*see* **Section 2.1.3**).
8. 37 and 30°C water baths.
9. Boiling water bath or heat block.
10. CIA and PCIA (*see* **Section 2.1.3**).
11. 10× reaction buffer: 200 mM Tris-acetate, pH 7.9, 100 mM $MgCl_2$, 500 mM KCl, and 10 mM DTT (*see* **Note 6**).
12. BSA stock solution: 10 mg/ml BSA, 20 mM KPO_4, pH 7.0, 50 mM NaCl, 0.1 mM EDTA, 5% glycerol.
13. 0.8% agarose gel (*see* **Section 2.1.1**).
14. Transilluminator UV light and gel documentation system.
15. Liquid scintillation vials (volume ~10 ml), liquid scintillation cocktail (EcoLume, MP Biologicals, Inc.), scintillation counter (Beckman LS6500 multipurpose scintillation counter).
16. QIAGEN QIAquick nucleotide removal kit (#28304) or illustra MicroSpin G-25 columns (#27-5325-01) from GE Healthcare.

2.4. Detection of Topoisomerase Contaminations

2.4.1. Topoisomerase I

1. Commercial RF I ΦX174 (NEB, N3021L) (*see* **Note 5**).
2. *E. coli* topoisomerase I (NEB, M0301S) as control enzyme.
3. 2× topoisomerase I reaction buffer: 50 mM Tris-acetate pH 7.9, 20 mM $MgCl_2$, 100 mM potassium acetate, 200 mM NaCl, 2 mM DTT, 1 mM EDTA.
4. Proteinase K solution: 0.71% SDS, 0.357 M EDTA, and 4.2 mg/ml proteinase K.
5. BSA stock solution: 10 mg/ml BSA, 20 mM KPO_4, pH 7.0, 50 mM NaCl, 0.1 mM EDTA, 5% glycerol.
6. 30°C water bath.
7. 0.8% agarose gel made with TAE, TAE running buffer, and staining buffer (*see* **Section 2.1.1**).
8. 10× nucleic acid agarose gel loading buffer (*see* **Section 2.1.1**).
9. Transilluminator UV light and gel documentation system.

2.4.2. Topoisomerase II

The same as **Section 2.4.1** except

1. Human topoisomerase II (Inspiralis, HT201, 10 units/μl) as control enzyme.
2. 2× topoisomerase II reaction buffer: 40 mM Tris-HCl, pH 7.9, 20 mM $MgCl_2$, 2 mM DTT, 100 mM KCl, 100 mM NaCl, 1 mM EDTA, 2 mM ATP.

3. Methods

3.1. Detection of Nucleic Acid Contaminations

3.1.1. Agarose Gel Electrophoresis

Agarose gel electrophoresis is a robust method to visualize and separate nucleic acids based on their size and topology. Staining by ethidium bromide or alternative dyes provides a simple way to detect nucleic acid contaminations in protein preparations (*see* **Note 2**).

1. Set up the gel tray and make sure the surface is level. Prepare 0.8% agarose LE solution with TAE or TBE. Carefully heat the solution with microwave until the agarose is completely dissolved. Pour the gel.
2. Add 1/10 volume of 10× nuclei acid gel loading buffer to each sample (final volume up to 30 μl). Load the sample on the gel. A DNA ladder is loaded as a size standard.
3. Run gel at 90 V (~5 V/cm) for 1.5–2 h.
4. Stain the gel in 1 μg/ml ethidium bromide. Destain the gel in ddH_2O for 10–20 min.
5. Visualize nucleic acids with UV transilluminator and document.

3.1.2. Spectrophotometry with NanoDrop

Protein and nucleic acids show UV absorption maxima at 280 and 260 nm, respectively. The OD_{260}/OD_{280} ratio is typically used to evaluate the purity of nucleic acid preparations, ensuring the absence of protein. Likewise, nucleic acid contamination in protein preparations can be detected the same way. The NanoDrop is ideal for this purpose for its sparing use of sample (as little as 2 μl).

1. Start the NanoDrop application on your computer. Follow the screen prompt to initialize the machine.
2. Blank the machine using 2 μl storage buffer used for storing the protein sample. Wipe the buffer from the upper and low pedestals.
3. Establish a UV spectrum (220–350 nm) of the sample by applying a 2 μl volume. A peak or shoulder at 260 nm and a ratio of A_{260}/A_{280} greater than 1 indicate nucleic acid contamination in the protein sample.

3.1.3. DNA Phosphorylation Methods (^{32}P)

During protein extraction it is likely that all nucleic acids have been sheared to their linear form. This facilitates their detection by 5′-end labeling with polynucleotide kinase after removing any potential phosphates using a phosphatase. This method is significantly more sensitive than the electrophoretic or spectrophotometric approaches and can be performed directly with a sample of the purified protein (start at step 10) or after extraction of the nucleic acid (steps 1–9), as tight binding to the protein may prevent access by the phosphatase/kinase. We recommend doing both (*see* **Note 4**).

1. Start with a sample ≥0.2 ml in a 1.5 ml Eppendorf tube. Add an equal volume of PCIA.
2. Vortex vigorously for about 10 s. Spin 1 min at 16,000×*g* in a table-top centrifuge.
3. Carefully transfer the top aqueous phase to a new 1.5 ml Eppendorf tube. Repeat the PCIA extraction process one to two times.
4. Extract the aqueous phase 1× with CIA to eliminate residual phenol.
5. Add 1/10 volume of 3 M sodium acetate, pH 5.2 to the aqueous solution, mix well.
6. Add 2.5 volume of ice-cold 100% ethanol. Vortex. Keep the sample at −80°C for 30 min or −20°C overnight.
7. Spin the tube at 4°C (16,000×*g*) for 20 min. Carefully remove the supernatant.
8. Wash the pellet in 0.5 ml ice-cold 70% ethanol. Remove the supernatant completely after centrifugation. Dry the

nucleic acid pellet by leaving the tube open on bench for at least 30 min or use a Speedvac (*see* **Note 4**).

9. Resuspend nucleic acids in 12.5 μl ddH_2O.
10. Add 1.5 μl 10× Antarctic phosphatase reaction buffer and 1 μl of Antarctic phosphatase. Incubate the mixture at 37°C for 1 h.
11. Incubate the mixture at 70°C for 10 min to inactivate Antarctic phosphatase.
12. Add 3 μl 10× PNK buffer, 10 units (1 μl) T4 polynucleotide kinase, and 3 μl [γ-^{32}P]-ATP. Adjust final volume to 30 μl by adding 8 μl ddH_2O. Incubate the reaction at 37°C for 1 h.
13. Add 5 μl proteinase K solution and incubate at 30°C for 20 min.
14. Load 20–30 μl of the mixture into one well of a 0.8% agarose gel and run the gel with 5 V/cm, for 2 h.
15. Dry the gel and expose it to a phosphorimager and quantify with ImageQuant.

3.2. Detection of ATPase Contamination

Hundreds of cellular enzymes hydrolyze ATP to generate energy in support of different functions. Indeed many proteins involved in homologous recombination display or are predicted to exhibit ATPase activity, including homologs or paralogs of RecA, DNA helicases, and DNA translocases. Of practical importance are also molecular chaperones that use the energy of ATP hydrolysis to support folding and often co-purify with overexpressed recombinant proteins. Contaminations by DNA helicases or DNA translocases can lead to changes in the designed substrates, potential intermediates, or reaction products in reconstituted recombination reactions, severely compromising potential interpretations. These enzymes often display DNA-dependent ATPase activity (ssDNA or dsDNA) and require Mg^{2+}. Hence, these cofactors must be provided to detect such contaminations. Here, we provide protocols for two simple ATPase assays (*see* **Note 7**).

3.2.1. Charcoal Assay

Activated charcoal specifically binds nucleotides but not phosphate, allowing to monitor the hydrolysis of ATP using [γ-^{32}P]-ATP by measuring the accumulation of [^{32}P] in solution (4). Any radioactivity above background indicates ATPase activity.

1. Add 25 μl 2X ATPase buffer, 1 μg ssDNA, and 1 μg dsDNA in a 1.5 ml Eppendorf tube.
2. Add an aliquot of protein preparation (up to 24 μl) or 1 μg RecA as positive control.
3. Add 1 μl [γ-^{32}P]-ATP solution (containing 0.1 μCi ^{32}P).
4. Adjust the final volume to 50 μl with ddH_2O.

5. Two control reactions are needed, background and input. Both of them contain buffer and [γ-^{32}P]-ATP only.
6. Incubate at 30°C for 30 min.
7. Add 0.5 ml charcoal solution (agitate during adding), vortex, incubate on ice for 10 min for reaction and background control. For input control, add the same volume of ddH_2O instead. Vortex once in the middle of incubation. Spin at 16,000×*g* at 4°C for 10 min.
8. Transfer 0.4 ml of supernatant to a scintillation vial containing 4 ml liquid scintillation cocktail. Mix well.
9. Determine radioactivity in supernatant by scintillation.

3.2.2. Thin-Layer Chromatography Assay

Thin-layer chromatography on polyethyleneimine plates allows the separation of ATP, ADP, AMP, and phosphate. The protocol provided here is simpler than the original version (5), as the labeled phosphate is separated more easily from the [γ-^{32}P]-ATP than [α-^{32}P]-ADP from [α-^{32}P]-ATP.

1. Assemble the reactions as in the charcoal assay (**Section 3.2.1**).
2. Add 2.5 μl [γ-^{32}P]-ATP mixture to a final volume of 50 μl.
3. Incubate the reaction at 30°C.
4. Withdraw 2.5 μl from the reaction at 15, 30, and 60 min and mix with 1.25 μl stop solution.
5. 1 μl of the mixture from step 4 is spotted on the cellulose PEI-F plate close to the bottom edge.
6. Run the PEI-F plate in the TLC running solution in a TLC tank until the front of the solution almost reaches the top of the plate.
7. Remove the plate from the tank and let it dry in fume hood or using a hair dryer.
8. Expose overnight.
9. Scan and quantify with ImageQuant software. Calculate the free phosphate increase above background.

3.3. Detection of Nuclease/Phosphatase Contaminations

Nuclease contaminations significantly compromise the integrity of the substrates in reconstituted recombination reactions, and it is important to confirm the absence of ssDNA and dsDNA endo- and exonuclease activities. Using 5′-end-labeled substrates for exonuclease assays concomitantly monitors also for contaminations by phosphatases.

3.3.1. Endonuclease

The conversion of circular to linear DNA provides a simple reaction to detect endonuclease activity by electrophoretic mobility differences on agarose gels and commercially available circular ssDNA and dsDNA provide easy access to substrates.

1. Incubate 20 μl reaction containing 2 μl 10× buffer, 2 μl BSA stock solution, 1–5 μg purified protein with 0.5 μg of form I dsDNA or circular ssDNA for 2 h at 30°C.
2. Use DNase I (NEB, M0303S) as positive control, adding 2 units/reaction, and no protein added as negative control.
3. Add 2.5 μl proteinase K solution in each reaction tube and incubate at 30°C for 20 min.
4. Add 1/10 volume 10× nucleic acid agarose gel loading buffer to each sample.
5. Run the sample on a 0.8% agarose gel with TAE buffer system, 5 V/cm, 2 h.
6. Stain the gel with 1 μg/ml ethidium bromide and visualize with a transilluminator UV light. Any change in band intensity and migration position of form I or circular ssDNA bands compared to the negative control signals potential endonuclease activity (*see* also **Section 3.4**).

3.3.2. Exonuclease/ Phosphatase

Exonucleases require a terminus of a polynucleotide chain to hydrolyze dsDNA and/or ssDNA. Using 5′-[^{32}P]-labeled linearized ΦX174 dsDNA as a substrate provides a sensitive assays to detect contaminating exonuclease and phosphatase activities.

3.3.2.1. Linearization of Circular dsDNA and 5′-End Labeling

1. 30 μg of ΦX174 RF I dsDNA in 100 μl volume with 10 μl 10× NEBuffer #4, 10 μl BSA stock solution, and 50 units *Xho*I at 37°C for 1 h. Check for completion of linearization by agarose gel electrophoresis.
2. Remove restriction enzyme by PCIA extraction (*see* **Section 3.1.3**), precipitate with ethanol plus NaOAc, and dissolve in 80 μl ddH_2O.
3. Take 15 μl the linearized dsDNA from step 2 and add 5 μl 10× Antarctic phosphatase buffer, 28 μl ddH_2O, and 1.5 μl (7.5 units) Antarctic phosphatase. Incubate at 37°C for 1 h.
4. Inactivate the Antarctic phosphatase by heating the sample at 70°C for 10 min.
5. Add ddH_2O to 200 μl. Remove the phosphatase by PCIA extraction and ethanol precipitation (steps 1–9 of **Section 3.1.3**). Dissolve the DNA in 40 μl ddH_2O.
6. Add 5 μl 10× PNK buffer, 10 units (1 μl) T4 polynucleotide kinase, and 4 μl [γ-^{32}P]-ATP (10 μCi/μl). Incubate the reaction at 37°C for 1 h.
7. Inactivate the enzyme by heating at 75°C for 10 min. Then remove the unused [γ-^{32}P]-ATP using QIAGEN QIAquick nucleotide removal kit or an illustra MicroSpin G-25 column (GE Healthcare).

8. Analyze 1 μl of the labeled DNA on an agarose gel to monitor the efficiency of labeling, stain with ethidium bromide and visualize under UV, then dry gel and analyze using a phosphorimager. Determine radioactivity of 1 μl of the labeled DNA by liquid scintillation.

3.3.2.2. Detection of Exonuclease/ Phosphatase

1. Set up four parallel sets of reactions, two for dsDNA and two for ssDNA to be analyzed either by scintillation or by gel electrophoresis. The dsDNA can be used directly from **Section 3.3.2.1** for ssDNA, heat denature the linear dsDNA by boiling in a heat block for 1 min and placing the tube directly on ice.
2. Each 20 μl reaction contains 0–10 μg purified protein, 2 μl 10× reaction buffer, 0.2 μl BSA stock solution, and 0.5–1 μl of 5′-[^{32}P] labeled linearized ΦX174 dsDNA or ssDNA.
2. Incubate the reaction at 30°C for 2 h.
3. Add 4 μl proteinase K solution. Continue the incubation at 30°C for 20 min.
4. In the set for scintillation detection, add 36 μl ddH_2O and 12 μl 30% TCA to a final concentration of 5%. Incubate the tubes on ice for 30 min. Then spin the samples at 16,000×*g* for 10 min at 4°C. Transfer the supernatant into 2 ml EcoLume. Mix well and determine radioactivity in scintillation counter. Any increase in radioactivity in the supernatant above the no protein control signals potential exonuclease or phosphatase activity.
5. In the set for gel electrophoresis, add 2.2 μl agarose gel loading buffer to each tube. Run the sample on 0.8% agarose gel with TAE at 5 V/cm. Visualize the DNA under a transilluminator UV light.
6. Dry the gel and expose the gel to a phosphorimager screen for 12 h. Scan and quantify the gel.

Loss of signal compared to the no protein control indicates exonuclease or phosphatase activity. Shortening of the labeled DNA indicates 3′–5′ exonuclease activity or endonuclease activity, depending on the results with **Section 3.3.1.**

3.4. DNA Topoisomerase

Relaxation of naturally negatively supercoiled DNA into a ladder of more relaxed topological isomers provides a simple assay to detect type I and type II DNA topoisomerases. While type I enzymes catalyze the reaction independent of ATP, type II enzymes require ATP, making for easy distinction. The activity of bacterial gyrase is not detected in this assay, as it only relaxes positive but not negative supercoils. Gyrase, however, is an ATPase and should be identified as a contamination by **Section 3.2.** More detail is found in the following sources (6, 7).

3.4.1. Topoisomerase I

1. 30 μl reaction mixture contains 15 μl 2× topoisomerase I buffer, 0.5 μg form I dsDNA, 100 μg/ml BSA, 2.5 units of *E. coli* Topo I, or 0–10 μg purified protein. Adjust the final volume to 30 μl with ddH_2O.
2. Incubate at 30°C for 1–2 h.
3. Stop the reaction by adding 6 μl protease K solution. Incubate at 30°C for 20 min.
4. Add 4 μl 10× DNA gel loading buffer, mix. Load on 0.8% agarose gel made with 1× TAE buffer. Run the gel at 5 V/cm for 2 h in 1× TAE.
5. Stain the gel with 1 μg/ml ethidium bromide solution for 30 min and destain the gel with ddH_2O for 20 min with agitation, and visualize bands with gel documentation system.

3.4.2. Topoisomerase II

1. Set up 30 μl reactions including 15 μl 2× topoisomerase II reaction buffer containing ATP, 0.5 μg form I dsDNA, 100 μg/ml BSA, 2 units of human topoisomerase II, or 0–10 μg purified protein.
2. Incubate the reactions at 30°C for 1–2 h and proceed as described for topoisomerase I (**Section 3.4.1**).

In both assays, topoisomerase contamination is signaled by the appearance of a ladder of more relaxed topological isoforms running above the negatively supercoiled input DNA.

4. Notes

1. Several methods have been developed to remove nucleic acids during the early stages of a protein purification protocol. Nucleic acids can be removed by binding to a positively charged substance, such as polyethyleneimine (PEI) (8). If the target protein is acidic, elevated salt concentrations avoid precipitation of the target protein (9). Nucleic acids can also be eliminated by enzymatic digestion using DNase I (10) or Benzonase, a nuclease available from many commercial sources (e.g., Merck) which degrades linear or circular DNA and RNA in their single-stranded or double-stranded form, producing 5′-monophosphate terminating oligonucleotides of two to five bases. Alternatively, nucleic acids can be removed from protein samples by chromatography on strong anion exchange resins such as DEAE cellulose or Q-Sepharose (11, 12) or DNA-binding proteins can be selectively captured by affinity chromatography including media such as DNA cellulose (13), Affi-Gel Blue

(Bio-Rad) (14), or Fast Flow Cibacron Blue 3GA (Sigma) (15). A caveat for using DNA cellulose is the observation that DNA slowly leaches off the column, introducing contaminating DNA.

2. The detection limit for ethidium bromide staining is ~20 ng/band of dsDNA, while the method is less sensitive for RNA or single-stranded DNA. Ethidium bromide is a known mutagen and appropriate caution should be taken for handling and disposal. More sensitive alternatives to ethidium bromide have been developed, for example, SYBR dyes (Molecular Probes, Invitrogen) increase the sensitivity to ~1 ng/band, but are significantly more expensive. Ethidium bromide can be added directly into the gel solution before casting or the gel can be stained after electrophoresis. The latter staining method gives clearer background and limits the liquid waste volume.
3. Basic equipment includes a Geiger counter, shields, mask or safety glasses, shielded liquid and solid waste containers. A user should be trained and follow the common and local rules for isotope usage.
4. Phenol extraction is used to remove protein from the sample, which may interfere with the labeling process. The concentration of the nucleic acid contamination in the sample is likely to be low. Hence, the DNA pellet may be invisible. Mark the expected position of the pellet on the wall of the tube. Make sure that the pellet will not be lost. The minimal start volume of phenol extraction should be 0.2 ml to ease handling. Antarctic phosphatase (NEB) is used to dephosphorylate the 5′-ends of nucleic acids. This enzyme can be inactivated by incubation at 65°C for 5 min. Labeling may also be performed directly without PCIA extraction.
5. Commercial DNA is adequate for this purpose, although the quality of the batches varies in the proportion of form I supercoiled DNA for dsDNA (as opposed to form II nicked and form III linear DNA) and the proportion of circular DNA for ssDNA (as opposed to linear ssDNA). However, sufficient circular DNA for endonuclease assays and supercoiled DNA for topoisomerase assays must be present. Be aware that frequent freeze/thaw cycles generate nicks in DNA.
6. The buffer should provide conditions under which most nucleases exhibit at least some activity. If a nuclease contamination is suspected, the important parameters (Mg^{2+} concentration, buffer type, salt type, and concentration) can be varied to enhance sensitivity. The standard buffers provided with restriction enzymes provide a useful range of reaction conditions. These are 10× concentrations. NEBuffer 1:

100 mM Bis-Tris-propane-HCl, pH 7.0, 100 mM $MgCl_2$, 10 mM DTT. NEBuffer 2: 100 mM Tris-HCl, pH 7.9, 100 mM $MgCl_2$, 500 mM NaCl, 10 mM DTT. NEBuffer 3: 500 mM Tris-HCl, pH 7.9, 100 mM $MgCl_2$, 1,000 mM NaCl, 10 mM DTT. NEBuffer 4: 200 mM Tris-acetate, pH 7.9, 100 mM magnesium acetate, 500 mM potassium acetate, 10 mM DTT.

7. The charcoal and TLC assays have similar sensitivities. The volumes in the TLC assay can be reduced for more sparing use of protein sample. In addition, the TLC assays provide information on the reaction products of the ATPase activity.

Acknowledgments

We thank Clare Fasching, Erin Schwartz, Kirk Ehmsen, Shannon Ceballos, and William Wright for helpful comments on the manuscript. Our work is supported by the NIH (GM58015, CA92276), the DoD (BC083684), and a Susan G. Komen Breast Cancer Foundation postdoctoral fellowship (PDF403213) to XPZ.

References

1. Wood, R.D., Mitchell, M., and Lindahl, T. (2005) Human DNA repair genes, 2005. *Mutat Res* **577**, 275–283.
2. Game, J.C. (2000) The *Saccharomyces* repair genes at the end of century. *Mutat Res* **451**, 277–293.
3. Kornberg, A. (2003) Ten commandments of enzymology, amended. *Trends Biochem Sci* **28**, 515–517.
4. Zimmerman, S.B., and Kornberg, A. (1961) Deoxycytidine di- and triphosphate cleavage by an enzyme formed in bacteriophage-infected *Escherichia coli*. *J Biol Chem* **236**, 1480–1486.
5. Scott, J.F., Eisenberg, S., Bertsch, L.L., and Kornberg, A. (1977) A mechanism of duplex DNA replication revealed by enzymatic studies of phage phi X174: catalytic strand separation in advance of replication. *Proc Natl Acad Sci USA* **74**, 193–197.
6. Stewart, L., and Champoux, J.J. (2001) Assaying DNA topoisomerase I relaxation activity. In *DNA topoisomerase protocols: enzymology and drugs*, N. Osheroff, M.-A. Bjornsti eds., Vol. 95, Methods in Molecular Biology. (Totowa, NJ: Humana Press), pp. 1–11.
7. Fortune, J.M., and Osheroff, N. (2001) Topoisomerase II-catalyzed relaxation and catenation of plasmid DNA. In *DNA topoisomerase protocols: enzymology and drugs*, N. Osheroff, M.-A. Bjornsti eds., Vol. 95, Methods in Molecular Biology. (Totowa, NJ: Humana Press) pp. 275–281.
8. Burgess, R.R. (1991) Use of polyethyleneimine in purification of DNA-binding proteins. *Methods Enzymol* **208**, 3–10.
9. Burgess, R.R. (2009) Protein precipitation techniques. *Methods Enzymol* **463**, 331–342.
10. Burgess, R.R. (1969) A new method for the large scale purification of *Escherichia coli* deoxyribonucleic acid-dependent ribonucleic acid polymerase. *J Biol Chem* **244**, 6160–6167.
11. Mangel, W.F. (1974) Initial steps in the large-scale purification of *Escherichia coli* deoxyribonucleic acid-dependent ribonucleic acid polymerase. *Arch Biochem Biophys* **163**, 172–177.
12. Berthold, W., and Walter, J. (1994) Protein purification: aspects of processes for pharmaceutical products. *Biologicals* **22**, 135–150.

13. Alberts, B.M., Amodio, F.J., Jenkins, M., Gutmann, E.D., and Ferris, F.L. (1968) Studies with DNA-cellulose chromatography. I. DNA-binding proteins from *Escherichia coli*. *Cold Spring Harb Symp Quant Biol* **33**, 289–305.
14. Deutscher, M.P. (2009) Affi-gel blue for nucleic acid removal and early enrichment of nucleotide binding proteins. *Methods Enzymol* **463**, 343–345.
15. Zhang, X.P., Lee, K.I., Solinger, J.A., Kiianitsa, K., and Heyer, W.D. (2005) Gly-103 in the N-terminal domain of *Saccharomyces cerevisiae* Rad51 protein is critical for DNA binding. *J Biol Chem* **280**, 26303–26311.

Chapter 20

Assays for Structure-Selective DNA Endonucleases

William D. Wright, Kirk T. Ehmsen, and Wolf-Dietrich Heyer

Abstract

Structure-selective nucleases perform DNA strand incisions crucial to the repair/resolution of branched DNA molecules arising during DNA replication, recombination, and repair. From a combination of genetics and in vitro nuclease assay studies, we are just beginning to understand how these enzymes recognize their substrates and to identify their in vivo DNA structure targets. By performing nuclease assays on a variety of substrates meant to mimic cellular intermediates, structural features of branched DNA molecules that are important for robust catalysis can be defined. However, since these enzymes often are capable of cleaving a range of DNA structures, caution must be taken not to overemphasize the significance of incision of a certain structure before a careful and detailed kinetic analysis of a variety of DNA substrates with different polarities and structural features has been completed. Here, we provide protocols for the production of a variety of oligo-based DNA joint molecules and their use in endonuclease assays, which can be used to derive the kinetic parameters K_M and k_{cat}. Determination of these values for a variety of substrates provides meaningful comparisons that allow inferences to be made regarding in vivo DNA structure target(s).

Key words: DNA joint molecule, endonuclease, flap, incision site, Mus81–Mms4/Eme1, recombination, XPF paralogs, kinetic analysis, Michaelis–Menten analysis, Holliday junction.

1. Introduction

Mus81–Mms4 (MUS81–EME1), Slx1–Slx4, Rad1–Rad10 (XPF-ERCC1), Yen1 (GEN1), and Rad27 (FEN1) are all examples of endonucleases which base their selectivity for incision of branched DNA molecules on structural features. These features can include the presence of double or single-strand DNA "arms" in a specific orientation relative to a branch point, DNA strand end(s), bubble, gap, or other features that direct the active site for

H. Tsubouchi (ed.), *DNA Recombination*, Methods in Molecular Biology 745,
DOI 10.1007/978-1-61779-129-1_20, © Springer Science+Business Media, LLC 2011

incision on a specific strand. Each of these nucleases have unique biochemical and genetic properties; because of lack of space for discussion here, we refer the reader to the literature (1–5). The exact structures of in vivo targets of these enzymes are in many cases unknown and occur within a chromatin context that can only be minimally approximated in vitro by synthetic DNA structures. Nevertheless, experiments using a variety of branched DNA molecules meant to mimic replication or recombination-dependent DNA structures found in the cell have proved fruitful for describing the basic biochemical properties of structure-selective endonucleases. The methods described below have been developed through our studies of the *Saccharomyces cerevisiae* Mus81–Mms4 protein complex (6, 7). The reader is advised to refer to these references for more examples of data that can be generated through this type of analysis.

We refer to these endonucleases as structure-selective, not structure-specific, because they almost invariably will cleave a range of substrates meant to mimic various replication or recombination-associated structures formed in vivo. Consequently, demonstration that an endonuclease can cleave a certain DNA structure has little meaning until a thorough analysis comparing the kinetic parameters of multiple different DNA structures has been conducted. For example, Mus81–Mms4 will catalytically incise most of the structures shown in **Table 20.1**, with poorly cleaved structures requiring conditions where there is excess enzyme to DNA substrate molecules (**Fig. 20.1a**, **c**). Meaningful distinctions between substrates can be made after meticulous kinetic analysis under non-saturating (catalytic) conditions where the enzyme concentration is less than the DNA substrate concentration. Using Michaelis–Menten analysis, kinetic parameters can be determined for the structures under comparison. The Michaelis constant, K_M, is the concentration of substrate at precisely one-half of the enzyme's maximal velocity and is a measure of the substrate concentration required for efficient catalysis to occur. Mus81–Mms4 has low K_M values in the range of 1–7 nM for DNA structures that are cut well, as expected for an enzyme that targets a single substrate entity within a cell, corresponding to an in vivo concentration on the order of 1 nM. The k_{cat}, or the turnover number, is another useful value which gives the number of substrate molecules turned over per enzyme molecule per unit time at maximal velocity and hence is a direct measure of catalytic efficiency, which includes product release. Discussions of how to determine these parameters using Michaelis–Menten kinetic analysis can be found in various biochemistry texts (8–10). Here, it is our purpose to describe the methods associated with producing purified branched DNA structures, the basic setup of nuclease assay time courses useful in determining kinetic parameters K_M and k_{cat}, and protocols for mapping incision sites on these structures.

Table 20.1a
DNA joint molecule structure schematics and descriptions

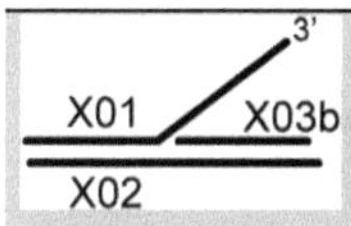

3' Flap
3'-flaps can result from over-synthesis in the Synthesis-Dependent Strand Annealing (SDSA) pathway

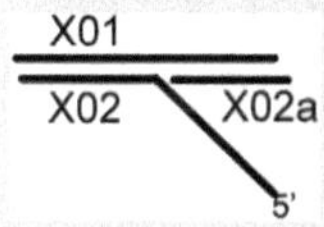

5' Flap
5'-flaps arise as intermediates of lagging strand synthesis resulting from RNA primer displacement. In vivo, 5'-flaps can isoenergetically convert to 3'-flaps and vice versa

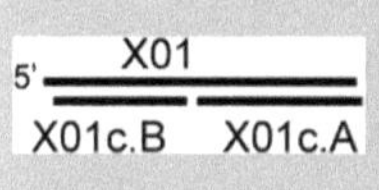

Nicked Duplex
Control for determining the importance of a branchpoint relative to flap structures. It is also a ligation control used in the incision site mapping protocol

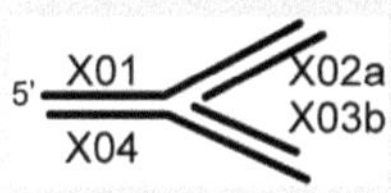

Replication Fork-like
Replication fork mimic

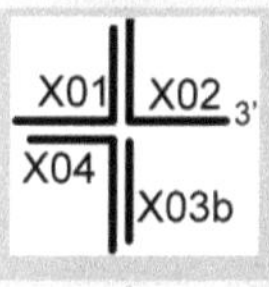

Partial X012-3'
Mimics a structure that could occur at a replication fork if the regressed leading strand were longer than the lagging strand

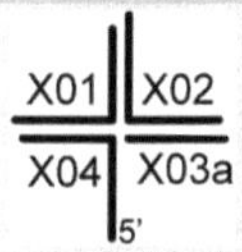

Partial X012-5'
Mimics a structure that could occur at a replication fork if the regressed lagging strand were longer than the leading strand

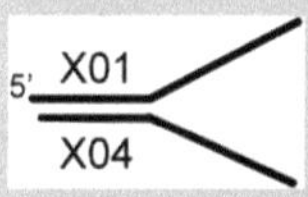

Simple Y
Minimal fork substrate to test the importance of duplex arms flanking the branchpoint

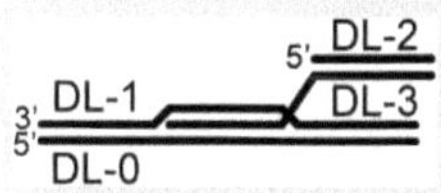

D-loop
Displacement-loop mimic, a key synaptic homologous recombination intermediate

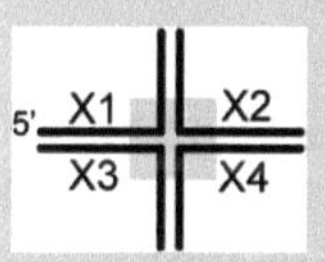

X12
Holliday junction mimic, a key post-synaptic recombination intermediate. The central core has 12 bp of homology which allows branch migration

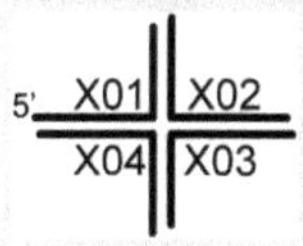

X012
Holliday junction mimic with a non-mobile core

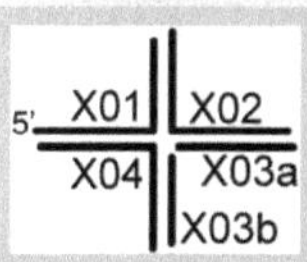

Nicked X012
The nicked Holliday junction tests the influence of a DNA nick in the vicinity of the branchpoint

Table 20.1b
Oligo names and sequences

X1	5′-gACgCTgCCgAATTCTggCTTgCTAggACATCTTTgCCCACgTTgACCCg-3′
X2	5′-CgggTCAACgTgggCAAAgATgTCCTAgCAATgTAATCgTCTATgACgTC-3′
X3	5′-gACgTCATAgACgATTACATTgCTAggACATgCTgTCTAgAgACTATCgC-3′
X4	5′-gCgATAgTCTCTAgACAgCATgTCCTAgCAAgCCAgAATTCggCAgCgTC-3′
X01	5′-CAACgTCATAgACgATTACATTgCTACATggAgCTgTCTAgAggATCCgA-3′
X02	5′-gTCggATCCTCTAgACAgCTCCATgATCACTggCACTggTAgAATTCggC-3′
X03	5′-TgCCgAATTCTACCAgTgCCAgTgATggACATCTTTgCCCACgTTgACCC-3′
X04	5′-TgggTCAACgTgggCAAAgATgTCCTAgCAATgTAATCgTCTATgACgTT-3′
X02a	5′-gTCggATCCTCTAgACAgCTCCATg-3′
X03a	5′-TgCCgAATTCTACCAgTgCCAgTgAT-3′
X03b	5′-ggACATCTTTgCCCACgTTgACCC-3′
X01c.A	5′-gTCggATCCTCTAgACAgCTCCATgT-3′
X01c.B	5′-AgCAATgTAATCgTCTATgACgTT-3′
DL-0	5′-gACgCTgCCgAATTCTACCAgTgCCTTgCTAggACATCTTTgCCCACCTgCAgg TTCACCC-3′
DL-1	5′-gggTgAACCTgCAggTgggCggCTgCTCATCgTAggTTAgTTggTAgAATTCggCAg CgTC-3′
DL-2	5′-TAAgAgCAAgATgTTCTATAAAAgATgTCCTAgCAAggCAC-3′
DL-3	5′-TATAgAACATCTTgCTCTTA-3′

2. Materials

2.1. Branched DNA Substrate Production

1. Oligonucleotides, PAGE-purified for greater than 50-mers.
2. 6× annealing buffer: 0.9 M NaCl, 90 mM sodium citrate.
3. Thermocycler or microwave.
4. 10 × 10 cm (e.g., Hoefer Mighty Small/GE model SE260) or 20 × 10 cm (e.g., OWL model 73.1020 V) PAGE gel apparatus, 1.5 mm gel spacers, and gel comb with large ~1 cm^2 wells.
5. Polyacrylamide gel electrophoresis (PAGE) solutions: 29:1 acrylamide:bisacrylamide solution, Tris–acetate EDTA (TAE) buffer (40 mM Tris–acetate, 1 mM EDTA), *N*,*N*,*N*′,*N*′-tetramethyl-ethylenediamine (TEMED), 10% ammonium persulfate.
6. 10× DNA loading dye for native DNA gels: 50% glycerol, 0.05% bromophenol blue (pH 8.0).
7. Short-wave hand-held UV light source.

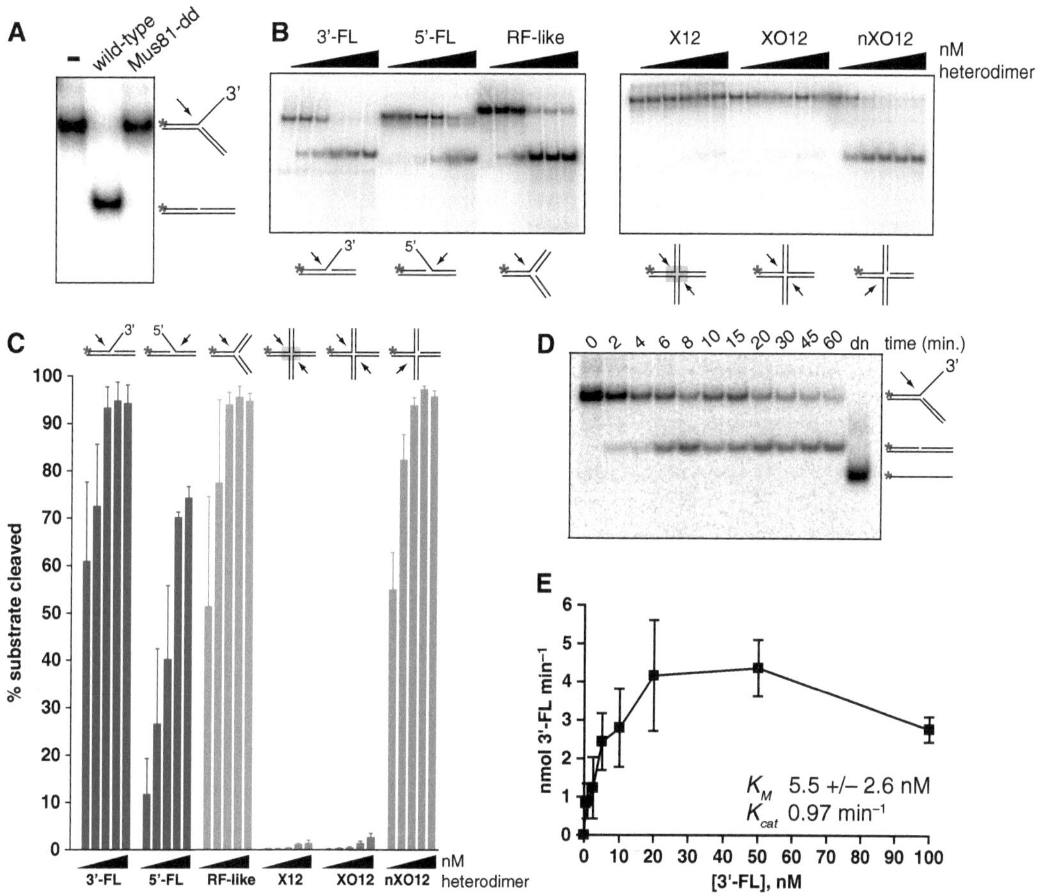

Fig. 20.1. Comparison of nuclease activity on DNA joint molecules and kinetic analysis. (**a**) *Saccharomyces cerevisiae* Mus81–Mms4 incises model DNA joint molecules such as a 3′-flapped DNA. Incision is shown to depend on the nuclease activity of the endonuclease, as a purified mutant complex cannot cut DNA (Mus81-dd is Mus81–D414A, D415A). (**b**) Fixed time point Mus81–Mms4 nuclease assays for several DNA joint molecules are shown at fixed substrate concentration (50 nM) and a titration of Mus81–Mms4 from limiting concentration (5 nM) to excess concentration (100 nM). (**c**) Incision proficiency on different DNA joint molecules can be quantitated and graphically represented, as in this quantitation of the data in (**b**). (**d**) To perform a reaction time course, aliquots from an ongoing nuclease reaction are removed at determined intervals and stopped. Here, a 3′-FL time course is shown. "dn" represents heat-denatured substrate, demonstrating that the incision product is specific to the enzyme and not time-dependent denaturation of the substrate. When percent substrate cleaved versus time is plotted (reaction progress curves), the initial rate of the reaction can be extrapolated from early points in the time course over the interval when the reaction rate is linear. (**e**) With initial rates determined over a range of substrate concentrations, a plot of initial velocity versus substrate concentration can be used to determine the Michaelis concentration (K_M) and catalytic turnover (k_{cat}) of the nuclease on substrates it incises.

8. Thin-layer chromatography (TLC) paper (for UV shadowing). We use polyethyleneimine sheets with a fluorescence enhancer (PEI-F, JT Baker #447400).
9. Scalpel or razor blade.
10. Autoradiography film, cassette, and developer machine.
11. Small light box.

12. Radiation work area and safety equipment (*see* **Note 1**).
13. Electroelution device. We use the Centrilutor (Millipore). However, this device has been discontinued and a suitable replacement system must be used (*see* **Note 2**).
14. Scintillation counter, vials, and scintillation cocktail (e.g., EcoLume, MP Biomedicals).
15. Nanodrop® spectrophotometer or alternative unit that can read the A_{260} of small volumes of DNA in solution.
16. T4 polynucleotide kinase (PNK) (New England Biolabs, NEB).
17. γ^{32}P-ATP (specific activity 6,000 Ci/mmol).
18. Size exclusion spin columns, e.g., GE MicroSpin™ G-25 or G-50.

2.2. Nuclease Assays

1. Purified endonuclease of interest. Please see Chapter 19 by Zhang and Heyer in this volume for information on establishing the quality of protein preparations.
2. Purified DNA structures, including radioactive (hot) reaction spikes and unlabeled (cold) structures at higher concentration (see below).
3. Reaction buffer mixture (1×): 25 mM HEPES, pH 7.5, 100 mM NaCl, 3 mM $Mg(OAc)_2$, 0.1 mM dithiothreitol (DTT), 100 μg/ml BSA. Prepare at 1.67× (multiply the 1× concentrations by this factor).
4. Reaction stop mix: 2.5 mg/ml proteinase K, 2.5% SDS, 125 mM EDTA.
5. Water bath and/or heat block.
6. 10× DNA loading dye for native DNA gels: 50% glycerol, 0.05% bromophenol blue (pH 8.0). For denaturing DNA gels 1× is 0.005% bromophenol blue dissolved in formamide.
7. 20×10 cm gel electrophoresis unit (e.g., OWL model 73.1020 V) with 0.75 mm gel spacers and 25–36 well comb.
8. Two ~25 well, 10×20 cm, 10% TBE PAGE gels required; depending on the substrate and incision site, a denaturing gel (+7 M urea) may be required (*see* **Section 3**).
9. Gel equilibration solution: 20% methanol, 5% glycerol.
10. Gel dryer with vacuum pump.
11. Phosphorimaging screen and scanner (e.g., Storm 860 by Molecular Dynamics, now GE Healthcare).
12. Whatman® filter paper 3.

13. ImageQuant™ software (GE Healthcare) or equivalent program.

2.3. Incision Site Mapping

1. As for nuclease assays, plus the following:
2. Oligonucleotides, PAGE-purified and radiolabeled.
3. DNA sequencing gel apparatus (e.g., OWL model S3S).
4. T4 DNA ligase (optional).
5. 60 mM $Mg(OAc)_2$/10 mM ATP solution.

3. Methods

The reaction DNA substrate concentration is defined with unlabeled structures, while an otherwise identical radiolabeled substrate spike of negligible concentration is used to "report" on the cleavage of the entire substrate population. Using this strategy, the substrate concentration can be titrated without saturating the signal on phosphorimaging screens at higher concentrations. Also, the concentrations of unlabeled structures can be determined much more accurately using A_{260} values. When first testing an endonuclease on any branched DNA structure, it is necessary to determine on which strand(s) incision takes place so as to know which strand to end-label to monitor substrate cleavage. Initially, the labeled strand of radiolabeled structures will need to be varied and the DNA analyzed on denaturing gels in order to determine which strand is incised. Depending on which strand is incised and where, native DNA PAGE may be sufficient to resolve cleaved from uncleaved radiolabeled structures. For example, in the case of Mus81–Mms4, the 3′-flap structure is incised in a manner that removes the ssDNA flap, and the product can be resolved from uncleaved substrate by native gel electrophoresis (**Fig. 20.1a**), while the D-loop structure is incised on a strand that requires denaturing gel electrophoresis to observe. When first working with a new enzyme, it is best to optimize reaction buffer conditions for such parameters as types and concentrations of divalent cation and salt, pH, type of buffer. Optimizing these parameters from the start is much easier than re-collecting data to incorporate a change in reaction conditions.

A nuclease assay can be performed either as a fixed time point assay or as a kinetic time course. If fixed time point assays are intended to become the basis for comparison of various branched DNA substrates, caution should be taken in their interpretation. As single data points, they provide much less information than determination of kinetic values like K_M and k_{cat}. If nuclease entities are in excess of substrate molecules, this can mask differences that could have otherwise been noted. **Figure 20.1b, c** gives examples of this type of data. Fixed time point experiments are useful for determining strand incision sites and to screen different sets of reaction conditions. They may also be useful, though caution must be taken as mentioned above, in demonstrating large differences in substrate preference in a simple gel figure. Kinetic time courses follow individual reaction progresses with time by withdrawing a portion of the reaction for analysis at closely spaced intervals after the nuclease is added (e.g., **Fig. 20.1d**). These data are used to derive the initial (linear) reaction velocities. Initial velocity data are plotted over a wide range of substrate concentrations (all at one specific enzyme concentration) in a Michaelis–Menten plot, from which the parameters K_M and k_{cat} can be easily derived (e.g., **Fig. 20.1e**). Determination of these parameters requires more experimentation than fixed time point assays, but they give a set of values that can be used to compare different branched DNA structures. In the case of Mus81–Mms4, the enzyme's broad selectivity profile did not allow us to assign a probable single structure of the in vivo DNA molecule it cleaves. However, we were able to identify features of the substrate that are essential for efficient catalysis. For instance, Mus81–Mms4 requires two double-strand DNA "arms" to flank a three (or four)-way branch point, in which the third branch can be double- or single-strand DNA (6). Further, binding and catalysis appear to be influenced by the presence of a DNA end/nick at a branch point where dsDNA transitions to ssDNA, although it is solely the position of the flap adjacent to this discontinuity which directs the active site to the position of the cleavage (7).

3.1. Production of Oligonucleotide-Based Joint Molecule Substrates

3.1.1. Non-radiolabeled Structures

1. Dilute component oligonucleotides to 100 pmol/μl in TE (10 mM Tris–HCl, 1 mM EDTA, pH 7.5).
2. In 1× annealing buffer, add 600 pmol 50-mers or 1,200 pmol 25-mers in a total volume of 60 μl or less.
3. Using a thermocycler, step down the temperature as follows (*see* **Note 3**):
   a. 95°C, 3 min
   b. 65°C, 10 min
   c. 55°C, 10 min
   d. 45°C, 10 min
   e. 35°C, 10 min
   f. 4°C thereafter
4. Add native DNA loading dye to samples to 1× final concentration.

5. Pour a 10% 29:1 acrylamide:bisacrylamide, 1× TAE gel with large ~1 cm^2 wells and pre-run for 15 min at 100 V.
6. Load samples and run at 100 V for 1–2 h.
7. Carefully separate the glass gel plates and transfer gel onto a piece of plastic wrap.
8. Place gel on top of a sheet of PEI-F TLC paper. UV illumination of the gel will reveal the shadows of the product bands, which are usually the slowest migrating DNA species (*see* **Note 4**). Use a scalpel to carefully excise the gel slice containing the target DNA band and transfer product slices to 1.5 ml microcentrifuge tubes.
9. Electroelute the product DNAs from the gel slices. We use YM-10 Centricon™ units together with the centrilutor™ electroelution system (Millipore) (*see* **Note 1**). Assemble the unit with the gel slice in the sample tube as per manufacturer's instructions. Electroelute in degassed TAE buffer at 100 V for at least 2 h.
10. Concentrate the DNA in the Centricon device by centrifugation in a Beckman JA-20.1 rotor at 5,000×*g* maximum for 1 h to 1 h 45 min at 4°C. A final volume of ≤300 μl is desirable.
11. The sample can be left in TAE buffer or here dialyzed into TE or any other desired buffer using a Tube-o-dialyzer Medi tube, MWCO 15 kDa (GenoTech Inc.).
12. Measure the A_{260} using a Nanodrop® spectrophotometer. Convert to micromolar DNA molecules (for sample calculation, *see* **Note 5**).

3.1.2. Radiolabeled Structures (See **Note 1**)

Radiolabeled structures are produced in much the same way as the cold substrates, with a few changes to the protocol, as follows:

1. Radiolabel the 5′-end of the diagnostic strand (to-be-cleaved strand).
   a. Combine 200 pmol oligo to be labeled with 5 μl γ^{32}P-ATP (specific activity 6,000 Ci/mmol) and 1 μl T4 polynucleotide kinase (T4-PNK) in 20 μl total volume 1× T4 PNK buffer.
   b. Incubate 30–60 min at 37°C.
   c. Separate the labeled oligo from the unincorporated radionucleotides using an appropriate size exclusion spin column (GE MicroSpin™ G-25 or G-50).
2. In 1× annealing buffer, add 20 pmol radiolabeled oligo (one-tenth post-spin column volume; ~3 μl) and for the other non-radiolabeled strands add 100 pmol 50-mers and 200 pmol 25-mers in a total volume of 40 μl or less.

3. Anneal and separate structures on native PAGE as for cold structures (see above).
4. After completing electrophoresis, wrap gel in plastic wrap, minimizing wrinkles and ensuring that the radioactive moisture is contained within.
5. Place the gel in a standard autoradiography cassette and expose to regular (the sensitivity is not critical) autoradiography film for about 10 min. Develop the film. Optimal exposure for clean extraction is long enough to faintly see the outline of the edges/wells of the gel, but not so long that the signal of the product bands becomes undefined.
6. Place the film on top of a standard white light box and line up the gel with its image on the film below. Excise the product bands with a clean scalpel and dispose of the rest of the gel in ^{32}P dry waste.
7. Electroelute and concentrate as for cold oligo structures (see above).
8. Measure the activity of 1–2 μl of the recovered structures. Greater than 10,000 cpm/μl is desirable but much less is still workable as a reaction spike, for some time before decay renders it unusable (*see* **Note 6**). The DNA concentration of a structure prepared in this way is negligible and can be ignored in comparison to nanomolar and higher non-radiolabeled branched DNA concentrations.

3.2. Nuclease Assay Time Course Protocol

This protocol describes how to perform a set of reaction time courses that can be used to determine kinetic parameters for one branched DNA substrate by producing a Michaelis–Menten plot. The following protocol has been optimized for Mus81–Mms4. Optimal buffer and substrate concentrations, as well as other conditions, will need to be determined for other nucleases. At a minimum, we recommend using time points of 0, 3, 6, 10, 15, 20, and 30 min at each [substrate] and substrate concentrations of 2.5, 5, 10, 20, 50, 100 nM. As described, each substrate would therefore require a total of 42 time points (and gel lanes).

3.2.1. Preparation

1. Aliquot 0.5 μl of reaction stop mix to forty-two 0.5 ml appropriately labeled reaction tubes. Keep these at 4°C if not to be used immediately (to slow proteinase K self-digestion).
2. Prepare a 10× (50 nM in this case) stock of the nuclease and keep on ice. We dilute Mus81–Mms4 with 10 mM Tris–HCl pH 7.5, 0.5 mg/ml BSA.
3. Prepare 100 μl of a 1.67× stock of the reaction buffer mixture (*see* **Section 2.1**), keep on ice. Always make this fresh.

4. Make small volumes of 5× stocks of the above substrate concentrations by diluting cold substrate with TE; 500 nM is therefore the lowest workable stock concentration, in this case.
5. Measure or calculate new activity (after decay) of your reaction spike stock. If necessary, dilute to an activity such that the desired amount of counts per reaction is delivered in 1.06 μl/reaction (*see* **Note 6**).

3.2.2. Performing the Assay

1. According to **Table 20.2**, add buffer mix, substrate, and spike to 9.5 μl in a 0.5 ml microcentrifuge tube.
2. Place this tube in a 30°C (or other chosen assay temperature) water bath and incubate 5 min before taking a 0.5 μl "zero" time point before the nuclease is added.
3. Place this and all subsequent 0.5 μl time point withdrawals in the pre-aliquoted and labeled tubes containing 0.5 μl stop mix (thaw tubes with stop mix shortly beforehand in the water bath). Incubate the stop mix with quenched reaction withdrawals in the water bath, allowing all time points at least 30 min for proteinase K digestion (even though the reaction stops immediately this aids in resolution of the cleaved products by PAGE later).
4. Add the nuclease (1 μl of a 50 nM stock is suggested), and start a timer.
5. Take all subsequent 0.5 μl reaction withdrawals at the appropriate times. By staggering the start times of two or more reactions, multiple time courses can be performed simultaneously.

Table 20.2
Time course reaction additions/withdrawals

Additions	Withdrawals	Concentration factor (× final concentration)	Reaction volume
6.33 μl 1.67× buffer		1.67×	6.33
2.11 μl 5× substrate		–	8.44
1.06 μl substrate spike		1.11×	9.5
	0.5 μl for "zero" time point (t.p.)	1.11×	9.0
1 μl of 10× nuclease (start timer)		1×	10
	0.5 μl for 3 min t.p.	1×	9.5
	0.5 μl for all other t.p.	1×	9.0, 8.5, 8.0,...

6. At the end of the reaction time course, add 9 μl of 1× appropriate (native or denaturing) DNA loading dye to each time point sample and spin briefly in a microcentrifuge.
7. Pour a 10% 10 cm tall native (or denaturing) TBE polyacrylamide gel. Pre-run the gel for 15–20 min at 100 V.
8. Load reaction time points and run at 100 V for 65 min.
9. If the gel is a *native gel*, it can be transferred directly to Whatman filter paper and dried. Separate the glass plates of the gel, and with the gel still adhered to one glass plate, press the filter paper against the gel such that it sticks to the paper and will transfer from the plate to the filter paper without cracking or tearing. Cover the gel with plastic wrap and place in the gel dryer, apply the vacuum to the gel dryer and heat for 1 h at 80°C.
   If the gel is a *denaturing gel*, remove the glass plates and place the gel in gel equilibration solution to remove the urea. After 30–60 min of gentle rocking at room temperature, place the gel face down on a plastic tray, blot off excess moisture with a paper towel, and finally press a piece of Whatman paper on the gel to transfer it to the paper. Dry as above.
10. Tape the filter paper upon which the gel is dried to the inside of a phosphorimaging screen cassette and expose to the screen according to the guidelines given in **Note 6**. If the desired signal is not achieved, the screen can simply be exposed again for an alternative length of time.
11. Develop the phosphorimaging screen using a Storm 860 (Molecular Dynamics, now GE) or equivalent scanner (excitation = 635 nm, emission = 390 nm).
12. To quantitate incision of joint molecule substrates using ImageQuant™ software, draw a vertical line through the uncleaved and cleaved bands and adjust the width to include the entire width of the bands within that lane. Next, generate an intensity graph, define the two peaks, and generate an area report. This gives the percent of total intensity for each defined peak. For the product (cleaved species) band, this is the same as the percent substrate cleaved.
13. Determine kinetic parameters K_M and k_{cat} by constructing a Michaelis–Menten plot, which relates initial velocity as a function of substrate concentration. Initial velocities are derived by drawing tangents to the near-linear slopes of the early part of individual reaction progress curves (the raw data, % substrate cleaved versus time) at each substrate concentration. Then, construct a Michaelis plot by

graphing the reaction rate (e.g., in units of nmol/min) versus substrate concentration (e.g., in units of nanomolar). The asymptote of the Michaelis plot is V_{max} (nmol/min). The k_{cat} is simply V_{max}/nmol enzyme present in the reaction and has units of time^{-1}. K_M is the concentration of substrate at ½ V_{max}. Please refer to various biochemistry textbooks for further discussion of determining these kinetic values and elaboration of their meaning (8–10).

3.3. Incision Site Mapping Protocol

This protocol extends the nuclease assay described earlier by offering a way to identify the phosphodiester bond(s) hydrolyzed by a nuclease in model DNA substrate molecules. Properties of a DNA structure relevant to determining incision sites can be defined by mapping incision sites on substrates with varied structural features, allowing inference of branch point properties that a nuclease most strongly uses as a reference to define where it delivers hydrolysis. Whether or not nuclease incision generates products that can be re-ligated can also be determined (meaning the incision position occurs such that adjacent nicks can be sealed by DNA ligase). In general, denaturing polyacrylamide gel electrophoresis is used to resolve oligonucleotide lengths to nucleotide resolution. Direct comparison to a nested set of oligonucleotide length markers allows identification of the phosphodiester bond hydrolyzed in the incised strand. The sequences of these marker oligonucleotides are identical to the incised strand in model substrates but shortened by single-nucleotide intervals flanking the structure's branch point and function as standards from which the incised strand length can be directly determined.

Incision site mapping can be performed on substrates processed by the nuclease assay protocol described in **Section 3.2**, with the following modifications:

1. If oligonucleotides were ordered without 5′-phosphorylation, phosphorylate 5′-ends that may be relevant to substrate processing when annealed into a DNA joint molecule. 5′-Phosphorylate oligonucleotides by incubating 250 pmol oligonucleotide with 10 U T4 PNK in T4 DNA ligase buffer (containing 1 mM ATP) in a 50 μl volume for 30 min at 37°C followed by 10 min incubation at 65°C. Recover oligonucleotides using a Qiaquick® Nucleotide Removal kit (Qiagen) or Microspin G-25 Sepharose columns (GE Healthcare). If necessary, confirm phosphorylation by denaturing urea-PAGE, which confirms a greater electrophoretic mobility after addition of the negatively charged phosphate group.
2. Anneal structures as described in **Section 3.1**. Perform nuclease reactions as described in **Section 3.2**, with the exception that reaction volumes may be increased to 20 μl,

with enzyme and substrate at equimolar concentrations (e.g., 50 nM enzyme:50 nM substrate) or at limiting enzyme:substrate concentrations (e.g., 10 nM enzyme:50 nM substrate). Incubate reactions at 30°C or other appropriate optimal temperature for 30 min or other appropriate times.

3. To assay the fraction of incised material that can be re-ligated after nuclease incision, incubate approximately half of the nuclease reaction volume with T4 DNA ligase following the nuclease assay. Remove 9 μl reaction volume and transfer to a new 500 μl Eppendorf tube. Add 0.5 μl 60 mM $Mg(OAc)_2$, 10 mM ATP plus 0.5 μl (200 U) T4 DNA ligase. Incubate at room temperature for 15 min. In parallel, verify ligase activity by performing nicked duplex ligation controls. Use 50 nM nicked duplex substrate (prepared with or without phosphate at the internal nick), incubated with T4 DNA ligase under the same conditions.
4. Stop all reactions by denaturation at 95°C for 2 min, followed by transfer to ice. Normalize samples for activity (total cpm to be added per well of an analytical gel); add formamide/bromophenol blue to a volume of 2–3 μl and load onto an analytical 8–12% acrylamide/8 M urea denaturing PAGE gel. In lanes flanking the nuclease reactions, run an oligonucleotide size ladder that will serve as a migration standard from which the incision site in the incised DNA joint molecule strand can be determined (**Fig. 20.2**).
5. Oligonucleotide size ladders can be prepared by designing oligonucleotides of defined lengths that represent potential incision products along the incised DNA strand. Order these PAGE-purified or PAGE-purify yourself on a denaturing 12% polyacrylamide/8 M urea gel followed by band excision and oligonucleotide elution. Radiolabel the oligonucleotides separately from one another as described for oligonucleotide labeling in **Section 3.1** and remove unincorporated nucleotide using a Qiaquick® Nucleotide Removal Kit or Microspin™ G-25 Sepharose columns. Determine activities of each oligonucleotide by scintillation count and pool appropriate volumes of each labeled oligonucleotide to normalize their activities in a common ladder stock. In other words, pool the oligonucleotides according to cpm/μl of each oligonucleotide so that each oligoucleotide is of common intensity in the ladder regardless of the individual length and labeling efficiency.
6. Perform denaturing PAGE (8–12% acrylamide) for 3–5 h at 1,500 V. Transfer gel to Whatman paper, cover in Saran wrap, and dry at 80°C under vacuum for 1 h (*see* **Note 7**).

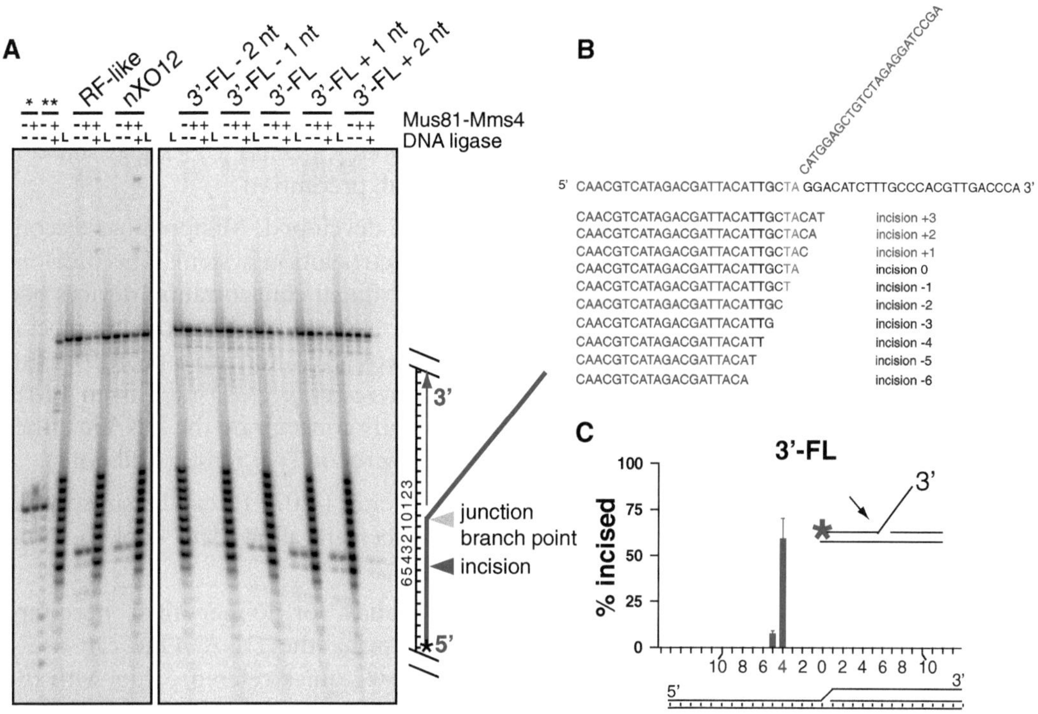

Fig. 20.2. Incision site mapping by direct comparison to an oligonucleotide size ladder. (**a**) A number of DNA joint molecules are represented on a denaturing PAGE gel, with oligonucleotide size markers ("L") representing a series of possible incision site products on the incised strand. (**b**) The oligonucleotides pooled as incision site markers flank the structure's branch point on the incised strand. (**c**) The fraction of molecules incised at a particular site can be graphed by quantitation of data in (**a**). For the 3'-FL shown, the majority of incision events occurred 4 NT 5' of the structure's branch point.

7. Transfer dried gel to a phosphorimager screen cassette and expose overnight or longer. Process using Storm Imager as described for nuclease assays in **Section 3.2**. Incision sites can be determined by direct comparison to oligonucleotide size ladders, because the component oligonucleotides represent a population of oligos whose length define single-nucleotide increments of potential incision site locations (*see* **Note 8**). Quantitation of band intensities in the nuclease incision products allows one to graph the preference for phosphodiester bond target sites relative to structural properties (as in **Fig. 20.2c**).

4. Notes

1. Basic equipment includes a Geiger counter, shields, mask or safety glasses, shielded liquid, and solid waste container. A user should be trained and follow the common and local

rules for isotope usage. When using ^{32}P, always remember to use appropriate shielding/protective eyewear and dispose of waste appropriately. This applies throughout substrate preparation; however, for the low activity actually used in spiked nuclease assays, working behind a Plexiglas shield is not a necessary, or practical, precaution.

2. Since these protocols were developed, Millipore has discontinued the Centrilutor electroelution system. The previous generation Centricon centrifugal concentration devices had the convenience of acting as vessel for both the electroelution and concentration steps. An alternative protocol is to use another electroelution vessel (e.g., D-tube from EMD Bioscience) and subsequently concentrate the DNA in either a Centricon ($\leq$2 ml) or Microcon ($\leq$0.5 ml) (Millipore).
3. A beaker of near-boiling water (~0.5 L) can be substituted for this step, in which tubes are allowed to cool to room temperature over ~1–2 h.
4. Avoid shadowing the bands for longer than necessary because UV light will damage the DNA. One can use a razorblade or scalpel to make quick reference cuts with the light on, then spend time transferring the slices to 1.5 ml tubes with the light off afterward. Also, avoid loading different structures with similar mobility in adjacent lanes to avoid possible cross contamination. To verify the correct and fully annealed structure by electrophoretic mobility, the different possible combinations of component strands (e.g., all strands versus only one to three strands) can be run out in adjacent lanes.
5. Conversion from A_{260} units to micromolar molecules:
   For *pure* double-stranded DNA (dsDNA) or single-stranded DNA, we convert from absorbance to μM nucleotides (NT), then divide by the number of NT per molecule (2× # bp) to give μM molecules:

$$\mu\text{M NT}/\#\text{NT per molecule} = \mu\text{M molecules}$$

For *dsDNA*, this conversion = $(A_{260} \times 150\,\mu\text{M NT } A_{260}^{-1})/(2 \times \#\text{bp})$.

For structures of mixed double- and single-strand DNA (ssDNA), we must treat the contributions of each type of DNA to A_{260} separately due to their different extinction coefficients. We can use the conversion that 1 NT of ssDNA absorbs 67% as much as 1 bp (2 NT in dsDNA), to express the ssDNA as bp equivalents of absorption. The following example illustrates this for the 3′-flap structure, which has 49 bp in dsDNA plus 27 NT in ssDNA:

$$\# \text{bp equivalents} = 49 + 27(0.67) = 67.1$$
$$\# \text{ds NT equivalents} = 134.2$$

Thus one would enter 134.2 for the "# of NT per molecule" value to convert to μM molecules using the conversion factor for dsDNA given above. Using this strategy, simple conversion factors can be calculated to convert A_{260} values to μM molecules for each structure. In the case of the 3′-flap, this conversion factor is simply (150)/134.2 = 1.12. Thus one A_{260} absorbance unit for the 3′-flap structure corresponds to 1.12 μM molecules. Of course, these are still rough conversions, as they assume average sequence composition and do not account for variable base stacking interactions within the ssDNA regions of different substrates.

The A_{260} of *radiolabeled substrates* should be too low to be measured with a Nanodrop® spectrophotometer or any other method of measuring DNA concentration. However, do not attempt to pool and further concentrate several decayed spike preparations of the same structure to avoid making a fresh preparation; the concentration of the spike will become high enough to become no longer negligible, as required.

6. A strong yet quantifiable signal of both substrate and product bands above ~1% can be obtained from a gel loaded with 500 cpm/lane radiolabeled substrate spike (in this case, ~10,000 cpm/reaction) after exposure to the phosphorimaging screen for several hours to overnight. However, as little as ten times less (50 cpm/lane; 1,000 cpm/reaction) can be used for successful quantitation with exposures of overnight to 2+ days. For endpoint assays, the entire reaction can be loaded in one lane, and therefore 20 times less cpm/μl reaction volume is required to spike the reactions than time course assay reactions.

7. Transferring a large sequencing gel to Whatman paper can be challenging because the gel is thin and easily torn. Keeping polyacrylamide concentration at or below 12% helps the gel adhere more readily to Whatman paper. Coating one inner surface of the glass plates with a thin film of Rain-X® can help glass plates separate more easily from the gel surface. To transfer the gel, a large sheet of Whatman paper can be gently pressed against the gel and then pulled up from one corner in a steady and swift motion. Alternatively, the second glass plate can be placed back on top of the Whatman paper such that the Whatman paper and gel are sandwiched between the two glass plates (bottom to top: glass plate, Whatman paper, gel, glass plate). Place the assembly near the edge of a solid counter surface. Leaving the lower

glass plate in place, pull the upper glass plate away from the counter edge so that the Whatman paper and gel begin to fall away from the upper glass plate by gravity. The initial adherence between the gel and Whatman paper, particularly at the edges of the gel, can be encouraged with a small stream of water from a squirt bottle.

8. An alternative to this direct comparison method is Maxam–Gilbert sequencing. In the case of Maxam–Gilbert sequencing, a correction for incision site location needs to be made because some functional groups are lost during chemical processing. This correction does not need to be made with the direct comparison method described here.

Acknowledgments

We thank Shannon Ceballos, Clare Fasching, Ryan Janke, Sucheta Mukherjee, Erin Schwartz, and Xiao-Ping Zhang for helpful comments on the manuscript. Our work is supported by the NIH (GM58015, CA92276), the DoD (BC083684), and a TRDRP predoctoral fellowship (17DT-0178) to W.W.

References

1. Ciccia, A., McDonald, N., and West, S.C. (2008) Structural and functional relationships of the XPF/MUS81 family of proteins. *Annu Rev Biochem* 77, 259–287.
2. Heyer, W.D., Ehmsen, K.T., and Solinger, J.A. (2003) Holliday junctions in the eukaryotic nucleus: resolution in sight? *Trends Biochem Sci* **28**, 548–557.
3. Hollingsworth, N.M., and Brill, S.J. (2004) The Mus81 solution to resolution: generating meiotic crossovers without Holliday junctions. *Genes Dev* **18**, 117–125.
4. Mimitou, E.P., and Symington, L.S. (2009) Nucleases and helicases take center stage in homologous recombination. *Trends Biochem Sci* **34**, 264–272.
5. Svendsen, J.M., and Harper, J.W. (2010) GEN1/Yen1 and the SLX4 complex: solutions to the problem of Holliday junction resolution. *Genes* **24**, 521–536.
6. Ehmsen, K.T., and Heyer, W.D. (2008) *Saccharomyces cerevisiae* Mus81-Mms4 is a catalytic, DNA structure-selective endonuclease. *Nucleic Acids Res* **36**, 2182–2195.
7. Ehmsen, K.T., and Heyer, W.D. (2009) A junction branch point adjacent to a DNA backbone nick directs substrate cleavage by *Saccharomyces cerevisiae* Mus81-Mms4. *Nucleic Acids Res* **37**, 2026–2036.
8. Parkin, K. (2002) Enzyme kinetics: a modern approach (Malden, MA: Wiley-Interscience).
9. Cornish-Bowden, A. (1995) Fundamentals of enzyme kinetics (London: Portland Press Limited).
10. Segel, I.H. (1993) Enzyme kinetics: behavior and analysis of rapid equilibrium and steady state enzyme systems (New York, NY: Wiley-Interscience).

Chapter 21

In Vitro Assays for DNA Pairing and Recombination-Associated DNA Synthesis

Jie Liu, Jessica Sneeden, and Wolf-Dietrich Heyer

Abstract

Homologous recombination (HR) is a high-fidelity DNA repair pathway that maintains genome integrity, by repairing double strand breaks (DSBs) and single-stranded DNA (ssDNA) gaps and by supporting stalled/collapsed replication forks. The RecA/Rad51 family of proteins are the key enzymes in this homology-directed repair pathway, as they perform DNA strand invasion and exchange, in concert with a host of ancillary factors. In vitro, the RecA/Rad51 family of proteins share similar enzymatic activities including catalyzing ssDNA-stimulated ATP hydrolysis, formation of displacement loops (D-loops), and DNA strand exchange. After successful DNA strand invasion, DNA synthesis restores the lost genetic information using an undamaged DNA template. In this chapter, we describe two well-established biochemical assays to investigate the signature DNA strand transfer activity of RecA/Rad51 family of proteins: the D-loop assay and the DNA strand exchange reaction. Moreover, we describe a D-loop extension assay coupling D-loop formation with DNA synthesis, which can be used to define the roles of DNA polymerases in HR. Additionally, we present a protocol to investigate protein-mediated DNA annealing, a critical step in the synthesis-dependent strand annealing (SDSA) and double-Holliday junction (dHJ) pathways as well as the single-strand annealing (SSA) pathway. The quality of supercoiled plasmid DNA is critical in reconstituted HR reactions, and a protocol describing the preparation of this DNA substrate is included.

Key words: D-loop, DNA polymerase, DNA strand exchange, DNA strand annealing, DNA synthesis, homologous recombination, Rad51, Rad52, RecA, supercoiled plasmid DNA.

1. Introduction

The RecA/Rad51 family of proteins are the central enzymes in homologous recombination (HR), a high-fidelity DNA repair pathway that processes DNA double strand breaks (DSB) and single-stranded DNA (ssDNA) gaps and also supports replication

H. Tsubouchi (ed.), *DNA Recombination*, Methods in Molecular Biology 745,
DOI 10.1007/978-1-61779-129-1_21, © Springer Science+Business Media, LLC 2011

forks in all domains of life (1, 2). These proteins form filaments on ssDNA and catalyze homology search, DNA strand invasion, and DNA strand exchange using homologous double-stranded DNA (dsDNA) as a template. In vitro, two different biochemical assays have been developed to demonstrate this recombinational activity: DNA strand exchange and displacement loop (D-loop) formation. In this chapter, we describe protocols for DNA strand exchange and D-loop formation, developed for the budding yeast Rad51 protein (**Figs. 21.1** and **21.2**). These assays have been critical to define the properties of the wild-type and mutant proteins. Moreover, subtle modifications to the assays allow testing the functions of additional protein factors, such as mediator proteins or anti-recombination helicases, involved in Rad51-dependent recombination (3–5).

After DNA strand invasion, the 3′-OH of the invading strand is positioned in the D-loop on an undamaged homologous template to initiate repair DNA synthesis. Our laboratory has recently developed a coupled reaction of D-loop formation and extension by DNA polymerase, using purified protein factors from *Saccharomyces cerevisiae* (**Fig. 21.2a**) (6). This assay provides a tool to test the functional interplay between recombination proteins and replication factors and to explore the roles of various DNA polymerases in the specific context of recombinational repair. In the D-loop formation and extension assays, supercoiled dsDNA serves as the homologous template, and its quality directly contributes to the minimization of experimental artifacts and interferences. Thus, we provide a protocol on how to prepare high-quality supercoiled plasmid dsDNA.

In vivo, DNA annealing is a late but critical step in two different subpathways of HR: the synthesis-dependent strand annealing (SDSA) and the double Holliday junction (dHJ) pathways (1).

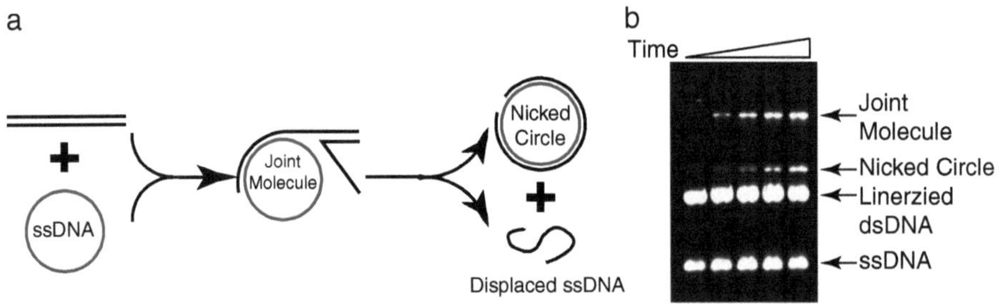

Fig. 21.1. DNA strand exchange reaction. (**a**) Reaction scheme for the DNA strand exchange assay. Homologous circular ssDNA and linearized dsDNA are the substrates. Joint molecules are intermediates. Nicked circles and displaced ssDNAs are the final products. (**b**) Time course of a Rad51-catalyzed DNA strand exchange assay. Rad51 (6.7 μM) was incubated with 20 μM (nt concentration) ϕX174 ssDNA for 15 min at 30°C, then 1.11 μM RPA was added and incubated for another 30 min. Then 20 μM (bp concentration) *Pst*I-linearized ϕX174 dsDNA was added to initiate the reaction. Samples were taken at 0, 30, 60, 90, 180 min and immediately quenched by stop buffer. An ethidium bromide-stained agarose gel is shown.

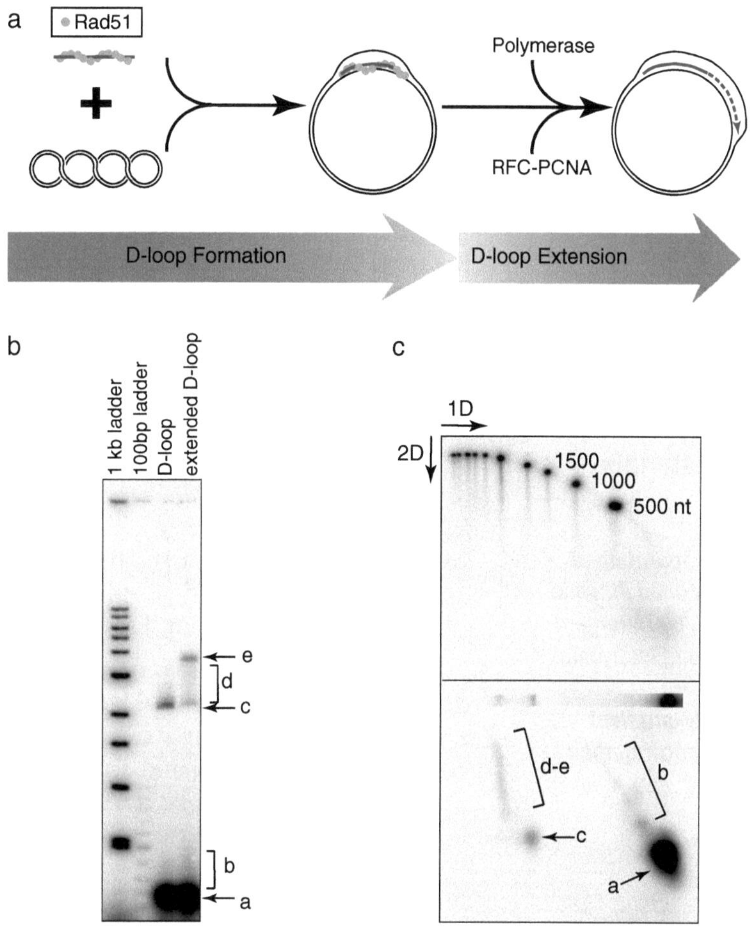

Fig. 21.2. D-loop formation and D-loop extension assays. (**a**) Reaction scheme for the D-loop formation and D-loop extension assay. Rad51 forms nucleoprotein filaments on a short oligo nucleotide, which catalyzes D-loop formation with supercoiled plasmid dsDNA. The extension assay is initiated by the addition of both DNA polymerase and its accessory factors PCNA-RFC, as well as dNTPs. NEB 1 kb ladder is 0.5, 1, 1.5, 2, 3, 4, 5, 6, 8, 10 kb. (**b**) D-loop formation and extension are monitored by analyzing DNA species on a 0.8% native agarose gel. (**c**) Two-dimensional native/denaturing agarose gel electrophoresis of a 1 kb ladder (*top panel*) and D-loop extension products (*bottom panel*). Labels in (**b**), (**c**): *a* free ssDNA, *b* short extension products, dissociated from D-loops, *c* unextended D-loops, *d* partially extended D-loops, *e* maximally extended D-loops.

Strand annealing is also the central reaction of the single strand annealing (SSA) pathway (2). We describe a DNA strand annealing assay protocol based on the yeast Rad52 protein under physiologically relevant conditions that include free Mg^{2+} and sufficient amounts of the ssDNA binding protein RPA to saturate the ssDNA (**Fig. 21.3**). This method can be used to test individual proteins, for example Rad52, and the modulation of their activity by other factors, such as Rad51 or Rad59 (7, 8).

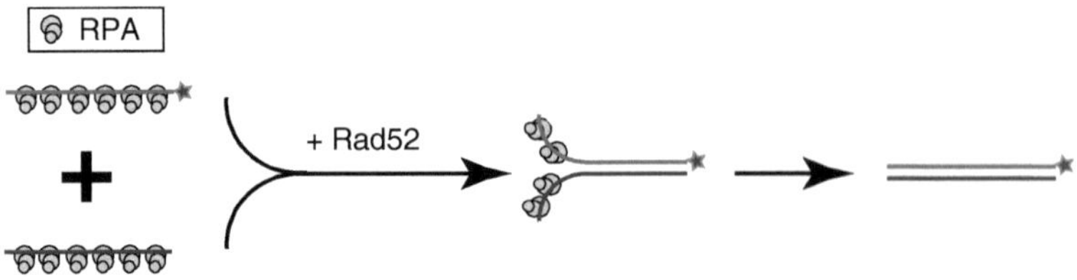

Fig. 21.3. DNA annealing assay. Reaction scheme for DNA strand annealing assay of RPA-covered ssDNA. Rad52 and its homolog in phage T4 (UvsY) and *E.coli* (RecO) can catalyze annealing between complementary ssDNA strands covered with RPA.

2. Materials

2.1. Preparation of Supercoiled Plasmid dsDNA by Detergent Lysis and Isopycnic CsCl density-gradient ultracentrifugation

1. Liquid LB media (1 l): 10 g tryptone, 5 g yeast extract, and 5 g NaCl. Suspend all solids into ddH_2O and autoclave it at 121°C for 25 min. Fresh media is preferred at this scale.
2. STE buffer: 10 mM Tris–HCl (pH 8.0), 0.1 M NaCl, and 1 mM EDTA (pH 8.0). Sterilize the solution by passing through a 25 mm syringe filter with 0.22 μm pore size (Fisher Scientific), and store at 4°C. All buffers described in this chapter can be sterilized in this manner. Use 200 ml for each 1 l cell culture.
3. Tris–sucrose buffer: 25% w/v sucrose and 50 mM Tris–HCl (pH 8.0). Sterilize by filtration and store at 4°C. Use 100 ml for each 1 l cell culture.
4. 0.5 M EDTA (pH 8.0), 10% SDS, and 5 M NaCl.
5. Lysozyme solution: 10 mg/ml hen egg white lysozyme, 25 mM Tris–HCl (pH 8.0). Prepare fresh solution each time before use.
6. Sodium iodide solution (100 ml): 7.6 M NaI, 40 mM Tris–HCl (pH 8.0), and 20 mM EDTA. Sterilize by filtration and store at 4°C.
7. 3 M NaOAc (pH 5.2), isopropanol, 100 and 70% ethanol.
8. TE buffer (10×): 100 mM Tris–HCl (pH 8.0) and 10 mM EDTA (pH 8.0). Sterilize by filtration and store at room temperature.
9. 10 mg/ml ethidium bromide solution and solid cesium chloride (CsCl).
10. Beckman Coulter Allegra^{TM} 6 Centrifuge: Swing-bucket benchtop centrifuge.
11. Beckman J2-MC Centrifuge and corresponding JA-20 and JA-14 rotors.

12. Beckman Optima™ LE-80 K Ultracentrifuge: for CsCl gradient centrifugation.
13. Quick-Seal polyallomer 13.5 ml capacity centrifuge tubes (Beckman catalog number: 342413).
14. ISO-TIP quick charge soldering iron (Beckman, catalog number: 7740).
15. Agarose gel running apparatus and power supply.
16. 10× TBE buffer: 0.89 M Tris base, 0.89 M H_3BO_3, and 25 mM EDTA. Prepare 1 l solution by dissolving 109 g of Tris base, 55 g of M H_3BO_3, and 50 ml of 0.5 M EDTA (pH 8.0) into 1 l ddH_2O.
17. Digital gel imaging system with UV illumination box.
18. Nanodrop® spectrophotometer or alternative unit that can read the A_{260} of small volumes of DNA in solution.

2.2. DNA Strand Exchange Reaction

1. DNA strand exchange buffer (5× SEB): 150 mM Tris–acetate (pH 7.5), 5 mM DTT, 250 μg/ml BSA, 12.5 mM ATP, 20 mM $Mg(OAc)_2$, and 100 mM phosphocreatine (*see* **Note 1**). Sterilize by filtration and store at –20°C.
2. Purified proteins from *S. cerevisiae*: Rad51 and RPA (*see* **Note 2**) (6).
3. 1 μg/μl creatine kinase stock solution, store at −20°C.
4. 100 mM spermidine (pH 7.4) stock solution, store at −20°C.
5. Stop buffer: 0.714% SDS, 357 mM EDTA, and 4.3 mg/ml proteinase K. Prepare 84 μl solution freshly by mixing 6 μl 10% SDS, 60 μl 0.5 M EDTA, and 18 μl 20 mg/ml Proteinase K.
6. DNA gel loading buffer (10×): 0.25% bromophenol blue (w/v), 50% glycerol (omit xylencyanol). Store at 4°C.
7. φX174 ssDNA (virion) is purchased from NEB (catalog number: N3023S) (*see* **Note 3**).
8. RFI φX174 dsDNA (RF I) is purchased from NEB (catalog number: N3021S) or prepared by the protocol in **Section 3.1**.
9. TBE–agarose gel running buffer, apparatus, and imaging system, as listed in **Section 2.1**.
10. ImageQuant software (version 5.1, GE Healthcare).

2.3. D-Loop Assay

1. T4 polynucleotide kinase reaction buffer (10×): 70 mM Tris–HCl (pH 7.6), 10 mM $MgCl_2$, and 5 mM DTT.
2. T4 polynucleotide kinase

3. D-loop buffer (5×): 150 mM Tris–acetate (pH 7.5), 5 mM DTT, 250 μg/ml BSA, 20 mM ATP, 25 mM $Mg(OAc)_2$, and 100 mM phosphocreatine (*see* **Note 1**). Sterilize by filtration and store at −20°C.
4. Purified proteins from *S. cerevisiae*: Rad51, Rad54, and RPA (*see* **Note 2**) (6).
5. Creatine kinase and spermidine: as described in **Section 2.2**.
6. Stop buffer and DNA loading buffer (10×): as described in **Section 2.2**.
7. 95-mer DNA oligonucleotides: The sequence of *Pst*I-95 mer is 5′-TgCAggCATgCAAgCTTggCgTAATCATggT CATAgCTgTTTCCTgTgTgAAATTgTTATCCgCTCACA ATTCCACACAACATACgAgCCggAAg-3′. It shares homology with pUC19 plasmid DNA.
8. pUC19 supercoiled plasmid DNA: Prepare according to the protocol in **Section 3.1** without alkaline cell lysis.
9. 10× TBE buffer: as described in **Section 2.1**.
10. Mini Spin Oligo Column (Roche Applied Science: catalog # 11814397001). These spin columns are used to remove unincorporated radioactive nucleotides from the labeled oligo substrates.
11. TBE–agarose gel running buffer, apparatus, and imaging system, as listed in **Section 2.1**.
12. DE81 paper (Whatman): a thin DEAE cellulose paper about 1 mm in width. It has weakly basic anion exchangers coupled with diethylaminoethyl groups, resulting in low retention of DNA during the gel drying process.
13. 3 M Whatman filter paper.
14. Gel dryer with vacuum pump, dedicated for use with radioisotopes.
15. Phosphorimaging screen and Storm 860 (Molecular Dynamics) PhosphorImaging system.
16. ImageQuant software (version 5.1, GE healthcare).

2.4. D-Loop Extension Assay

1. All the materials listed above in **Section 2.3**.
2. Stock solution of dNTP mix containing: 1 mM each dATP, dCTP, dTTP, dGTP.
3. Purified proteins from *S. cerevisiae*: RFC, PCNA, and polymerase δ (*see* **Note 2**) (6).
4. Denaturing agarose gel buffer: 50 mM NaOH, 1 mM EDTA.

5. DNA gel loading dye (6×): 1× TE, 18% ficoll (w/v), 0.15% bromocresol green (w/v), 50 mM EDTA.
6. [γ-^{32}P]-ATP-labeled 100 bp and 1 kb ladders.
7. 50°C water bath.

2.5. DNA Strand Annealing

1. Oligo A1 (5′-GCAATTAAGCTCTAAGCCATCCGCAAAAATGACCTCTTATCAAAAGGA) and Oligo A2 (5′-TCCTTTTGATAAGAGGTCATTTTTGCGGATGGCTTAGAGCTTAATTGC) either ordered PAGE-purified or purified by electrophoresis using 15% polyacrylamide gels containing 8 M urea. The nucleotide concentrations of oligo A1 and A2 can be measured using extinction coefficients of 1.0×10^4 and 9.6×10^3/M/cm at 260 nm, respectively.
2. DNA strand annealing buffer (5×): 150 mM Tris–acetate (pH 7.5), 25 mM magnesium acetate, 5 mM DTT. Sterilize by filtration and store at −20°C.
3. Purified proteins from *S. cerevisiae*: Rad52, RPA (*see* **Note 2**) (9).
4. Stop buffer: 1.54 μM unlabeled oligo A2, 0.77% SDS, and 1.54 mg/ml proteinase K. Make fresh stop buffer each time to reach this final concentration. Prepare 30 μl of fresh stop buffer by mixing 10 μl of 20 μM unlabeled oligo A2, 10 μl of 10% SDS, and 10 μl of 20 mg/ml proteinase K. Mix 3 μl fresh stop buffer with 10 μl sample aliquot rapidly. Store 10% SDS at room temperature and proteinase K at −20°C.
5. For radioactive samples: polyacrylamide gel running apparatus, power supply, and TBE buffer, as listed in **Section 2.1**.
6. For fluorescent dye method: 4′,6-diamidino-2-phenylindole (DAPI, Invitrogen). The stock concentration of DAPI is measured using a molar extinction coefficient of 3.3×10^4/M/cm at 345 nm.
7. For fluorescent dye method: an SLM8000 spectrofluorimeter or other similar spectrofluorimeter. Additionally, a 700 μl cuvette is needed, but other smaller cuvettes can be considered to minimize protein consumption.

3. Methods

Many biochemical assays including D-loop formation, DNA strand exchange, and DNA-binding use plasmid-based circular dsDNA as substrate. Commercial plasmid DNA is usually available as the product of an alkaline lysis method, the norm in industrial-scale plasmid preparation. However, alkaline treatment creates both nicked circles and locally melted regions in the dsDNA plasmid, which causes artifacts in these biochemical assays. For example, local melting of the dsDNA increases background for spontaneous D-loop formation, because of annealing between the ssDNA supplied to the assay as the invading strand and the ssDNA in the melted zones. The nicked circular form of dsDNA migrates very close to D-loops during agarose gel electrophoresis, interfering with data analysis. Equally relevant, an intact supercoiled dsDNA form will stabilize D-loops and therefore minimize D-loop dissociation during electrophoresis. Thus, a clean preparation of plasmid DNA in high-quality, supercoiled form provides a solid foundation for the success of D-loop formation and D-loop extension assays in **Sections 3.3** and **3.4.** In **Section 3.1**, we describe a protocol for the preparation of intact supercoiled plasmid dsDNA from *E. coli* that uses detergent lysis as an alternative to alkaline lysis. The supercoiled form is recovered by isopycnic centrifugation in CsCl gradients.

As the major ssDNA-binding protein in eukaryotes, RPA binds ssDNA rapidly with high affinity and saturates ssDNA, protecting it from nuclease digestion and removing inhibitory secondary structure. In vitro, when Rad51 is added before the addition of RPA, Rad51 cannot form a saturated nucleoprotein filament because of inhibitory secondary structures in ssDNA. The subsequent addition of RPA to partially formed Rad51–ssDNA filaments promotes the fully assembled Rad51–ssDNA filament via the slow displacement of RPA by Rad51. The end result is a direct increase of product yield in the DNA strand exchange reaction. This sequence is used in the protocols described in **Sections 3.2**, **3.3** and **3.4**. However, if the order of addition is reversed by adding RPA before Rad51, the strong and cooperative binding of RPA on ssDNA usually inhibits filament formation of Rad51 onto ssDNA, which results in an inhibition of Rad51-dependent DNA strand invasion. This order of addition is useful to demonstrate the function of mediator proteins, such as Rad52 protein, which facilitates Rad51–ssDNA filament formation under suboptimal conditions such as high salt, low Rad51 concentration, and RPA inhibition (5, 10). In general, mediator proteins may facilitate the nucleation or enhance stability of Rad51 filaments onto RPA-covered ssDNA. Consequently, Rad51 can propagate and form active nucleoprotein filaments to initiate homology search. For more details about this "order of addition" phenomenon, the reader is referred to references (5, 10, 11).

In vivo, D-loop formation represents the successful invasion of the ssDNA into a homologous duplex DNA; the stability and the processing of D-loops determine the outcome of competition between several HR subpathways. As the central intermediate of DNA strand exchange, the D-loop can be disrupted, migrated, and expanded, and/or it may lead to double Holliday junctions through second-end capture. Thus, the functional characterizations of various protein factors in Rad51-dependent D-loop formation will define the mechanisms that differentiate the SDSA, dHJ, and break-induced replication (BIR) pathways of HR. We describe a D-loop assay in **Section 3.3** that uses a short 95-mer as the ssDNA substrate and the homologous pUC19 plasmid dsDNA as the target of D-loop formation (**Fig. 21.2a**). Our laboratory has also developed a reconstituted D-loop extension assay, described in **Section 3.4** (**Fig. 21.2a**), to investigate the interplay between recombination proteins, DNA repair polymerases, and their accessory factors (6).

Protein-mediated annealing of complementary ssDNA molecules is a key step in the SDSA pathway and in second-end capture of the dHJ pathway. Moreover, annealing is the central reaction of the SSA pathway of DSB repair. Rad52 is a prototype protein for catalyzing annealing of complementary ssDNA under physiologically relevant conditions (ssDNA fully saturated with the cognate ssDNA-binding protein RPA, presence of free Mg^{2+} (9)). In **Section 3.5**, we describe two DNA annealing assays using purified Rad52 and RPA proteins.

For all assays described in this chapter, the reactions are terminated by a rapid inactivation of the proteins in the samples by metal ion chelation and SDS denaturation. This step is essential for reliable results that allow comparison of parallel conditions.

3.1. Preparation of Supercoiled pUC19 Plasmid DNA

1. Grow 1 l DH5α cells containing pUC19 plasmid in LB media with 100 μg/ml ampicillin, overnight at 37°C. Be sure to inoculate the culture with a single colony from a fresh transformation.
2. Collect the cells by centrifugation at 1,540×*g* for 15 min using a JA-14 rotor. Discard supernatant and resuspend pellets in 200 ml ice-cold STE buffer. Collect cells at 1,540×*g* for 15 min at 4°C using a JA-14 rotor. Pellet can be stored at −80°C.
3. Resuspend the cells in 100 ml ice-cold Tris–sucrose solution. Add 20 ml lysozyme solution (10 mg/ml) and 40 ml of 0.5 M EDTA (pH 8.0). Mix well by inverting gently and incubate on ice for 10 min.
4. Add 40 ml of 10% SDS and mix immediately but gently into the solution with a glass rod. Add 60 ml of 5 M NaCl (to 1 M final concentration). Mix with glass rod and incubate on ice for at least 1 h. The incubation time on ice can be extended for enhanced yield and purity.
5. Centrifuge at 13,870×*g* for 30 min at 4°C using a JA-14 rotor. Then transfer the supernatant into four ultracentrifuge tubes and centrifuge at 4°C, 70,400×*g* for 30 min using a Ti45 rotor.

6. Add 2 volumes of 100% ethanol to the supernatant. Mix well and incubate at room temperature for 2 h.
7. Centrifuge at 13,870×*g* for 20 min at 4°C using a JA-14 rotor. Save the pellet and wash with 1 volume of 70% ethanol.
8. Centrifuge the resuspended pellet at 13,870×*g* for 20 min at 4°C using a JA-14 rotor. Air dry pellet and resuspend in 10 ml TE (pH 8.0) buffer. At this step, the sample can be stored at −20°C. Take OD_{260} reading to determine the DNA concentration using a molar extinction coefficient (bp) $\varepsilon_M = 6{,}500/M/cm$ for dsDNA. Analyze the sample on a 0.8% native agarose–TBE gel to establish the quality of the sample.
9. Add 1.28 volumes of NaI buffer and incubate for 5 min at room temperature.
10. Add 0.6–1.0 volume of isopropanol. Mix well and incubate for 10–15 min at room temperature.
11. Centrifuge at 13,870×*g* for 20 min at 4°C using a JA-14 rotor. Save the pellet and wash out the isopropanol with 1 volume of 70% ethanol.
12. Centrifuge at 13,870×*g* for 20 min at 4°C using a JA-14 rotor. Save the pellet and air dry.
13. Resuspend the dried pellet in 3 ml TE (pH 8.0) buffer. Add 10 mg/ml RNaseA to 3 ml sample in TE (pH 8.0) buffer to a final concentration of 50 μg/ml RNaseA and incubate at 37°C for 1 h. Centrifuge again at 13,870×*g* for 20 min at 4°C using a JA-14 rotor. Keep the clear supernatant and discard small white precipitates observed at the bottom of the tube.
14. Add 0.1 volume of 3 M NaOAc (pH 5.2). Mix well.
15. Add 0.6–1 volume of isopropanol. Mix well and incubate at room temperature for ~30 min. Centrifuge at 18,800×*g* for 30 min at 4°C using a JA-14 rotor.
16. Wash with 1 volume of 70% ethanol. Centrifuge at 4°C, 18,800×*g* for 20 min using a JA-14 rotor.
17. Air dry the pellet and resuspend in 1 ml TE (pH 8.0) buffer. Save 10 μl sample to be checked for by native agarose gel electrophoresis in later steps.
18. Add TE buffer to the sample from Step 17 to reach a total volume of 19 ml. Add 20.8 g CsCl into the solution and dissolve well. Once CsCl is dissolved, add 1 ml 10 mg/ml ethidium bromide into the solution. Weigh 1 ml of this solution; the mass should be ~1.55 g. Adjust the density with CsCl or TE, as necessary (*See* **Note 4**).

19. Vortex the tube to mix well and centrifuge at 2,280×*g* for 15 min using the Beckman Coulter Allegra ™6 Benchtop centrifuge. After centrifugation, take only the clear supernatant solution to fill in the Quick-Seal tubes in the next step. Discard the cloudy part at the top and the precipitants at the bottom of the tube.
20. Use an 18 gauge needle and a 5 ml syringe to transfer the supernatant into two Quick-Seal polyallomer 13.5 ml capacity tubes (Beckman cat. NO: 342413) up to the neck. Avoid any debris. Balance the tubes carefully and then seal the tube by melting the neck of the tube using ISO-TIP quick charge soldering iron (Beckman, catalog number: 7740). It is important to check the integrity of the seal, to prevent leakage and ethidium bromide contamination of the centrifuge.
21. Centrifuge at 160,400×*g* for 20–24 h (minimum for 10 h) at 20°C using a Ti65 rotor. This centrifugation step is critical to separate the intact supercoiled plasmid form from other forms of DNA.
22. Use a syringe with an 18 gauge needle to pierce the Quick-Seal tube and to extract the bright orange band corresponding to supercoiled DNA in the middle of the tube, about 2–4 ml from each tube. Apply the sample into a new centrifuge tube. Remove only the center of the band containing supercoiled plasmid DNA to avoid contamination by nicked circular DNA. Only use an 18 gauge needle, since a smaller needle might shear DNA.
23. Add 1 volume of *n*-Butanol saturated with TE to extract the ethidium bromide from the purified plasmid DNA. Vortex and centrifuge at 2,280×*g* for 15 min using the Beckman Coulter Allegra ™6 Benchtop centrifuge.
24. Discard the upper layer (*n*-Butanol + ethidium bromide) into a designated ethidium bromide hazardous waste container.
25. Repeat Steps 23–24 five times until the top organic phase is transparent.
26. Transfer the sample into a 35 ml centrifuge tube. Rinse the tube (Step 24 to 25) twice with 1 volume of sterile H_2O each time to wash any residual plasmid off the tube walls. Transfer the H_2O into the same 35 ml centrifuge tube with the sample.
27. Add 6 volumes of cold 100% ethanol (stored at −20°C). Vortex to mix and keep it at 4°C over night, since CsCl precipitates at −20°C.
28. Centrifuge the sample at 28,300×*g* for 1 h at 4°C using a JA-20 rotor.

29. Rinse the pellet with 10 ml of cold 70% ethanol. Centrifuge at 28,300×*g* for 1 h at 4°C using a JA-20 rotor.
30. Repeat Step 29. Air dry the pellet and resuspend it into 2 ml of TE buffer. A smaller amount of TE buffer can be used to dissolve the pellet to achieve a more concentrated DNA stock solution.
31. Take OD_{260} reading to determine the DNA concentration (bp) using a molar extinction coefficient (bp) ε_M = 6,500/M/cm for dsDNA. Run the saved samples in a 0.8% native agarose–TBE gel to establish the quality of the samples and to monitor the purification results.

3.2. DNA Strand Exchange Reaction

This assay describes a DNA strand exchange reaction catalyzed by the yeast Rad51 protein (**Fig. 21.1a**).

1. Prepare *Pst*I-linearized ϕX174 dsDNA:
   A. Set up a 100 μl restriction digestion reaction by incubating 26 μg ϕX174 RFI dsDNA with 50 units of *Pst*I, in NEB buffer #3 supplemented with 100 μg/ml BSA, at 37°C for 2 h.
   B. After incubation, remove *Pst*I enzyme using the Qiagen PCR Clean Up Kit as described in the manufacturer's instructions.
   C. Measure the nucleic acid absorbance at A_{260} and determine the sample concentration using the molar extinction coefficient (bp) ε_M = 6,500/M/cm at 260 nm.
2. Set up a 12.5 μl reaction by incubating 10 μM Rad51 protein (*see* **Note 5**) with 30 μM (in nucleotides) ϕX174 ssDNA (*see* **Note 6**) in 1× SEB buffer supplemented with 0.1 μg/μl creatine kinase (*see* **Note 7**) and 2.4 mM spermidine, at 30°C for 15 min. Notice that the total reaction volume will be 12.5 μl, and the total volume at this step should be 10 μl.
3. Dilute the RPA stock to add 1.8 μM RPA in 0.5 μl into the reaction. Incubate at 30°C for 30 min.
4. Initiate the reaction by adding 15 μM (in base pairs) *Pst*I-linearized ϕX174 dsDNA and incubate at 30°C for 4 h. If Rad54 is added, 0.2 μM Rad54 is added with the dsDNA. Dilute the stock solution of both ϕX174 dsDNA and Rad54 such that the total added volume is 2 μl. At this point, the reaction is fully assembled with all the substrates and proteins, and the total volume should be 12.5 μl.
5. Stop the reaction by adding 2 μl stop buffer and incubate at 30°C for 30 min.
6. Add 2 μl loading buffer and separate the samples by running a 0.8% agarose gel at a low voltage (25–30 V) for 12–20 h.

7. Stain gel with 0.5 μg/ml ethidium bromide solution for 20 min and destain with H_2O for 10 min (*see* **Note 8**).
8. Place gel on the UV light box in the digital gel imaging system and take image under 300 nm (UV) illumination. Avoid adding ethidium bromide into the agarose gel before pouring in Step 6, which creates uneven background and decreases the sensitivity needed.
9. After gel visualization, quantitate the intensity of substrate and product bands using densitometry with ImageQuant software (Version 5.1). As shown in **Fig. 21.1b**, there are joint molecule (JM) intermediates and nicked circle (NC) products. The yield of DNA strand exchange can be calculated through the equation: product % = (JM/1.5 + NC)/(JM/1.5 + NC + dsDNA).

3.3. D-Loop Formation Assay

This assay describes the formation of D-loops catalyzed by the *S. cerevisiae* Rad51 protein using a short oligo and a homologous supercoiled dsDNA substrate. D-loop formation by yeast Rad51 is dependent on Rad54. The D-loop yield is very time dependent and declines typically after 10 min, presumably caused by the motor activity of Rad54 protein (12).

1. To 5′ end label the 95-mer with ^{32}P, set up a labeling reaction as follows: 0.5 μg of gel-purified 95-mer, 2 μl of 10X T4 polynucleotide kinase reaction buffer, 25 units of T4 polynucleotide kinase, 5 μl of [γ-^{32}P]ATP (3,000 Ci/mmol), and add H_2O to a final volume of 20 μl. Incubate the reaction at 37°C for 1 h and inactivate the kinase by incubating at 65°C for 30 min (*see* **Note 9**).
2. Separate end-labeled oligonucleotides from unincorporated [γ-^{32}P]ATP through a Mini Spin Oligo Column (Roche Applied Science). These spin columns are ready-to-use, disposable, and microcentrifuge compatible. Prepare the column according to the manufacturer instructions and load the sample carefully into the center of the column bed. After centrifugation at 1,000×*g* for 4 min in a microcentrifuge, recover the eluate containing the end-labeled oligonucleotides. Quantitate ^{32}P-labeled 95-mer by counting 1 μl of a 1:10 diluted sample in a scintillation counter and expect between 20 and 100×10^6 cpm (Cerenkov).
3. Set up a 10 μl reaction by incubating 0.67 μM Rad51 (1:3 Rad51/nucleotide) (*see* **Note 6**) with 2 μM 95-mer ssDNA at 30°C for 10 min, in 1× D-loop buffer with 100 ng/μl creatine kinase.
4. Add 0.1 μM RPA into the reaction. Mix and incubate for 10 min at 30°C. The addition of RPA stabilizes and stimulates D-loop formation.

5. Add 100 nM Rad54 into the reaction with 56.6 μM (base pair) supercoiled pUC19 dsDNA to initiate D-loop formation. In the presence of Rad54, time points at 0, 2, 5, 10, and 20 min are usually taken. D-loop yield is very time sensitive (12). Thus, we recommend performing full time courses for each experiment.
6. Stop the reaction at desired time points through deproteinization by adding 2 μl of stop buffer into a 10 μl sample reaction and incubate at 37°C for 20 min.
7. Add 1.5 μl 10× loading dye into each sample and analyze by electrophoresis in a 1% agarose–TBE gel for 2.5 h at 100 V. Further details can be found in **Chapter 19** by Zhang and Heyer in this book.
8. Dehydrate the gel by placing it onto a DE81 membrane, on top of a piece of 3 M Whatman filter paper and a stack of paper towels. Cover the gel with a layer of plastic wrap and put a plastic plate on top with a 1 l bottle on top as extra weight to facilitate dehydration. Allow to dehydrate for no more than 1 h, otherwise bands will diffuse.
9. Place the wrapped gel with DE81 and filter paper underneath into the gel dryer to dry for 60 min at 80°C.
10. Put the dried gel with plastic wrap into the phosphorimage screen cassette for exposure. The exposure time is usually about 1–10 h, based on the specific activity of the isotope.
11. After exposure, place the screen face down into the PhosphorImager system, such as a Storm 860 (Molecular Dynamics), and scan the selected area to obtain the image.
12. Analyze and quantitate the joint molecule (D-loop) yield, using densitometry with a program such as ImageQuant (version 5.1). Calculate the yield as a percentage of the input ssDNA.

3.4. D-Loop Extension Assay

The D-loop extension assay tests the ability of DNA polymerases to prime DNA synthesis from the invading strand in a D-loop formed by Rad51, capturing the essence of recombinational DSB in vitro. The size of the newly synthesized DNA in the extended D-loop can be determined by two-dimensional gel electrophoresis (native/denaturing), as shown in **Fig. 21.2b** (native first dimension) and **c** (denaturing second dimension).

1. The initial steps of the D-loop extension assay are identical to the D-loop assay. Set up a 10 μl reaction by incubating 0.67 μM Rad51 (1:3 Rad51/nucleotide) with 2 μM 95-mer ssDNA at 30°C for 10 min, in 1× D-loop buffer supplemented with 100 ng/μl creatine kinase and 100 μM each of dATP, dGTP, dTTP, and dCTP (*see* **Note 10**).

2. Add 0.1 μM RPA into the reaction. Mix and incubate for 10 min at 30°C. Then add 72 nM Rad54 into the reaction with 56.6 μM (base pair) supercoiled pUC19 dsDNA and incubate for 2 min at 30°C.
3. Add 20 nM RFC and 20 nM PCNA into the reaction and incubate an additional 2 min.
4. To initiate DNA synthesis, add DNA polymerase to 20 nM final assay concentration. Typically, aliquots are sampled at 0, 2, 5, and 10 min.
5. Stop the reaction at desired time points through deproteinization by adding 2 μl of stop buffer into a 10 μl sample reaction and incubate at 30°C for 20 min.
6. As described in **Section 3.3**, analyze samples through electrophoresis using a 1% agarose–TBE gel.
7. For a two-dimensional gel involving denaturing electrophoresis in the second dimension (**Fig. 21.2c**), run the first dimension native agarose–TBE gel as described. After finishing, slice individual gel lanes carefully.
8. Prepare a 1.5% denaturing agarose gel by adding 360 ml H_2O to 6 g of agarose and heating to 95°C until the agarose is completely melted. Add 40 ml 10× denaturing agarose gel buffer and mix well. Equilibrate in a 50°C water bath.
9. In a 4°C cold room, place the cut gel slices at the top of a gel mold. Pour the agarose solution around the slices, taking care to ensure that the slices remain in the desired positions. Allow the gel to solidify.
10. Run the gel in a 4°C cold room at low voltage (about 2 V/cm), for several hours until the bromocresol green dye marker migrates about halfway through gel.
11. Repeat Steps 9–13 in **Section 3.3** for gel handling, image collection, and data analysis.

3.5. DNA Strand Annealing Assay

For DNA strand annealing, we provide two protocols: one is based on radioisotope-labeled substrates, the second on fluorescent dye intercalation (such as DAPI). Both methods will be described separately, and the scheme in **Fig. 21.3** depicts a radioactive substrate. Additionally, several different ssDNA oligos with various lengths and composition can be used as substrates, such as short oligos, poly dT and poly dA, or full-length heat-denatured *Pst*I-linearized pUC19 DNA. In this chapter, we will only discuss the annealing activity of Rad52 protein on short oligo substrates, since they are the most common substrates in use. Details on using longer DNA substrates can be found in reference (9).

3.5.1. For Radioactive Substrates

1. End label gel-purified oligo A2 using [γ-^{32}P]ATP as described in **Section 3.3**. Leave complementary oligo A1 unlabeled, as shown in **Fig. 21.3**.
2. Set up a 60 μl reaction by mixing 200 nM (nt) of radioactive oligo A2 and 200 nM (nt) of unlabeled oligo A1 in the 1× DNA strand annealing buffer containing 30 mM Tris–acetate (pH 7.5), 5 mM magnesium acetate, and 1 mM DTT, in a total current volume of 54 μl (*see* **Note 11**). Assemble the reaction on ice to minimize spontaneous annealing between the complementary oligos.
3. Add 30 nM RPA and incubate at 30°C for 15 min (*see* **Note 11**). The addition of saturating or over-saturating amounts of RPA decreases the spontaneous annealing between complementary short oligos efficiently. One RPA heterotrimer per 25 nts is considered to be a saturating amount.
4. Initiate the reaction by adding 20 nM Rad52 protein and incubate at 30°C. Keep the total volume added for RPA and Rad52 to 6 μl, thereby reaching a 60 μl final reaction volume.
5. At 2, 4, 6, 8, and 10 min, remove a 10 μl sample aliquot and quench it rapidly by mixing it with 3 μl stop buffer. Incubate the mixture for 15 min at 30°C. For the zero-time point sample, proteins and DNA are mixed in the stop buffer.
6. Mix the sample with 1 μl of 10× loading dye. Pre-run a 1 mm-thick 10% polyacrylamide gel in TBE buffer at 100 V for 20 min to eliminate ammonium persulfate (APS) in the gel, which might interfere with the electrophoresis of DNA oligonucleotides.
7. After the pre-run, load the samples into the wells and start electrophoresis at 100 V for 1–2 h to separate the radioactive annealed product from the substrates.
8. For the 1 mm-thick polyacrylamide gel, there is no need to dehydrate. Place the gel onto a DE81 membrane on top of a 3 M Whatman filter paper and cover the gel with a layer of plastic wrap. Place the wrapped gel with filter paper underneath into the gel dryer to dry for 60 min at 80°C.
9. Place the dried gel with plastic wrap into the phosphorimage screen cassette for exposure. The exposure time is usually about 1–10 h, based on the specific activity of the isotope.
10. After exposure, put the screen face down into the PhosphorImager system, such as a Storm 860 (Molecular Dynamics), and scan the selected area to obtain the image.

11. Analyze and quantitate the annealed duplex and ssDNA oligonucleotide substrates using densitometry with a program such as ImageQuant (version 5.1). Calculate the annealing efficiency as a percentage of the annealed duplex to the input ssDNA amount.

3.5.2. For Unlabeled Substrates Using DAPI Dye

1. This protocol describes a real-time and quantitative assay of DNA annealing by Rad52 protein, based on the unique fluorescent property of DAPI. DAPI binds to the minor groove of dsDNA specifically and exhibits an enhanced fluorescent signal at 467 nm, when excited at 345 nm (13, 14). The limitation of this method is the requirement of a spectrofluorimeter and the demand for a larger quantity of protein, compared to the assay as performed with radioactively labeled substrate.
2. Assemble a 400 μl reaction by adding 200 nM oligo A1 and 200 nM oligo A2 in the 1× DNA strand annealing buffer containing 30 mM Tris–acetate (pH 7.5), 5 mM magnesium acetate, and 1 mM DTT with 0.2 μM DAPI into the cuvette (*see* **Note 11**).
3. Set the excitation and emission wavelengths of the SLM8000 spectrofluorimeter to 345 and 467 nm, respectively. Set the slit widths for excitation and emission light to 1 and 4 mm, respectively. The DAPI fluorescence signal is proportional to the dsDNA concentration up to 10 μM (bp) at a 0.2 μM DAPI concentration (9).
4. Place the cuvette into the holder of the SLM8000 spectrofluorimeter and adjust the temperature control to 30°C.
5. Add 30 nM of RPA (*see* **Note 11**) and incubate at 30°C for 15 min. The addition of saturating amounts of RPA decreases the spontaneous annealing between complementary short oligonucleotides efficiently.
6. Initiate the reaction by adding 20 nM Rad52 protein to the reaction at 30°C and begin to record the signal at 10 or 20 s intervals continuously with the above setting for 10 min. Usually, there is a 3–5 s delay between mixing Rad52 into reaction solution and starting data collection. The total volume added for RPA and Rad52 should be equal to 10 μl to reach a 400 μl final reaction volume.
7. The increase in DAPI fluorescence signal reflects the annealing of complementary ssDNA oligonucleotides and the formation of dsDNA products. To calculate the percentage of annealing, divide the dsDNA formed by the total input DNA. Define background fluorescence signal on ssDNA substrates as 0%, and maximum fluorescence increases over ssDNA background on fully annealed dsDNA product as 100%.

8. Plot the percentage of annealing over time to demonstrate the annealing role of Rad52, compared to spontaneous annealing in the no protein control.

4. Notes

1. The presence of free Mg^{2+} is critical, since Mg^{2+}-ATP is the substrate of the RecA/Rad51-like proteins. For each enzyme, the optimum concentration of free Mg^{2+} is different. Ususally, a 1–5 mM free Mg^{2+} is used for the yeast and human Rad51 proteins. For the human RAD51 protein, the presence of Ca^{2+} stimulates the DNA strand exchange activity dramatically, compared to Mg^{2+} alone, by locking hRAD51 into the active, ATP-bound configuration (15). When using Ca^{2+}, its possible effect on other reaction components must be considered.
2. Store purified proteins at −80°C. Thaw and dilute immediately before use, to avoid loss of activity. Frequent freeze/thaw cycles can decrease protein activity. To minimize this loss, aliquot purified protein stocks into small volumes prior to storage. For protocols to ensure the absence of relevant contamination in preparations of HR proteins, *see* **Chapter 19** by Zhang and Heyer (this volume).
3. Yeast Rad51 prefers linearized ϕX174 dsDNA with 3′ overhang (e.g., *Pst*I-linearized) as the substrate in the DNA strand exchange assay, whereas human RAD51 prefers linearized ϕX174 dsDNA with 5′ overhangs (e.g., *Apa*L1-linearized). Furthermore, while *E. coli* RecA and bacteriophage T4 UvsX catalyze efficient DNA strand exchange using phage M13mp18 DNA substrates, the efficiency of DNA strand exchange with M13mp18 substrates is poor when using eukaryotic Rad51 proteins. The reasons for this substrate preference are unknown. Furthermore, the end products in DNA strand exchange for these proteins are quite different as well. RecA catalyzes the DNA strand exchange reaction highly efficiently, accumulating nicked circles as the final product. UvsX catalyzes the same reaction very fast and through multiple invasions, leading to the formation of large DNA networks (shown as aggregates in the well) as the major product. Yeast Rad51 catalyzes DNA strand exchange much less efficiently than RecA, accumulating joint molecule intermediates and fewer nicked circle products. Finally, human Rad51 protein is even less efficient, since nicked circle formation is very low and joint molecules are the major species.

4. After the addition of ethidium bromide, the sample tube should be wrapped with foil and work should be performed in low light until after the ethidium bromide is removed, to avoid nicking the DNA. Be sure to wear gloves for the steps involving ethidium bromide and CsCl. Ethidium bromide is a mutagen and carcinogen, and direct contact with it should be avoided. Cesium is a heavy metal, and exposure to it should be avoided. If there is a spill, rinse the contaminated area with a large amount of H_2O.
5. The addition of corresponding protein storage buffer to match the ionic strength in the no protein control to the ionic strength of the experimental samples is critical. The D-loop and DNA strand exchange assays catalyzed by RecA/Rad51 family proteins are very sensitive to the ionic strength. Artifacts can be avoided by ensuring that the actual buffer components in the reaction are identical.
6. The ratio between Rad51 protein and DNA is key for high yield in D-loop and DNA strand exchange reactions. The nucleotide binding site size of Rad51 is $n = 3$, which means one Rad51 binds three nucleotides in a saturated nucleoprotein filament. Higher or lower Rad51 to ssDNA ratios decrease the yields in D-loop and DNA strand exchange reactions. However, suboptimal conditions (non-optimal Rad51:ssDNA ratio, elevated salt concentration, presence of RPA inhibition, order of addition changes) can reveal the functions and mechanisms of ancillary proteins. For example, Rad54 can significantly boost the DNA strand exchange activity of sub-saturating amounts of Rad51 (Rad51/nucleotide = 1:14) (3). Unless otherwise specified, the nucleic acid concentrations in this protocol are given in nt for ssDNA and bp for dsDNA.
7. The ATP regenerating system is based on creatine kinase and phosphocreatine. Other regenerating systems can be used, such as pyruvate kinase and phosphoenolpyruvate (PEP). Commercial kinases are sometimes supplied as an ammonium sulfate suspension in buffer. This buffer might contain salts or chemicals that inhibit D-loop formation and DNA strand exchange activity. To avoid this, centrifuge a small aliquot of the kinase before each use and resuspend the kinase pellet in the corresponding reaction buffer.
8. The ethidium bromide staining solution is stable for several months at 4°C. If the dye concentration is low in this staining solution, add 10 μl of 10 mg/ml ethidium bromide solution into the staining solution before each new staining. In this way, ethidium bromide waste is minimized.

ssDNA does not stain well by ethidium bromide, and a brief destaining for only 10 min will enhance the ssDNA signal.

9. We recommend to order PAGE-purified oligonucleotides or to purify the oligonucleotide using 10–16% polyacrylamide gels containing 8 M urea. For all steps below involving radioactive materials, wear double gloves throughout and work on designated radioactive benches. All waste must be monitored. Discard all radioactive waste into the designated solid or liquid waste containers for ^{32}P.

   It is important to heat inactivate polynucleotide kinase, to avoid subsequent inhibition of the D-loop formation by the binding of kinase onto labeled 95-mer oligonucleotides and labeling of nicked or linearized plasmid during D-loop formation and synthesis.

10. In the D-loop extension assay, it is important to consider the K_M for dNTPs of the individual DNA polymerase. For example, translesion synthesis DNA polymerase η requires at least 100 μM each dNTP for efficient synthesis, while DNA polymerase δ can extend efficiently at a tenfold lower concentration. If [α-^{32}P]dCTP (0.22 μM, 6,000 uCi/mmol, Perkin Elmer) is used instead of labeled 95-mer to increase sensitivity, be aware that some DNA polymerases are highly sensitive to unbalanced dNTPs pools.

11. The presence of Mg^{2+} is key to condense the ssDNA structure and suppress DNA breathing at the dsDNA ends. Thus, fewer artifacts from DNA annealing will be introduced through the initial protein–DNA binding and the following deproteinization treatment. The presence of saturating amount of RPA (1 RPA/25 nts) better mimics the physiologically relevant situation and avoids artifacts caused by DNA condensation (protein-mediated aggregation) and spontaneous annealing.

Acknowledgments

We thank Kirk Ehmsen, William Wright, Clare Fasching, Ryan Janke, Erin Schwartz, Shannon Ceballos, Damon Meyer, Xiao-Ping Zhang, and Margarita Alexeeva for helpful comments on the manuscript. Our work is supported by the NIH (GM58015, CA92276), the DoD (BC083684), an NIH training grant fellowship (5T32CA108459) to J.S., and a TRDRP Postdoctoral fellowship (17FT-0046) to J.L.

References

1. Li, X., and Heyer, W.D. (2008) Homologous recombination in DNA repair and DNA damage tolerance. *Cell Res* **18**, 99–113.
2. Krogh, B.O., and Symington, L.S. (2004) Recombination proteins in yeast. *Annu Rev Genet* **38**, 233–271.
3. Solinger, J.A., Kiianitsa, K., and Heyer, W.-D. (2002) Rad54, a Swi2/Snf2-like recombinational repair protein, disassembles Rad51:dsDNA filaments. *Mol Cell* **10**, 1175–1188.
4. Prakash, R., Satory, D., Dray, E., Papusha, A., Scheller, J., Kramer, W., Krejci, L., Klein, H., Haber, J.E., Sung, P., and Ira, G. (2009) Yeast Mph1 helicase dissociates Rad51-made D-loops: implications for crossover control in mitotic recombination. *Genes Dev* **23**, 67–79.
5. New, J.H., Sugiyama, T., Zaitseva, E., and Kowalczykowski, S.C. (1998) Rad52 protein stimulates DNA strand exchange by Rad51 and replication protein A. *Nature* **391**, 407–410.
6. Li, X., Stith, C.M., Burgers, P.M., and Heyer, W.-D. (2009) PCNA is required for initiating recombination-associated DNA synthesis by DNA polymerase . *Mol Cell* **36**, 704–713.
7. Sugiyama, T., Kantake, N., Wu, Y., and Kowalczykowski, S.C. (2006) Rad52-mediated DNA annealing after Rad51-mediated DNA strand exchange promotes second ssDNA capture. *EMBO J* **25**, 5539–5548.
8. Wu, Y., Kantake, N., Sugiyama, T., and Kowalczykowski, S.C. (2008) Rad51 protein controls Rad52-mediated DNA annealing. *J Biol Chem* **283**, 14883–14892.
9. Sugiyama, T., New, J.H., and Kowalczykowski, S.C. (1998) DNA annealing by Rad52 Protein is stimulated by specific interaction with the complex of replication protein A and single-stranded DNA. *Proc Natl Acad Sci USA* **95**, 6049–6054.
10. Shinohara, A., and Ogawa, T. (1998) Stimulation by Rad52 of yeast Rad51-mediated recombination. *Nature* **391**, 404–407.
11. Sung, P. (1997) Function of yeast Rad52 protein as a mediator between replication protein A and the Rad51 recombinase. *J Biol Chem* **272**, 28194–28197.
12. Bugreev, D.V., Hanaoka, F., and Mazin, A.V. (2007) Rad54 dissociates homologous recombination intermediates by branch migration. *Nature Struct Mol Biol* **14**, 746–753.
13. Kapuscinski, J., and Szer, W. (1979) Interactions of 4′, 6-diamidine-2-phenylindole with synthetic polynucleotides. *Nucleic Acids Res* **6**, 3519–3534.
14. Kubista, M., Akerman, B., and Norden, B. (1987) Characterization of interaction between DNA and 4′,6-diamidino-2-phenylindole by optical spectroscopy. *Biochemistry* **26**, 4545–4553.
15. Bugreev, D.V., and Mazin, A.V. (2004) Ca^{2+} activates human homologous recombination protein Rad51 by modulating its ATPase activity. *Proc Natl Acad Sci USA* **101**, 9988–9993.

Chapter 22

An In Vitro Assay for Monitoring the Formation and Branch Migration of Holliday Junctions Mediated by a Eukaryotic Recombinase

Yasuto Murayama and Hiroshi Iwasaki

Abstract

DNA strand exchange is a core reaction of homologous recombination directly catalyzed by Rad51/Dmc1 RecA family recombinases in eukaryotes. This reaction proceeds through the formation of several DNA intermediates. The X-shaped four-way DNA structure known as a Holliday junction (HJ) is a central intermediate in homologous recombination. Genetic and biochemical studies indicate that the HJ is important for the production of crossover-type recombinants, which are reciprocal exchange products. According to a recombination model for the repair of DNA double-strand breaks, the formation of HJs requires a reciprocal duplex–duplex DNA exchange known as the DNA four-strand exchange reaction. In vitro analyses using purified recombination proteins and model DNA substrates provide a mechanistic insight into the DNA strand exchange reaction, including the steps leading to the formation and branch migration of Holliday junctions.

Key words: Homologous recombination, Holliday junction, Rad51 recombinase, DNA strand exchange, gel electrophoresis.

1. Introduction

The eukaryotic RecA family recombinases Rad51 and Dmc1 bind single-stranded DNA and form a nucleoprotein filament known as the presynaptic filament in an ATP-dependent manner (1–3). In this process, several accessory proteins known as recombination mediators stimulate the assembly of the filament (4, 5). The filament pairs with its homologous, intact duplex DNA and exchanges with its complementary strand, forming the first-strand

H. Tsubouchi (ed.), *DNA Recombination*, Methods in Molecular Biology 745,
DOI 10.1007/978-1-61779-129-1_22, © Springer Science+Business Media, LLC 2011

exchange intermediate known as the D-loop. The ssDNA-binding protein complex replication protein A (RPA) facilitates the reaction, both by eliminating secondary structures of ssDNA and by preventing the reverse reaction of strand exchange (6). The 3′-end of invaded ssDNA is used as the primer for repair DNA synthesis (7). Since ssDNA regions exist at the end of duplex DNA (or double-stranded DNA; dsDNA), the initial DNA strand exchange finally reaches the ss–ds junction of the invading DNA molecule. A Holliday junction (HJ) is formed if the DNA strand exchange proceeds over the junction and is converted from a ss–dsDNA exchange (three-strand exchange) to a duplex–duplex reciprocal exchange (four-strand exchange) (**Fig. 22.1**). This chapter describes one example of an in vitro assay for DNA strand exchange mediated by the fission yeast *Schizosaccharomyces pombe* Rad51 ortholog Rhp51 recombinase.

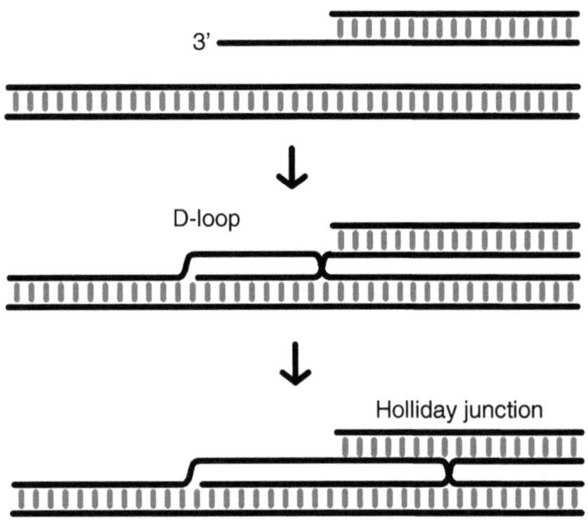

Fig. 22.1. A model of HJ formation in homologous recombination-mediated DNA DSBs repair. DNA strand exchange is initiated between ssDNA generated by the processing of nucleases and its homologous, intact duplex DNA. An HJ is formed if the strand exchange proceeds over the ss–ds DNA junction and is converted to duplex–duplex reciprocal exchange.

2. Materials

2.1. Culture Media and Supplements

1. M9 minimal medium: 0.6% (w/v) Na_2HPO_4, 0.3% KH_2PO_4, (w/v), 0.05% NaCl, 0.1% NH_4Cl, 0.1 mM $MgSO_4$, 0.1 mM $CaCl_2$, 0.2% (w/v) glucose. To make the solid medium, 1.5% (w/v) agar is added.
2. LB medium: 1% (w/v) tryptone, 0.5% (w/v) yeast extract, 0.5% (w/v) NaCl, adjust to pH 7.0 with NaOH. To make the solid medium, 1.5% (w/v) agar is added.

3. 2× YT medium: 1.6% (w/v) tryptone, 1% (w/v) yeast extract, 0.5% (w/v) NaCl adjusted to pH 7.5.
4. Ampicillin: A stock solution (100 mg/ml) is prepared by dissolving ampicillin powder in sterilized water. The working concentration consists of 1/1,000 of the culture volume.
5. Isopropyl β-D-1-thiogalactopyranosid (IPTG): A stock solution (1 M) is prepared by dissolving IPTG powder in sterilized water.

2.2. Protein Expression Plasmids (See Notes 1 and 2)

1. Rhp51: pET11b-Rhp51 [*rhp51*$^+$] cDNA is inserted into pET11b (Novagen) (8).
2. Swi5–Sfr1: pBKN220-Swi5–Sfr1 (*swi5*$^+$ and *sfr1*$^+$ cDNAs are inserted into a pET21a derivative high copy number plasmid, pBKN220) (9).
3. RPA: pET11b-RPA (the three cDNAs of the fission yeast RPA complex are inserted into pET11b as an operon array lined with subunits 2, 3, and 1) (9).

2.3. Protein Purification

1. P buffer: 20 mM potassium phosphate (pH 7.5), 1 mM dithiothreitol (DTT), 0.5 mM EDTA, 10% (w/v) glycerol.
2. R buffer: 20 mM Tris-HCl (pH 8.0), 1 mM DTT, 0.5 mM EDTA, 10% (w/v) glycerol.
3. 0.5% (w/v) polyethyleneimine (PEI): Dissolve 50% (w/v) polyethyleneimine (Sigma) in water (*see* **Note 3**) and adjust to pH 8.0 with HCl. Filter with a 0.22 μm pore membrane and store at 4°C.
4. A complete protease inhibitor cocktail (Roche Diagnostics).
5. An SP Sepharose FF column: Gel slurry of SP Sepharose FF (GE Healthcare) is packed into an XK16/20 or XK26/20 column (GE Healthcare) following the manufacturer's instructions.
6. A Q Sepharose FF column: Gel slurry of Q Sepharose FF (GE Healthcare) is packed into an XK16/20 or XK26/20 column following the manufacturer's instructions.
7. A prepacked Resource Q column (1 ml) (GE Healthcare).
8. A prepacked HiTrap heparin HP column (5 ml) (GE Healthcare).
9. A HiLoad 16/60 Superdex 200 pg column (GE Healthcare).
10. Dialysis membrane with a molecular weight cutoff (MWCO) of 10,000 (*see* **Note 4**).

2.4. Preparation of Substrate DNAs

1. PEG/NaCl solution: 20% (w/v) PEG6000, 2.5 M NaCl.
2. TE buffer: 10 mM Tris-HCl (pH 7.5), 0.1 mM EDTA.

3. CsCl: ultra-centrifugation analysis grade.
4. Protease K: 20 mg/ml protease K (Roche Diagnostics).
5. 10% SDS solution.
6. TE-saturated phenol.
7. Phenol/chloroform/isoamylalcohol (25:24:1).
8. Solution I: 25 mM Tris-HCl (pH 8.0), 10 mM EDTA, 50 mM glucose.
9. Solution II: 0.2 N NaOH, 1% (w/v) SDS.
10. Solution III: mix 60 ml of 5 M potassium acetate, 11.5 ml glacial acetic acid, and 28.5 ml H_2O.
11. NaCl-saturated isopropanol.
12. 5× annealing buffer: 125 mM Tris-HCl (pH 8.0), 1 M NaCl, 10 mM $MgCl_2$.
13. TAE buffer (pH 8.0): 40 mM Tris-acetate (pH 8.0), 1 mM EDTA.
14. 6× DNA loading buffer: 25 mM Tris-HCl (pH 7.5), 0.25% (w/v) bromophenol blue, 0.25% (w/v) xylene cyanol, 15% (w/v) ficol.
15. Ultrafiltration membrane: Microcon Ultracel Y-100 (Millipore).
16. SYBR Gold (Molecular Probes).

2.5. DNA Four-Strand Exchange Reaction Assay

1. 5× Reaction buffer F: 150 mM Tris-HCl (pH 7.5), 5 mM DTT, 750 mM NaCl, 50 mM $MgCl_2$, 12.5% (w/v) glycerol, 10 mM ATP, 40 mM creatine phosphate, 40 U/ml creatine kinase. The final concentrations are as follows: 30 mM Tris-HCl (pH 7.5), 1 mM DTT, 150 mM NaCl, 10 mM $MgCl_2$, 2.5% (w/v) glycerol, 2 mM ATP, 8 mM creatine phosphate, 8 U/ml creatine kinase.
2. Stop solution: 120 mM Tris-HCl (pH 7.5), 120 mM EDTA, 3% (w/v) SDS, 4.8 mg/ml protease K.
3. TAE buffer (pH 7.6): 40 mM Tris-acetate (pH 7.6), 1 mM EDTA.
4. 6× DNA loading buffer: See above.
5. Psoralen: Stock solution is 6 mg/ml in ethanol. Store at –20°C under protection from light. A volume of 0.6 mg/ml psoralen solution is prepared by diluting the 6 mg/ml stock solution with water.

2.6. RuvC Cleavage

1. RuvC solution: RuvC protein (Bioacademia, Osaka, Japan) is diluted with an RuvC buffer.
2. RuvC buffer: 30 mM Tris-HCl (pH 7.5), 1 mM DTT, 15 mM $MgCl_2$, 2.5% (w/v) glycerol.

3. G-25 Spin Column: Illusta MicroSpin G-25 Column (GE Healthcare) is pre-equilibrated with RuvC buffer.

3. Methods

Rhp51 is the *S. pombe* ortholog of Rad51 recombinase, which functions in both mitosis and meiosis (10). The Swi5–Sfr1 complex functions as a mediator of Rhp51 by stabilizing and activating the Rhp51 filament (9, 11, 12). This complex does not facilitate the formation and branch migration of HJ by Rhp51, but it stimulates the formation of initial three-stranded intermediates named sigma structures (see below) (8). RPA is needed for full efficiency of the three-strand exchange reaction by the mechanism mentioned in **Section 1** (6, 9). Bacterial ssDNA-binding protein (SSB) can be used as a substitute for RPA in the in vitro three-strand exchange reaction. RuvC from *Escherichia coli* is a DNA endonuclease that specifically resolves the Holliday structure (13, 14) and it is used as an indicator of HJ formation in this assay (8).

3.1. Protein Expression

1. An *E. coli* BL21-CodonPlus (DE3)-RIPL strain is transformed with the expression plasmids for Rhp51 or the Swi5–Sfr1 complex. BL21 (DE3) strain is also transformed with the RPA expression plasmid.
2. Approximately 20 fresh transformant colonies are scratched from the medium plates and inoculated into LB liquid medium containing 100 μg/ml ampicillin (1 l × 5 cultures).
3. The cultures are grown at 37°C with shaking (130–200 rpm).
4. When the optical density of the culture, measured at 600 nm, reaches 0.5, protein expression is induced by the addition of IPTG to a final concentration of 1 mM for Swi5–Sfr1 or RPA, and 0.5 mM for Rhp51.
5. The cultures are then incubated at 18°C with shaking at 130–200 rpm overnight (12–18 h) (*see* **Note 5**).
6. The cells are collected by centrifugation and washed with 0.9% NaCl. The cell pellets are frozen in liquid nitrogen and can be stored at –80°C for several months before use.

3.2. Rhp51 Purification (See Note 6)

1. Resuspend the cell pellets from a 5 l culture in 100 ml of P buffer containing 0.3 M KCl + a complete protease inhibitor cocktail (Roche Diagnostics).
2. Disrupt the cells thoroughly by sonication (*see* **Note 7**).

3. Centrifuge the crude cell extract at 40,000×g for 1 h.
4. Add ammonium sulfate to the supernatants at a final saturation concentration of 35% and stir for 1 h.
5. Centrifuge at 15,000×g for 30 min to precipitate the proteins in the cell lysate.
6. Discard the supernatant.
7. Dissolve the pellets by adding 20 ml of P buffer containing 0.5 M KCl. After dissolving the pellets completely, dilute the protein solution with 80 ml of P buffer to a final concentration of 0.1 M KCl.
8. Apply the protein sample to an SP Sepharose FF column (20 ml bed volume) (*see* **Note 8**) that is pre-equilibrated with P buffer containing 0.1 M KCl (*see* **Note 9**).
9. Apply the flow-pass fraction to a Q Sepharose FF column (30 ml bed volumes) that is pre-equilibrated with P buffer containing 0.1 M KCl.
10. Wash the column with 30 ml (1-bed volume) of P buffer containing 0.1 M KCl.
11. Elute the proteins with a linear gradient of 0.1–0.8 M KCl in P buffer (10-bed volume, 300 ml). Rhp51 is eluted at approximately 0.45 M KCl.
12. Pool the peak top fractions (20 ml) and dilute them with 70 ml of P buffer (4.5-fold) to a final concentration of 0.1 M KCl.
13. Apply the diluted peak top fraction to a HiTrap Heparin HP column (5 ml bed volume).
14. Wash the column with 25 ml (5-bed volumes) of P buffer containing 0.1 M KCl.
15. Elute the proteins with a linear gradient of 0.1–0.8 M KCl in P buffer (75 ml or 15-bed volumes). Rhp51 is eluted at approximately 0.35 M KCl.
16. Pool the peak top fractions (5 ml) and dilute them with 12.5 ml of P buffer (3.5-fold) to a final concentration of 0.1 M KCl.
17. Apply the diluted peak top fractions to a Resource Q column (1 ml) (GE Healthcare) that is pre-equilibrated with P buffer containing 0.1 M KCl.
18. Wash the column with 10 ml (10-bed volumes) of P buffer containing 0.1 M KCl.
19. Elute the proteins with a linear gradient of 0.1–0.6 M KCl in P buffer (50-bed volumes, 50 ml). Rhp51 is eluted at approximately 0.45 M KCl.
20. Dialyze the peak top fractions (2 ml) against 2 l of P buffer containing 0.2 M KCl for 5 h, with two buffer changes.

21. Determine the Rhp51 concentration by measuring absorbance at 280 nm with an extinction coefficient of 1.86×10^4 M/cm. Approximately 5 mg of Rhp51 are routinely obtained with this procedure.
22. Freeze aliquots in liquid nitrogen and store at –80°C.
23. The progression of Rhp51 purification is summarized in **Fig. 22.2**.

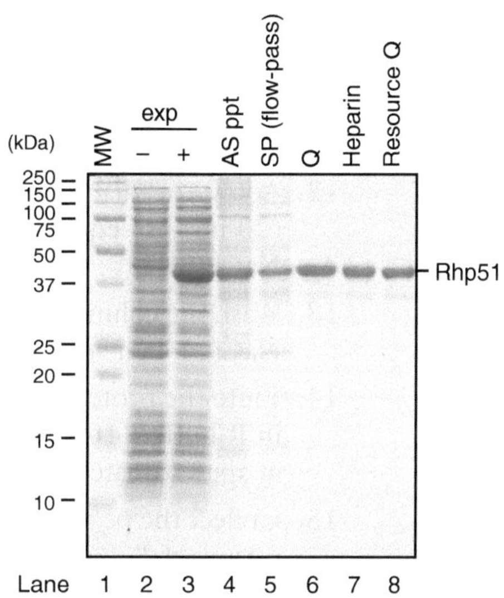

Fig. 22.2. Overproduction and purification of Rhp51 protein. Samples were separated by 12.5% SDS-PAGE and stained with CBB. *Lane 1*, molecular size markers. *Lane 2*, total extracts of IPTG-induced cells of BL21-CodonPlus (DE3)-RIPL harboring the empty vector pET11b. *Lane 3*, total extracts of IPTG-induced cells of BL21-CodonPlus (DE3)-RIPL harboring the Rhp51$^+$ plasmid on pET11b (pET11b-Rhp51). *Lane 4*, a protein sample after step 7 (~5 μg), which corresponds to a fraction after ammonium sulfate precipitation. *Lane 5*, the flow-pass fraction (~2 μg) of SP Sepharose FF chromatography (a sample after step 8). *Lane 6*, the fraction (~2 μg) after Q Sepharose FF chromatography (a sample after step 12). *Lane 7*, the fraction of HiTrap heparin chromatography (a sample after step 16). *Lane 8*, the final fraction (~2 μg) after Resource Q chromatography.

3.3. Purification of the Swi5–Sfr1 Complex

1. Resuspend the cell pellets from 2 l cultures in 100 ml of R buffer containing 0.2 M NaCl and a protease inhibitor cocktail (Complete Inhibitor Cocktail, Roche Diagnostics).
2. Disrupt the cells thoroughly by sonication.
3. Centrifuge the crude cell extract at 40,000*g* for 1 h.
4. Gradually add a 0.5% (w/v) PEI solution to the supernatants with a stirring bar to a final concentration of 0.05% (w/v) and continue stirring gently for 1 h.
5. Centrifuge at 15,000×*g* for 30 min.

6. Add ammonium sulfate to the supernatants at a final saturation concentration of 35% and stir for 1 h.
7. Centrifuge at 15,000×*g* for 30 min to precipitate proteins in the cell lysate.
8. Discard the supernatant.
9. Wash the pellets with R buffer containing ammonium sulfate (50% saturation) and 0.1 M NaCl.
10. Dissolve the pellets in 10 ml of R buffer containing 1 M NaCl.
11. Dialyze the protein sample against 1 l of R buffer + 0.1 M NaCl for 3 h, followed by a second dialysis against 1 l of R buffer containing 0.05 M NaCl for 3 h.
12. Apply the dialyzed sample to a Q Sepharose FF column (20 ml) pre-equilibrated with R buffer containing 0.05 M NaCl.
13. Wash the column with 20 ml of R buffer containing 0.05 M NaCl (1-bed volume).
14. Elute the proteins with a linear gradient of 0.05–1 M NaCl in R buffer (10-bed volumes, 200 ml). Swi5–Sfr1 is eluted at approximately 0.3 M KCl.
15. Collect the peak top fractions (20 ml) and dilute them with 60 ml of R buffer (threefold) to a final concentration of 0.1 M NaCl.
16. Apply the diluted peak top fractions to a HiTrap Heparin HP column (5 ml).
17. Wash the column with 25 ml of R buffer containing 0.1 M NaCl (5-bed volumes).
18. Elute the proteins with a linear gradient of 0.1–1 M NaCl in R buffer (20-bed volumes, 100 ml). The Swi5–Sfr1 complex is eluted at approximately 0.4 M NaCl.
19. Apply the peak top fractions (5 ml) to a HiLoad 16/60 Superdex 200 pg column (120 ml) pre-equilibrated with R buffer containing 1.5 M NaCl and elute with the same buffer.
20. Dialyze the peak top fractions (4 ml) against 2 l of 0.2 M NaCl in R buffer for 5 h.
21. Determine the concentration of the Swi5–Sfr1 complex by measuring absorbance at 280 nm with the extinction coefficient 1.44×10^4 M/cm. This procedure yields approximately 5 mg of Swi5–Sfr1.
22. Freeze the aliquots in liquid nitrogen and store at –80°C.
23. The progression of Swi5–Sfr1 complex purification is summarized in **Fig. 22.3**.

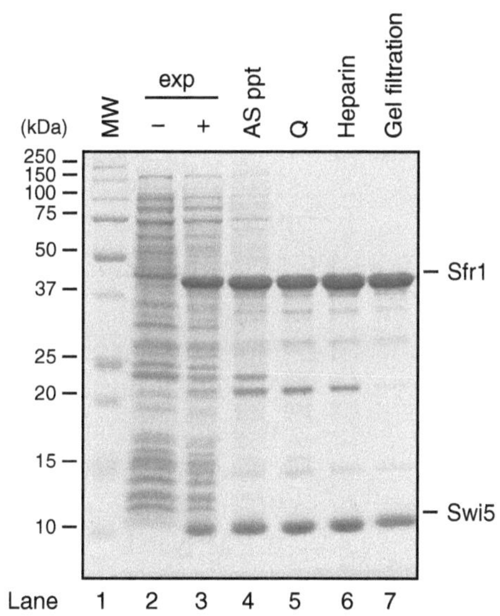

Fig. 22.3. Overproduction and purification of Swi5–Sfr1 protein complex. Samples were separated by 12.5% SDS-PAGE and stained with CBB. *Lane 1*, molecular size markers. *Lane 2*, total extracts of IPTG-induced cells of BL21-CodonPlus (DE3)-RIPL harboring the empty vector pBKN220. *Lane 3*, total extracts of IPTG-induced cells of BL21-CodonPlus (DE3)-RIPL harboring the Swi5–Sfr1 plasmid on pBKN220 (pBKN220-Swi5–Sfr1). *Lane 4*, a protein sample after step 11 (~5 μg), which corresponds to a fraction after ammonium sulfate precipitation. *Lane 5*, the fraction (~2 μg) after Q Sepharose FF chromatography (a sample after step 15). *Lane 6*, the fraction (~2 μg) after HiTrap heparin chromatography (a sample after step 18). *Lane 7*, the final fraction (~2 μg) after a HiLoad 16/60 superdex 200 pg chromatography.

3.4. Purification of RPA

1. Resuspend the cell pellets from 5 l cultures in 100 ml of R buffer containing 0.3 M NaCl and a protease inhibitor cocktail (Complete).
2. Disrupt the cells thoroughly by sonication.
3. Centrifuge the crude cell extract at 40,000×*g* for 1 h.
4. Gradually add 0.5% (w/v) PEI solution to the supernatants with a stirring bar to a final concentration of 0.05% (w/v) and continue to stir gently for 1 h.
5. Centrifuge at 15,000×*g* for 30 min.
6. Add ammonium sulfate to the supernatants at a final saturation concentration of 40% and stir for 1 h.
7. Centrifuge at 15,000×*g* for 30 min to precipitate proteins in the cell lysate.
8. Discard the supernatant.
9. Wash the pellets with R buffer containing ammonium sulfate (50% saturation) and 0.1 M NaCl.
10. Dissolve the pellets in 10 ml of R buffer containing 0.5 M NaCl.

11. Dialyze the protein sample against 1 l of R buffer containing 0.1 M NaCl for 3 h, followed by a second dialysis against 1 l of R buffer containing 0.05 M NaCl for 3 h.
12. Apply the dialyzed solution to an SP Sepharose FF column (50 ml) that is pre-equilibrated with R buffer containing 0.05 M NaCl.
13. Wash the column with 50 ml of R buffer containing 0.05 M NaCl (1-bed volume).
14. Elute the proteins with a linear gradient of 0.05–0.6 M NaCl in R buffer (10-bed volumes, 500 ml). RPA is eluted at approximately 0.3 M NaCl.
15. Collect the peak top fractions (30 ml) and dilute them with 90 ml of R buffer (threefold) to a final concentration of 0.1 M NaCl.
16. Apply the diluted peak top fractions to a HiTrap Heparin HP column (5 ml).
17. Wash the column with 25 ml of R buffer containing 0.1 M NaCl (5-bed volumes).
18. Elute the proteins with a linear gradient of 0.1–0.8 M NaCl in R buffer (15-bed volumes, 75 ml). RPA is eluted at approximately 0.5 M NaCl.
19. Apply the peak top fractions (5 ml) to a HiLoad 16/60 Superdex 200 pg column (120 ml) pre-equilibrated with R buffer containing 1 M NaCl and elute with the same buffer.
20. Dialyze the peak top fractions (4 ml) against 2 l of R buffer containing 0.1 M NaCl for 5 h with two buffer changes.
21. Determine the concentration of the RPA complex by measuring absorbance at 280 nm with an extinction coefficient of 9.98×10^4 M/cm. This procedure yields approximately 5 mg of RPA.
22. Freeze the aliquots in liquid nitrogen and store at –80°C.
23. The progression of RPA purification is summarized in **Fig. 22.4**.

3.5. Preparation of DNA Substrates

Circular single-stranded DNA (cssDNA) of pSKsxAS+ is used for the preparation of gapped circular DNA (gDNA) for the DNA four-strand exchange assay. pSKsxAS+ is a 4.3 kbp plasmid DNA that is a pBluscript SKII derivative (Stratagene) made by inserting the *ade6*$^+$ gene fragment (8).

3.5.1. Preparation of Circular Single-Stranded DNA (cssDNA)

1. Introduce pSKsxAS+ into the *E. coli* strain JM103 to allow colony formation on LB solid medium containing 100 μg/ml ampicillin (*see* **Note 10**).

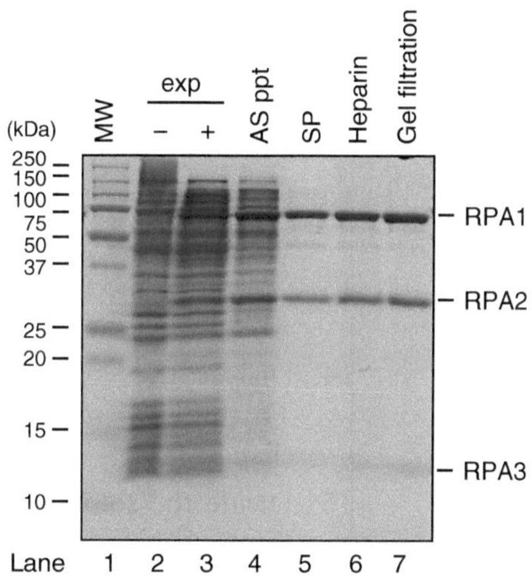

Fig. 22.4. Overproduction and purification of RPA complex. Samples were separated by 15% SDS-PAGE and stained with CBB. *Lane 1*, molecular size markers. *Lane 2*, total extracts of IPTG-induced cells of BL21(DE3) harboring the empty vector pET11b. *Lane 3*, total extracts of IPTG-induced cells of BL21(DE3) harboring an RPA plasmid on pET11b (pET11b-RPA). *Lane 4*, a protein sample after step 11 (~5 μg), which corresponds to a fraction after ammonium sulfate precipitation. *Lane 5*, the fraction (~2 μg) of SP Sepharose FF chromatography (a sample after step 15). *Lane 6*, the fraction (~2 μg) after HiTrap heparin chromatography (a sample after step18). *Lane 7*, the final fraction (~2 μg) after a HiLoad 16/60 superdex 200 pg chromatography.

2. Inoculate a single colony into 10 ml of 2× YT liquid medium containing 150 μg/ml ampicillin and incubate the culture overnight with vigorous shaking at 37°C.
3. Add M13KO7 helper phage to an m.o.i. =10 and incubate at 37°C for 1 h with gently shaking.
4. Take 5 ml of the cell suspension to inoculate 1.5 l of 2× YT liquid medium containing 150 μg/ml ampicillin and 50 μg/ml kanamycin and incubate overnight with vigorous shaking at 37°C.
5. Remove *E. coli* cells by centrifugation at 3,000×*g* (5,000 rpm) for 15 min in a JLA10.500 rotor (Beckman Coulter).
6. To the supernatant containing the phage particles, add 1/4 volume of 20% PEG/NaCl solution and stir at 4°C for at least 3 h.
7. Collect the phage particles by centrifugation at 3,000×*g* for 15 min (*see* **Note 11**).
8. Resuspend the particles with 200 ml of TE buffer.
9. Remove the residual *E. coli* cells by centrifugation and collect the phage particles by PEG/NaCl precipitation as described in **steps 6–7**.

10. Resuspend the phage particles in 6 ml TE buffer.
11. Add 3.5 ml of 1.32 g/ml CsCl to a 13×51 mm UC tube (Beckman Coulter) and layer 1 ml of the phage solution, followed by a 300 μl layer of TE buffer (*see* **Note 12**).
12. Centrifuge at 22,000×*g* (48,000 rpm) at 20°C for 15 h in an SW55Ti rotor (Beckman Coulter).
13. Collect phage particles (top band) by withdrawing with a syringe.
14. Dialyze the collected phage solution against 500 ml of TE buffer at room temperature for 2 h with two buffer changes.
15. Divide the solution into 400 μl aliquots into 1.5 ml centrifuge tubes, add 20 μl of 20 mg/ml protease K to each tube, and incubate at 37°C for 30 min.
16. Add 48 μl of 10% SDS and incubate at 37°C for an additional 30 min, followed by 5 min at 55°C; cool on ice for 15 min.
17. Remove the protein contaminates with TE-saturated phenol and subsequently with phenol/chloroform/isoamylalcohol (25:24:1) treatment.
18. Recover the cssDNA by ethanol precipitation (*see* **Note 13**).
19. Dissolve the phage single-stranded DNA with 100 μl of TE buffer in each tube and dialyze against 500 ml of TE buffer at 4°C for 2 h. Determine the DNA concentration by measuring A_{260} using 33 μg/ml per A_{260}.

3.5.2. Preparation of Duplex Circular DNA

1. Introduce the pSKsxAS+ plasmid (or pSKsxAS-HP used for preparation of gDNA) into an *E. coli* DH5α strain (*see* **Note 14**) to allow colony formation on LB plates containing 100 μg/ml ampicillin.
2. Inoculate a single colony into 1 l of LB liquid medium containing 100 μg/ml ampicillin and incubate the culture overnight with vigorous shaking at 37°C.
3. Collect cells by centrifugation at 4,500×*g* for 10 min.
4. Resuspend the cells completely with 40 ml of solution I.
5. Add 80 ml of solution II and immediately mix gently and completely. Incubate at room temperature for 5 min.
6. Add 60 ml of ice-cold solution III and mix gently and completely.
7. Centrifuge at 5,000×*g* at room temperature for 15 min. Transfer the supernatant to a new centrifuge tube. Add 120 ml of isopropanol and mix completely.

8. Centrifuge at 10,000×g at 4°C for 20 min. Wash the pellets with 70% ethanol and dry under vacuum.
9. Dissolve the pellets with 10 ml of TE containing 20 μg/ml RNase A and incubate at 37°C for 10 min.
10. Remove the protein contaminates by TE–phenol treatment and subsequently with phenol/chloroform/isoamylalcohol (25:24:1) treatment.
11. Recover the DNA by ethanol precipitation and dissolve with 8 ml of TE buffer. Add 8.64 g of CsCl (1.59 g/ml) to the DNA solution and dissolve completely. Add 640 μl of 5 mg/ml ethidium bromide and mix completely.
12. Transfer the solution into 13×51 mm heat-sealable tubes (Quick-Seal Centrifuge Tubes, Beckman Coulter) and centrifuge at 230,000×g (60,000 rpm) at 20°C for 8–18 h using an NVT100 rotor (Beckman Coulter).
13. Visualize the DNA species by UV irradiation. Two bands are usually observed. The topmost layer contains nicked circular DNA and genomic DNA. The lower band mostly includes intact, closed circular plasmid DNA. Collect the lower layer by withdrawing with a syringe.
14. Pool the DNA solution and combine with an equal volume of NaCl-saturated isopropanol, mixing very gently. Discard the top isopropanol layer and repeat this procedure until both layers become colorless.
15. Add 2 volumes of TE buffer and treat with phenol/chloroform/isoamylalcohol (25:24:1).
16. Recover the plasmid DNA by ethanol precipitation.
17. Dissolve the DNA in 500 μl of TE buffer and dialyze the DNA solution against 500 ml of TE buffer at 4°C for 2 h. Determine the DNA concentration by measuring A_{260} using 50 μg/ml per A_{260}.

3.5.3. Preparation of Linear Duplex DNA (dsDNA)

Linear duplex DNA (ldsDNA) substrates are generated by digestion of the pSKsxAS+ plasmid DNA with the restriction enzymes *Pst*I or *Hin*dIII, which are used for the DNA strand exchange assay of the 3′-end or the 5′-end homology reactions, respectively. As described below, *Pst*I- and *Hin*dIII-double-digested pSKsxAS-HP is used for the preparation of gDNA.

1. Digest 150 μg of plasmid with the indicated restriction enzyme in 400 μl of reaction mixture (*see* **Note 15**).
2. Remove the restriction enzyme by phenol/chloroform/isoamylalcohol treatment.
3. Recover the digested DNA by ethanol precipitation.
4. Dissolve the DNA with 200 μl of TE buffer and dialyze the DNA solution against 500 ml of TE buffer at 4°C for 2 h.

Determine the DNA concentration by measuring A_{260} using 50 μg/ml per A_{260}.

3.5.4. Preparation of Gapped Circular DNA (gDNA)

Gapped DNA (gDNA) is a circular duplex DNA that contains 0.6 kb single-stranded DNA gap region. This DNA substrate is generated by heat-annealing the pSKsxAS+ cssDNA and the *Pst*I/*Hin*dIII-digested fragment of pSKsxAS-HP (**Fig. 22.5**). The pSKsxAS-HP plasmid is homologous to pSKsxAS+, but the sequence between the *Pst*I and *Hin*dIII sites is replaced by a 0.3 kb length non-homologous DNA fragment. The *Pst*I/*Hin*dIII-digested mixture prepared as in the instructions of **Section 3.5.3** can therefore be used without further treatment.

1. Mix 125 μg of pSKsxAS+ cssDNA and 75 μg of pSKsxAS-HP fragment in a 1.5 ml centrifuge tube and bring to 1 ml with water. Add 250 μl of 5× annealing buffer and mix completely.
2. Divide the mixture into 250 μl aliquots and transfer to new 0.6 ml tubes.
3. Layer 100 μl of mineral oil on top of the mixture.
4. Incubate the tubes at 98°C for 5 min, and then immediately incubate at 65°C for 5 min using a block incubator. Power off the block incubator to gradually cool the heating block to room temperature.
5. Combine all aliquots together and reduce the volume to one half with an ultrafiltration membrane (Microcon Ultracel Y-100, Millipore).

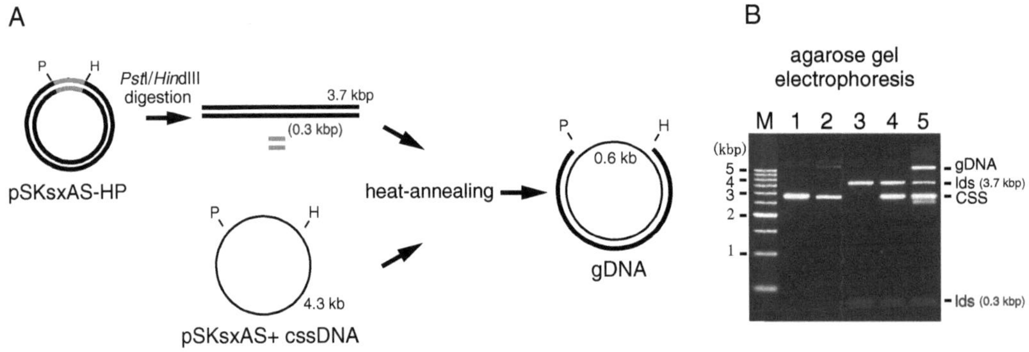

Fig. 22.5. (**a**) The preparation scheme for gDNA. A gDNA containing a 0.6 kb ssDNA region is generated by heat-annealing of pSKsxAS+ cssDNA (4.3 kb) and the 3.7 kb complementary strand derived from *Pst*I–*Hin*dIII-digested fragment of pSKsxAS-HP. The products are purified by agarose gel electrophoresis. The gray region located between *Pst*I and *Hin*dIII sites of pSKsxAS-HP is non-homologous to pSKsxAS+. P and H denote the *Pst*I and *Hin*dIII sites of each DNA molecule, respectively. (**b**) Gel image of the annealing products. The heat-annealing is carried out with 1 μg of each DNA substrate in a 10 μl reaction mixture. *Lane 1*, pSKsxAS+ cssDNA. *Lane 2*, pSKsxAS-HP plasmid. *Lane 3*, *Pst*I–*Hin*dIII-digested pSKsxAS-HP. *Lanes 4* and *5*, mixture of pSKsxAS+ cssDNA and *Pst*I–*Hin*dIII-digested pSKsxAS-HP before and after heat treatment, respectively.

6. Add 1/5 volume of 6× DNA loading buffer.
7. Separate the annealing products by 1.15% agarose gel electrophoresis with TAE (pH 8.0) buffer. The gel electrophoresis is performed using APPOLO 2025R (Continental Lab Products Inc.), with a gel volume of 500 ml at 150 V, at 4°C for 15 h. The gDNA products migrate close behind the xylene cyanol dye.
8. Cut the agarose gel at a width of 10 cm around the xylene cyanol dye.
9. Stain the gel with Syber Gold (Molecular Probes) at 1/10,000 in TAE buffer. Visualize the band corresponding to the gDNA products using a UV illuminator and cut out the region of interest.
10. Put the gel slice into a dialysis bag, add 5 ml of TAE buffer (pH 8.0), and completely remove air bubbles.
11. The gDNA products are eluted by electrophoresis performed at 70 V, at 4°C for 12 h. Reverse the current for 1 min to peel the DNA molecules attached to the dialysis bag.
12. Withdraw the solution containing the eluted gDNA with a sterile Pasteur pipette.
13. Recover the gDNA by ethanol precipitation.
14. Dissolve the gDNA in 500 μl of TE buffer and dialyze the DNA solution against 500 ml of TE buffer at 4°C for 2 h. Determine the DNA concentration by measuring A_{260} using 50 μg/ml per A_{260}.

3.6. DNA Four-Strand Exchange Reaction Assay

The DNA four-strand exchange reaction assay monitors the reciprocal DNA strand exchange that occurs between two duplex DNA molecules (**Fig. 22.6a**) (3). This exchange proceeds through the formation and branch migration of Holliday junctions. In this reaction, gDNA and its homologous ldsDNA are used as DNA substrates. The reaction is initiated by the pairing between the single-stranded DNA region of gDNA and its homologous ldsDNA, yielding the first joint molecule intermediate called a sigma structure. The three-strand exchange is then converted to duplex–duplex reciprocal exchange if the DNA strand exchange proceeds over the single–double-stranded DNA junction of gDNA following the formation of a single Holliday junction. This second intermediate is called an alpha structure. The completion of the strand exchange yields a nicked circular DNA (NC) and linear duplex DNA containing a single-stranded DNA tail (tailed DNA). These DNA species can be separated by agarose gel electrophoresis (**Fig. 22.6b**).

The polarity preference of the DNA four-strand exchange can also be determined using different types of ldsDNA (8, 15).

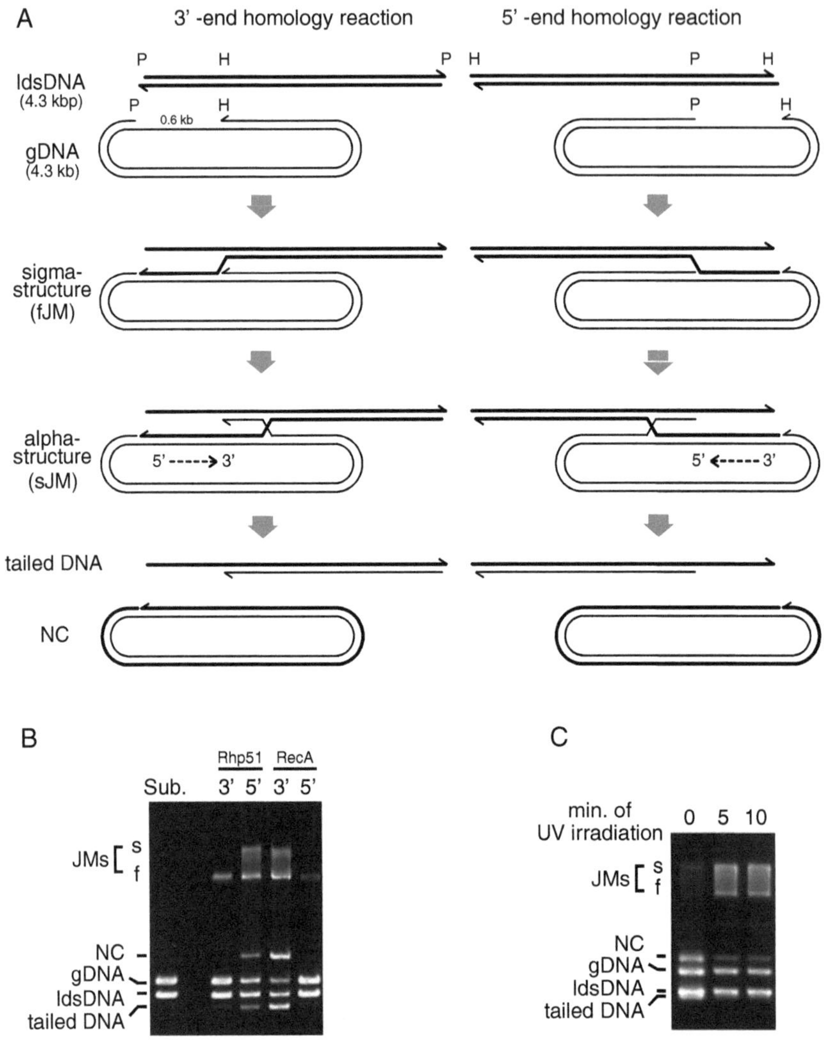

Fig. 22.6. (**a**) Schematic representation of the in vitro DNA four-strand exchange reaction. See details in the text. (**b**) Gel image of the products of the DNA four-strand exchange reaction mediated by Rhp51 (4 h) and RecA (30 min). The RecA-mediated reaction was performed as described in (8). The agarose gel electrophoresis was performed at 4°C. sJMs include sigma structures and fJMs mainly consist of alpha structures. 3′ and 5′ denote 3′- and 5′-end homology reactions, respectively. Sub., DNA substrates. (**c**) JMs are stabilized by psoralen crosslinking. The Rhp51-mediated 5′-end homology reaction mixture (4 h) was mixed with psoralen and irradiated with UV. The agarose gel electrophoresis was performed at room temperature.

As shown in **Fig. 22.6a**, the duplex–duplex reciprocal exchange only occurs when the DNA strand exchange proceeds in the 5′–3′ direction relative to the direction of the single-stranded DNA region of gDNA in a 3′-end homology reaction using *PstI*-linearized ldsDNA. On the other hand, a 3′–5′ polarity is required for the reciprocal exchange in the 5′-end homology reaction using *Hin*dIII-linearized ldsDNA.

The concentrations of all reaction components are stated as the final concentrations. The DNA concentrations are indicated in terms of total nucleotides.

1. Mix Rhp51 (7.5 μM) and Swi5–Sfr1 (0.3 μM) in reaction buffer F, with a final volume of 7 μl on ice.
2. Add 1 μl of gDNA (15 μM) and incubate at 37°C for 10 min.
3. Add 1 μl of RPA (0.3 μM) and incubate at 37°C for 10 min.
4. Initiate the strand exchange reaction by the addition of 1 μl of ldsDNA (14 μM) and incubate at 37°C for 4 h.
5. Add 2 μl of stop solution and incubate at 25°C for 10 min.
6. Add 2.4 μl of 6X DNA loading dye.
7. The samples are analyzed by 1.1% agarose gel electrophoresis with TAE (pH 7.6) at 4°C, at 5 V/cm for 10 h (*see* **Note 16**).
8. Stain the DNA species by dipping the gel in Syber Gold at 1/30,000 in TAE at 4°C for 4–12 h. Wash the gel with TAE buffer at room temperature for 1 h. Scan the gel using Fuji LAS-4000 (Fuji Photo Film Co.).

3.6.1. Psoralen Crosslinking (Option) (See Note 17)

DNA crosslinking is required to analyze the four-strand exchange products by agarose gel electrophoresis at room temperature. DNA crosslinking prevents the spontaneous branch migration of Holliday junctions, thus stabilizing the joint molecule products and allowing the agarose gel electrophoresis to occur at room temperature (**Fig. 22.6c**).

1. After step 5 described above, add 0.5 μl of psoralen solution (0.6 mg/ml) to the reaction mixture for a final concentration of 30 μg/ml.
2. Incubate the mixture on ice for 5 min.
3. Open the lid of the reaction tube and irradiate UV to the mixture for 5 min on ice (365 nm, 0.3 W/cm^2) (*see* **Note 18**).
4. Add 2.1 μl of stop solution and incubate at 25°C for 10 min.
5. Follow the subsequent steps as described in the above section, except that gel electrophoresis is performed at room temperature.

3.7. RuvC Cleavage of Holliday Junction

E. coli RuvC protein is a homodimeric endonuclease that specifically introduces symmetric incisions into a Holliday junction, resulting in two nicked duplex DNAs as products (**Fig. 22.7**) (13, 14). The alpha structure formed during the four-strand exchange is also a target for RuvC cleavage, allowing the confirmation of Holliday junction formation. In principle, two types of cleavage products are generated according to the different cleavage sites. As shown in **Fig. 22.7a**, cleavage at A–C yields a linear dimer product of twice the length of ldsDNA. On the other hand, B–D cleavage products are nicked circular and nicked tailed

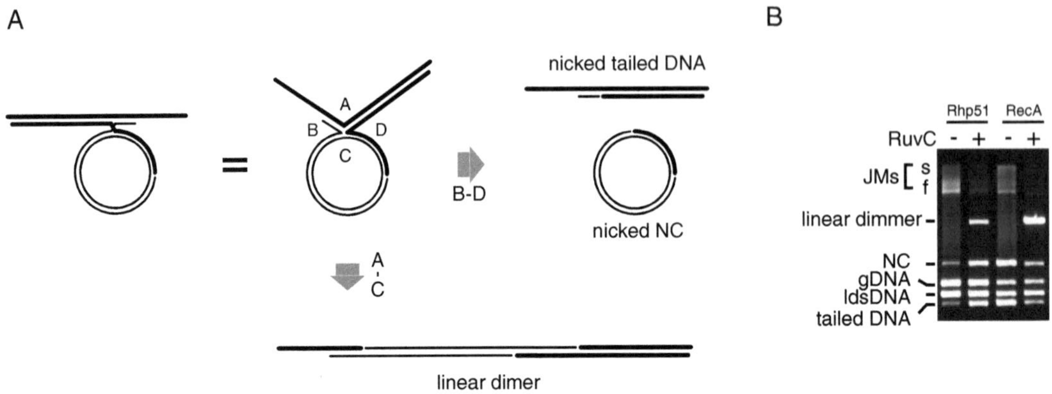

Fig. 22.7. (**a**) RuvC resolves the alpha structure at the HJ and generates two types of cleavage products. (**b**) Gel image showing the RuvC cleavage products. The RecA-mediated reaction is carried out as described in (8). The agarose gel electrophoresis was performed at 4°C.

DNAs. These two products migrate to a very similar position to that of the final products of the DNA four-strand exchange reaction. The B–D products are therefore not used as indicators of RuvC cleavage (**Fig. 22.7b**).

3.7.1. Ongoing Cleavage of the Holliday Junction

1. Initiate the DNA four-strand exchange reaction (15 μl) as described in **Section 3.6** and incubate for 3 h.
2. Withdraw 7 μl of the reaction mixture and aliquot into two new tubes. Add 0.5 μl of 125 mM $MgCl_2$ to each tube for a final concentration of 15 mM. Add RuvC (10 ng in 0.5 μl) to one tube and mock buffer (0.5 μl) to the other tube (*see* **Note 19**).
3. Incubate at 37°C for 1 h.
4. Add 1.6 μl of stop solution and incubate at 25°C for 10 min.
5. The DNA species are analyzed as described in **Section 3.6**.

3.7.2. RuvC Cleavage with Protein-Free DNA Substrate

1. Initiate the DNA four-strand exchange reaction in a volume of 50 μl and incubate for 3 h.
2. Add 1 μl of 20 mg/ml protease K to the reaction and incubate at 25°C for 5 min.
3. Treat the mixture with an equal volume of phenol/chloroform/isoamylalcohol (25:24:1). Remove the aqueous layer and save for the next step.
4. Exchange the reaction buffer of the aqueous layer with RuvC buffer by passing the solution through a G-25 spin column following the manufacturer's instructions (*see* **Note 20**).
5. Add 0.5 μl of RuvC (10 ng) to 15 μl of each sample and incubate at 37°C for 30 min.
6. Add 3 μl of stop solution and incubate at 25°C for 10 min.
7. The DNA species are analyzed as described in **Section 3.6**.

4. Notes

1. All target proteins used in this chapter are expressed using the T7 promoter expression system (Novagen).
2. All of the expressed target proteins and their tracks at each purification step in this chapter can be detected by a standard Coomassie brilliant blue (CBB) staining of SDS-PAGE gels.
3. Unless stated otherwise, all solutions should be prepared in Milli-Q water (Millipore) or an equivalent with a resistivity of 18.2 MΩ cm and a total organic content of less than 5 ppb. This standard is referred to as "water" in this text.
4. We use Slide-A-Lyzer 10 K MWCO Dialysis Cassettes, Extra strength (Thermo Scientific) following the manufacturer's instructions.
5. The protein solubility is significantly increased when the culture is incubated at 18°C. We use an Inova™ incubator shaker 44R (NBS) for protein expression. When the temperature is shifted down from 37 to 18°C, we change the temperature setting to 18°C on the same shaker in which the 37°C pre-culture was incubated. By this method, the gradual temperature shiftdown ensures a balance between the optimal temperature for protein expression and that for proper protein folding.
6. All procedures for protein purification are performed at 4°C. Samples are treated on an ice bucket and all solutions are pre-cooled at 4°C.
7. We use a model XL2020 sonicator (Misonix) with a standard horn of the 0.5 in. probe tip (catalog # 200). Sonication conditions, which are routinely applied for protein purifications in our laboratory, are as follows: The power setting of the amplitude control knob is 7, which corresponds to 30% of the maximum output. One cycle contains 16 time pulses (1 s on and 1 s off, total "on" time 30 s). Five cycles are applied to disrupt *E. coli* cells at 3-min interval on ice.
8. All column systems are connected to an AKTA Prime Plus or an AKTA explore purification system (GE Healthcare).
9. A fraction of Rhp51 binds to SP Sepharose resins under these conditions. The bound fraction should be discarded because SP Sepharose-bound Rhp51 protein is in an inactive form.
10. The genotype of *E. coli* JM103 is as follows: *endA1 glnV44 sbcBC rpsL thi-1* Δ(*lac-proAB*) *F′* [*traD36 proAB*$^+$ *lacI*^{q}

lacZΔM15]. Similar strains with F′, such as JM101, JM109, and TG1, can be substituted for JM103 as a host.

11. Remove the PEG/NaCl solution completely by wiping the inside of the tube with a clean kimwipe.
12. CsCl gradient centrifugation is an essential step to obtain an extra pure ssDNA for the strand exchange assay. This step can remove cell debris and separate phage particles from PEG. If this step is omitted, PEG would interfere, by a direct binding, with purification of ssDNA that comes out in solution after deproteination of phage particle.
13. The addition of a salt, such as Na acetate, should not be overlooked. Without a salt, a recovery of ssDNA by ethanol precipitation would not work well. The addition of a salt addition is much more effective to precipitation of ssDNA than that of dsDNA.
14. The genotype of *E. coli* DH5α is as follows: *F⁻ Φ80 dlacZΔM15 Δ(lacZYA-argF)U169 deoR recA1 endA1 hsdR17(rK⁻, mK⁺) phoA supE44 λ⁻ thi-1 gyrA96 relA1*.
15. The reaction mixtures should always be checked by agarose gel electrophoresis for the complete digestion of the plasmid DNA before proceeding to the next steps.
16. The electrophoresis must be performed at 4°C because joint molecule products are fragile and readily collapse when the electrophoresis is performed at room temperature (compare the third lane from the right in **Fig. 22.6b** and the leftmost lane in **Fig. 22.6c**). However, the electrophoresis can be performed at room temperature when the products are fixed by psoralen crosslinking (**Fig. 22.6c**).
17. Psoralen is a photoreactive interstrand DNA crosslinker that preferentially crosslinks pyrimidines at a TA sequence when irradiated with near-UV light.
18. A 5-min irradiation is sufficient for DNA crosslinking (**Fig. 22.6c**).
19. RuvC should be used immediately after the dilution of the stock solution with RuvC buffer.
20. G-25 spin columns should be pre-equilibrated by passing through an excess amount of RuvC buffer.

Acknowledgments

We thank K. Ito and T. Koizumi for preparation of the manuscript. This study was supported in part by grants in aid for Scientific Research on Priority Areas from the Ministry of

Education, Culture, Sports, Science, and Technology (MECSST) of Japan and for scientific research (A) to HI and for young scientist (B) to YM from the Japan Society for the Promotion of Science (JSPS), and by the Uehara Memorial Foundation to HI.

References

1. Masson, J.Y., and West, S.C. (2001) The Rad51 and Dmc1 recombinases: a non-identical twin relationship. *Trends Biochem Sci* **26**, 131–136.
2. Neale, M.J., and Keeney, S. (2006) Clarifying the mechanics of DNA strand exchange in meiotic recombination. *Nature* **442**, 153–158.
3. Cox, M.M. (2007) Motoring along with the bacterial RecA protein. *Nat Rev Mol Cell Biol* **8**, 127–138.
4. Haruta, N., Akamatsu, Y., Tsutsui, Y., Kurokawa, Y., Murayama, Y., Arcangioli, B., and Iwasaki, H. (2008) Fission yeast Swi5 protein, a novel DNA recombination mediator. *DNA Repair (Amst)* **7**, 1–9.
5. Sung, P., and Klein, H. (2006) Mechanism of homologous recombination: mediators and helicases take on regulatory functions. *Nat Rev Mol Cell Biol* **7**, 739–750.
6. Sugiyama, T., Zaitseva, E.M., and Kowalczykowski, S.C. (1997) A single-stranded DNA-binding protein is needed for efficient presynaptic complex formation by the Saccharomyces cerevisiae Rad51 protein. *J Biol Chem* **272**, 7940–7945.
7. Symington, L.S. (2002) Role of RAD52 epistasis group genes in homologous recombination and double-strand break repair. *Microbiol Mol Biol Rev* **66**, 630–670.
8. Murayama, Y., Kurokawa, Y., Mayanagi, K., and Iwasaki, H. (2008) Formation and branch migration of Holliday junctions mediated by eukaryotic recombinases. *Nature* **451**, 1018–1021.
9. Haruta, N., Kurokawa, Y., Murayama, Y., Akamatsu, Y., Unzai, S., Tsutsui, Y., and Iwasaki, H. (2006) The Swi5-Sfr1 complex stimulates Rhp51/Rad51- and Dmc1-mediated DNA strand exchange in vitro. *Nat Struct Mol Biol* **13**, 823–830.
10. Muris, D.F., Vreeken, K., Carr, A.M., Broughton, B.C., Lehmann, A.R., Lohman, P.H., and Pastink, A. (1993) Cloning the RAD51 homologue of Schizosaccharomyces pombe. *Nucleic Acids Res* **21**, 4586–4591.
11. Akamatsu, Y., Dziadkowiec, D., Ikeguchi, M., Shinagawa, H., and Iwasaki, H. (2003) Two different Swi5-containing protein complexes are involved in mating-type switching and recombination repair in fission yeast. *Proc Natl Acad Sci USA* **100**, 15770–15775.
12. Kurokawa, Y., Murayama, Y., Haruta-Takahashi, N., Urabe, I., and Iwasaki, H. (2008) Reconstitution of DNA strand exchange mediated by Rhp51 recombinase and two mediators. *PLoS Biol* **6**, e88.
13. Connolly, B., Parsons, C.A., Benson, F.E., Dunderdale, H.J., Sharples, G.J., Lloyd, R.G., and West, S.C. (1991) Resolution of Holliday junctions in vitro requires the *Escherichia coli ruvC* gene product. *Proc Natl Acad Sci USA* **88**, 6063–6067.
14. Iwasaki, H., Takahagi, M., Shiba, T., Nakata, A., and Shinagawa, H. (1991) *Escherichia coli* RuvC protein is an endonuclease that resolves the Holliday structure. *EMBO J* **10**, 4381–4389.
15. West, S.C., Cassuto, E., and Howard-Flanders, P. (1982) Postreplication repair in E. coli: strand exchange reactions of gapped DNA by RecA protein. *Mol Gen Genet* **187**, 209–217.

Chapter 23

Reconstituting the Key Steps of the DNA Double-Strand Break Repair In Vitro

Matthew J. Rossi, Dmitry V. Bugreev, Olga M. Mazina, and Alexander V. Mazin

Abstract

Double-stranded DNA breaks (DSB), the most harmful type of DNA lesions, cause cell death and genome instability. Homologous recombination repairs DSB using homologous DNA sequences as templates. Here we describe a set of reactions that lead to reconstitution of the double-stranded DNA break repair process in vitro employing purified human homologous recombination proteins and DNA polymerase η. Reconstitution of critical steps of DSB repair in vitro may help to better understand the mechanisms of recombinational DNA repair and the role of various human homologous recombination proteins in this process.

Key words: Homologous recombination, DNA strand exchange, branch migration, Holliday junction, joint molecules, D-loops.

1. Introduction

The Rad51 protein plays a key role in homologous recombination and DSB repair in eukaryotes (1, 2). It shares high homology with its prokaryotic and archaeal homologs, RecA and RadA. Rad51 possesses a unique activity – DNA strand exchange. Following resection of broken DNA ends, Rad51 is loaded onto the ssDNA to form a contiguous helical nucleoprotein filament, which searches for an intact homologous dsDNA template (3). Once the homologous sequence is found, Rad51 promotes the exchange of DNA strands that leads to formation of joint molecules (D-loops).

H. Tsubouchi (ed.), *DNA Recombination*, Methods in Molecular Biology 745,
DOI 10.1007/978-1-61779-129-1_23, © Springer Science+Business Media, LLC 2011

Following joint molecule formation, the 3′ ssDNA tails of the broken chromosome are extended by DNA polymerase, retrieving the information lost at the break site. Afterward, it is currently thought that the joint molecules formed during the initial steps of homologous recombination continue down one of two pathways (4–6). Either they dissociate, leading to rejoining of the broken chromosome through synthesis-dependent strand annealing (SDSA), or they proceed via the double-strand break repair (DSBR) by capture of the second processed DNA end, producing Holliday junctions which are resolved by structure-specific endonucleases. These basic mechanisms of homologous recombination appear to have been conserved throughout all forms of life. To better understand the role of different proteins in homologous recombination, here we describe a method for reconstituting individual steps and the entire process of DSB repair in vitro via the SDSA pathway (**Fig. 23.1**) using purified human proteins: RAD51, DNA polymerase η, RAD52, RAD54, and RPA (7).

2. Materials

2.1. DNA Molecules

Oligonucleotides are obtained from Integrated DNA Technologies (www.idtdna.com) and purified by electrophoresis in denaturing polyacrylamide gels containing 50% urea.

1. dsDNA comprised of two complementary ssDNA oligonucleotides:

   AVM #25, 48-mer: GCA ATT AAG CTC TAA GCC ATC CGC AAA AAT GAC CTC TTA TCA AAA GGA;

   AVM #26, 48-mer: TCC TTT TGA TAA GAG GTC ATT TTT GCG GAT GGC TTA GAG CTT AAT TGC.

2. Tailed dsDNA #1 comprised of two ssDNA oligonucleotides:

   AVM #199, 36-mer: CAC TGC TAA TAG CGT CCG GTA AGT AAA ATG AGA ATT;

   AVM #209, 100-mer: AAT TCT CAT TTT ACT TAC CGG ACG CTA TTA GCA GTG GGT GAG CAA AAA CAG GAA GGC AAA ATG CCG CAA AAA AGG GAA TAA GGG CGA CAC GGA AAT GTT G.

3. Tailed dsDNA #2 comprised of two oligonucleotides:

   AVM #199, 36-mer (shown above);

   AVM #320, 100-mer: AAT TCT CAT TTT ACT TAC CGG ACG CTA TTA GCA GTG CAA CAT TTC CGT GCC GCC CTT ATT CCC TTT TTT GCG GCA TTT TGC CTT CCC GTT TTT GCT CAC C.

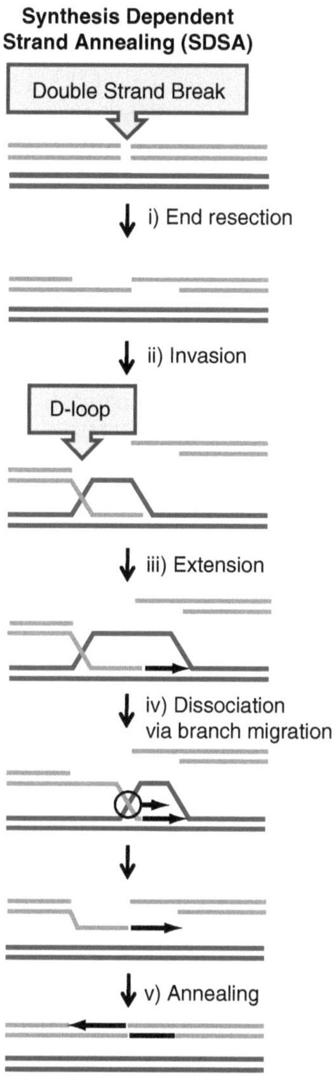

Fig. 23.1. DSB repair through the SDSA mechanism. In the SDSA model of double-strand break repair, (*i*) the DNA ends are processed at the site of the break to produce tailed dsDNA with 3′-ssDNA extensions. RAD51 forms a nucleoprotein filament on these 3′-ssDNA and promotes the initial stages of homologous recombination (*ii*) including homology searching, invasion, and hetero-duplex extension that leads to D-loop formation. Then, DNA polymerase (*iii*) extends the invading ssDNA strand to recover the genetic information lost at the DNA break site. Following extension, D-loop dissociation (*iv*) occurs via branch migration (indicated by *black circle* and *arrow*). Finally, the dissociated DNA ends are (*v*) annealed and ligated to reproduce an intact chromosome.

4. Tailed dsDNA #2* comprised of two oligonucleotides:

   AVM #199, 36-mer (shown above);

   AVM #231, 100-mer: AAT TCT CAT TTT ACT TAC CGG ACG CTA TTA GCA GTG GCT CAT GAG ACA ATA ACC CTG ATA AAT GCT TCA ATA ATA TTG AAA AAG GAA GAG TAT GAG TAT T.

5. ssDNA:
   (a) AVM #90, 90-mer: CGG GTG TCG GGG CTG GCT TAA CTA TGC GGC ATC AGA GCA GAT TGT ACT GAG AGT GCA CCA TAT GCG GTG TGA AAT ACC GCA CAG ATG CGT;
   (b) AVM #214: CTA GAT AAA AAT ATT ATA TAT TTA AAT TAT TAA AAA ATT TAT TTA TAA AAT TAT.
6. Supercoiled pUC19 plasmid DNA purified from *Escherichia coli* using alkaline lysis method (Qiagen) followed by CsCl–ethidium bromide equilibrium centrifugation (8). To produce plasmid DNA with a sufficient level of superhelicity, we recommend *E. coli* host strains with the intact gyrase gene, e.g., HB101.

2.2. Proteins and Reaction Buffers

1. T4 polynucleotide kinase and T4 polynucleotide kinase buffer (New England Biolabs).
2. [γ-^{32}P]ATP (10 mCi/ml, PerkinElmer).
3. "Ca^{2+} buffer" for Rad51 filament formation: 25 mM Tris–acetate (pH 7.5), 1 mM ATP, 1 mM magnesium acetate, 2 mM $CaCl_2$, 2 mM DTT, BSA (100 μg/ml), 20 mM phosphocreatine, and creatine phosphokinase (30 U/ml).
4. "Mg^{2+} buffer" for Rad51 filament formation: 25 mM Tris–acetate (pH 7.5), 2 mM ATP, 3 mM magnesium acetate, 2 mM DTT, BSA (100 μg/ml), 20 mM phosphocreatine, creatine phosphokinase (30 U/ml).
5. Phosphocreatine (Sigma) and creatine phosphokinase (Calbiochem, from rabbit skeletal muscle).
6. PCR-grade Proteinase K, recombinant (Roche).
7. Loading buffer for gel electrophoresis: 70% glycerol, 0.1% bromophenol blue.
8. Equipment for agarose and polyacrylamide gel electrophoresis.
9. TAE buffer: 40 mM Tris–acetate (pH 8.0) and 1 mM EDTA.
10. DE81 chromatography paper (Whatman).
11. PhosphorImager system (GE Healthcare) or any similar device.
12. ssDNA annealing buffer: 14 mM Tris–HCl (pH 7.5), 2 mM magnesium chloride, 1 mM DTT, 121 mM sodium chloride, and 12.1 mM sodium citrate.
13. D-loop dissociation buffer: 25 mM Tris–acetate (pH 7.5), 1 mM ATP, 1 mM magnesium acetate, 2 mM DTT, BSA (100 μg/ml), 20 mM phosphocreatine, and creatine phosphokinase (30 U/ml).
14. *Dde*I restriction endonuclease (New England Biolabs).

15. Polyacrylamide, Bis-*N*,*N'*-methylene-bis-acrylamide, ammonium persulfate, and TEMED (BioRad).
16. TBE buffer: 90 mM Tris–borate (pH 8.3) and 1 mM EDTA.
17. S-400 Spin columns (GE Healthcare).
18. Deoxyribonucleoside triphosphate set, PCR grade (Roche).
19. SDSA reconstitution mixture: 25 mM Tris–acetate (pH 7.5), 1 mM ATP, 2 mM magnesium acetate, 2 mM $CaCl_2$, 2 mM DTT, 100 mg/ml BSA, 20 mM phosphocreatine, 30 U/ml creatine phosphokinase, 1.5 ng/ml DNA polymerase η, dNTPs (dATP, dTTP, and dCTP; 100 μM each).
20. Human proteins, RAD51, RAD52, RAD54, RPA, DNA polymerase η, and BLM were purified as described in (9–14).

3. Methods

All of the methods presented in this chapter begin with formation of a RAD51 nucleoprotein filament on ssDNA. Relying on its DNA strand exchange activity, the RAD51 filament is used to study different aspects of the homologous recombination pathway. These protocols have all been optimized for human proteins. Proteins from other organisms require additional optimization of the reaction conditions. For instance, we have found that human RAD51 requires Ca^{2+} for efficient D-loop formation; however, *Saccharomyces cerevisiae* Rad51 does not (15).

The defining step of the SDSA model of DSB repair is the dissociation of D-loops via branch migration after the completion of template-dependent DNA synthesis (**Fig. 23.1**, Step iv). This model provides a mechanism of DSB repair through which non-crossover recombinants are faithfully produced, diminishing the rate of mitotic crossovers that may lead to loss of heterozygosity and ultimately cancer (16). Here we describe a number of assays that can be used in order to investigate various steps of the SDSA pathway including nucleoprotein filament formation and disassembly, D-loop formation, and their dissociation through branch migration. Then utilizing these individual assays, we present a protocol developed for in vitro reconstitution of the SDSA pathway (7).

3.1. The RAD51–ssDNA Filament Disassembly Assay

In order to promote DNA strand exchange, monomers of RAD51 polymerize on ssDNA forming an active nucleoprotein filament. The disassembly of the RAD51–ssDNA protein filament may represent an important step in the regulation of homologous recombination. Several proteins were shown to disassemble the RAD51 filament (14, 17). Here, we tested the ability of BLM to disrupt the RAD51–ssDNA filament by monitoring the inhibition of cleavage by a restriction endonuclease on a dsDNA probe protected by RAD51 monomers when they reassemble following disruption of the initial filament (14) (**Fig.23.2a**).

1. Label oligonucleotide AVM #25 using [γ-^{32}P]ATP and T4 polynucleotide kinase (NEB). For labeling, mix 2 μl of the oligonucleotide at a concentration of 500 μg/ml, 2 μl of [γ-^{32}P]ATP (10 mCi/ml; Perkin Elmer), 2 μl of 10× T4 polynucleotide kinase buffer, and 1 unit of T4 polynucleotide kinase in a total volume of 20 μl. Incubate the reaction mixture for 1 h at 37°C and then inactivate T4 polynucleotide kinase by heating the mixture for 10 min at 75°C. Store labeled oligonucleotides at –20°C. Note that

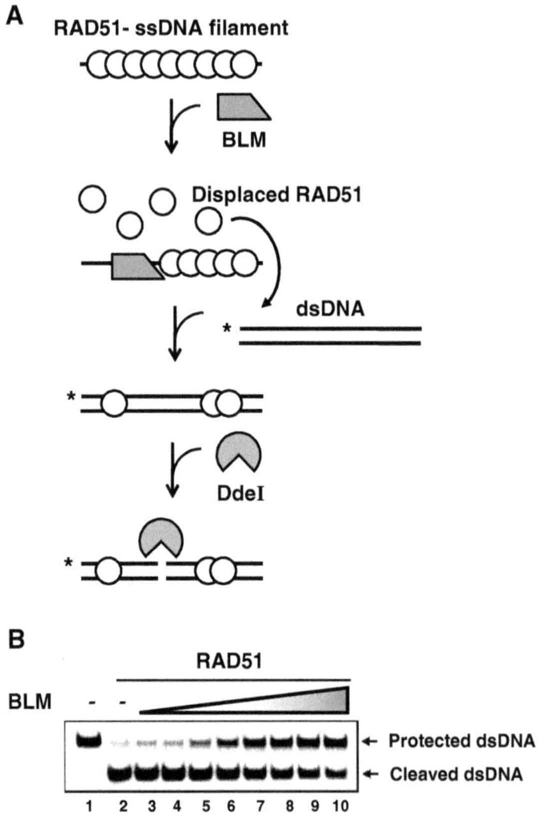

Fig. 23.2. RAD51 disassembly. (**a**) Experimental scheme: RAD51 displaced from ssDNA by BLM binds to dsDNA probe and protects it against the cleavage with *Dde*I restriction endonuclease. The *asterisk* indicates the ^{32}P label. (**b**) BLM disrupts the RAD51 filament in a concentration-dependent manner. *Lanes 2–10*, the fraction of protected DNA increases with the increase in BLM concentration from 0 to 200 nM. *Lane 1* shows the original dsDNA fragment in the absence of proteins. No protection was observed in the absence of RAD51 (14).

although not necessary for this procedure, if desired, the labeled oligonucleotides can be purified from unincorporated [γ-^{32}P]ATP by passing the labeling reaction through spin columns, e.g., Micro Bio-Spin 6 (Bio-Rad).

2. Prepare the dsDNA fragment by mixing ^{32}P-labeled ssDNA oligonucleotide (AVM #25) (usually 100 μg/ml) with an equimolar amount of the complementary ssDNA (AVM #26) in annealing buffer (*see* **Note 1**). Heat the reaction for 3 min at 95°C, followed by annealing for 1 h at the optimal hybridization temperature (T_h). Calculate T_h using the formula $T_h = 1.24T_m$–43.8°C, where T_m is the melting temperature of the double-stranded DNA. Calculate the T_m using a tool from the Promega website (www.promega.com/biomath/calc11.htm).
3. Incubate RAD51 (1 μM) with ssDNA (AVM #90; 60 nM) in "Mg^{2+} buffer" for RAD51 filament formation for 15 min at 37°C.
4. Assay for filament disruption activity by adding BLM (in a range of concentrations from 0 to 250 nM) to the RAD51–ssDNA filament and incubate for 15 min at 37°C.
5. Then, add ^{32}P-labeled dsDNA (AVM #25/#26) (15 nM) and incubate for an additional 10 min to allow RAD51 time to bind to dsDNA.
6. Digest the dsDNA by adding *Dde*I restriction endonuclease (0.2 units/μl) and incubating for 10 min at 37°C.
7. Stop the reaction and deproteinize DNA products by adding SDS to 1% and proteinase K to 800 μg/ml and by incubating the mixture for 15 min at 37°C.
8. Analyze the DNA products by electrophoresis in a 10% polyacrylamide–TBE gel (**Fig. 23.2b**). Mix the aliquots with a 1/10 volume of loading buffer and load them onto the gel. Dry the gel on DE81 chromatography paper (Whatman), visualize, and quantify the products of branch migration using a PhosphorImager system (GE Healthcare) or another appropriate device.

3.2. Formation of Single and Double D-Loops

The D-loop is generated by RAD51-promoted invasion of ssDNA molecule that mimics one end of broken DNA into covalently closed plasmid supercoiled DNA (**Fig. 23.3a**). The double D-loop is produced from the single D-loop by capturing the second broken DNA end and annealing it to the displaced ssDNA strand (**Fig. 23.3b**). The annealing reaction is promoted by RAD52 (18).

3.2.1. Formation of D-Loops with RAD51

1. Prepare the tailed dsDNA #1 by ^{32}P-labeling of ssDNA oligonucleotide (AVM #209) and then mix this ^{32}P-labeled

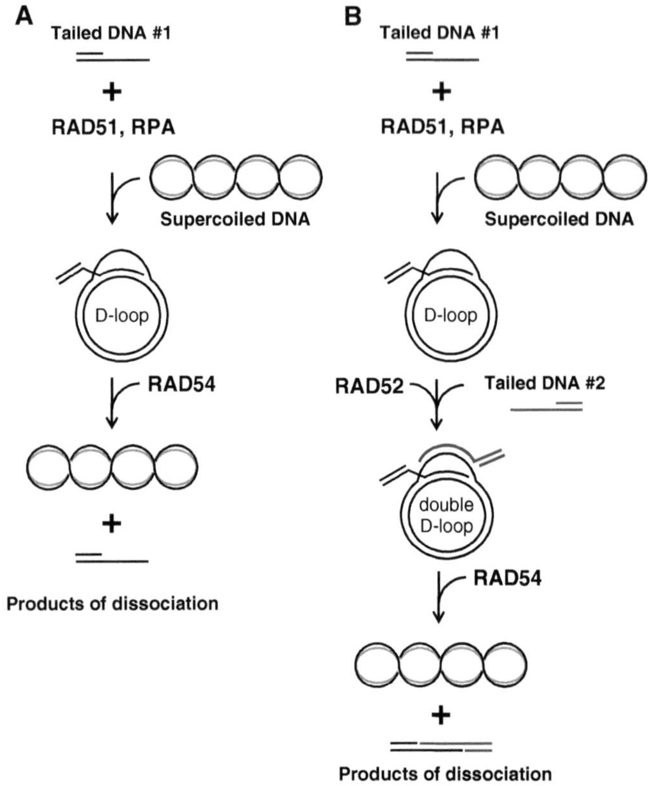

Fig. 23.3. Formation and dissociation of single and double D-loops. Rad54 promotes dissociation of single D-loops produced by RAD51 (**a**) or double D-loops produced by RAD51 and RAD52 (**b**).

ssDNA (usually 100 μg/ml) with an equimolar amount of the complementary ssDNA (AVM #199) in annealing buffer.

2. Incubate RAD51 (1 μM) with tailed dsDNA (AVM #199/#209) (30 nM) (*see* **Note 2**) in "Ca^{2+} buffer" for 15 min at 37°C.
3. Initiate D-loop formation by adding pUC19 dsDNA (50 μM, nucleotides) and continue to incubate for 15 min at 37°C.
4. Stop the reaction and deproteinize DNA products by adding SDS to 1% and proteinase K to 800 μg/ml, incubate the mixture for 15 min at 37°C.
5. Mix the aliquots with a 1/10 volume of loading buffer and analyze D-loop formation by electrophoresis in a 1% agarose–TAE gel. Dry the gel on DE81 chromatography paper (Whatman), visualize, and quantify the D-loops using a PhosphorImager system (GE Healthcare) or another appropriate device.

6. After confirming formation of D-loops, purify them by passing the mixture (from Step 4) twice through S-400 Spin columns (GE Healthcare) equilibrated with 25 mM Tris–acetate (pH 7.5) and 25 mM NaCl.

3.2.2. Formation of Double D-Loops with RAD51 and RAD52

1. To form double D-loops, begin by forming D-loops as described in **Section 3.2.1**, Steps 1–4.
2. Purify deproteinized D-loops by passing the reaction twice through S-400 Spin columns (GE Healthcare) equilibrated with 25 mM Tris–acetate (pH 7.5) and 25 mM NaCl.
3. In a separate tube incubate RAD52 (1.5 μM) with tailed dsDNA #2 (AVM #199/#320) (90 nM) in "Ca^{2+} buffer" for 15 min at 37°C in 0.5× reaction volume relative to the volume of the single D-loop mixture.
4. Initiate double D-loop formation by addition of the RAD52/tailed dsDNA complexes to the single D-loop mixture.
5. Deproteinize DNA by adding SDS to 1% and proteinase K to 800 μg/ml to the reaction mixture followed by incubation for 15 min at 37°C and purify double D-loops through a 1% agarose gel as described in (8).
6. Alternatively, to study non-deproteinized double D-loops, instead of Step 5, deplete Ca^{2+} by the addition of 2 mM EGTA followed by a 5-min incubation (*see* **Note 3**).
7. As a control, incubate the D-loops reaction in the same buffer with *Eco*RI endonuclease (400 U/ml) for 5 min at 37°C to enable specific detection of double D-loops, which, in contrast to single D-loops, are stable after plasmid DNA linearization with a restriction enzyme (7).

3.3. Dissociation of D-Loops with RAD54 or BLM

Dissociation of D-loops appears to be a crucial step in advancing along the SDSA recombination pathway (**Fig. 23.1**, Step iv; **Fig. 23.3a**). Previously, it was shown that RAD54 and BLM can catalyze D-loop dissociation (7, 14). The double D-loop, a highly stable structure, was thought to be dissolved by concerted action of BLM and topoisomerase or processed via the DSBR pathway of homologous recombination (19, 20). However, we recently showed that double D-loops can also be dissociated by RAD54 through the mechanism that couples D-loop dissociation with rejoining of the broken DNA ends obviating formation of ssDNA intermediates and their reannealing (**Fig. 23.3b**) (7).

The D-loop dissociation assay described below can be used for identification of proteins that specifically interact with joint molecules and dissociate them, as it was previously shown for BLM (14) and RAD54 (7). The assay may also help in determining the conditions that either facilitate or inhibit D-loop dissociation.

3.3.1. Dissociation of Single D-Loops

1. Purify deproteinized D-loops as described in **Section 3.2.1**, Step 6.
2. Alternatively, to study non-deproteinized D-loops, instead of deproteinizing D-loops as in **Section 3.2.1**, Step 4, deplete Ca^{2+} by the addition of 2 mM EGTA followed by a 5-min incubation (*see* **Note 3**). Do not pass non-deproteinized D-loops through the S-400 spin columns as it may lead to protein dissociation and losses.
3. Incubate D-loops (0.25 nM) with BLM (100 nM) or RAD54 (100 nM) in D-loop dissociation buffer at 37°C. Determine the kinetics of D-loop dissociation by withdrawing aliquots at various time points, typically from 0 to 30 min.
4. Stop the reaction by adding SDS to 1% and proteinase K to 800 μg/ml. Incubate the mixture for an additional 15 min at 37°C to deproteinize the DNA products.
5. Analyze the DNA products by electrophoresis in 1% agarose–TAE gels. Dry the gels on DE81 chromatography paper (Whatman), visualize, and quantify the D-loops using a PhosphorImager system (GE Healthcare) or another appropriate device.

3.3.2. Dissociation of Double D-Loops

1. Perform dissociation of double D-loops (0.25 nM) with the RAD54 (20 nM) and RPA (50 nM) in D-loop dissociation buffer at 37°C. Determine the kinetics of D-loop dissociation by withdrawing aliquots at various time points, typically from 0 to 30 min.
2. Stop the reaction by adding SDS to 1%, proteinase K to 800 μg/ml, and unlabeled ssDNA oligonucleotide (AVM #209) (100 nM) that is used to prevent spontaneous annealing of the tailed dsDNAs during deproteinization. Incubate the mixture for an additional 15 min at 37°C to deproteinize the DNA products.
3. Visualize the products of D-loop dissociation by electrophoresis in a 1% agarose–TAE gel (to visualize D-loops) and an 8% polyacrylamide–TBE gel (to visualize the oligonucleotide products of dissociation), in parallel. Dry the gels on DE81 chromatography paper (Whatman), visualize, and quantify the D-loops using a PhosphorImager system (GE Healthcare) or another appropriate device.

3.4. Reconstitution of the Key Steps of the SDSA Pathway In Vitro

Here we describe a reconstitution assay that incorporates all of the major steps required to repair the DSB via the SDSA pathway: nucleoprotein filament formation, homology searching, invasion, DNA strand exchange, DNA replication, second-end capture, D-loop dissociation, and reannealing (**Fig. 23.4a**).

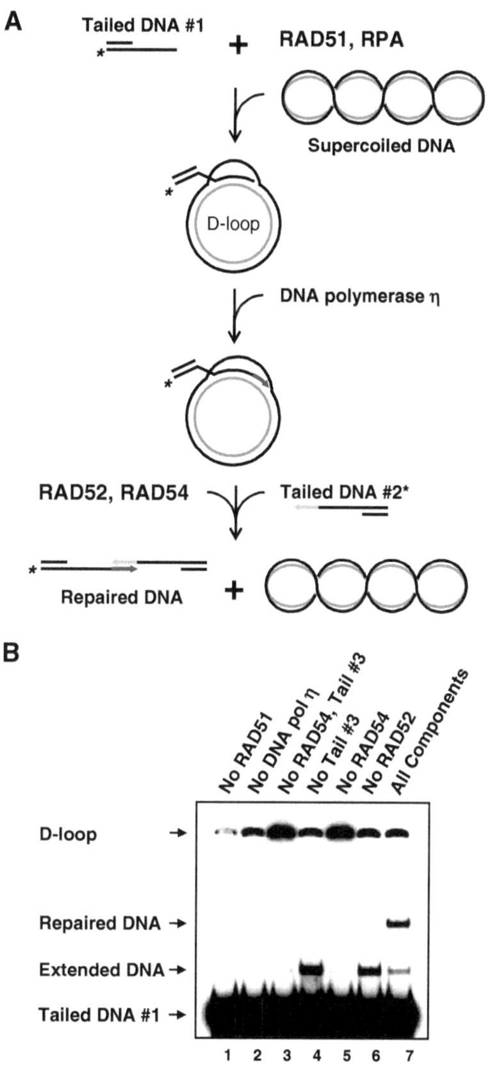

Fig. 23.4. In vitro reconstitution of DSB repair. (**a**) Experimental scheme. The *asterisk* denotes ^{32}P label. *Right arrow* represents extension of tailed dsDNA #1 by DNA polymerase η; *left arrow* denotes the complementary segment of tailed dsDNA #2* that mimics the second end of a broken DNA. Extension by polymerization stops at the first cytosine in the template sequence, as dGTP was omitted. (**b**) A representative reconstitution reaction. In *lanes 1–6*, one or more indicated components of the reaction mixture have been omitted (*lane 1*: RAD51, *lane 2*: DNA polymerase η, *lane 3*: Rad54 and tailed dsDNA #2*, *lane 4*: tailed dsDNA #2*, *lane 5*: RAD54, and *lane 6*: RAD52). The expected product, "repaired DNA," is only present when all the components were included (*lane 7*). "Extended DNA" refers to tailed dsDNA #1 that has been extended by DNA polymerase η and dissociated from the supercoiled DNA.

The reconstitution experiment begins with RAD51-promoted D-loop formation between tailed dsDNA #1 and supercoiled pUC19 DNA, representing the processed DNA end at the break site and the homologous, intact chromosome, respectively. Then using tailed dsDNA #1 as a primer and pUC19 as a template, DNA polymerase η performed DNA synthesis. To control the length of the DNA extension product, the synthesis was performed in the presence of only three dNTPs (dGTP was omitted) resulting in a 32 nucleotide extension of the invading DNA ("extended DNA"; **Fig. 23.4b**, lane 4). Next tailed dsDNA #2*, which is only complementary to the extended DNA, is added along with RAD52 and RAD54. Rad54 promotes dissociation of the D-loop structure, while RAD52 facilitates the annealing of the two-tailed dsDNA molecules, yielding the expected product ("repaired DNA"; **Fig. 23.4b**, lane 7) (7). These reactions directly demonstrate the feasibility of the SDSA mechanism of DSB repair and provide a useful method of determining how and where other proteins may fit into the framework of homologous recombination.

1. Begin by forming D-loops with RAD51 as described in **Section 3.2.1**, Steps 1–4, except replacing D-loop formation buffer with SDSA reconstitution mixture, and the addition of RAD51 is preceded by a 5 min incubation with RPA (225 nM). Note that RAD51, when added after RPA, can efficiently displace RPA from tailed dsDNA, but not ssDNA.
2. Following D-loop formation, deplete Ca^{2+} by the addition of 2 mM EGTA. Add tailed dsDNA #2* (30 nM) and incubate for 5 min at 37°C.
3. Finally, add RAD54 (200 nM) and RAD52 (1.5 μM) for 30 min at 37°C.
4. Terminate the reactions by adding SDS to 1%, proteinase K to 800 μg/ml, and an ssDNA oligonucleotide (AVM #214) (1.2 μM). Incubate the mixture for an additional 15 min at 37°C to deproteinize the DNA products. AVM #214, a 32-base oligonucleotide complementary to tailed dsDNA 2*, was added with the stop buffer to prevent protein-independent annealing of the extension product with tailed dsDNA #2* during DNA deproteinization.
5. Analyze the products of dissociation by electrophoresis in a 1% agarose–TAE gel and an 8% polyacrylamide–TBE gel, in parallel (**Fig. 23.4b**; agarose gel is not shown). Dry the gels on DE81 chromatography paper (Whatman), visualize, and quantify the D-loops using a PhosphorImager system (GE Healthcare) or another appropriate device.

4. Notes

1. Oligonucleotide-based substrates are generally prepared to contain only 10% of the radiolabeled oligonucleotide and

90% unlabeled in order to limit the total amount of radioactivity used in each experiment.

2. DNA concentrations are given in units of molecules, unless otherwise indicated.
3. Ca^{2+} depletion destabilizes the RAD51 filament, causing it to undergo a transition from an active to an inactive filament (15).

Acknowledgments

This work was supported by the NIH Grant CA100839, MH084119, and the Leukemia and Lymphoma Society Scholar Award 1054-09 (to AVM) and NIH Grant F31 AG033484-01 (to MJR).

References

1. San Filippo, J., Sung, P., and Klein, H. (2008) Mechanism of eukaryotic homologous recombination. *Annu Rev Biochem* 77, 229–257.
2. Krogh, B.O., and Symington, L.S. (2004) Recombination proteins in yeast. *Annu Rev Genet* **38**, 233–271.
3. Kowalczykowski, S.C. (2008) Structural biology: snapshots of DNA repair. *Nature* **453**, 463–466.
4. Pâques, F., and Haber, J.E. (1999) Multiple pathways of recombination induced by double-strand breaks in Saccharomyces cerevisiae. *Microbiol Mol Biol Rev* **63**, 349–404.
5. Allers, T., and Lichten, M. (2001) Differential timing and control of noncrossover and crossover recombination during meiosis. *Cell* **106**, 47–57.
6. Hunter, N., and Kleckner, N. (2001) The single-end invasion: an asymmetric intermediate at the double-strand break to double-holliday junction transition of meiotic recombination. *Cell* **106**, 59–70.
7. Bugreev, D.V., Hanaoka, F., and Mazin, A.V. (2007) Rad54 dissociates homologous recombination intermediates by branch migration. *Nat Struct Mol Biol* **14**, 746–753.
8. Sambrook, J., and Russell, D.W. (2001) Molecular cloning: a laboratory manual, 3rd Edition (Cold Spring Harbor, NY: Cold Spring Harbor Laboratory Press).
9. Sigurdsson, S., Trujillo, K., Song, B., Stratton, S., and Sung, P. (2001) Basis for avid homologous DNA strand exchange by human Rad51 and RPA. *J Biol Chem* **276**, 8798–8806.
10. Kumar, J.K., and Gupta, R.C. (2004) Strand exchange activity of human recombination protein Rad52. *Proc Natl Acad Sci USA* **101**, 9562–9567.
11. Mazina, O.M., and Mazin, A.V. (2004) Human Rad54 protein stimulates DNA strand exchange activity of hRad51 protein in the presence of Ca^{2+}. *J Biol Chem* **279**, 52042–52051.
12. Henricksen, L.A., Umbricht, C.B., and Wold, M.S. (1994) Recombinant replication protein A: expression, complex formation, and functional characterization [published erratum appears in J Biol Chem 1994 Jun 10;269(23):16519]. *J Biol Chem* **269**, 11121–11132.
13. Masutani, C., Kusumoto, R., Iwai, S., and Hanaoka, F. (2000) Mechanisms of accurate translesion synthesis by human DNA polymerase eta. *EMBO J* **19**, 3100–3109.
14. Bugreev, D.V., Yu, X., Egelman, E.H., and Mazin, A.V. (2007) Novel pro- and anti-recombination activities of the Bloom's syndrome helicase. *Genes Dev* **21**, 3085–3094.
15. Bugreev, D.V., and Mazin, A.V. (2004) Ca^{2+} activates human homologous recombination protein Rad51 by modulating its ATPase

activity. *Proc Natl Acad Sci USA* **101**, 9988–9993.
16. Richardson, C. (2005) RAD51, genomic stability, and tumorigenesis. *Cancer Lett* **218**, 127–139.
17. Sommers, J.A., Rawtani, N., Gupta, R., Bugreev, D.V., Mazin, A.V., Cantor, S.B., and Brosh, R.M., Jr. (2009) FANCJ uses its motor ATPase to destabilize protein-DNA complexes, unwind triplexes, and inhibit RAD51 strand exchange. *J Biol Chem* **284**, 7505–7517.
18. Sugiyama, T., Kantake, N., Wu, Y., and Kowalczykowski, S.C. (2006) Rad52-mediated DNA annealing after Rad51-mediated DNA strand exchange promotes second ssDNA capture. *EMBO J* **25**, 5539–5548.
19. Wu, L., and Hickson, I.D. (2003) The Bloom's syndrome helicase suppresses crossing over during homologous recombination. *Nature* **426**, 870–874.
20. Raynard, S., Bussen, W., and Sung, P. (2006) A double Holliday junction dissolvasome comprising BLM, topoisomerase IIIalpha, and BLAP75. *J Biol Chem* **281**, 13861–13864.

Chapter 24

Biochemical Studies on Human Rad51-Mediated Homologous Recombination

Youngho Kwon, Weixing Zhao, and Patrick Sung

Abstract

Rad51-mediated pairing between homologous DNA sequences during homologous recombination (HR) plays pivotal roles in DNA double-strand break repair. The multi-step process occurs through cooperation of Rad51 and a number of accessory protein factors. The development of various biochemical analyses with the requisite purified factors provides an opportunity to understand the molecular mechanisms of HR. In this chapter, we describe detailed procedures of in vitro assays using human Rad51, a polypeptide derived from the BRCA2 protein, and the Hop2–Mndl complex, to examine (1) homologous DNA pairing, (2) Rad51 targeting to single-stranded DNA, (3) stabilization of the Rad51 nucleoprotein filament, and (4) duplex capture by the Rad51 nucleoprotein filament. These methods are invaluable for delineating the functional interplay of HR factors.

Key words: Rad51, BRCA2, Hop2–Mndl, homologous recombination, presynaptic filament, homologous, DNA pairing.

1. Introduction

Rad51 is the eukaryotic orthologue of the *Escherichia coli* RecA recombinase. Like RecA, Rad51 catalyzes the homologous DNA pairing reaction that links homologous chromatids during HR, and it is active in both mitotic and meiotic cells (1, 2). During HR, Rad51 is loaded onto 3′ single-stranded (ss) DNA tails, produced via nucleolytic resection of DNA ends at double-strand breaks (DSBs) (**Fig. 24.1**). Biochemical analyses have shown that Rad51 assembles onto ssDNA to form a right-handed helical filament in a process that requires ATP binding but not its hydrolysis (3, 4). The Rad51-ssDNA filament, also called the presynaptic

H. Tsubouchi (ed.), *DNA Recombination*, Methods in Molecular Biology 745,
DOI 10.1007/978-1-61779-129-1_24, © Springer Science+Business Media, LLC 2011

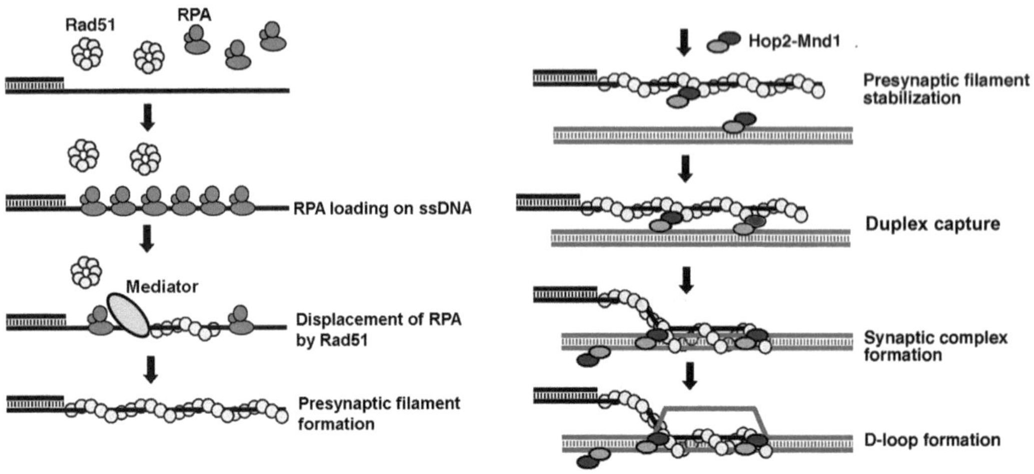

Fig. 24.1. Formation of the presynaptic filament and displacement loop. The RPA molecules coat on a ssDNA tail generated from DNA end resection. Recombination mediators (a BRCA2-derived polypeptide, Rad51B–Rad51C complex, Rad52 and Rad55–Rad57 complex) help Rad51 displace RPA, thus facilitating Rad51 presynaptic filament assembly. The filament is further stabilized by accessory factors, including Hop2–Mnd1 and Rad54, captures duplex DNA, and searches for homologous sequence. HR accessory factors including Rad54, Rad54B, Rdh54, and Hop2–Mnd1 promote formation of a DNA joint, known as the displacement (D)-loop.

filament, possesses the ability to promote invasion of the ssDNA into a homologous dsDNA target (5). Since Rad51 possesses considerable affinity for dsDNA, presynaptic filament assembly is inhibited by double-stranded (ds) DNA (5). We have found that formation of the presynaptic filament is promoted by the BRCA2-derived polypeptide BRC3/4-DBD, containing BRC3 and BRC4 repeats (BRC3/4) and DNA binding domain (DBD) of BRCA2, by targeting Rad51 to ssDNA even when an excess of dsDNA is present (6).

Once assembled, the presynaptic filament probes homology in the target duplex and, once homology is located, mediates invasion of the target to form a DNA joint known as the displacement loop (D-loop, *see* **Fig. 24.1**). RPA and recombination mediators are important factors that facilitate the formation of the presynaptic filament (1, 7). RPA is a conserved ssDNA binding protein consisting of 14, 32, and 70 kDa subunits (8), and it facilitates presynaptic filament assembly by minimizing secondary structure in ssDNA which can interfere with the growth of the presynaptic filament (9, 10). However, owing to the high affinity of RPA for ssDNA, the transition from RPA-bound ssDNA to a Rad51-ssDNA filament requires additional protein factors, known as recombination mediators, such as Rad52 (11, 12) and the Rad55–Rad57 complex (13) in *Saccharomyces cerevisiae*, BRCA2 (6) and the Rad51B–51C complex (14) in humans. These mediator proteins thus help ensure the timely assembly of the Rad51 presynaptic filament during HR (**Fig. 24.1**).

As alluded to above, assembly of catalytically active presynaptic filaments only requires ATP binding (15). Thus, even a non-hydrolyzable ATP analogue, such as AMP-PNP, is sufficient for activating the recombinase activity of Rad51 (16, 17). Interestingly, since ATP hydrolysis prompts the turnover of the presynaptic filament, suppressing ATP hydrolysis by (1) mutating Lys to Arg in the Walker A motif in Rad51 (16, 17), (2) adding Ca^{2+} ion (18), or (3) using a non-hydrolyzable ATP analogue (15) leads to presynaptic filament stabilization and an enhancement of the ability of Rad51 to catalyze homologous DNA pairing. In addition, the presynaptic filament is also stabilized by the Hop2–Mnd1 complex (19) and Rad54 protein (20).

Prior to the formation of a stable joint between the recombining ssDNA and dsDNA, the DNA molecules are homologously aligned in a paranemically linked joint, known as the "synaptic complex," within the presynaptic filament (1). Importantly, Hop2–Mnd1 functions in conjunction with the presynaptic filament to capture duplex DNA, thus facilitating synaptic complex formation (19, 21) (**Fig. 24.1**). Strand switching in the synaptic complex then generates the nascent D-loop, in which the DNA strands are topologically intertwined, or plectonemically linked. The size of the D-loop is expanded by continual uptake of the ssDNA into the duplex donor molecule (1).

This chapter provides detailed protocols for various in vitro analyses germane for examination of different phases of the homologous DNA pairing reaction: (1) oligonucleotide-based homologous DNA pairing, (2) specific targeting of hRad51 to ssDNA, (3) stabilization of the Rad51 presynaptic filament, and (4) duplex capture by the presynaptic filament.

2. Materials

2.1. Protein Preparation

Published protocols were used for the purification of human Rad51 (hRad51) (9), human RPA (hRPA) (9), the BRCA2-derived peptide-BRC3/4-DBD (6), and the Hop2–Mnd1 complex (19).

2.2. DNA Preparation

1. DNA substrates: The sequences of oligonucleotides are listed in **Table 24.1**
2. Micro Bio-Spin 6 chromatography column (Bio-Rad)
3. Micro-dialysis kit (GeBAflex, 3 ml, IBI)

Table 24.1
Oligonucleotide sequences

	Length	Sequence	Use
Oligo 1	40 nt	5′-TAATACAAAATAAGTAAATGAATAAACAGAGAAAA TAAAG-3′	Section 3.2
Oligo 2	40 nt	5′-CTTTATTTTCTCTGTTTATTCATTTACTTATTTTGT ATTA-3′	Section 3.2
Oligo 3	150 nt	5′-TCTTATTTATGTCTCTTTTATTTCATTTCCTATATT TATTCCTATTATGTTTTATTCATTTACTTATTCTTT ATGTTCATTTTTTATATCCTTTACTTTATTTTCTCT GTTTATTCATTTACTTATTTTGTATTATCCTTATCT TATTTA-3′	Section 3.2
Oligo 4	60 nt	5′-TGTCTCTTTTATTTCATTTCGTTTTATTCACTATAT TTTTTATTTCTATTTTACTTATTT-3′	Section 3.3
Oligo 5	60 nt	5′-AAATTAAGTAAAATAGAAATAAAAAATATAGTGAAT AAAACGAAATGAAATAAAGAGACA-3′	Section 3.3
Oligo 6	80 nt	5′-AAATAAGTAAAATAGAAATAAAAAATATAGTGAAT AAAACGAAATGAAATAAAAGAGACA-3′	Section 3.5
Oligo-7	80 nt	5′-ATGAACATAATTGAAATAAGGATCCGGCTAATACA AAATAAGTAAAAGGTTAAACATAGAATTCAAAGTA AAGGATATAA-3′	Section 3.5

4. Denaturing gel loading buffer (2×): 94% formamide, 20 mM Tris–HCl, pH 7.5, 2 mM EDTA, 0.05% bromophenol blue, and 0.05% xylene cyanol
5. Native gel loading buffer (4×): 30 mM Tris–HCl (pH 7.5), 50% glycerol, and 0.1% orange G
6. T4 polynucleotide kinase (NEB) and PNK buffer (10×) (supplied from the manufacturer)
7. γ^{32}P-ATP, 10 μCi/μl, 6,000 Ci/mmol (Perkin Elmer)
8. TE buffer (1×): 10 mM Tris–HCl, pH 7.5, and 1 mM EDTA
9. TAE (1×): 40 mM Tris–acetate, pH 7.5, and 0.5 mM EDTA
10. Buffer H: 50 mM Tris–HCl, pH 7.5, 10 mM $MgCl_2$, and 100 mM NaCl
11. Hand-held UV lamp
12. Amicon Ultra-4 (10 K NMWL, Millipore)

2.3. Preparation of DNA Magnetic Beads

1. Streptavidin-coated magnetic beads (Roche Molecular Biochemicals)

2. Buffer B: 10 mM Tris–HCl, pH 7.5, 100 mM NaCl, and 1 mM EDTA
3. Buffer W: 10 mM Tris–HCl, pH 7.5, 1,000 mM NaCl, and 1 mM EDTA
4. Magnetic separator (with 1.5 ml microfuge tube holders)
5. Rotary mixer

2.4. Homologous DNA Pairing

1. 0.5 μM 150-mer ssDNA (Oligo 3)
2. 0.5 μM 40-bp dsDNA (annealing product of Oligo 2 and ^{32}P-labeled Oligo 1)
3. Buffer E (5×): 250 mM Tris–HCl, pH 7.5, 5 mM $MgCl_2$, and 5 mM dithiothreitol (DTT)
4. 10 μg/μl BSA
5. Buffer T300: 25 mM Tris–HCl, pH 7.5, 10% glycerol, 0.5 mM EDTA, 1 mM DTT, and 300 mM KCl
6. 12.5 mM ATP (*see* **Note 1**)
7. 25 μM hRad51, 7.5 μM hRPA, 12.5 μM BRC3/4-DBD
8. 50 mM spermidine (*see* **Note 2**)
9. 5% SDS
10. 10 mg/ml proteinase K (*see* **Note 3**)

2.5. Targeting of hRad51 to ssDNA

1. ssDNA magnetic beads (*see* **Section 3.1.3**)
2. 20 μM 60-mer dsDNA (annealing product of Oligo 4 and Oligo 5)
3. Buffer R (5×): 250 mM Tris–HCl, pH 7.5, 300 mM KCl, 5 mM $MgCl_2$, and 5 mM DTT
4. Buffer T (1×): 25 mM Tris–HCl, pH 7.4, 10% glycerol, 0.5 mM DTT, and 1 mM $MgCl_2$
5. Magnetic separator
6. 54 μM hRad51, 12.5 μM BRC3/4-DBD
7. 100 mM ATP, 2% SDS, 10 mg/ml proteinase K, 10% Igepal (*see* **Note 4**)
8. SDS-PAGE loading buffer (2×): 100 mM Tris–HCl, pH 6.8, 0.2 M DTT, 4% SDS, 20% glycerol, and 0.01% bromophenol blue
9. Native gel loading buffer (4×)

2.6. Stabilization of the Rad51 Nucleoprotein Filament

1. ssDNA magnetic beads (*see* **Section 3.1.3**)
2. 40 μM ^{32}P-labeled poly dT_{83}
3. Buffer S (5×): 175 mM Tris–HCl, pH 7.5, 250 mM KCl, 5 mM $MgCl_2$, and 5 mM DTT

4. Magnetic separator
5. 54 μM hRad51, 45 μM Hop2–Mnd1
6. 100 mM ATP, 2% SDS, 10 mg/ml proteinase K, 10% Igepal
7. SDS-PAGE loading buffer (2×), native gel loading buffer (4×)

2.7. dsDNA Capture by the Rad51 Nucleoprotein Filament

1. ssDNA magnetic beads (*see* **Section 3.1.3**)
2. 0.5 μM ^{32}P-labeled 80-mer ssDNA (Oligo 6)
3. 0.5 μM ^{32}P-labeled 80-bp dsDNA (annealing product of Oligo 7 and ^{32}P-labeled Oligo 6)
4. Buffer C (5×): 175 mM Tris–HCl, pH 7.5, 250 mM KCl, 5 mM DTT, and 5 mM $MgCl_2$
5. Buffer N: 35 mM Tris–HCl, pH 7.5, 50 mM KCl, 100 ng/μl BSA, 1 mM DTT, 2 mM AMP-PNP, and 1 mM $MgCl_2$
6. 54 μM hRad51, 18 μM Hop2–Mnd1
7. Magnetic separator
8. 100 mM AMP-PNP, 2% SDS, 10 mg/ml proteinase K, native gel loading buffer (4×)

2.8. Electrophoresis and Detection (Other Similar Systems Can Be Used)

1. Protean II (Bio-Rad) gel electrophoresis kit
2. Mini-Protean 3 (Bio-Rad) gel electrophoresis kit
3. Denaturing gel for purification of oligonucleotides (20 × 16 cm, 1.5 mm thick, containing 7 M urea, 10% polyacrylamide, and 1 × TAE, prepared according to instructions of Protein II system)
4. Native polyacrylamide gel for purification of dsDNA (20 × 16 cm, 1.5 mm thick, containing 10% polyacrylamide (*see* **Note 5**) in 1 × TAE, prepared according to instructions of Protein II system)
5. Native polyacrylamide gel for assays (10 × 7 cm, 1.5 mm thick, containing 10% polyacrylamide in 1 × TAE, prepared according to instructions of Mini-Protean 3)
6. Personal FX phosphorimager and the Quantity One software (Bio-Rad)
7. DE81 DEAE paper (Whatman), chromatography paper, 3MM Chr (Whatman)
8. Gel dryer
9. SDS-PAGE gel for protein analysis (10 × 7 cm, 0.75 mm thick, 10% polyacrylamide, 1% SDS, prepared according to instructions of Mini-Protean 3)

3. Methods

3.1. Preparation of DNA Substrates

3.1.1. Gel Purification of Oligonucleotides

1. Dissolve 200 μg chemically synthesized oligonucleotide (listed in **Table 24.1**) in 100 μl TE buffer and mix with an equal volume of the denaturing gel loading buffer.
2. Incubate the sample at 95°C for 5 min and then chill on ice for 5 min. Load on the denaturing gel containing 7 M urea, 10% acrylamide, and 1 × TAE. Run the gel at 150 V for 4 h at 55°C.
3. After the electrophoresis, disassemble the gel apparatus and place the gel on Saran wrap over a sheet of white paper. Locate DNA band in the gel by shadowing the gel under a hand-held UV lamp. Excise the gel piece containing DNA with a razor blade, place it in a micro-dialysis tube, and electro-elute in 1 × TAE for 2 h, 100 V at 4°C. Finish eluting by reversing the current for 2 min.
4. Transfer the DNA solution in the dialysis tube into the filter unit of Amicon Ultra-4 and concentrate by centrifugation at 6,000 × *g* until the volume of the buffer is reduced to about 100 μl. Add 500 μl of fresh 1 × TAE and repeat centrifugation to remove residual urea and acrylamide. Repeat the step twice.
5. Transfer the buffer into a microfuge tube and measure the concentration of DNA from absorbance at 260 nm using a UV spectrometer, make ~10 μl aliquots, and store at –20°C.

3.1.2. Preparation of the 5′-End Radiolabeled dsDNA

1. Incubate 5 μg of a purified oligonucleotide in 70 μl of 1 × PNK buffer supplemented with 50 μCi γ^{32}P-ATP, 1 μl of PNK at 37°C for 1 h. Stop the reaction by incubating at 65°C for 15 min and remove unincorporated γ^{32}P-ATP by using a Micro Bio-Spin 6 chromatography column according to the manufacturer's protocol.
2. Mix the 5′-end-labeled oligomer and the complementary sequence in an equal molar ratio in 70 μl of buffer H and anneal them using slow cooling from 90 to 4°C (**Note 6**).
3. Add 30 μl of native gel loading buffer and load on the native polyacrylamide gel. Run the gel with 1 × TAE at 150 V for 4 h at 4°C. The double-stranded DNA band can be identified by exposing the gel to UV as described in **Section 3.1.1**.
4. Excise the gel piece containing the dsDNA from the gel, extract DNA by electro-elution, and concentrate DNA as described in **Section 3.1.1**.

3.1.3. Preparation of DNA Magnetic Beads

1. Mix streptavidin-coated magnetic beads (Roche Molecular Biochemicals) thoroughly and transfer 400 μl of the slurry to a 1.5 ml microfuge tube. Place the tube in a magnetic separator and wait until the magnetic beads are completely separated from buffer.
2. Remove the buffer using a pipette, and wash beads three times with 400 μl buffer B.
3. Add 15 μg of biotinylated dT_{83} purified as described in **Section 3.1.1** and mix for 4 h at 25°C with gentle mixing using a rotary mixer.
4. Remove the supernatant and wash the beads twice with 800 μl buffer W. Resuspend the beads in 400 μl buffer B and measure the concentration of DNA from absorbance at 260 nm with a UV spectrometer using mock-treated beads as a control.
5. Store the beads at 4°C (*see* **Note 7**).

3.2. Homologous DNA Pairing

In this assay, DNA strand exchange between 150-mer ssDNA and homologous 40-bp dsDNA is performed by hRad51 in the presence of ATP (*Standard homologous DNA pairing*). The DNA pairing product, 40/150-mer, is identified by native gel electrophoresis, followed by phosphor imaging of the gel. In *restoration of homologous DNA pairing reaction* (**Fig. 24.2**), preincubation of RPA with ssDNA significantly reduces formation of the product, due to high binding affinity of RPA for ssDNA. A mediator (BRC3/4-DBD) restores Rad51-mediated strand exchange in the presence of PRA.

3.2.1. Standard Homologous DNA Pairing Reaction with hRad51

1. A 12.5 μl reaction is set up as follows:
   2.5 μl buffer E (5×),
   0.125 μl 10 μg/μl BSA,
   1 μl 12.5 mM ATP,
   1 μl 0.5 μM Oligo 3 (150-mer),
   2.875 μl H_2O (to make final volume 12.5 μl).

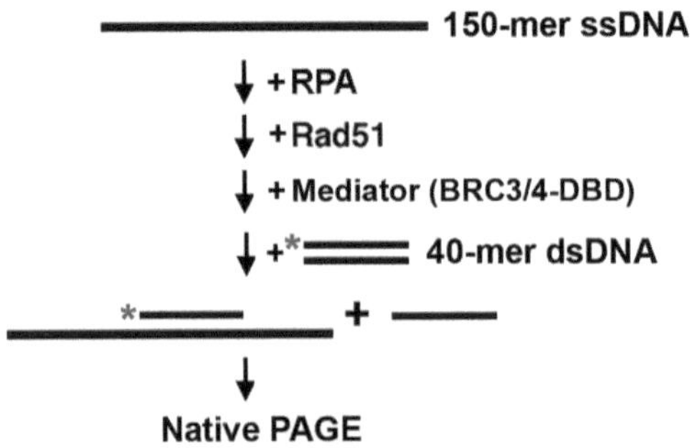

Fig. 24.2. Homologous DNA pairing assay.

2. Add buffer T300 to adjust final KCl concentration of the reaction to 75 mM.
3. Add 1 μl of 25 μM hRad51 (*see* **Note 8**) to the mixture and mix gently while avoiding generating air bubbles.
4. Incubate at 37°C for 5 min.
5. Add 1 μl of 0.5 μM 40-bp dsDNA and 1 μl of 50 mM spermidine. Incubate at 37°C for 30 min.
6. Stop the reaction with 1.25 μl 5% SDS and 0.7 μl of 10 mg/ml proteinase K.
7. Incubate the sample for 15 min at 37°C.
8. Add 5 μl of native gel loading buffer and load the sample onto a 10% native polyacrylamide gel in 1 × TAE.
9. Run the gel at 120 mA for 90 min in 1 × TAE.
10. Disassemble the gel apparatus and dry the gel on a sheet of DEAE paper to prevent the loss of DNA under vacuum using a gel dryer.
11. Expose the dried gel to a phosphorimaging screen for an appropriate time and analyze it in a Personal FX phosphorimager using the Quantity One software (Bio-Rad).

3.2.2. Restoration of Homologous DNA Pairing Reaction by BRC3/4-DBD

12. Set up the reaction in Step 1.
13. Add 1 μl of 7.5 μM hRPA (*see* **Note 9**) and incubate at 37°C for 5 min.
14. Add 1 μl of 25 μM hRad51 and incubate at 37°C for 10 min.
15. Add 0.2 μl of 12.5 μM BRC3/4-DBD (*see* **Note 10**) and incubate at 37°C for 5 min.
16. Follow Steps 5–10 described in the standard reaction.

3.3. Targeting of hRad51 to ssDNA by BRC3/4-DBD

hRad51 is able to form nucleoprotein filaments on both ssDNA and dsDNA, but only the ssDNA filament is active for strand exchange. Previously we demonstrated that a BRCA2-derived peptide BRC3/4-DBD increases hRad51's loading on ssDNA. In this assay, the effect of BRC3/4-DBD on partitioning hRad51 between dsDNA and ssDNA immobilized on streptavidin magnetic beads is examined (**Fig. 24.3a**).

1. Set up a 20 μl reaction in a 1.5 ml microfuge tube as follows:
   4 μl buffer R (5×),
   0.2 μl 10 μg/μl BSA,
   0.4 μl 5 mM ATP,
   7.68 μl H_2O (to make final volume 20 μl).
2. Add 1 μl of 54 μM hRad51 and 0.32 μl of 12.5 μM BRC3/4-DBD. Incubate at 25°C for 10 min.

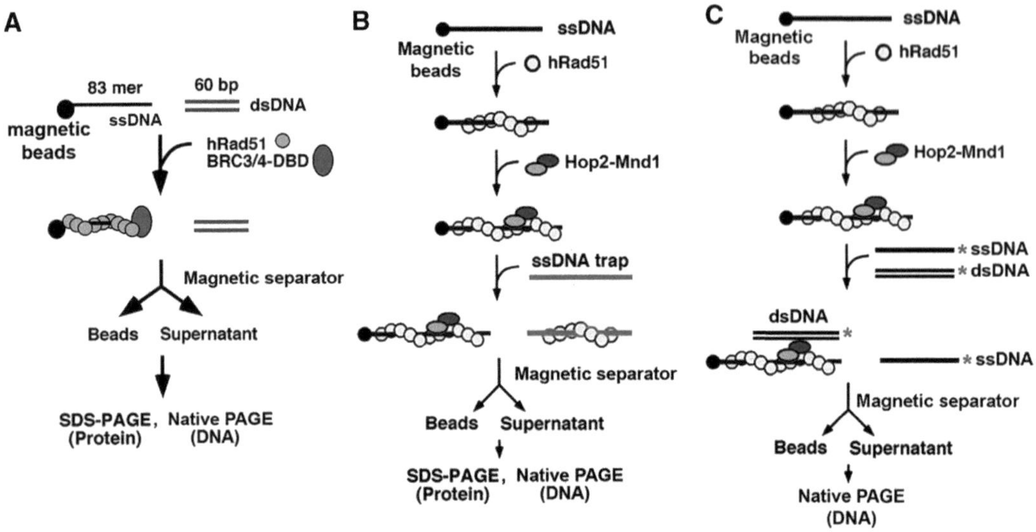

Fig. 24.3. Schematic of magnetic beads-based assays. (**a**) Targeting of hRad51 to ssDNA. (**b**) Stabilization of the hRad51 presynaptic filament. (**c**) Duplex capture by the hRad51 presynaptic filament.

3. In a separate tube, mix 1.2 μl of 20 μM 60-mer dsDNA and 4 μl of ssDNA magnetic beads containing biotinylated poly dT_{83}.
4. Add 5.2 μl of the DNA mixture to the tube containing Rad51 and BRC3/4-DBD.
5. Incubate at 37°C for 10 min with gentle mixing every 30 s.
6. Collect the beads with a magnetic separator and transfer the supernatant to a separate tube.
7. Wash the beads twice with 20 μl of buffer T supplemented with 0.01% Igepal and 0.1 mM ATP. (ATP is included in the washing buffer to maintain the assembled Rad51 filament on biotinylated poly dT_{83}.)
8. Add 20 μl of 2% SDS to the bead, vortex the tube, and separate the SDS eluate and beads with spinning.
9. Add 3 μl SDS-PAGE loading buffer to 10 μl of the SDS eluate and supernatant from Step 6. Run SDS-PAGE at 200 V for 40 min.
10. Stain the gel with Coomassie staining to visualize proteins (*see* **Note 11**).
11. Add 0.5 μl of 10 mg/ml proteinase K to 10 μl of the SDS eluate and supernatant and incubate at 37°C for 15 min.
12. Add 4 μl of native gel loading buffer and load the sample onto a 10% native polyacrylamide gel in 1 × TAE. Run native gel electrophoresis at 120 mA for 90 min.
13. Stain the gel with ethidium bromide to visualize DNA species.

3.4. Stabilization of the Rad51 Nucleoprotein Filament

In this assay, the Rad51 nucleoprotein filament is assembled in the presence of ATP on biotinylated ssDNA (poly dT_{83}) that is linked to streptavidin magnetic beads. The filament is challenged with an excess of free ssDNA (poly dT_{83}) with/without Hop2–Mnd1. The bead and supernatant fractions are separated with a magnetic separator, followed by identification of proteins and DNA using gel electrophoresis (**Fig. 24.3b**).

1. Set up a 20 μl reaction in a 1.5 ml microfuge tube as follows:
   4 μl buffer S (5×),
   0.2 μl 10 μg/μl BSA,
   4 μl of ssDNA magnetic beads,
   0.4 μl 100 mM ATP,
   9.4 μl H_2O (to make final volume 20 μl).
2. Add 1 μl of 54 μM hRad51 and incubate at 37°C for 5 min.
3. Add 0.4 μl of 45 μM Hop2–Mnd1 (*see* **Note 12**) and incubate another 5 min with gentle mixing.
4. To trap dissociated protein from beads, add 0.6 μl of 40 μM ^{32}P-labeled poly dT_{83} and gently mix for 10 min at 37°C.
5. Separate the magnetic beads and supernatant with a magnetic separator.
6. Transfer the supernatant to a microfuge tube for later analysis.
7. Wash the beads twice with 40 μl buffer S (1×) supplemented with 2 mM ATP and 0.01% Igepal.
8. Remove the buffer and add 20 μl SDS-PAGE loading buffer (1×) to the beads.
9. Vortex the tube and separate the SDS eluate and beads with spinning.
10. Load 10 μl of the SDS eluate and supernatant from Step 6 on a SDS-PAGE gel and electrophorese at 200 V for 40 min.
11. Visualize proteins with Coomassie staining (*see* **Note 13**).
12. Add 0.5 μl of 10 mg/ml proteinase K to the 10 μl SDS eluate and supernatant from Step 6 and incubate at 37°C for 15 min.
13. Add 4 μl of native gel loading buffer and load the samples onto a 10% native polyacrylamide gel in 1 × TAE. Run native gel electrophoresis at 120 mA for 90 min.
14. Visualize DNA species as described in **Section 3.2.1**.

3.5. dsDNA Capture by the Rad51 Nucleoprotein Filament

An engaging step of the presynaptic filament and homologous dsDNA is required for strand exchange, and protein factor(s), e.g., Hop2–Mnd1, that accelerates this step stimulates DNA joint formation. The ability of the presynaptic filament to capture duplex DNA can be examined in the following assay (**Fig. 24.3c**):

1. Set up a 20 μl reaction in a 1.5 ml microfuge tube as follows:
   4 μl buffer C (5×),
   0.2 μl 10 μg/μl BSA,
   4 μl of ssDNA magnetic beads,
   0.4 μl 100 mM AMP-PNP (*see* **Note 14**),
   10.4 μl H_2O (to make final volume 20 μl).
2. Add 1 μl of 54 μM hRad51 and incubate at 37°C for 5 min.
3. Separate the magnetic beads from supernatants with a magnetic separator. Remove supernatant and wash beads with 20 μl of buffer N and then resuspend the beads in 19 μl of buffer N.
4. Add 1 μl of 18 μM Hop2–Mnd1 to the mixture and incubate at 30°C for 5 min. The magnetic beads are collected with a magnetic separator. Then, wash the beads with 20 μl of buffer N and resuspend in 18 μl of buffer N.
5. Add 2 μl each of 0.5 μM ^{32}P-labeled 80-mer ssDNA and 0.5 μM ^{32}P-labeled 80-mer dsDNA to the beads and incubate at 30°C for 10 min with gentle mixing every 30 s.
6. Collect the beads with a magnetic separator and set aside the supernatant for analysis later. Wash the beads twice with 20 μl of buffer N containing 0.01% Igepal. The bound proteins and radiolabeled DNA are eluted with 20 μl of 2% SDS.
7. Add 1 μl of 10 mg/ml proteinase K to the SDS eluates and the supernatants and incubate at 37°C for 15 min.
8. Add 7 μl of native gel loading buffer and load the sample onto a 10% native polyacrylamide gel in 1 × TAE.
9. Run the gel at 120 mA for 90 min and follow Steps 9–10 in **Section 3.2.1**.

4. Notes

1. The concentration of the ATP stock solution is 100 mM. To make the stock, dissolve ATP in H_2O and adjust pH to 7.5 with 10 N NaOH. Store small aliquots at –80°C. Dilute to the desirable concentration with H_2O before use.
2. To make the 50 mM stock of spermidine, dissolve 0.1273 g of spermidine (Sigma) in 10 ml of 50 mM Tris–HCl, pH 7.5. Store small aliquots at –80°C.

3. Dissolve 100 mg of proteinase K (Roche) in 10 ml of 20 mM Tris–HCl, pH 7.5. Store small aliquots at –80°C.
4. Tenfold dilution of Igepal CA-630 (Sigma) with H_2O.
5. The polyacrylamide gels are made with an acrylamide/ *N,N'*-methylene-bis-acrylamide concentration of 30/0.8. However, a higher concentration of *N,N'*-methylene-bis-acrylamide (30/1.5) is used for purification of 40-mer dsDNA in order to separate the dsDNA from ssDNA.
6. Using a thermal cycler, incubate the tube containing two complementary oligos at 95°C for 10 min. Then, gradually decrease temperature to 25°C over a period of 140 min and then incubate at 25°C for 30 min. Store the tube at 4°C. Alternatively, put the mixture in a 1.5 ml microfuge tube and float the tube in a 1 l beaker filled with boiling water. Place the beaker in a styrofoam box and let it cool slowly to room temperature. Then, store the tube at 4°C.
7. Concentration of ssDNA (poly dT_{83}) in the beads should be close to 20 ng/μl. Otherwise, adjust the volume of the solution with buffer B. Store the beads in an amber (light protection) colored tube on ice. The ssDNA magnetic beads can be used within 1 month.
8. Dilute the hRad51 stock solution with buffer T300 (25 mM Tris–HCl, pH 7.5, 10% glycerol, 0.5 mM EDTA, 1 mM DTT, and 300 mM KCl) accordingly and store on ice. The activity of Rad51 does not significantly change on ice for a week. The precise concentration of Rad51 to generate maximum homologous DNA pairing should be optimized by titration.
9. Dilute concentrated hRPA with buffer T300. The amount of RPA sufficient to abrogate homologous DNA pairing of Rad51 should be determined by titration.
10. Appropriate control reactions should be carried out in parallel. The BRC3/4-DBD protein is prone to degradation. Refreezing is undesirable.
11. Without BRC3/4-DBD, 20–30% of hRad51 is present in the ssDNA magnetic beads fraction. The amount of hRad51 increases up to ~90% by addition of 100–200 nM BRC3/4-DBD.
12. Dilute Hop2–Mnd1 with buffer T300 to obtain proper concentration. The activity of Hop2–Mnd1 does not significantly change on ice for 1 week.
13. Appropriate control reactions should be carried out in parallel. Without Hop2–Mnd1, ~30% of hRad51 remains in ssDNA magnetic beads, whereas Hop2–Mnd1 increases hRad51 in the beads up to ~70%.

14. Use AMP-PNP as a nucleotide cofactor instead of ATP, because hRad51 forms a stable presynaptic filament with AMP-PNP. Alternatively, use hRad51 K133R (change of lysine 133 to arginine) mutant and ATP as a nucleotide cofactor.

Acknowledgments

The studies described in this chapter have been supported by research grants from the US National Institutes of Health.

References

1. San Filippo, J., Sung, P., and Klein, H. (2008) Mechanism of eukaryotic homologous recombination. *Annu Rev Biochem* **77**, 229–257.
2. Symington, L.S. (2002) Role of RAD52 epistasis group genes in homologous recombination and double-strand break repair. *Microbiol Mol Biol Rev* **66**, 630–670.
3. Sung, P., Krejci, L., Van Komen, S., and Sehorn, M.G. (2003) Rad51 recombinase and recombination mediators. *J Biol Chem* **278**, 42729–42732.
4. Sheridan, S.D., Yu, X., Roth, R., et al. (2008) A comparative analysis of Dmc1 and Rad51 nucleoprotein filaments. *Nucleic Acids Res* **36**, 4057–4066.
5. Sung, P., and Robberson, D.L. (1995) DNA strand exchange mediated by a RAD51-ssDNA nucleoprotein filament with polarity opposite to that of RecA. *Cell* **82**, 453–461.
6. San Filippo, J., Chi, P., Sehorn, M.G., Etchin, J., Krejci, L., and Sung, P. (2006) Recombination mediator and Rad51 targeting activities of a human BRCA2 polypeptide. *J Biol Chem* **281**, 11649–11657.
7. Wang, X., and Haber, J.E. (2004) Role of Saccharomyces single-stranded DNA-binding protein RPA in the strand invasion step of double-strand break repair. *PLoS Biol* **2**, E21.
8. Bochkarev, A., and Bochkareva, E. (2004) From RPA to BRCA2: lessons from single-stranded DNA binding by the OB-fold. *Curr Opin Struct Biol* **14**, 36–42.
9. Sigurdsson, S., Trujillo, K., Song, B., Stratton, S., and Sung, P. (2001) Basis for avid homologous DNA strand exchange by human Rad51 and RPA. *J Biol Chem* **276**, 8798–8806.
10. Sugiyama, T., Zaitseva, E.M., and Kowalczykowski, S.C. (1997) A single-stranded DNA-binding protein is needed for efficient presynaptic complex formation by the Saccharomyces cerevisiae Rad51 protein. *J Biol Chem* **272**, 7940–7945.
11. Shinohara, A., and Ogawa, T. (1998) Stimulation by Rad52 of yeast Rad51-mediated recombination. *Nature* **391**, 404–407.
12. Sung, P. (1997) Function of yeast Rad52 protein as a mediator between replication protein A and the Rad51 recombinase. *J Biol Chem* **272**, 28194–28197.
13. Sung, P. (1997) Yeast Rad55 and Rad57 proteins form a heterodimer that functions with replication protein A to promote DNA strand exchange by Rad51 recombinase. *Genes Dev* **11**, 1111–1121.
14. Sigurdsson, S., Van Komen, S., Bussen, W., Schild, D., Albala, J.S., and Sung, P. (2001) Mediator function of the human Rad51B-Rad51C complex in Rad51/RPA-catalyzed DNA strand exchange. *Genes Dev* **15**, 3308–3318.
15. Robertson, R.B., Moses, D.N., Kwon, Y., et al. (2009) Structural transitions within human Rad51 nucleoprotein filaments. *Proc Natl Acad Sci USA* **106**, 12688–12693.
16. Chi, P., Van Komen, S., Sehorn, M.G., Sigurdsson, S., and Sung, P. (2006) Roles of ATP binding and ATP hydrolysis in human Rad51 recombinase function. *DNA Repair* **5**, 381–391.

17. Sung, P., and Stratton, S.A. (1996) Yeast Rad51 recombinase mediates polar DNA strand exchange in the absence of ATP hydrolysis. *J Biol Chem* **271**, 27983–27986.
18. Bugreev, D.V., and Mazin, A.V. (2004) Ca^{2+} activates human homologous recombination protein Rad51 by modulating its ATPase activity. *Proc Natl Acad Sci USA* **101**, 9988–9993.
19. Chi, P., San Filippo, J., Sehorn, M.G., Petukhova, G.V., and Sung, P. (2007) Bipartite stimulatory action of the Hop2-Mnd1 complex on the Rad51 recombinase. *Genes Dev* **21**, 1747–1757.
20. Mazin, A.V., Alexeev, A.A., and Kowalczykowski, S.C. (2003) A novel function of Rad54 protein. Stabilization of the Rad51 nucleoprotein filament. *J Biol Chem* **278**, 14029–14036.
21. Pezza, R.J., Voloshin, O.N., Vanevski, F., and Camerini-Otero, R.D. (2007) Hop2/Mnd1 acts on two critical steps in Dmc1-promoted homologous pairing. *Genes Dev* **21**, 1758–1766.

Chapter 25

Studying DNA Replication Fork Stability in *Xenopus* Egg Extract

Yoshitami Hashimoto and Vincenzo Costanzo

Abstract

A crucial process to ensure cell survival and genome stability is the correct replication of the genome. DNA replication relies on complex machinery whose mechanisms are being elucidated using different model systems. A major aspect of this process, which is an intense subject of investigation, is what happens when replication forks encounter obstacles impairing their progression such as modified bases, pausing sites, and single strand breaks. The detailed biochemical analysis of DNA replication in the presence of DNA damage has been impeded by the lack of a cell-free system recapitulating DNA replication. Here we describe assays based on the vertebrate *Xenopus laevis* egg extract to study the biochemical aspects of replication fork stability.

Key words: DNA replication, *Xenopus laevis*, DNA repair, DNA damage, fork restart, replication inhibitors and chromatin.

1. Introduction

Replication forks frequently encounter DNA lesions. Among these, interruptions of DNA template continuity represent a major threat to the stability of replication forks. For example, single strand breaks occur quite frequently, although they are repaired very efficiently (1). However, even a single un-repaired nick can lead to a chromosome break at the passage of a replication fork. In this case the resulting chromosome break needs to be rapidly detected and repaired to promote the restart of the damaged fork and ensure complete replication.

Cell-free systems based on the *Xenopus laevis* egg extract are particularly useful to study DNA replication fork stability as they

H. Tsubouchi (ed.), *DNA Recombination*, Methods in Molecular Biology 745,
DOI 10.1007/978-1-61779-129-1_25,

allow extensive biochemical analysis and can reproduce basic cell cycle events such as chromatin formation, nuclear assembly, semi-conservative DNA replication, and DNA damage checkpoint activation. The high degree of genetic conservation between *Xenopus* and mammalian organisms facilitates the biochemical study of large proteins that can be easily isolated and characterized.

An important tool available for exploitation of this system is the possibility of depleting native proteins and protein complexes from the egg extract using specific antibodies. This approach can be considered the equivalent of a "biochemical knockout". Protein depletion can lead to the complete and rapid removal of a specific protein.

Using egg extract we have monitored responses to a variety of aberrant DNA structures that mimic damaged DNA and examined subsequent progression into S-phase.

For example, to study the role of ATM and ATR in promoting fork recovery we have designed assays in which replicating forks arrested by DNA damaging agents that impair fork progression are then allowed to restart in extracts lacking these agents. Using these assays we have shown that ATM, ATR, and the MRN complex promote restart of disrupted replication forks (2). Here we present some methodologies to study replication fork behaviour in the presence of DNA damage using *Xenopus* egg extract.

2. Materials

1. Interphase egg extract and demembranated sperm chromatin (**Note 1**).
2. Aphidicolin (Sigma) is dissolved in ethanol at 1 mg/ml or in DMSO at 10 mg/ml, stored at –20°C.
3. Camptothecin (Sigma) is dissolved in chloroform/methanol (4:1) at 4 mg/ml, stored at –20°C.
4. Solution of methyl methan sulfonate (MMS) from Sigma, stored at 4°C.
5. Cisplatin (Sigma) is dissolved in dH_2O at 5 mg/ml, stored at –20°C.
6. EB buffer: 50 mM Hepes–KOH (pH 7.5), 100 mM KCl, 2.5 mM $MgCl_2$, stored at R.T.
7. Replication stop buffer (neutral agarose gel): 8 mM EDTA, 80 mM Tris–HCl (pH 8.0), 0.13% phosphoric acid, 10% ficoll, 5% SDS, 0.2% bromophenol blue, stored at R.T.
8. Replication stop buffer (alkaline agarose gel): 20 mM Tris–HCl (pH 8.0), 5 mM EDTA, 200 mM NaCl, 0.5% SDS, stored at R.T.

9. Alkaline loading buffer: 1.25% ficoll, 0.0125% bromophenol blue, 1 mM EDTA, 50 mM NaOH (NaOH should be added just before using), stored at R.T.
10. Mounting solution (for immunofluorescence): 3.7% formaldehyde, 2 μg/ml Hoechst 33342, 80 mM KCl, 15 mM NaCl, 15 mM PIPES–KOH (pH 7.2), 50% glycerol, stored at −20°C.
11. Roscovitine (Calbiochem) is dissolved in DMSO at 100 mM, stored at −20 or −80°C.

3. Methods

3.1. Agents Inducing Stalled Replication Forks

When replication forks encounter DNA lesions on the template, fork progression is inhibited and the replication checkpoint is activated. Many agents are known to induce stalled replication forks. The most commonly used drugs and treatments in egg extracts are aphidicolin, camptothecin, ultraviolet light (UV), MMS, and cisplatin. Aphidicolin and camptothecin are inhibitors of DNA polymerases α/δ/ε and topoisomerase I, respectively, and can be added to egg extracts at various concentrations. UV, MMS, and cisplatin can instead be used to treat sperm chromatin and to induce damage before sperm chromatin is added to egg extract. UV treatment, which induces mainly pyrimidine dimers, requires a UV crosslinker such as the Stratalinker (Stratagene). To treat sperm chromatin with UV an aliquot of sperm nuclei is layered onto parafilm on ice in the Stratalinker and irradiated with 254 nm UV light (typically at 100–1,000 J/m^2). For MMS treatment, which methylates DNA, an aliquot of sperm chromatin is mixed with 0.2–1% MMS and incubated for 30 min at R.T. For cisplatin treatment, which induces intrastrand crosslinking of DNA, an aliquot of sperm chromatin is mixed with 1–10 μg/ml cisplatin and incubated for 2 h at R.T.

3.2. Analysing Chromatin and Nuclear Proteins by Western Blotting

Sucrose density centrifugation can be used to isolate nuclear and chromatin fractions from egg extract. Chromatin fractions can be isolated by disrupting nuclear membrane with detergent. With regard to stalled fork-related events, nuclear fraction isolation can be useful for the detection of Chk1 kinase phosphorylation, a central component of the replication checkpoint pathway. Chromatin fractions can be instead isolated to monitor the association and dissociation of replication fork proteins to DNA and to study fork restart.

It should be noted that it is sometimes difficult to separate chromatin bound proteins from egg extract due to the fact that egg extracts contain a large maternal protein stockpile. A useful control to make sure that the separation has been effective is to use a mock sample without sperm nuclei to be subjected to centrifugation through sucrose. This sample can be used as negative control to compare the Western blot signals obtained without sperm DNA with the ones obtained with sperm chromatin. In this protocol, the buffer used to prepare nuclear and chromatin fractions contains 100 mM KCl. It should be mentioned that replication fork proteins become soluble when the salt concentration is increased up to 0.5 M NaCl (7).

3.2.1. Nuclei Isolation

1. After incubation for an appropriate time of 3–4,000 sperm nuclei/μl in egg extract, re-suspend 50 μl of sample with 10 volumes of EB, and layer it on top of 10 volume of EB + 30% sucrose in 1.5 ml eppendorf tubes.
2. Spin at 6,000–10,000×*g* for 3 min at 4°C using swinging bucket rotor.
3. Carefully remove the supernatant together with the sucrose layer.
4. Add 1 ml of EB + 30% sucrose without disrupting the pellet. Spin at 6,000–10,000×*g* for 1 min at 4°C. Carefully remove the buffer.
5. Re-suspend sample with SDS-PAGE sample buffer and perform Western blotting.

3.2.2. Chromatin Isolation

1. After incubation for an appropriate time of 3–4,000 sperm nuclei/μl in egg extract, re-suspend each sample with 20 volume of EB + 0.2% triton X-100 (or NP40), and layer it on top of 200 μl of EB + 0.2% triton X-100 (or NP40) + 30% sucrose.
2. Spin at 10,000×*g* for 5 min at 4°C using a swinging bucket rotor.
3. Carefully remove the supernatant together with the sucrose layer.
4. Add 300 μl EB without disrupting the pellet. Spin at top speed (more than 10,000×*g*) for 1 min at 4°C. Carefully remove the buffer.
5. Suspend with SDS-PAGE sample buffer and perform Western blotting.

3.3. Analysing Replication Activity by Neutral or Denatured Agarose Gel Electrophoresis

To measure the relative bulk replication activity of egg extract, the easiest way is neutral agarose gel electrophoresis separation of ^{32}P-labelled replication products (**Fig. 25.1**). Alkaline agarose gel electrophoresis under denaturing condition is instead suitable for monitoring the maturation of nascent DNA strands whose sizes are less than 10 kb (**Fig. 25.2**). It should be noted that these

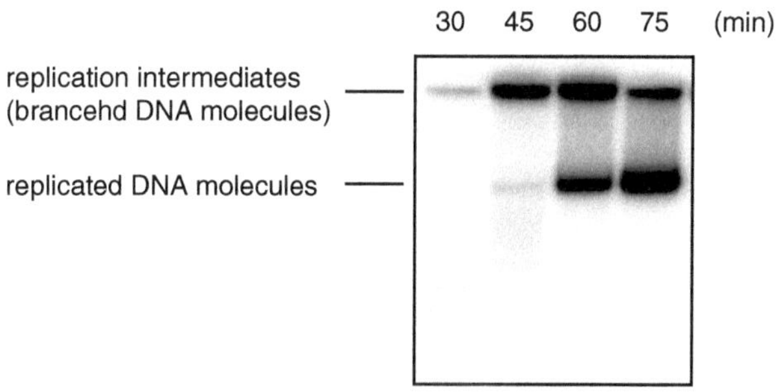

Fig. 25.1. Replication assay using neutral agarose gel. Sperm nuclei (4,000/μl) were incubated in 40 μl of egg extract with ^{32}P-dATP. After 30 min, 10 μl of extract was sampled at every 15 min. The replication products were resolved with 0.8% TAE agarose gel, followed by autoradiography. Two distinct bands are visible; the *upper band* corresponds to replication intermediates or branched DNA molecules, while the *lower band* corresponds to unbranched or fully replicated DNA molecules.

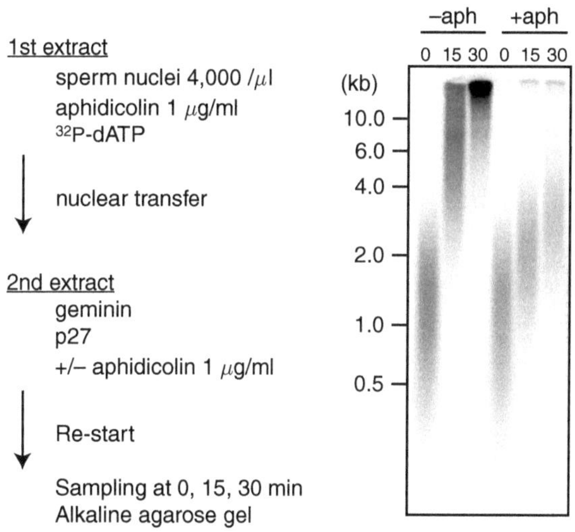

Fig. 25.2. Monitoring the maturation of nascent strand DNA pre-labelled with ^{32}P-dATP after fork restart. Sperm nuclei (4,000/μl) were incubated for 40 min in the first egg extract with ^{32}P-dATP and aphidicolin (1 μg/ml). Then, the sample was equally divided into two tubes and the nuclear fractions were isolated. Unincorporated ^{32}P-dATP is washed out and samples are re-suspended with the second extracts plus or minus aphidicolin (1 μg/ml) in the presence of geminin and p27. The replication products were resolved with 0.8% alkaline agarose gel, followed by autoradiography.

methods are not used for determining the absolute amount of replicated DNA, for which TCA precipitation must be used (8).

3.3.1. Neutral Agarose Gel

1. Use 1 μl of α-^{32}P-dATP (3,000 mCu/mmol) per 40–100 μl of egg extracts.
2. After incubation for an appropriate time of 3–4,000 sperm nuclei/μl in egg extract, mix the sample with one volume of

replication stop buffer. It is possible to freeze the samples at –20°C at this point.

3. Add 0.5 mg/ml of proteinase K and 50 μg/ml of RNase A in each sample and incubate for 2 h at 37°C (**Note 2**).
4. Half volumes of each sample are loaded onto 0.8% TAE agarose gel and run at 100 V for 1–2 h.
5. Wash the gel with dH_2O for 5 min, gently shaking three times.
6. The gel is dried on filter paper with a gel dryer and subjected to autoradiography.

3.3.2. Alkaline Agarose Gel

1. Steps 1–3 are similar to the neutral protocol. The samples are subjected to phenol/chloroform extraction and ethanol precipitation, and the pellets are air-dried and re-suspended with alkaline loading buffer.
2. Half volumes of each sample are loaded into 0.8% alkaline agarose gel and run at 2 V/cm for 14–16 h (**Note 3**).
3. Fix the gel with 10% methanol + 10% acetic acid for 30 min with gentle shaking.
4. Wash the gel with dH_2O for 5 min with gentle shaking three times.
5. The gel is dried on filter paper with a gel dryer and subjected to autoradiography.

3.4. Analysing Chromatin or Nuclear Localization by Immunofluorescence

1. For nuclear localization experiments, re-suspend samples with 100 μl of EB + 3.7% formaldehyde and incubate for 10 min at R.T. For chromatin localization, re-suspend with 90 μl of EB + 0.2% triton X-100 (or NP40). Then, add 11 μl of 37% formaldehyde and incubate for 10 min at R.T.
2. Add 400 μl of EB to stop fixation by formaldehyde.
3. Collect nuclei (or chromatin) on poly-L-lysine coated cover slip through EB + 30% sucrose using centrifugation at 500×*g* for 5 min.
4. Wash the cover slip with PBS-T (PBS + 0.05% Tween20).
5. Incubate with 20–30 μl of the first antibody solution appropriately diluted with 10% skim milk in PBS-T at R.T. for 2–3 h (or at 4°C, overnight).
6. Wash three times with PBS-T.
7. Incubate with 20–30 μl of the secondary antibody solution appropriately diluted with 10% skim milk in PBS-T at R.T. for 2 h.
8. Wash with PBS-T, PBS, and dH_2O sequentially and mount on slide glass using 3–5 μl of mounting solution.

3.5. Replication Restarting Assay

The replication restart assay using chromatin/nuclear transfer is a powerful tool to understand how replication resumes after fork stall or collapse. The basic idea is that replication is started in a first extract in the presence of replication blocking agents to induce stalled forks. Chromatin containing stalled forks is then isolated and transferred to another extract which does not contain the stalling agents to allow the restart of the stalled forks. Replication in the second extract can be followed as described (**Fig. 25.2**).

Although there is no obvious difference between early and late replication origins during DNA replication in egg extracts, many potential origins exist at each chromosome and only a small portion of them is actually utilized at any given time (9–11). These extra origins can be used to replicate regions affected by stalled forks. Therefore, unless these compensative origin firing events are inhibited, it is difficult to address the stalling and restarting of each replication fork. To overcome this problem, geminin, p27, and roscovitine can be used. Geminin, an inhibitor of the formation of pre-RC (pre replication complex), is a cell-cycle regulated protein and prevents re-replication during G2 phase. The replication activity is completely inhibited in egg extract pre-incubated with more than 80 nM of recombinant geminin. However, when geminin is added during early stage of pre-RC formation (3–5 min after the start of incubation), the number of pre-RC assembled on chromatin can be reduced. This condition is called "minimal licensing" (12). This treatment can be used to reduce the number of replication origins and study the effects of DNA damaging agents on DNA replication in the absence of redundant pathways affecting replication activity (8). In the replication restart protocol geminin can be used to block relicensing in the second extract when nuclear membrane is disrupted (see below). CDK inhibitors such as p27 and roscovitine are also useful to inhibit replication from origins that have assembled in the first extract and that can fire in the second one. In this case 40 μg/ml of recombinant p27 protein or 0.5 mM roscovitine can almost completely inhibit the replication initiation, but not the elongation. It is also possible to suppress the activation of additional replication origin firing by adding p27 or roscovitine after the initiation reaction (typically around 30 min after the start of incubation).

3.5.1. Chromatin/Nuclear Transfer Experiment

Chromatin or nuclear transfer is a useful method when it is necessary to change the extract conditions from one state to another (e.g., from extract treated with DNA damaging agents to an extract in which these agents are absent or from an extract depleted of specific component to an undepleted extract). Chromatin transfer obtained after permeabilization of the nuclear membrane is useful to assess the effect of factors depleted from the first extract that are present in the second extract. In contrast to chromatin intact nuclei are not completely exposed to the proteins of the second extract. It is important to point out that complete disruption of nuclear structure makes the chromatin difficult to replicate unless a limited concentration of detergent is used during the chromatin transfer experiment. The permeabilized nuclear membrane is repaired in the second extract.

1. Demembranated sperm chromatin (typically 1,000–4,000 nuclei/μl) is mixed with appropriate amount of interphase egg extract (20–50 μl per each sample).
2. Incubate at 23°C for appropriate time (**Note 4**).
3. For nuclear transfer, the sample is gently re-suspended with 20 volume of EB buffer and underlayered with 200 μl of EB buffer/30% sucrose. For chromatin transfer, EB and EB + 30% sucrose supplemented with 0.002–0.01% Triton X-100 or NP40 are used (more than 0.01% makes the chromatin difficult to replicate). (**Note 5**)
4. Spin at 10,000×*g* for 5 min at 4°C.
5. Carefully remove the supernatant together with the sucrose layer.
6. (Optional) Add 300 μl EB without disrupting the pellet. Spin at 10,000×*g* for 1 min at 4°C. Carefully remove the buffer.
7. Gently re-suspend with appropriate amount of egg extract (usually the same amount used for the first incubation) for the second incubation (**Note 6**).

4. Notes

1. There are many variations of the methods to prepare interphase egg extracts (3–5). Our favourite method has been previously described (6).
2. Since RNase A is also degraded by proteinase K, it may be better to do RNase A treatment for 20–30 min before adding proteinase K. Insufficient treatment of RNase A keeps the sample very viscous, which is hard to load into the gel.
3. It is possible to increase the voltage up to 4 V/cm as long as the gel and the buffer are kept refrigerated.
4. Usually replication initiates at 20–30 min and finishes at 60–90 min in the absence of exogenous DNA damage.
5. It is possible to add 2 mM DTT and 1 mM EDTA to the buffer for stabilization of nuclei.

6. A cut-tip should be used for re-suspension to avoid destruction of nuclear structure. It is necessary to repeat pipetting at least 10 times to solubilize the pellets.

References

1. Caldecott, K.W. (2007) Mammalian single-strand break repair: mechanisms and links with chromatin. *DNA Repair (Amst)* **6**, 443–453.
2. Trenz, K., Smith, E., Smith, S., and Costanzo, V. (2006) ATM and ATR promote Mre11 dependent restart of collapsed replication forks and prevent accumulation of DNA breaks. *EMBO J* **25**, 1764–1774.
3. Kornbluth, S., Yang, J., and Powers, M. (2006) Analysis of the cell cycle using Xenopus egg extracts, *Curr Protoc Cell Biol* **Chapter 11**, p. Unit 11 11.
4. Lupardus, P.J., Van, C., and Cimprich, K.A. (2007) Analyzing the ATR-mediated checkpoint using Xenopus egg extracts. *Methods* **41**, 222–231.
5. Powers, M., Evans, E.K., Yang, J., and Kornbluth, S. (2001) Preparation and use of interphase Xenopus egg extracts. *Curr Protoc Cell Biol* **Chapter 11**, p. Unit 11 10.
6. Trenz, K., Errico, A., and Costanzo, V. (2008) Plx1 is required for chromosomal DNA replication under stressful conditions. *EMBO J* **27**, 876–885.
7. Lee, J., Gold, D.A., Shevchenko, A., and Dunphy, W.G. (2005) Roles of replication fork-interacting and Chk1-activating domains from Claspin in a DNA replication checkpoint response. *Mol Biol Cell* **16**, 5269–5282.
8. Chong, J.P., Thommes, P., Rowles, A., Mahbubani, H.M., and Blow, J.J. (1997) Characterization of the Xenopus replication licensing system. *Methods Enzymol* **283**, 549–564.
9. Blow, J.J., and Ge, X.Q. (2009) A model for DNA replication showing how dormant origins safeguard against replication fork failure. *EMBO Rep* **10**, 406–412.
10. Ge, X.Q., Jackson, D.A., and Blow, J.J. (2007) Dormant origins licensed by excess Mcm2-7 are required for human cells to survive replicative stress. *Genes Dev* **21**, 3331–3341.
11. Woodward, A.M., Gohler, T., Luciani, M.G., Oehlmann, M., Ge, X., Gartner, A., Jackson, D.A., and Blow, J.J. (2006) Excess Mcm2-7 license dormant origins of replication that can be used under conditions of replicative stress. *J Cell Biol* **173**, 673–683.
12. Oehlmann, M., Score, A.J., and Blow, J.J. (2004) The role of Cdc6 in ensuring complete genome licensing and S phase checkpoint activation. *J Cell Biol* **165**, 181–190.

Chapter 26

Supported Lipid Bilayers and DNA Curtains for High-Throughput Single-Molecule Studies

Ilya J. Finkelstein and Eric C. Greene

Abstract

Single-molecule studies of protein–DNA interactions continue to yield new information on numerous DNA processing pathways. For example, optical microscopy-based techniques permit the real-time observation of proteins that interact with DNA substrates, which in turn allows direct insight into reaction mechanisms. However, these experiments remain technically challenging and are limited by the paucity of stable chromophores and the difficulty of acquiring statistically significant observations. In this protocol, we describe a novel, high-throughput, nanofabricated experimental platform enabling real-time imaging of hundreds of individual protein–DNA complexes over hour timescales.

Key words: Single molecule, TIRF microscopy, nanofabrication, DNA curtains, nucleosome, DNA motors.

1. Introduction

In recent years, fluorescence-based single-molecule imaging techniques have been used to follow the action of macromolecular machines along a DNA substrate. Direct observations of DNA replication (1–4), transcription (5–7), and repair (8–10) at the single-molecule level are continuing to offer fresh insights into these complex, multi-step reactions. The knowledge gained from these studies typically could not be accessed using traditional biochemical or biophysical approaches. Single-molecule experiments offer the advantage of being able to study rare or short-lived intermediates that can be obfuscated among the often-heterogenous populations of molecules that are studied in traditional biochemical assays.

H. Tsubouchi (ed.), *DNA Recombination*, Methods in Molecular Biology 745,
DOI 10.1007/978-1-61779-129-1_26, © Springer Science+Business Media, LLC 2011

Despite decades of intense technique development, single-molecule observations of protein–DNA interactions continue to be experimentally challenging. The relatively short fluorescent lifetimes of most organic dyes significantly limit the accessible reaction timescales. The molecules under investigation usually must be anchored to a surface that is inherently different from the normal environments encountered within cells. In addition, experimental platforms that manipulate the target DNA via optical or magnetic tweezers are typically carried out on single DNA molecules in a serial fashion (i.e., one molecule at a time), and this low data throughout often limit the scope of the experimental results. In the protocol presented here, we describe a method for rapid, real-time imaging of hundreds of individual protein–DNA complexes over extended, biological timescales within a biologically friendly microenvironment. The method is flexible and can be used to address a number of different biological problems. We have successfully applied this experimental approach to observe the diffusion and translocation of DNA repair proteins (10, 11), the localization of nucleosomes along an intrinsic DNA-binding energy landscape (12), and to follow the polymerization activity of recombinases on double-stranded DNA (13–15).

In this protocol, we describe a nanofabricated, micro-fluidic system for simultaneous imaging of hundreds of DNA molecules in real time (**Fig. 26.1**). The DNA molecules are organized into "DNA curtains" on the surface of a micro-fluidic sample chamber that is otherwise coated with a fluid lipid bilayer. Various aspects of the DNA curtains technology have been presented previously (16–19). Briefly, the experimental system consists of a total internal reflection fluorescence (TIRF) microscope built

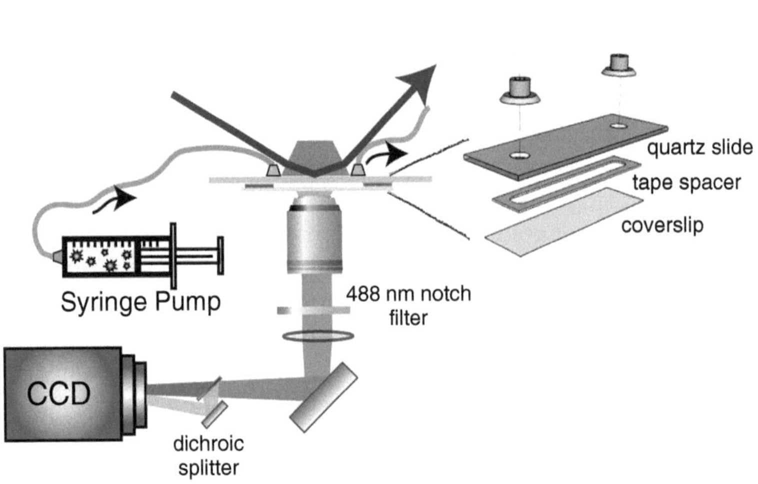

Fig. 26.1. Schematic of the fluorescence microscope setup. The flowcell is placed on a microscope stage in an inverted configuration. A 488 nm laser impinges on a DOVE prism that rests atop the flowcell. Fluorescent signal is collected by a high N.A. objective and is passed through a 488 nm notch filter and a DualView beam splitter before being imaged on a 512 × 512 pixel EM-CCD. A syringe pump delivers continuous buffer flow through the flowcell inlet port.

around an inverted Nikon TE2000 microscope. Laser illumination is provided by a ~200 mW 488 nm diode laser. The laser beam impinges on a DOVE prism atop a flowcell constructed from a silica microscope slide containing nanofabricated barriers to lipid diffusion (**Fig. 26.2**). An evanescent wave is generated at the water–silica interface, illuminating a shallow observation volume at the flowcell surface. Fluorescence from molecules immobilized at the surface (see below) is collected by a 60×, N.A. 1.2 water immersion objective. The signal is passed through a holographic 488 nm notch filter and imaged on a back-thinned 512 × 512 pixel EM-CCD. For multi-color fluorescence imaging, the signal is passed through a DualView beam splitter and each color imaged on one half of the CCD chip.

The surface of the flowcell is passivated by a fluid lipid bilayer (20–22). DNA is immobilized at the lipid bilayer by a streptavidin–biotin linkage and extended into the evanescent wave via shear buffer flow delivered by a syringe pump. The fluidity of the lipid bilayer permits organization of individual DNA molecules at nanofabricated diffusion barriers (**Fig. 26.3**). The spacing, density, and orientation of DNA molecules relative to one another may be controlled by appropriately designed diffusion barriers (17). Recently, we have also extended the DNA curtain technology to generate DNA arrays that are immobilized at both ends (16).

For our studies we label the proteins with highly fluorescent semiconductor nanocrystal quantum dots (QDs). Quantum dots

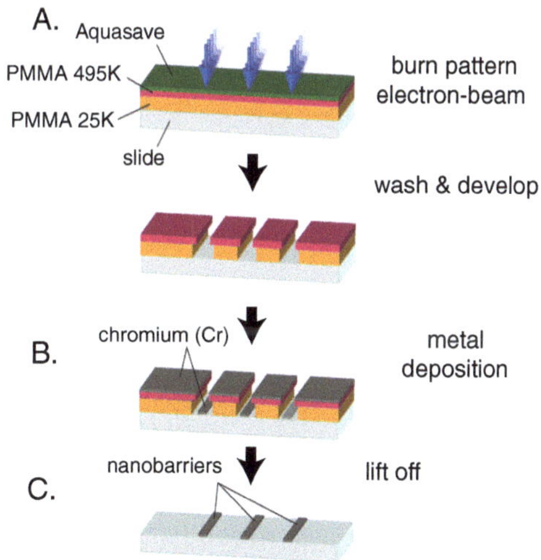

Fig. 26.2. Overview of electron beam lithography. (**a**) For e-beam lithography, the slide is first coated with PMMA, and a layer of Aquasave, and an electron beam rastered across the surface to burn through these layers creating a pattern that defines the shapes of the diffusion barriers. (**b**) Chromium (Cr) is deposited on the entire surface, and (**c**) the remaining PMMA is lifted off, leaving behind the nanofabricated barriers.

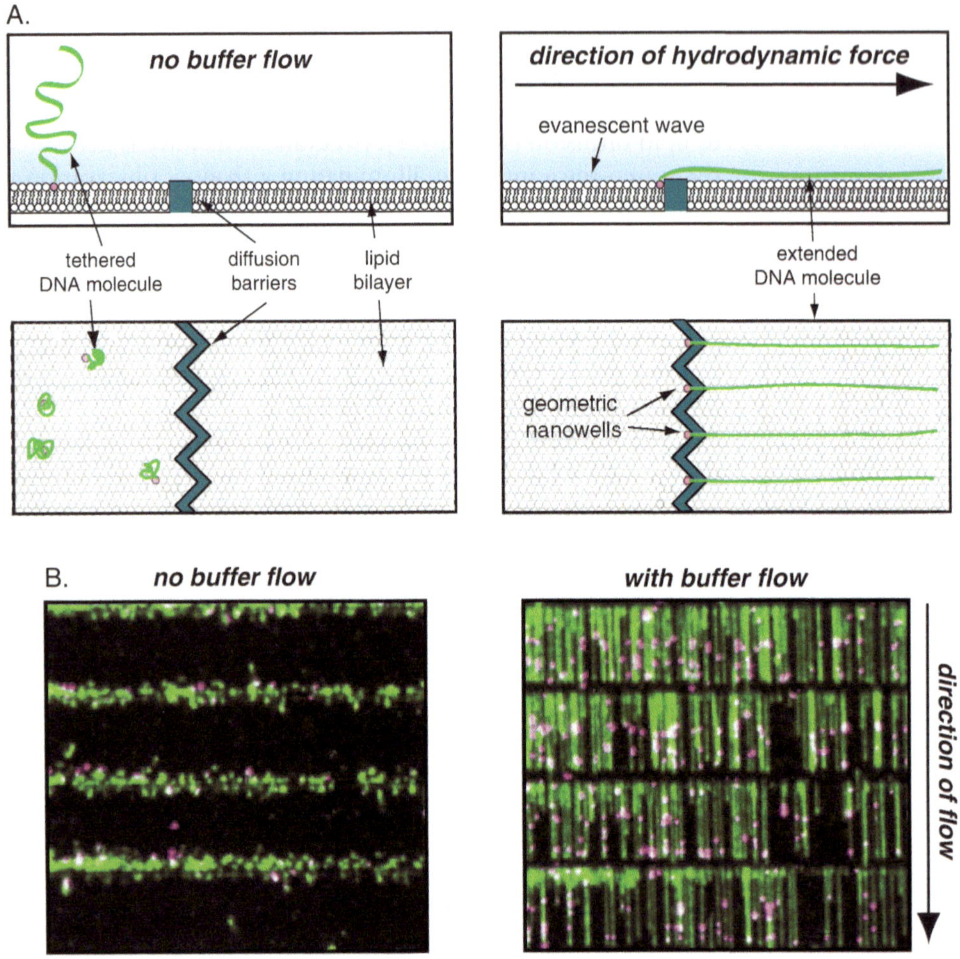

Fig. 26.3. Assembly of DNA curtains. (**a**) A schematic illustration of DNA molecules assembled into DNA curtains on a fluid lipid bilayer. DNA is tethered to the bilayer by a streptavidin–biotin linkage. In the presence of buffer flow, individual DNA molecules are pushed through the lipid bilayer until the molecules assemble at nanofabricated diffusion barriers. (**b**) A single field of view permits the observation of up to four rows of assembled λ-DNA curtains in the absence (*left panel*) and presence (*right panel*) of buffer flow. The DNA molecules have been decorated with QD-labeled RNA polymerase.

are relatively small (~10–20 nm diameter) nanoparticles that display broad excitation spectra, narrow emission peaks, large Stokes shifts, large absorbance cross sections, and very high quantum yields (23). Individual QDs can be readily visualized at data collection rates of 100 frames/s and the QDs do not bleach even after prolonged illumination (23–26). This allows imaging for extended periods (up to hours) without risk of photobleaching the sample. To specifically label a protein of interest, an epitope tag is engineered into the protein. Antibodies raised against the epitope tag are chemically linked to QDs and the QD–antibody complex is conjugated with the protein prior to visualization on the DNA curtain.

The DNA can be viewed by staining with very low concentrations (1–2 nM) of the intercalating dye YOYO1. To avoid rapid photobleaching of YOYO1 and concomitant DNA damage due to the reactivity of excited fluorophores with molecular oxygen, we employ an enzymatic oxygen scavenging system. The coupled activity of glucose oxidase and catalase in a buffer containing millimolar amounts of glucose significantly reduces DNA breaks and permits observations of individual molecules for tens of minutes (27). Although this approach does not inhibit the biochemical activity of many enzymes (9, 10), care should be taken to biochemically assay all the protein–DNA interactions in the presence of YOYO1, as well as all additional buffer components. If necessary, YOYO1 may be used to stain the DNA at the beginning of an experiment and subsequently flushed out by washing the flowcell with a high-salt (500 mM NaCl or 10 mM $MgCl_2$) buffer. In addition, alternative labeling procedures that employ recognition of digoxigenin (DIG)-labeled DNA by anti-DIG antibody–QD conjugates have also been developed (13, 14, 28). These labeling methods leave the duplex almost completely unperturbed and do not require intercalating DNA dyes or an oxygen scavenging system to visualize the DNA curtains.

2. Materials

2.1. TIRF Microscope

1. Laser: 488 nm, ~200 mW cw diode laser (Coherent).
2. Microscope: TE2000 Eclipse Inverted Microscope (Nikon).
3. Objective: Plan Apo 60X 1.2, W.I. 0.22 WD (Nikon).
4. Dual View Imaging System (Optical Insights).
5. Holographic 488 nm Notch Filter (Semrock).
6. DOVE Prism (ESCO).
7. Syringe pump (KD Scientific).
8. EM-CCD (Photometrix).
9. NIS-Elements Imaging Software (Nikon).

2.2. Nanofabricated Silica Slides

1. Nanofabrication facility: electron beam (e-beam) lithography and e-beam evaporator apparatus.
2. Spin-coater (Laurell Technologies).
3. Bath sonicator (VWR).
4. NanoStrip solution (CyanTek).
5. Polymethylmethacrylate (PMMA): MW of 25 kDa, 3% in anisole, and MW of 495 kDa, 1.5% in anisole (MicroChem).

6. AquaSave conducting polymer (Mitsubishi Rayon).
7. Resist Developer: 3:1 solution of 2-propanol and methyl isobutyl ketone (MicroChem).
8. Acetone.
9. 2-Propanol.

2.3. Buffer Solutions

1. PBS buffer: 10 mM phosphate, 138 mM NaCl, 2.7 mM KCl, pH 7.2. Autoclaved and stored at room temperature.
2. TE buffer: 10 mM Tris-HCl, pH 8, 1 mM EDTA. Autoclaved and stored at room temperature.
3. Lipids buffer: 10 mM Tris-HCl, pH 7.8, 100 mM NaCl. Filter through a 0.22 μm syringe filter and store at room temperature.
4. Imaging buffer: 40 mM Tris-HCl, pH 7.8, 1 mM DTT, 1 mM $MgCl_2$, 0.2 mg/ml bovine serum albumin (BSA; fraction VI, Sigma-Aldrich). Filter through a 0.22 μm syringe filter and use the same day as experiment.

2.4. DNA Substrate

1. λ-Phage DNA (New England Biolabs).
2. T4 ligase (New England Biolabs).

2.5. Antibody-Quantum Dots Conjugate Preparation

1. Quantum Dot Antibody conjugation kit (Invitrogen).
2. Appropriate antibody stock, stored as directed by manufacturer (e.g., monoclonal anti-FLAG M2 antibody, Sigma-Aldrich).

2.6. Lipid Bilayers

1. Micro-tip sonicator.
2. Chloroform, 99%.
3. 18:1 PC (*cis*) 1,2-dioleoyl-*sn*-glycero-3-phosphocholine (DOPC; Avanti Polar Lipids): prepared as a 20 mg/ml stock in chloroform.
4. 16:0 biotinyl cap PE 1,2-dipalmitoyl-*sn*-glycero-3-phosphoethanolamine-*N*-cap biotinyl, sodium salt (DOPE-biotin; Avanti Polar Lipids): purchased as 10 mg/ml solution in chloroform.
5. 18:1 PEG550 PE: 1,2-dioleoyl-*sn*-glycero-3-phosphoethanolamine-*N*-[methoxy(polyethylene glycol)-550] ammonium salt (DOPE-mPEG; Avanti Polar Lipids): stored as 10 mg/ml solution in chloroform.
6. Streptavidin: stored as 1 mg/ml solution in water at –20°C.
7. Glass screw-cap vials with Teflon liner (Avanti Polar Lipids).
8. Glass syringes (100 and 500 μl) with Teflon plungers (Hamilton).

2.7. Flowcell Assembly

1. Quartz microscope slides (1 in. × 3 in. × 1 mm; G. Finkenbeiner).
2. Scotch double-sided tape (3 M).
3. Microscope slide cover glass (Fisher Scientific).
4. Nanoport assembly system (Upchurch Scientific).
5. Tubing, Teflon PFA (Upchurch Scientific).
6. Medium size binder clips.
7. Low-temperature melting glue and glue gun.
8. Disposable Luer lock connector syringes (BD Scientific).

2.8. Imaging DNA Curtains

1. YOYO1 dye: stored as 1 mM stock in DMSO at –20°C (Invitrogen).
2. Glucose oxidase (Sigma-Aldrich).
3. Catalase (Sigma-Aldrich).

2.9. Data Processing Software

For experiments requiring tracking of QD-labeled proteins, the pointspread function of the fluorescent QD signal is fit to a 2D Gaussian for each frame of the multi-frame particle trajectory (10). The strong fluorescence signal from QDs and the resulting high *S*/*N* ratio offer a high precision fit to within several nanometers (29, 30). In practice, accuracy of the particle trajectory is limited due to the Brownian motion fluctuations of the double-stranded DNA (31, 32). Our lab has developed several in-house particle tracking programs that were written in MATLAB and IgorPro, and excellent commercial and free software tools for routine particle tracking have also been reported (33).

3. Methods

3.1. Nanofabrication of Lipid Diffusion Barriers

Nanofabricated barrier patterns are made by electron beam (e-beam) lithography, which yields uniform barrier patterns of high quality with nanometer precision (**Fig. 26.2**). The general process of e-beam lithography involves first coating the microscope slide with a thin polymer film (a bilayer of polymethylmethacrylate (PMMA), followed by a layer of Aquasave conducting polymer). An electron beam is then used to "burn" a desired pattern into the polymer film and expose the underlying slide surface. Metal (we typically use chromium) is vaporized under vacuum and deposited over the entire surface, including the exposed slide surface and the PMMA. The remaining polymer is then peeled away in a process called "liftoff," leaving behind the metal pattern on the microscope slide, which acts as lipid diffusion barriers when assembling the DNA curtains.

1. Drill quartz slides on a high-speed drill press using fresh diamond-tipped drills (*see* **Note 1**).
2. Clean slides in NanoStrip solution for ~30 min (*see* **Note 2**).
3. The NanoStrip solution may be reused for several rounds of cleaning.
4. Exhaustively rinse slides with water to remove all NanoStrip.
5. Wash slides individually with acetone. Wash with 2-propanol before acetone dries. Blow-dry using filtered, ultrapure N_2 (*see* **Note 3**).
6. Spin-coat a layer of 25 kDa PMMA on the freshly cleaned slide. Spin-coat the high MW, 495 kDa PMMA on top of the low MW layer. Finally, spin-coat the slide with a few drops of AquaSave conducting polymer. Each layer is spun at 4,000 rpm for 45 s using a ramp rate of 300 rpm/s (*see* **Note 4**).
7. Linear barriers are written by e-beam lithography using an FEI Sirion scanning electron microscope equipped with a pattern generator and lithography control system (J. C. Nabity, Inc., Bozeman, MT, USA).
8. After patterning, gently rinse the slide with water to remove the AquaSave layer.
9. Develop the resist by placing the slide in the MIBK developer and sonicating in a bath sonicator for 1 min at 4°C.
10. Immediately after developing, wash the slide with 2-propanol and blow-dry. Extended immersion in developer will over-develop the PMMA resist (*see* **Note 5**).
11. A semicore e-beam evaporator is used to deposit a 25 nm layer of chromium.
12. Lift off the PMMA/chromium by soaking the slides in boiling acetone for 30 min, followed by a 5-min sonication in acetone.
13. Following liftoff, samples are rinsed with acetone to remove stray chromium flakes and dried with N_2.

Microscope slides with nanofabricated diffusion barriers produce reproducible, highly controlled DNA curtains. Nanofabricated slides are durable and can be reused for tens of experiments, but require access to a nanofabrication facility. Simple diffusion barriers can also be made by scratching the microscope slide with a glass scribe or diamond-tipped drill. These barriers are easy to make but are highly variable in quality and spacing (18, 19).

3.2. Preparation of λ-Phage DNA Substrates

Genomic DNA from the bacteriophage λ is 48.5 kb long, commercially available, and contains 12 nucleotide 5′-ssDNA overhangs that are used to ligate a biotinylated synthetic oligonucleotide.

1. In a 1.5 ml Eppendorf tube, combine 100 μl of 10× T4 ligase buffer, 100–500 μg of λ-DNA (*see* **Note 6**), and the 3′-biotinylated and 5′-phosphorylated complementary oligonucleotide to a final concentration of 1 μM (*see* **Note 7**). Add water to bring the total volume to 990 μl. Gently mix, warm to 65°C, and cool slowly to room temperature.
2. Once the solution has cooled, add 10 μl of T4 ligase (400 U/μl) and place in a 42°C bath for 4 h to overnight. After the ligation reaction is complete, heat inactivate the ligase according to manufacturer's recommendations.
3. Filter the reaction through a Sephacryl S-200 or similar gel-filtration FPLC column at 4°C in TE + 150 mM NaCl running buffer to remove excess oligonucleotide and other reaction components. The 48.5 kb λ-DNA comes out in the void volume of the column.
4. Dilute λ-DNA fractions are pooled and stored at 4°C or divided into 100 μl aliquots and stored frozen at –20°C.

3.3. Preparation of Antibody–Quantum Dot Conjugates

1. Quantum dot (QD)–antibody conjugates are prepared according to the manufacturer-provided protocol with a minor modification in the final gel-filtration step.
2. After the conjugation reaction is quenched with 2-mercaptoethanol (manufacturer protocol), the QD–antibody mixture is passed through a Superose 6 FPLC column. The QDs come out in the void volume while the antibodies and other reaction components enter the Superose matrix.
3. QD conjugates are concentrated with MW 50 kDa microcentrifuge concentrators and stored for up to 3 months at 4°C in PBS at a final concentration of 100–500 nM.

3.4. Liposome Stock Solution Preparation

1. Rinse a new 2 ml glass vial (*see* **Note 8**) with water and ethanol. Dry the vial in an oven at 120°C under vacuum for about 20 min.
2. Warm the lipid stock solutions (in chloroform) to room temperature.
3. Combine 1 ml of 20 mg/ml DOPC, 160 μl of 10 mg/ml DOPE–mPEG, and 10 μl of 10 mg/ml DOPE–biotin chloroform stocks in the dry 2 ml glass vial.
4. This mixture of chloroform stocks is evaporated by carefully blowing a weak stream of nitrogen gas while rotating the vial

continuously to form a thin, uniform layer of dried lipid on the side walls. The glass vial is then placed under vacuum for at least 2 h (may be left overnight) to ensure complete removal of all residual chloroform.

5. Add 2 ml of lipids buffer to the dried lipid mixture and allow it to hydrate for at least 2 h to overnight.
6. Vortex the hydrated lipid mixture for 2–3 min to form large multilamellar vesicles. At this point, the liposomes should appear as an opaque, cloudy white solution.
7. Transfer the liposome solution to a 5 ml polypropylene culture tube for sonication. Sonicate the liposomes using a probe tip sonicator to form small unilamellar vesicles under the following settings (VirSonic micro-tip sonicator): set the power output to 15%. Sonicate for 1.5 min with 2-min intervals on ice. Repeat this cycle two more times. After sonication, the liposome solution should clear considerably and become translucent.
8. Filter the liposome solution using a 0.22 μm nylon syringe filter into 1.5 ml Eppendorf tubes. Store the liposome solution at 4°C for up to 2 weeks (*see* **Note 9**). Do not freeze. The final concentration of liposome solution is 10 mg/ml DOPC with 0.5% (w/w) biotinylated DOPE and 8% (w/w) mPEG(550) DOPE.

3.5. Construction of Flowcells

The flowcell is constructed from a segment of double-sided tape sandwiched between a silica microscope slide and a glass coverslip (**Fig. 26.1**). Tube ports are attached to pre-drilled holes in the silica slides. In practice, 5–10 flowcells are assembled simultaneously and stored in a vacuum dessicator for up to a week, but prolonged storage compromises surface quality, which can prevent deposition of a fluid bilayer.

1. Nanofabricated slides (*see* **Section 3.1**) are rinsed in filtered MilliQ H_2O, gently agitated in 2% Hellmanex cleaning solution for 1 h, rinsed thoroughly in H_2O, soaked 1 h in 1 M NaOH, and rinsed again with water and 100% methanol.
2. The slides are dried under a dry nitrogen steam and baked at 120°C in a vacuum oven for an hour. Cleaned silica slides that have not been assembled into flowcells may be stored in the vacuum oven indefinitely.
3. To assemble the flowcell, mask off a segment of double-sided tape with a narrow strip of paper. This paper will eventually be cut out, and the resulting channel in the tape will form the observation chamber for microscopy experiments.
4. Place the double-sided tape over the silica slide. Make sure that both ends of the paper cover the drilled holes and that the paper template covers the chrome diffusion barriers.

5. Cut out the paper template using a razor. Keep close to the template, ensuring that the tape does not cover the holes.
6. Place a glass coverslip over the double-sided tape. Remove excess tape that is not covered by the coverslip.
7. Sandwich the flowcell between two clean glass microscope slides, apply even pressure with four small binder clips, and bake in a vacuum oven at 120°C for up to an hour.
8. The nanoports are attached with a low-temperature melting hot-glue gun to the silica side of the flowcell assembly.
9. The assembled flowcells may be stored at room temperature under vacuum for up to a week without significant degradation to the flowcell surface and lipid bilayer fluidity.

3.6. Preparation of DNA Curtains

1. Attach a syringe with 10 ml of H_2O to one end of the flowcell. Rinse the flowcell with water, while tapping gently. Tapping the flowcell loosens and flushes out all air bubbles within the system. Air bubbles must be avoided, as even a small bubble will ruin the lipid bilayer surface. All subsequent syringes must be attached to the system by making drop-to-drop Luer lock connections.
2. Wash the flowcell with 2–3 ml lipids buffer, contained in a 3 ml Luer lock syringe that is attached to the second nanoport. Alternating between the two flowcell ports reduces the chance of injecting air bubbles into the tubing.
3. Dilute 40 μl of stock liposome solution (*see* **Section 3.4**) with 960 μl of lipids buffer. Inject 1 ml of the diluted liposome solution into the flowcell as a series of three injections with a 5- to 10-min incubation time between injections.
4. Rinse the flowcell with 2–3 ml lipids buffer. Incubate for 30 min to promote vesicle fusion and bilayer growth along the silica surface.
5. Slowly inject 1 ml BSA buffer and incubate for 10 min to allow BSA to block remaining exposed surfaces.
6. Inject 300 μl of 0.1 mg/ml streptavidin in BSA buffer.
7. Rinse flowcell with 2–3 ml BSA buffer to flush out free streptavidin.
8. Dilute 5–50 μl stock biotinylated λ-DNA (*see* **Section 3.2**) into 1 ml BSA buffer. Slowly inject this solution into the flowcell and incubate 5–10 min to allow for DNA binding to the lipid bilayer surface. The amount of injected DNA may be adjusted to obtain the desired surface–DNA density.
9. Rinse the DNA out with 2–3 ml BSA buffer. At this point, the flowcell is ready for imaging experiments and should be transferred to the microscope syringe pump system.

3.7. Imaging of Flowcells

1. Attach the flowcell to the syringe pump system pre-rinsed with imaging buffer. With moderate buffer flow, the DNA molecules will be pushed along the surface and will align at the diffusion barriers, a process that may take several minutes.
2. Mount the flowcell atop the microscope objective and place the DOVE prism on top of the silica slide.
3. Focus the objective at the slide surface by adjusting the focus knob until the DNA fluorescence signal is maximized. Adjust the 488 nm laser beam and total internal reflection angle to maximize the fluorescence signal (*see* **Note 10**). Experiments that use YOYO1-stained DNA must include an oxygen scavenging system for extended imaging (27).
4. For experiments that utilize QD-tagged proteins, inject the protein of interest in the appropriate reaction buffer to initiate the experiment.

4. Notes

1. The silica slides tend to shatter if drilled too quickly or with blunted drill bits. Work slowly, under a constant stream of running water, and change the drill bits frequently.
2. NanoStrip consists of a mixture of concentrated sulfuric acid and hydrogen peroxide, is extremely corrosive, and should be handled with care in a proper acid hood.
3. At this point, care should be taken to keep the cleaned slides away from dust.
4. The PMMA may be filtered through a 0.22 μm syringe filter to produce uniform, dust-free polymer layers.
5. It is generally possible to see the developed pattern using a white light illumination optical microscope and a 50× objective. It is generally a good idea to check the quality of the developed pattern before continuing with the next e-beam evaporation step.
6. Pre-warm the λ-DNA stock at 65–75°C for a few minutes before pipetting. Pre-warming the λ-DNA melts the cohesive ssDNA ends, which anneal at the high stock concentration. Care must be taken when working with any long DNA molecule to avoid shearing. Avoid multiple freeze–thaw cycles, pipette minimally with a wide-hole pipette tip, and mix all solutions gently by inverting and tapping the Eppendorf tube.

7. If both ends of the λ-DNA need to be functionalized, include the second oligo at a concentration of 1 μM at this step. Although the two oligos are complimentary and will anneal, the large excess of oligos over λ-DNA ends and efficient ligation at 42°C (near the oligo melting temperature) yields a majority of λ-DNA molecules with both ends functionalized.
8. Most laboratory plastics are susceptible to chloroform. Care must be taken to avoid all plastic lab wares when working with chloroform solutions. All stocks are stored in glass vials with Teflon-sealed caps. Lipids are transferred using glass Hamilton syringes with Teflon plungers that have been pre-rinsed thoroughly with chloroform.
9. Over time, the small unilamellar vesicles formed during initial sonication slowly fuse to make larger, more stable structures. Aged liposome stocks yield patchy bilayers with substantially reduced fluidity and poor surface blocking properties.
10. For experiments that do not use fluorescently stained DNA, it is possible to focus on the nanofabricated pattern or random imperfections on the slide surface.

Acknowledgments

We thank the many members of the Greene Laboratory who have worked on developing the DNA curtain experimental platform, in particular, Teresa Fazio for establishing the nanofabrication process. The Greene Laboratory is supported by the Howard Hughes Medical Institute, the National Institutes of Health, the National Science Foundation, the Susan G. Komen Foundation, and the Irma T. Hirschl Trust. IJF is supported by the NIH Fellowship #F32GM80864. We apologize to any colleagues whose work we were not able to cite due to length limitations.

References

1. Hamdan, S.M., Loparo, J.J., Takahashi, M., Richardson, C.C., and van Oijen, A.M. (2009) Dynamics of DNA replication loops reveal temporal control of lagging-strand synthesis. *Nature* **457**, 336–339.
2. van Oijen, A.M. (2007) Single-molecule studies of complex systems: the replisome. *Mol Biosyst* **3**, 117–125.
3. Perumal, S.K., Yue, H., Hu, Z., Spiering, M.M., and Benkovic, S.J. (2010) Single-molecule studies of DNA replisome function. *Biochim Biophys Acta* **1804**, 1094–1112.
4. Yao, N.Y., Georgescu, R.E., Finkelstein, J., and O'Donnell, M.E. (2009) Single-molecule analysis reveals that the lagging strand increases replisome processivity but

slows replication fork progression. *Proc Natl Acad Sci USA* **106**, 13236–13241.
5. Bai, L., Santangelo, T.J., and Wang, M.D. (2006) Single-molecule analysis of RNA polymerase transcription. *Annu Rev Biophys Biomol Struct* **35**, 343–360.
6. Hodges, C., Bintu, L., Lubkowska, L., Kashlev, M., and Bustamante, C. (2009) Nucleosomal fluctuations govern the transcription dynamics of RNA polymerase II. *Science* **325**, 626–628.
7. Herbert, K.M., Greenleaf, W.J., and Block, S.M. (2008) Single-molecule studies of RNA polymerase: motoring along. *Annu Rev Biochem* 77, 149–176.
8. Finkelstein, I.J., and Greene, E.C. (2008) Single molecule studies of homologous recombination. *Mol Biosyst* **4**, 1094–2104.
9. Spies, M., Amitani, I., Baskin, R.J., and Kowalczykowski, S.C. (2007) RecBCD enzyme switches lead motor subunits in response to chi recognition. *Cell* **131**, 694–705.
10. Gorman, J., Chowdhury, A., Surtees, J.A., Shimada, J., Reichman, D.R., Alani, E., and Greene, E.C. (2007) Dynamic basis for one-dimensional DNA scanning by the mismatch repair complex Msh2-Msh6. *Mol Cell* **28**, 359–370.
11. Kwon, Y., Seong, C., Chi, P., Greene, E.C., Klein, H., and Sung, P. (2008) ATP-dependent chromatin remodeling by the Saccharomyces cerevisiae homologous recombination factor Rdh54. *J Biol Chem* **283**, 10445–10452.
12. Visnapuu, M.L., and Greene, E.C. (2009) Single-molecule imaging of DNA curtains reveals intrinsic energy landscapes for nucleosome deposition. *Nat Struct Mol Biol* **16**, 1056–1062.
13. Robertson, R.B., Moses, D.N., Kwon, Y., Chan, P., Zhao, W., Chi, P., Klein, H., Sung, P., and Greene, E.C. (2009) Visualizing the disassembly of S. cerevisiae Rad51 nucleoprotein filaments. *J Mol Biol* **388**, 703–720.
14. Robertson, R.B., Moses, D.N., Kwon, Y., Chan, P., Chi, P., Klein, H., Sung, P., and Greene, E.C. (2009) Visualizing the disassembly of S. cerevisiae Rad51 nucleoprotein filaments. *Proc Natl Acad Sci USA* **106**, 12688–12693.
15. Prasad, T.K., Yeykal, C.C., and Greene, E.C. (2006) Visualizing the assembly of human Rad51 filaments on double-stranded DNA. *J Mol Biol* **363**, 713–728.
16. Gorman, J., Fazio, T., Wang, F., Wind, S., and Greene, E.C. (2009) Nanofabricated racks of aligned and anchored DNA substrates for single-molecule imaging. *Langmuir* **26**, 1372–1379.
17. Visnapuu, M.L., Fazio, T., Wind, S., and Greene, E.C. (2008) Parallel arrays of geometric nanowells for assembling curtains of DNA with controlled lateral dispersion. *Langmuir* **24**, 11293–11299.
18. Fazio, T., Visnapuu, M.L., Wind, S., and Greene, E.C. (2008) DNA curtains and nanoscale curtain rods: high-throughput tools for single molecule imaging. *Langmuir* **24**, 10524–10531.
19. Graneli, A., Yeykal, C.C., Prasad, T.K., and Greene, E.C. (2006) Organized arrays of individual DNA molecules tethered to supported lipid bilayers. *Langmuir* **22**, 292–299.
20. Visnapuu, M.L., Duzdevich, D., and Greene, E.C. (2008) The importance of surfaces in single-molecule bioscience. *Mol Biosyst* **4**, 394–403.
21. Groves, J.T., Ulman, N., and Boxer, S.G. (1997) Micropatterning fluid lipid bilayers on solid supports. *Science* **275**, 651–653.
22. Richter, R.P., Bérat, R., and Brisson, A.R. (2006) Formation of solid-supported lipid bilayers: an integrated view. *Langmuir* **22**, 3497–3505.
23. Jaiswal, J.K., Mattoussi, H., Mauro, J.M., and Simon, S.M. (2003) Long-term multiple color imaging of live cells using quantum dot bioconjugates. *Nat Biotechnol* **21**, 47–51.
24. Medintz, I.L., Uyeda, H.T., Goldman, E.R., and Mattoussi, H. (2005) Quantum dot bioconjugates for imaging, labeling and sensing. *Nat Mater* **4**, 435–446.
25. Ebenstein, Y., Gassman, N., Kim, S., Kim, Y., Ho, S., Samuel, R., Michalet, X., and Weiss, S. (2009) Lighting up individual DNA binding proteins with quantum dots. *Nano Lett* **9**, 1598–1603.
26. Pinaud, F., Michalet, X., Bentolila, L.A., Tsay, J.M., Doose, S., Li, J.J., Iyer, G., and Weiss, S. (2006) Advances in fluorescence imaging with quantum dot bio-probes. *Biomaterials* **27**, 1679–1687.
27. Rasnik, I., McKinney, S.A., and Ha, T. (2006) Nonblinking and long-lasting single-molecule fluorescence imaging. *Nat Methods* **3**, 891–893.
28. Escude, C., Geron-Landre, B., Crut, A., and Desbiolles, P. (2009) Multicolor detection of combed DNA molecules using quantum dots. *Methods Mol Biol* **544**, 357–366.
29. Thompson, R.E., Larson, D.R., and Webb, W.W. (2002) Precise nanometer localization analysis for individual fluorescent probes. *Biophys J* **82**, 2775–2783.

30. Yildiz, A., and Selvin, P.R. (2005) Fluorescence imaging with one nanometer accuracy: application to molecular motors. *Acc Chem Res* **38**, 574–582.
31. Gueroui, Z., Freyssingeas, E., Place, C., and Berge, B. (2003) Transverse fluctuation analysis of single extended DNA molecules. *Eur Phys J E Soft Matter* **11**, 105–108.
32. Quake, S.R., Babcock, H., and Chu, S. (1997) The dynamics of partially extended single molecules of DNA. *Nature* **388**, 151–154.
33. Carter, B.C., Shubeita, G.T., and Gross, S.P. (2005) Tracking single particles: a user-friendly quantitative evaluation. *Phys Biol* **2**, 60–72.

Chapter 27

FRET-Based Assays to Monitor DNA Binding and Annealing by Rad52 Recombination Mediator Protein

Jill M. Grimme and Maria Spies

Abstract

During homologous recombination and homology-directed repair of broken chromosomes, proteins that mediate and oppose recombination form dynamic complexes on damaged DNA. Quantitative analysis of these nucleoprotein assemblies requires a robust signal, which reports on the association of a recombination mediator with its substrate and on the state of substrate DNA within the complex. Eukaryotic Rad52 protein mediates recombination, repair, and restart of collapsed replication forks by facilitating replacement of ssDNA binding protein replication protein A (RPA) with Rad51 recombinase and by mediating annealing of two complementary DNA strands protected by RPA. The characteristic binding mode whereby ssDNA is wrapped around the Rad52 ring allowed us to develop robust and sensitive FRET-based assays for monitoring Rad52 interactions with protein-free DNA and ssDNA–RPA complexes. By reporting on the configuration of ssDNA dually labeled with Cy3 and Cy5 fluorescent dyes, solution-based FRET is used to analyze Rad52–RPA–DNA interactions under equilibrium binding conditions. Finally, FRET between Cy3 and Cy5 dyes incorporated into two homologous ssDNA molecules can be used to analyze interplay between Rad52-mediated DNA strand annealing and duplex DNA destabilization by RPA.

Key words: Annealing protein, fluorescence, Föster resonance energy transfer (FRET), homologous recombination, Rad52, recombination mediator, RPA.

1. Introduction

1.1. Rad52 Protein

Homologous genetic recombination (HR) is a genome maintenance process vital for error-free repair of the most deleterious DNA lesions affecting both strands of the duplex and for the re-initiation of replication at stalled or collapsed replication forks (1, 2). Proteins orchestrating HR and homology-directed DNA repair (HDR) in eukaryotes belong to the so-called Rad52-epistasis group, owing its name to the yeast Rad52 protein.

H. Tsubouchi (ed.), *DNA Recombination*, Methods in Molecular Biology 745,
DOI 10.1007/978-1-61779-129-1_27, © Springer Science+Business Media, LLC 2011

The central step in eukaryotic HR is assembly of the nucleoprotein filament formed by multiple Rad51 recombinases binding along the ssDNA. The assembled presynaptic filament then mediates the search for homology and exchanges the DNA strands resulting in the formation of DNA joints between two recombining molecules. Assembly of the Rad51 filament is the most important step in HR and is a very delicate process since both defective and excessive recombination can cause genomic instability and lead to cancer or cell senescence. Nucleation of the recombinase filament on ssDNA is a slow process and is tightly regulated. Rad51 fails to compete for binding to ssDNA with eukaryotic single-stranded DNA binding protein RPA unless nucleation is assisted by a recombination mediator (RM) protein (reviewed in (3)). Rad52 (4) is one such RM and has additional functions in promoting annealing of complementary DNA strands as well as annealing between ssDNA and RPA complexes (5, 6). Both the mediator function and the annealing activity play important roles in classic double-stranded break repair (DSBR) and synthesis-dependent strand annealing (SDSA) pathways of HR that confer the highest fidelity of DNA repair (7–12). Annealing activity is also indispensable for single-strand annealing (SSA) pathway of mutagenic DNA repair (13).

A unique feature of Rad52 among eukaryotic recombination mediators is its oligomeric structure characterized by a ring-shaped morphology. Rad52 interacts with ssDNA primarily through the binding site located in a deep, narrow, positively charged groove spanning the outside surface of the predominantly heptameric ring of the full-length protein or the undecameric ring of N-terminal DNA binding domain of Rad52 (5, 14–18). The conserved N-terminal domain of Rad52 also contains an additional DNA binding site important for the RM function (19).

With exception of some plants and invertebrates, Rad52 homologues are found across all eukaryotic organisms. In addition to Rad52, yeast contain an additional homologue, Rad59, capable of annealing homologous ssDNA, but not ssDNA–RPA complexes (20, 21). A number of bacterial and bacteriophage RMs and single-strand annealing proteins (SSAPs) display similar ring-shaped morphology: bacteriophage SSAP Sak forms undecameric rings and is believed to be both structural and functional homologue of the N-terminal DNA binding domain of Rad52 protein (22); bacteriophage T4 RM UvsY forms hexameric rings, which bind ssDNA as well as ssDNA bound by gp32 (functional equivalent of eukaryotic RPA) complexes in a wrapped configuration (23), while Redβ protein whose DNA binding domain is related to that of Rad52 forms broken ring structures (24).

1.2. FRET-Based DNA Binding and Annealing Assays as Alternatives to EMSA and SPR

Binding of yeast and human Rad52 to ssDNA has been traditionally evaluated using electrophoretic mobility shift assays (EMSA or gel-shift assays). Although robust, these assays are not true equilibrium techniques and therefore commonly underestimate binding affinity. Because Rad52 forms dynamic complexes by rapidly binding to and dissociating from DNA, EMSA requires cross-linking the protein to ^{32}P-labeled or fluorescently labeled DNA prior to separation. EMSA performed using radiolabeled DNA allows using substrate concentrations in a low nanomolar range and evaluation of high-affinity nucleoprotein complexes. The use of fluorescently labeled DNA substrates for EMSA eliminates the safety hazards associated with handling of ^{32}P-labeled materials and time-consuming visualization step. Additionally, because fluorescent dyes such as Cy3 or Cy5 emit light in the visible range that can be detected through glass, one can carry out electrophoresis in very low density gels and distinguish various high molecular weight complexes (25). Affinity of human Rad52 for ssDNA obtained using this technique ranged from sub-nM to 100 nM (18, 26, 27).

Surface plasmon resonance (SPR) presents a laborious alternative to EMSA. Under the right conditions, however, it allows measuring both kinetic association and dissociation constants and therefore provides a measure of equilibrium dissociation constant (26). Drawbacks of SPR measurements include possible artifacts associated with surface tethering of the substrate and technical difficulty of measuring both kinetic constants in the same experiment under the same conditions (28).

Here we describe a robust fluorescence-based assay for monitoring binding of Rad52 to ssDNA and ssDNA–RPA complex under a wide range of experimental conditions (25). This assay exploits the difference in the DNA conformation in solution, in complex with RPA, and in the stoichiometric complex with Rad52. Föster resonance energy transfer (FRET) (29) between two fluorescent dyes Cy3 (FRET donor) and Cy5 (FRET acceptor) incorporated into the synthetic oligonucleotide at a distance of 25–50 nucleotides from one another allows differentiation of the conformational states of the substrate molecule. The binding of RPA to ssDNA straightens ssDNA and extends it to the contour length. This results in a decrease in the FRET between Cy3 and Cy5 dyes. In contrast, wrapping of ssDNA by Rad52 brings the two dyes in close proximity resulting in high FRET signal. The main advantages of the FRET-based assay are that it measures binding under true equilibrium conditions and that it can be carried out at low (sub-nM) concentrations of substrate.

The FRET-based DNA binding assay can be complemented by the annealing assay, which takes advantage of the donor (Cy3) and acceptor (Cy5) fluorophores incorporated into two complementary oligonucleotides (25, 30). Duplex formation brings Cy3 and Cy5 dyes in a close proximity resulting in an increase in the FRET signal, which can be monitored in real time and allows determination of both the rate and the extent of annealing. This is in contrast to measuring annealing using gel-based methods, which report only on the extent of the annealing reaction. The kinetics of annealing can be also monitored by following fluorescence of nucleic acid binding dyes, such as DAPI, which has high fluorescence yield when bound to dsDNA (21). Advantages of a DAPI-based assay are that it does not require labeling of DNA substrates and that it can be applied to short oligonucleotides or long plasmid length DNA. Due to lower sensitivity, however, higher concentrations of DNA are required, while sub-nanomolar concentrations of Cy3- and Cy5-labeled nucleotides can be successfully visualized in the FRET-based assays. Another advantage of using Cy3 and Cy5 dyes is that they emit light in the visible range allowing one to add BSA to the reaction mixture to prevent nonspecific adhesion of the enzymes to the cuvette walls.

2. Materials and Equipment

2.1. Equipment

1. Protein purification system(s): AKTA prime and AKTA FPLC (GE Healthcare Life Sciences) or BioLogic DuoFlow system (Bio-Rad).
2. UV spectrophotometer: We use Cary BIO 300 (Varian). Alternatively, a spectrophotometer from Agilent Technologies, Eppendorf, Thermo Scientific, Shimadzu, Beckman Coulter, PerkinElmer, or NanoDrop can be used to measure protein and DNA concentrations.
3. Fluorescence spectrophotometer equipped with a temperature controller and capable of simultaneously detecting fluorescence in two channels: We used Cary Eclipse (Varian). Alternatively, a spectrofluorimeter from ISS or Shimadzu can be used. The assays described below also can be adapted for a multiwell plate format for use with a plate reader.
4. Quartz or optical glass cuvettes for binding titrations: 5 mm square cuvette (500 μl volume) (Starna or Helma).
5. Micro-cuvette for annealing reactions: 100–150 μl minimum volume cuvette configured for measuring fluorescence (Starna or Helma)

2.2. Proteins

Purified human RPA and Rad52 proteins. Human RPA protein, a heterotrimer of RPA70, RPA35, and RPA14 subunits, is purified as described in (25, 31). Typical purification yields 30 μM protein stock in storage buffer containing 30 mM HEPES (pH 7.8), 0.1 mM DTT, 0.25 mM EDTA, 0.25 mM inositol, 0.01% Nonidet P40, 300 mM KCl, and 10% glycerol (*see* **Notes 1** and **2**).

Human Rad52 is expressed and purified as described in (32, 33) (*see* **Notes 3** and **4**). Typical purification results in 50–200 μM Rad52 stock in the storage buffer containing 50 mM Tris-HCl (pH 7.5), 1 mM EDTA, 0.1 mM DTT, 250 mM KCl, and 10% glycerol.

2.3. Buffers

1. Protein storage buffers are described above.
2. Protein–DNA binding and annealing buffer: 30 mM Tris-acetate [pH 7.5] and 1 mM DTT.

2.4. Oligonucleotides

Synthetic oligonucleotides used in the binding and annealing assays described here are obtained commercially. Among other commercial sources, Integrated DNA Technologies (IDT) can incorporate Cy3 or Cy5 at the 3′ of the oligonucleotide, internally or at the 5′-terminus. Due to synthetic limitations, it is advantageous to design substrates that have one label at or near the 5′-terminus of oligonucleotide while having the second label positioned not further than 50–60 nucleotides from the 5′-end. If necessary, long oligonucleotide (more than 50 nt in length) with a fluorescent dye at or near the 3′-terminus can be produced by ligating two shorter oligonucleotides. Concentrations of oligonucleotides can be determined spectrophotometrically using extinction coefficients provided by the manufacturer.

2.5. Fluorimeter Settings for FRET-Based Binding and Annealing Assays

1. Cy3 excitation wavelength at 530 nm and emission wavelength at 565 nm.
2. Cy5 excitation wavelength at 630 nm and emission wavelength at 660 nm.
3. Excitation and emission slit widths set to 10 nm each.
4. PMT voltage set, so that the amplitude of the Cy3 signal is below 60% of the available scale (*see* **Note 5**).
5. Both binding and annealing assays are carried out in kinetics mode. The time base for binding titrations is 1 s. Time base in annealing reactions is 0.1 s.

3. Methods

3.1. Calculating FRET Efficiency

1. These instructions assume the use of a Cary Eclipse fluorescence spectrophotometer (Varian) with a temperature controller set to 25°C and the following instrument settings: Cy3 donor excitation 530 nm and emission 565 nm, Cy5 acceptor emission 660 nm.

2. The FRET efficiency (E_{FRET}) is calculated as a fraction of donor acceptor intensity adjusted by correction factors (*see* **Note 6**). For Cary Eclipse fluorescence spectrophotometer (Varian) E_{FRET} between Cy3 and Cy5 dyes can be calculated using

$$E_{FRET} = \frac{4.2 \times I_{Cy5}}{4.2 \times I_{Cy5} + 1.7 \times I_{Cy3}}, \qquad [1]$$

where I_{Cy5} is the averaged acceptor intensity and I_{Cy3} is the averaged donor intensity (*see* **Note 7**).

Example calculation: At 8 nM Rad52 (**Fig. 27.3**), $E_{FRET} = 4.2(36.3)/[4.2(36.3) + 1.7(37.5)] = 0.71$. This number is the peak E_{FRET} observed for Cy3 and Cy5 upon Rad52 binding.

3.2. Assays for Determining RPA–ssDNA Binding Stoichiometry

1. The binding experiments are carried out using Cary Eclipse fluorescence spectrophotometer (Varian) with the setting described in **Section 2.5** in freshly prepared protein–DNA binding/annealing buffer (*see* **Note 8** and **Section 2.3**) (**Fig. 27.1**).
2. RPA binding is analyzed using square 5 mm cuvette. The reaction volume is 500 μl. Measurement is initiated by recording background fluorescence of protein–DNA binding buffer in the absence of fluorescently labeled DNA (indicated by the first arrow in **Fig. 27.2a**) in the Cy3 and Cy5 channels.
3. The Cy3/Cy5-labeled ssDNA substrate (1 nM) is then added to the cuvette (second arrow in **Fig. 27.2a**). Pipette solution to mix thoroughly.
4. A substantial increase in the fluorescence intensities of both dyes should be seen and recorded until the Cy3 and Cy5 signals stabilize.
5. RPA is then added in aliquots to incrementally increase the total protein concentration from 0 to 5 nM (arrows in **Fig. 27.2a**). After each addition, the reaction mixture in the cuvette is thoroughly mixed (*see* **Note 9**).
6. The fluorescence of the Cy3 and Cy5 dyes is allowed to equilibrate and is recorded and averaged for at least 1 min (inset in **Fig. 27.2a**).
7. The average background fluorescence is subtracted from the averaged fluorescence recorded at each RPA concentration.
8. Resulting values for Cy3 and Cy5 fluorescence can be plotted as functions of RPA concentration (**Fig. 27.2b**) and converted into E_{FRET} values using equation [1] (**Fig. 27.2c**) (*see* **Note 10**).

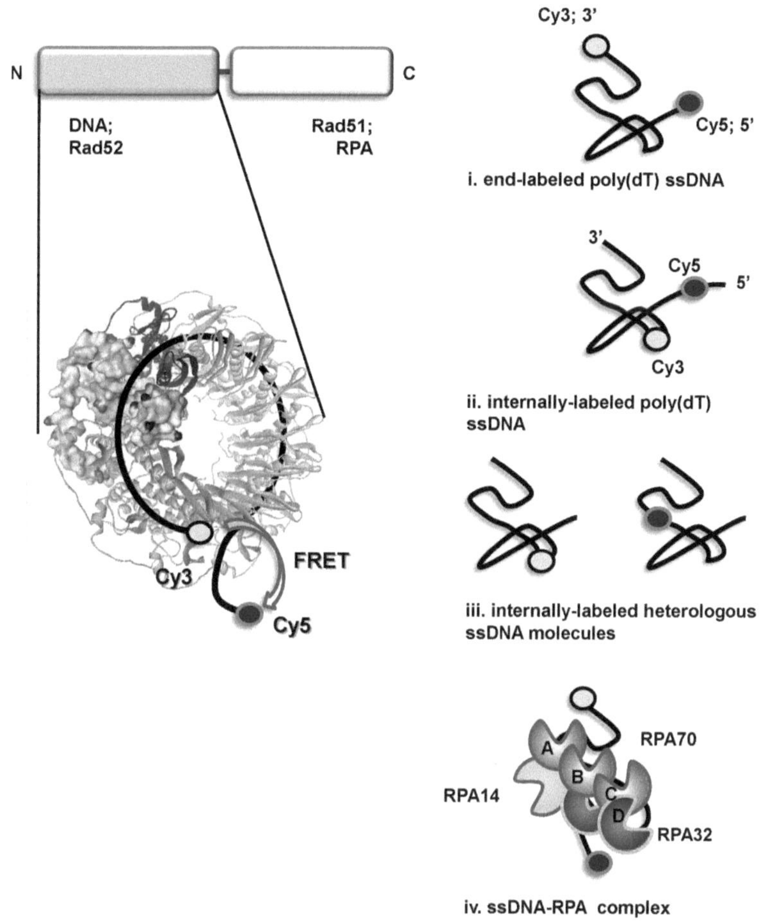

Fig. 27.1. The Rad52 oligomer forms a ring structure with ssDNA binding site spanning along the perimeter: structure of the undecameric ring of the DNA binding domain of human Rad52 protein (coordinates were taken from PDB: 2H1I (16)). One subunit is shown as an electrostatic potential surface. The adjacent subunit (*dark gray*) displays residues important for DNA binding (*ball* and *stick* representation) (18). The *right panel* shows examples of substrates used in the FRET-based binding assays: (*i–iii*) Cy3- and Cy5-labeled ssDNA, (*iv*) ssDNA–RPA complex. Seven OB folds of RPA protein are depicted as packman shapes. OB folds involved in ssDNA binding are marked as A–D.

3.3. FRET-Based Equilibrium Binding Assays for Rad52 Interaction with RPA-Coated ssDNA

1. The binding experiments are carried out using Cary Eclipse fluorescence spectrophotometer (Varian) with the setting described in **Section 2.5** in freshly prepared protein–DNA binding/annealing buffer (*see* **Note 8** and **Section 2.3**).
2. Five millimeter square cuvette is used in these measurements. The reaction volume is 500 μl.
3. Measurements are initiated by recording background fluorescence in the Cy3 and Cy5 channels of protein–DNA binding buffer containing 2 nM RPA (first arrow in **Fig. 27.3a**).

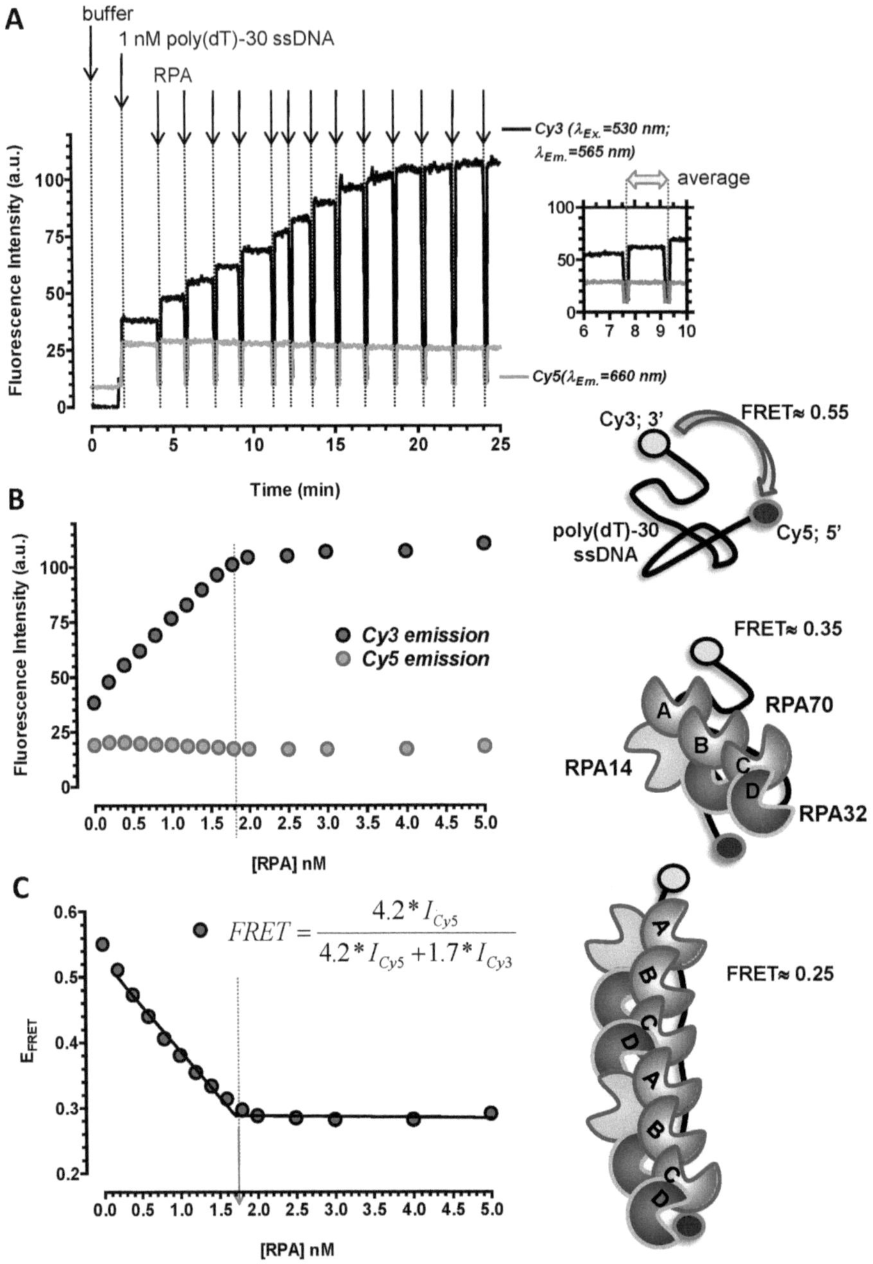

Fig. 27.2. RPA–ssDNA binding experiment: (**a**) Raw data from a representative experiment. Fluorescence intensity of Cy3 (*black*) and Cy5 (*gray*) are recorded in kinetic mode every 1 s. *Arrows above the graph* indicate time points when the cuvette was removed from the fluorimeter to add DNA or RPA protein. The *inset* shows the fragment of the trace used to obtain the average Cy3 and Cy5 intensities for a particular RPA concentration. (**b**) Averaged intensities of Cy3 and Cy3 dyes are plotted as a function of RPA concentration. (**c**) Calculated FRET efficiency (E_{FRET}) as a function of RPA concentration. Data shown in these figures were originally published in *NAR (Oxford University Press)*; Grimme et al. (25).

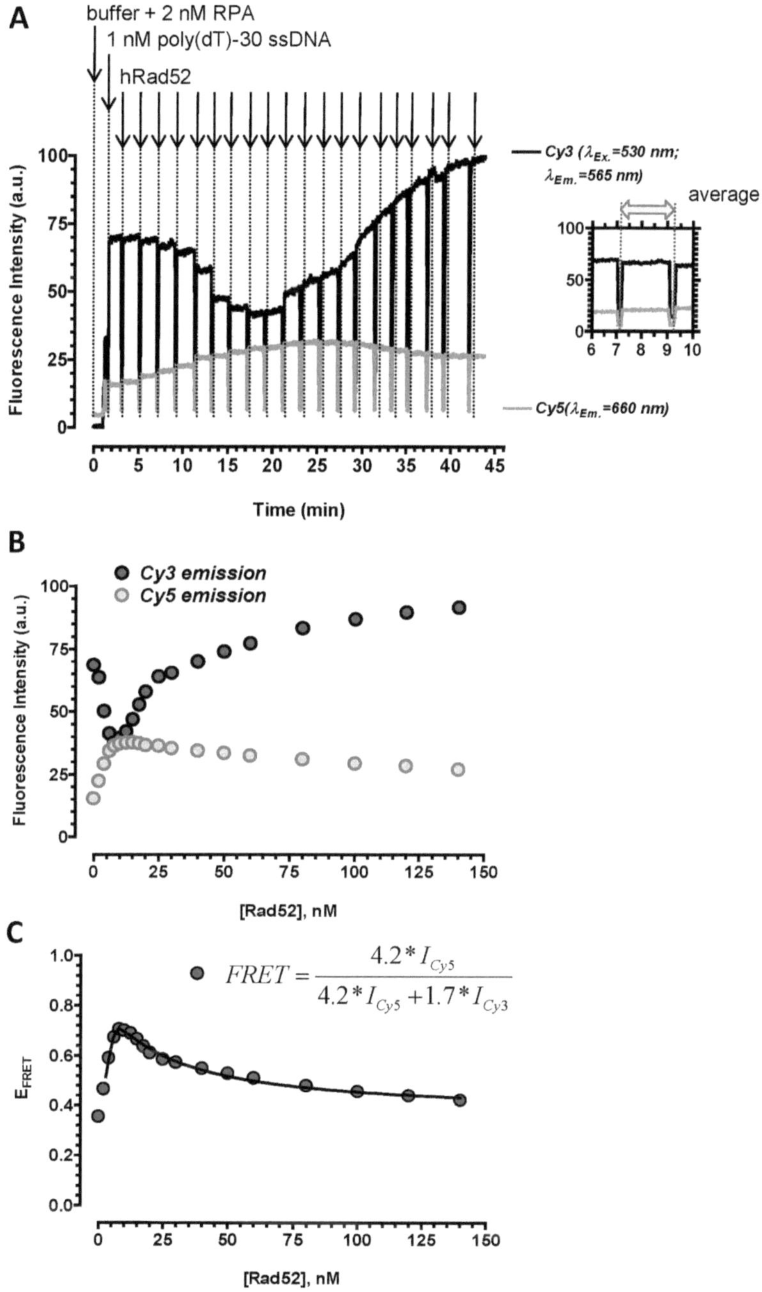

Fig. 27.3. Rad52 binding to the RPA–ssDNA complex: (**a**) Raw data from a representative experiment. Fluorescence intensity of Cy3 (*black*) and Cy5 (*gray*) are recorded in kinetic mode every 1 s. *Arrows above the graph* indicate time points DNA or Rad52 protein were added to the cuvette. The *inset* shows the fragment of the trace used to obtain the average Cy3 and Cy5 intensities for a particular Rad52 concentration. (**b**) Averaged intensities of Cy3 and Cy3 dyes plotted as a function of Rad52 concentration. (**c**) Calculated FRET efficiency (E_{FRET}) as a function of Rad52 concentration. Data shown in these figures were originally published in *NAR (Oxford University Press)*; Grimme et al. (25).

4. The ssDNA substrate (1 nM poly(dT)-30) is then added to the cuvette and the solution mixed well (second arrow in **Fig. 27.3a**). The signal for ssDNA–RPA complex is allowed to equilibrate and is averaged for 1 min (*see* **Note 11**).
5. The aliquots of Rad52 protein are then titrated into the reaction mixtures (addition of each subsequent aliquot is indicated by an arrow in **Fig. 27.3a**). The fluorescence of the donor (Cy3) and the acceptor (Cy5) is recorded as in step 4 (*see* **Note 12**).
6. The data are plotted by subtracting the background from each recorded and averaged value (**Fig. 27.3b**).
7. The calculated E_{FRET} is then plotted against the Rad52 concentrations (in monomers) (**Fig. 27.3c**).

3.4. FRET-Based Equilibrium Binding Assays for Monitoring Rad52 Interactions with ssDNA Substrates of Various Lengths

1. The assays are *carried out using bare ssDNA dually labeled with Cy3 and Cy5 fluorophores* as described in **Section 3.3**, but with no RPA present in the reaction mixture.
2. Separate binding experiments are performed using 1 nM end-labeled poly(dT)-22, poly(dT)-30, poly(dT)-39, poly(dT)-50 ssDNA, as well as an internally labeled poly(dT)-60 ssDNA (**Fig. 27.4a**) (*see* **Notes 13**).
3. After subtracting the background from each recorded and averaged value, the calculated E_{FRET} is plotted against the Rad52 concentrations (in monomers) for each substrate (**Fig. 27.4a**) (*see* **Note 14**).
4. The binding titration assays for each substrate should be performed in triplicate and the calculated E_{FRET} values are averaged and plotted against Rad52 concentration (in monomers). A representative assay is shown in **Fig. 27.4b** for Rad52 binding to the poly(dT)-30 ssDNA substrate (*see* **Note 15**).

3.5. Analysis of the Rad52–ssDNA (±RPA) Binding Isotherms

1. Binding of Rad52 to ssDNA or ssDNA–RPA complex typically produces a biphasic isotherm. At sub-saturating concentrations, binding of Rad52 to the poly(dT)-30 or poly(dT)-30 -RPA complex causes an increase in E_{FRET}, which is attributed to the growing fraction of ssDNA molecules wrapped around the Rad52 ring and designated below as Complex1 (High FRET phase in **Fig. 27.4b**).
2. Increase in E_{FRET} continues until the Rad52 concentration reaches approximately stoichiometric amounts (1 oligonucleotide per 7–8 Rad52 monomers).
3. The difference between maximal E_{FRET} amplitude corresponding to the completely wrapped oligonucleotide and E_{FRET} value characteristic to Rad52-free ssDNA (E_0) is shown in **Fig. 27.4b** as ΔE_1. The appearance of wrapped

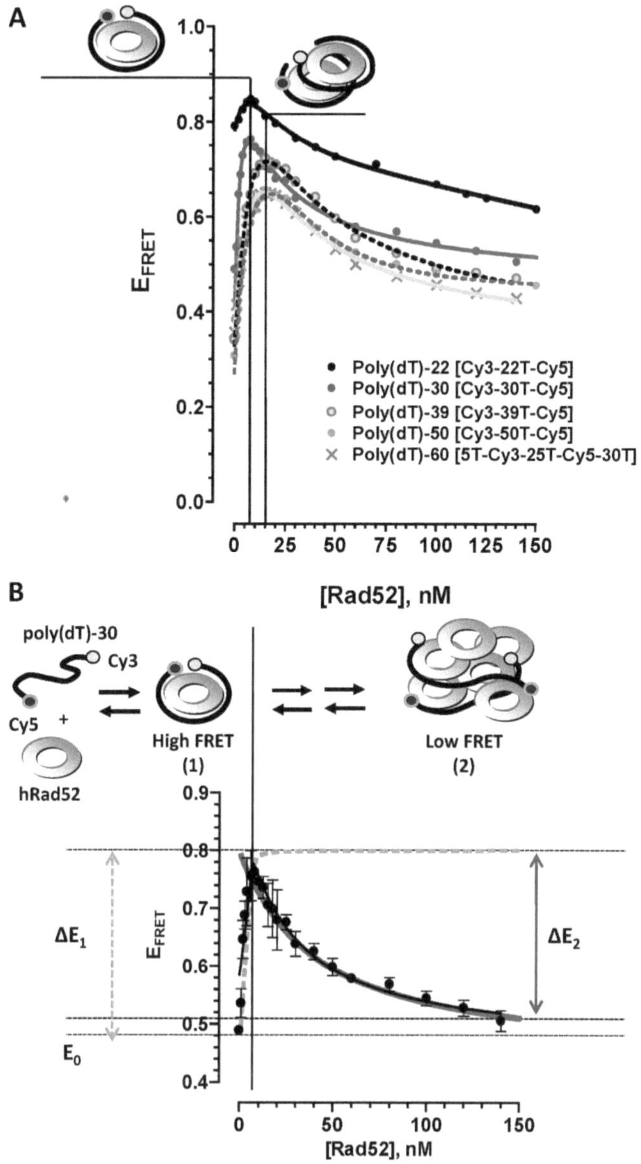

Fig. 27.4. Binding of Rad52 protein to oligonucleotides of different lengths. (**a**) Binding titrations with human Rad52 and 1 nM ssDNA of various lengths were performed and the FRET efficiency (E_{FRET}) was calculated as depicted in **Fig. 27.3**. The substrates used were Cy3–Cy5 end-labeled poly(dT)-22, poly(dT)-30, poly(dT)-39, poly(dT)-50, and internally labeled poly(dT)-60. Cartoons over the graph depict the DNA conformation and DNA/Rad52 stoichiometry corresponding to the highest FRET efficiency. (**b**) Interpretation of FRET data. Data points represent averages and standard deviations for three independent binding titrations with poly(dT)-30. ΔE_1 and ΔE_2 represent the change in the E_{FRET} value due to formation of wrapped complex (*see* equation [2]) and extended complex (*see* equation [3]), respectively. Data shown in these figures were originally published in *NAR (Oxford University Press)*; Grimme et al. (25).

species as a function of Rad52 concentration can be described as follows:

$$\text{Complex 1} = \frac{K_{d1} + [\text{DNA}] + \frac{[\text{Rad52}]}{7} - \sqrt{\left(K_{d1} + [\text{DNA}] + \frac{[\text{Rad52}]}{7}\right)^2 - 4 \times [\text{DNA}] \times \frac{[\text{Rad52}]}{7}}}{2}, \quad [2]$$

where [DNA] is the total substrate concentration (in molecules, 1 nM).

[Rad52] is the concentration of Rad52 monomers (respectively, concentration of Rad52 heptamers is [Rad52]/7) and K_{d1} is the equilibrium dissociation constant for this species. "Complex 1" represents one molecule of ssDNA wrapped around one heptameric ring of Rad52. Equation [2] is derived from a simplified 1:1 binding scheme, $\text{DNA}_{\text{FREE}} + \text{Rad52}^{\text{HEP}}_{\text{FREE}} \overset{K_{d1}}{\Leftrightarrow} \text{Complex 1}$ which assumes that free DNA and free Rad52 heptamers exist in equilibrium with "Complex 1."

4. Further increase in the Rad52 concentration results in a decrease in E_{FRET}. This represents the unwrapping of the ssDNA due to distributing the substrate molecule between multiple Rad52 oligomers (Low FRET phase in **Fig. 27.4b**).
5. The characteristic amplitude of this phase is ΔE_2, which represents difference in the E_{FRET} value between wrapped and extended substrates. Compared to homogenous "Complex 1" the FRET signal in this phase represents a heterologous mix of species. Nevertheless, these species can be collectively depicted as "Complex 2," which exists in equilibrium with "Complex 1" and free Rad52 defined by the equilibrium dissociation constant K_{d2}:

$$\text{Complex 2} = \frac{K_{d2} + \text{Complex 1} + \frac{[\text{Rad52}]}{7} - \sqrt{\left(K_{d2} + \text{Complex 2} + \frac{[\text{Rad52}]}{7}\right)^2 - 4 \times \text{Complex 2} \times \frac{[\text{Rad52}]}{7}}}{2}, \quad [3]$$

where K_{d2} is an apparent equilibrium dissociation constant for conversion of "Complex 2" to "Complex 1."

6. The E_{FRET} value at each Rad52 concentration then is the weighted sum of signals derived from free DNA, "Complex 1," and "Complex 2" and obeys the following relation:

$$E_{\text{FRET}} = E_0 + \Delta E_1 \times \frac{\text{Complex 1}}{[\text{DNA}]} - \Delta E_2 \times \frac{\text{Complex 2}}{\text{Complex 1}} \quad [4]$$

Note that while the overall shape of representative binding isotherms shown in **Figs. 27.3** and **27.4** obeys the above description, these assays were carried out under stoichiometric binding conditions for the first phase of the curves, which precludes quantitative evaluation of binding constants. To measure the two binding constants, one needs to carry out binding assays under conditions where the concentration of DNA substrate is significantly below the expected K_d. This can be achieved either by lowering the DNA concentration below 0.1 nM or by selecting conditions that impede Rad52 binding without affecting ssDNA flexibility. To accurately determine the four unknown parameters (K_{d1}, K_{d2}, ΔE_1, and ΔE_2), the binding experiments need to be carried out at several concentrations of the DNA substrate and then globally fit (*see* **Note 16**).

3.6. Kinetics of Rad52-Mediated Annealing of Short Oligonucleotides

1. The solution conditions and instrument settings are the same as described for binding assays.
2. We suggest using the following DNA substrates (25): a "target" molecule 28 nucleotides in length, T-28: 5′-ATAGTTATGGTGAGGACCC/**iCy3**/CTTTGTTTC-3′, where iCy3 indicates the position of the Cy3 dye and a Cy5-labeled complementary "probe" molecule, P-28: 5′-GAAACAAAGGGGTCC/**iCy5**/ TCACCATAACTAT-3′, where iCy5 marks the position of the Cy5 dye (*see* **Note 17**).
3. To minimize reaction volume, annealing assays can be carried out in the micro-cuvette (minimum volume 150 μl).
4. Fluorescence of Cy3 and Cy5 dyes is recorded in kinetics mode simultaneously over 400–600 s with the time resolution of 0.1 s. This time resolution is required to collect maximal possible number of data points during the initial phase of annealing reaction and to accurately define the initial rate of annealing.
5. **Figure 27.5a** shows raw data from a representative annealing reaction and **Fig. 27.5b** schematically depicts the change in dye positions and FRET states upon annealing of T-28 and P-28.
6. To ensure that annealing reactions are carried out under the same conditions as binding assays, it is advisable to premix Rad52 and RPA (if present) proteins in 300 μl of binding/annealing buffer and then separate the mixture into two half reactions. The first half reaction is directly placed in the cuvette.
7. After measuring the background fluorescence, the Cy3-labeled T-28 oligo is added to the cuvette and thoroughly mixed (**Fig. 27.5a**).
8. While recording baseline, the Cy5-labeled P-28 oligo is mixed in the second half reaction kept in the Eppendorf tube on the bench top.

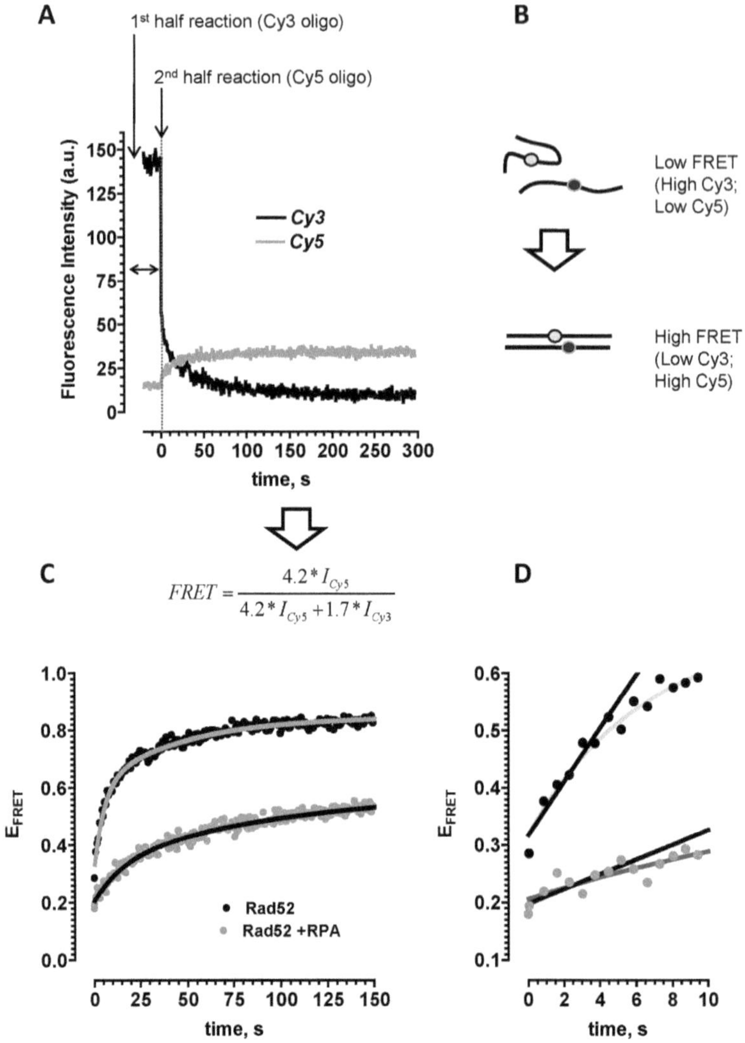

$$FRET = \frac{4.2 * I_{Cy5}}{4.2 * I_{Cy5} + 1.7 * I_{Cy3}}$$

Fig. 27.5. Kinetics of Rad52-mediated strand annealing: (**a**) Representative raw data from a typical annealing experiment. Fluorescence intensity of Cy3 (*black*) and Cy5 (*gray*) are recorded in kinetics mode every 0.1 s. *Arrows above the graph* indicate time points when the first and second half reactions were added to the cuvette. (**b**) Schematic representation of the experiment. (**c**) Calculated FRET efficiency (E_{FRET}) for annealing reaction in the absence (*gray* curve over *black* data points) and presence (*black* curve over *gray* data points) of RPA is fitted to double exponential function to yield the extent of reaction. (**d**) The initial linear portion of the FRET trajectory can be fitted to a straight line to yield the initial rate of annealing. Data shown in these figures were originally published in *NAR (Oxford University Press)*; Grimme et al. (25).

9. When the Cy3 signal from the first half reaction has stabilized, the second half reaction can be transferred into the cuvette and rapidly mixed. This initiates the annealing reaction (**Fig. 27.5a**) (*see* **Note 18**).

10. Data extracted from Cy3 and Cy5 progress curves can be converted into E_{FRET} values by first subtracting background and then using equation [1] (**Fig. 27.5c**).
11. To ensure reproducibility, the change in E_{FRET} is calculated by averaging the E_{FRET} value for three independent annealing reactions for each tested condition.
12. Averaged E_{FRET} can be converted to a fraction, a percentage, or a concentration of annealed DNA molecules. The E_{FRET} value for dsDNA (0.81 under our conditions) is set to 100% annealed DNA, while the E_{FRET} value for fully ssDNA (0.18 under our conditions) is determined by mixing two heterologous Cy3- and Cy5-labeled oligonucleotides. (We used two identical sequences shown in **Table 27.1**, T28 labeled with Cy3 and P-28′ labeled with Cy5.)
13. The E_{FRET} progress curves (**Fig. 27.5c**) are then fitted to a double exponential whose combined amplitude is compared to E_{FRET} values for dsDNA and ssDNA to determine the extent of annealing reaction.
14. The initial rate of annealing is determined as the slope of the linear portion of the progress curve (5–20 s depending on the protein concentration) for each assay, divided by 0.63 (E_{FRET} difference between fully single-stranded and fully annealed DNA) and multiplied by the total amount of dsDNA present (0.5 nM) as shown in **Fig. 27.5d**.

3.7. Effect of Rad52 on RPA-Mediated Duplex Destabilization

1. The instrument settings, cuvette, and buffer conditions are identical to those described in **Section 3.3**.
2. The reaction is monitored in kinetics mode with time resolution of 1 s.

Table 27.1
DNA substrates recommended for binding and annealing assays

Oligonucleotide name	Oligonucleotide sequence
Poly(dT)-22	5′-Cy3-dT_{22}-Cy5-3′
Poly(dT)-30	5′-Cy3-dT_{30}-Cy5-3′
Poly(dT)-39	5′-Cy3-dT_{39}-Cy5-3′
Poly(dT)-50	5′-Cy3-dT_{50}-Cy5-3′
Poly(dT)-60	5′-dT_5-Cy3-dT_{25}-Cy5-dT_{30}-3′
Target T-28	5′-ATAGTTATGGTGAGGACCC/**iCy3**/CTTTGTTTC-3′
Probe P-28	5′-GAAACAAAGGGGTCC/**iCy5**/TCACCATAACTAT-3′
Heterologous P-28′	5′-ATAGTTATGGTGAGGACCC/**iCy5**/CTTTGTTTC-3′

3. After recording background fluorescence of protein–DNA binding buffer with (or without) Rad52, pre-annealed T-28/P-28 dsDNA is added into the micro-cuvette and thoroughly mixed.
4. The duplex destabilization reaction is initiated by addition of the first aliquot of RPA. RPA-mediated melting of the dsDNA is monitored by following a decrease in calculated E_{FRET} signal. After the signal equilibrates, the next aliquot of RPA is titrated into the cuvette and the reaction is thoroughly mixed. Titration is repeated until no change in Cy3 and Cy5 signals is observed upon addition of RPA. Typical titration proceeds between 0 and 15 nM RPA. The calculated E_{FRET} values are plotted against the RPA concentration and compared for different Rad52 concentrations.

4. Notes

1. It is extremely important that the proteins used in binding and annealing assays are free from nuclease contamination. The cation or anion exchange chromatography is used as the last step in purification of RPA and Rad52 proteins, respectively, to remove all residual nuclease activity that may co-purify with protein of interest. This step also provides a means of concentrating the proteins.
2. *Purification of active nuclease-free human RPA*: Human RPA protein is expressed in *E. coli* Rosetta strain using pET11b plasmid containing the DNA construct for expression of all three RPA subunits (31). RPA expressing cells are induced by addition of 0.25 mM IPTG at mid-log phase and grown for additional 4 h before harvesting. The protein is purified as described in (31) using HiTrap Blue, hydroxyl apatite, and high-resolution anion exchange chromatography (MonoQ (GE Healthcare) or UNO-Q (Bio-Rad) columns). We recommend storing the purified protein at –80°C in RPA storage buffer containing 10% glycerol (listed in **Section 2.2**). The extinction coefficient of 84,000 M^{-1} cm^{-1} is used to determine RPA concentration based on absorbance at 280 nm.
3. *Purification of active nuclease-free human Rad52 protein*: Human Rad52 is expressed in *E. coli* Rosetta strain using pET15b plasmid which allows expression of the 6-histidine-tagged protein. Rad52 is purified using Ni affinity, heparin affinity, and cation exchange (MonoS or UNO-S) chromatography as described in (32, 33). Store

purified protein at −80°C in the Rad52 storage buffer containing 10% glycerol. The extinction coefficient of 40,380 M^{-1} cm^{-1} is used for determining Rad52 concentration by measuring absorbance at 280 nm.

4. Rad52 is prone to irreversible aggregation if dialyzed at high protein concentrations. To avoid a dialysis step between Ni affinity and heparin columns, it is advisable to bind Rad52 protein to the Ni-charged column in high salt/low imidazole buffer (50 mM potassium phosphate pH 7.4, 500 mM KCl, 50 mM imidazole, 5 mM β-mercaptoethanol), then wash the column with low salt (50 mM KCl) buffer, and then elute with low salt/high imidazole (300 mM) buffer. Eluted protein can be immediately loaded onto heparin column equilibrated with 50 mM Tris-HCl (pH 7.5), 100 mM KCl, 1 mM EDTA, and 0.1 mM DTT and eluted with 100–500 mM KCl gradient. Human Rad52 protein elutes from the heparin column at KCl concentration between 250 and 270 mM and is usually sufficiently diluted, so it can be dialyzed overnight without the risk of precipitation. To avoid this dialysis step, Rad52 can be diluted with buffer containing no salt to achieve a KCl concentration of 100 mM. The dialyzed or diluted sample then can be subjected to a cation exchange chromatography. We recommend using high-resolution MonoS, ResourceS (GE Healthcare) or UNO-S (Bio-Rad) columns.
5. PMT voltage of the spectrofluorimeter can be increased to increase the signal range, but it also will increase the noise.
6. The correction factors used to calculate E_{FRET} values based on Cy3 and Cy5 emission are specific to the instrument and are assigned based on the fraction of the donor fluorescence in the acceptor channel and the fraction of acceptor fluorescence in the donor channel, respectively. In our case, these were 4.2 for the Cy3 dyes and 1.7 for the Cy5 dye. To determine these factors we separately recorded the emission spectra of two oligonucleotides labeled with either Cy3 or Cy5 dye. Upon exciting Cy3 dye at 530 nm, we compared its emission at 565 nm (Cy3 channel) and 660 nm (Cy5 channel). The Cy5 dye was directly excited at 630 nm.
7. The I_{Cy5} and I_{Cy3} values were calculated by averaging the measured fluorescence intensities for each dye over 2 min after the signal had equilibrated and subtracting the background fluorescence.
8. Low ionic strength buffers are required to distinguish free ssDNA from ssDNA in a wrapped configuration, but up to about 50 mM NaCl can be added. Higher ionic strength

or the presence of divalent cations (such as Mg^{2+}) will decrease repulsion between negatively charged phosphates in the DNA backbone by shielding or bridging them and therefore will decrease the average distance between donor and acceptor dyes on free ssDNA making it impossible to distinguish unbound and wrapped forms of ssDNA. Higher salt concentrations and divalent cations are compatible with FRET-based binding analysis when using RPA–ssDNA as a substrate for Rad52 binding.

9. Eukaryotic RPA is composed of three subunits: RPA70, RPA32, and RPA14 all containing oligonucleotide/oligosaccharide binding (OB) folds. Four of these OB folds (A–D) are involved in ssDNA binding (reviewed in (34)). Human RPA binds ssDNA with sub-nanomolar affinity (35–37) and occludes 25–30 nucleotides upon binding (38). Because the end-to-end distance of ssDNA increases upon RPA binding, ssDNA–RPA complex formation can be followed by measuring FRET changes between Cy3 and Cy5 dyes incorporated into the same ssDNA molecule (**Fig. 27.2**).
10. Under stoichiometric binding conditions, E_{FRET} decreases with increasing RPA concentration until ssDNA is saturated and maximally extended by RPA. This saturation point defines binding stoichiometry. Selecting the RPA concentration that is slightly above stoichiometric amounts (for example, 2 nM RPA/1 nM 30-mer ssDNA) yields the most profound difference between ssDNA extended by RPA and ssDNA–RPA complex wrapped around human Rad52 protein.
11. Each protomer of the Rad52 heptamer accommodates four nucleotides of ssDNA (39). Therefore, oligonucleotides with the length around 28 nt are expected to be able to make a full turn around the Rad52 ring under the stoichiometric binding conditions, bringing the donor and acceptor dyes in close proximity.
12. The assays described here are very sensitive to the protein concentration and to the fraction of active protein present in preparation. Established protocols for RPA (31) and Rad52 (32, 33) ensure that the purified proteins are 100% active. If loss of activity or inconsistency between replicate experiments is observed, this may be due to Rad52 or RPA inactivation due to adsorption on the walls of cuvette and/or microcentrifuge tubes used for protein dilution and storage. To avoid this all pipette tips and plastic tubes should be non-adsorbing and should not be autoclaved. The walls of the cuvette can be pretreated by washing the

cuvette with solution containing BSA. BSA can also be added to the reaction mixture.

13. Although similar binding trends were observed for the substrates of all tested lengths, the highest E_{FRET} values for poly(dT)-22 and poly(dT)-30 corresponded to binding of one Rad52 ring per one oligonucleotide molecule, while poly(dT)-39, poly(dT)-50, and poly(dT)-60 substrates were achieved at the protein concentrations indicative of wrapping these ssDNA molecules around two oligomeric rings.
14. In addition to reporting on conformation of the dually labeled DNA substrate, calculated E_{FRET} is less sensitive to variations from experiment to experiment than the absolute values of donor and acceptor fluorescence. Moreover, protein binding in proximity of Cy3 dye often increases quantum yield of this fluorophore (40, 41). In FRET-based assays, this increase in Cy3 fluorescence is offset by a respective increase in fluorescence of Cy5 dye.
15. Experimental variations do occur so titrations should be performed in triplicate and E_{FRET} values should be presented as averages for the three independent titrations. **Figure 27.4b** shows a representative binding curve depicting both averaged values and standard deviations for each Rad52 concentration.
16. It is also important to keep in mind that the FRET-based assays report only on the distance between the donor and acceptor fluorophores and not on the actual configuration of substrate DNA molecule. Appropriate controls and alternative methods are required to validate the conclusions of the FRET-based studies.
17. The ssDNA molecules used for annealing assays should be internally labeled with the donor Cy3 or the acceptor Cy5 such that when the annealed product is formed, the proximity of the dyes to each other (~5–6 nt) results in high E_{FRET} (≈ 0.81).
18. To unambiguously compare binding and annealing properties of Rad52, total DNA concentration in the annealing reaction should be the same as in the respective binding reactions. For example, if the binding reaction was carried out in the presence of 1 nM (molecules) poly(dT)-30, each annealing half reaction should contain 1 nM T-28 or P-28. Mixing of two half reactions will bring the concentration of each oligonucleotide to 0.5 nM and total concentration of DNA strands to 1 nM.

References

1. Couedel, C., Mills, K.D., Barchi, M., Shen, L., Olshen, A., Johnson, R.D., Nussenzweig, A., Essers, J., Kanaar, R., Li, G.C., Alt, F.W., and Jasin, M. (2004) Collaboration of homologous recombination and nonhomologous end-joining factors for the survival and integrity of mice and cells. *Genes Dev* **18**, 1293–1304.
2. Sung, P., and Klein, H. (2006) Mechanism of homologous recombination: mediators and helicases take on regulatory functions. *Nat Rev Mol Cell Biol* 7, 739–750.
3. Krogh, B.O., and Symington, L.S. (2004) Recombination proteins in yeast. *Annu Rev Genet* **38**, 233–271.
4. Mortensen, U.H., Lisby, M., and Rothstein, R. (2009) Rad52. *Curr Biol* **19**, R676–R77.
5. Mortensen, U.H., Erdeniz, N., Feng, Q., and Rothstein, R. (2002) A molecular genetic dissection of the evolutionarily conserved N terminus of yeast Rad52. *Genetics* **161**, 549–562.
6. Sugiyama, T., New, J.H., and Kowalczykowski, S.C. (1998) DNA annealing by RAD52 protein is stimulated by specific interaction with the complex of replication protein A and single-stranded DNA. *Proc Natl Acad Sci USA* **95**, 6049–6054.
7. Bugreev, D.V., Hanaoka, F., and Mazin, A.V. (2007) Rad54 dissociates homologous recombination intermediates by branch migration. *Nat Struct Mol Biol* **14**, 746–753.
8. Miyazaki, T., Bressan, D.A., Shinohara, M., Haber, J.E., and Shinohara, A. (2004) In vivo assembly and disassembly of Rad51 and Rad52 complexes during double-strand break repair. *EMBO J* **23**, 939–949.
9. Sugiyama, T., Kantake, N., Wu, Y., and Kowalczykowski, S.C. (2006) Rad52-mediated DNA annealing after Rad51-mediated DNA strand exchange promotes second ssDNA capture. *EMBO J* **25**, 5539–5548.
10. McIlwraith, M.J., and West, S.C. (2008) DNA repair synthesis facilitates RAD52-mediated second-end capture during DSB repair. *Mol Cell* **29**, 510–516.
11. Nimonkar, A.V., Sica, R.A., and Kowalczykowski, S.C. (2009) Rad52 promotes second-end DNA capture in double-stranded break repair to form complement-stabilized joint molecules. *Proc Natl Acad Sci USA* **106**, 3077–3082.
12. Paques, F., and Haber, J.E. (1999) Multiple pathways of recombination induced by double-strand breaks in Saccharomyces cerevisiae. *Microbiol Mol Biol Rev* **63**, 349–404.
13. Stark, J.M., Pierce, A.J., Oh, J., Pastink, A., and Jasin, M. (2004) Genetic steps of mammalian homologous repair with distinct mutagenic consequences. *Mol Cell Biol* **24**, 9305–9316.
14. Shinohara, A., Shinohara, M., Ohta, T., Matsuda, S., and Ogawa, T. (1998) Rad52 forms ring structures and co-operates with RPA in single-strand DNA annealing. *Genes Cells* **3**, 145–156.
15. Stasiak, A.Z., Larquet, E., Stasiak, A., Muller, S., Engel, A., Van Dyck, E., West, S.C., and Egelman, E.H. (2000) The human Rad52 protein exists as a heptameric ring. *Curr Biol* **10**, 337–340.
16. Singleton, M.R., Wentzell, L.M., Liu, Y., West, S.C., and Wigley, D.B. (2002) Structure of the single-strand annealing domain of human RAD52 protein. *Proc Natl Acad Sci USA* **99**, 13492–13497.
17. Kagawa, W., Kurumizaka, H., Ishitani, R., Fukai, S., Nureki, O., Shibata, T., and Yokoyama, S. (2002) Crystal structure of the homologous-pairing domain from the human Rad52 recombinase in the undecameric form. *Mol Cell* **10**, 359–371.
18. Lloyd, J.A., McGrew, D.A., and Knight, K.L. (2005) Identification of residues important for DNA binding in the full-length human Rad52 protein. *J Mol Biol* **345**, 239–249.
19. Kagawa, W., Kagawa, A., Saito, K., Ikawa, S., Shibata, T., Kurumizaka, H., and Yokoyama, S. (2008) Identification of a second DNA binding site in the human Rad52 protein. *J Biol Chem* **283**, 24264–24273.
20. Petukhova, G., Stratton, S.A., and Sung, P. (1999) Single strand DNA binding and annealing activities in the yeast recombination factor Rad59. *J Biol Chem* **274**, 33839–33842.
21. Wu, Y., Sugiyama, T., and Kowalczykowski, S.C. (2006) DNA annealing mediated by Rad52 and Rad59 proteins. *J Biol Chem* **281**, 15441–15449.
22. Ploquin, M., Bransi, A., Paquet, E.R., Stasiak, A.Z., Stasiak, A., Yu, X., Cieslinska, A.M., Egelman, E.H., Moineau, S., and Masson, J.-Y. (2008) Functional and structural basis for a bacteriophage homolog of human RAD52. *Curr Biol* **18**, 1142–1146.
23. Pant, K., Shokri, L., Karpel, R.L., Morrical, S.W., and Williams, M.C. (2008) Modulation of T4 gene 32 protein DNA binding activity by the recombination mediator protein UvsY. *J Mol Biol* **380**, 799–811.
24. Erler, A., Wegmann, S., Elie-Caille, C., Bradshaw, C.R., Maresca, M., Seidel, R.,

Habermann, B., Muller, D.J., and Stewart, A.F. (2009) Conformational adaptability of red[beta] during DNA annealing and implications for its structural relationship with Rad52. *J Mol Biol* **391**, 586–598.
25. Grimme, J.M., Honda, M., Wright, R., Okuno, Y., Rothenberg, E., Mazin, A.V., Ha, T., and Spies, M. (2010) Human Rad52 binds and wraps single-stranded DNA and mediates annealing via two hRad52-ssDNA complexes. *Nucleic Acids Res* **38**, 2917–2930.
26. Jackson, D., Dhar, K., Wahl, J.K., Wold, M.S., and Borgstahl, G.E. (2002) Analysis of the human replication protein A:Rad52 complex: evidence for crosstalk between RPA32, RPA70, Rad52 and DNA. *J Mol Biol* **321**, 133–148.
27. de Vries, F.A., Zonneveld, J.B., de Groot, A.J., Koning, R.I., van Zeeland, A.A., and Pastink, A. (2007) Schizosaccharomyces pombe Rad22A and Rad22B have similar biochemical properties and form multimeric structures. *Mutat Res* **615**, 143–152.
28. Majka, J., and Speck, C. (2007) Analysis of protein-DNA interactions using surface plasmon resonance. *Adv Biochem Eng Biotechnol* **104**, 13–36.
29. Clegg, R.M. (2002) FRET tells us about proximities, distances, orientations and dynamic properties. *J Biotechnol* **82**, 177–179.
30. Rothenberg, E., Grimme, J.M., Spies, M., and Ha, T. (2008) Human Rad52-mediated homology search and annealing occurs by continuous interactions between overlapping nucleoprotein complexes. *Proc Natl Acad Sci USA* **105**, 20274–20279.
31. Henricksen, L.A., Umbricht, C.B., and Wold, M.S. (1994) Recombinant replication protein A: expression, complex formation, and functional characterization. *J Biol Chem* **269**, 11121–11132.
32. Benson, F.E., Baumann, P., and West, S.C. (1998) Synergistic actions of Rad51 and Rad52 in recombination and DNA repair. *Nature* **391**, 401–404.
33. Reddy, G., Golub, E.I., and Radding, C.M. (1997) Human Rad52 protein promotes single-strand DNA annealing followed by branch migration. *Mutat Res* **377**, 53–59.
34. Fanning, E., Klimovich, V., and Nager, A.R. (2006) A dynamic model for replication protein A (RPA) function in DNA processing pathways. *Nucleic Acids Res* **34**, 4126–4137.
35. Gomes, X.V., Henricksen, L.A., and Wold, M.S. (1996) Proteolytic mapping of human replication protein A: evidence for multiple structural domains and a conformational change upon interaction with single-stranded DNA. *Biochemistry* **35**, 5586–5595.
36. Gomes, X.V., and Wold, M.S. (1996) Functional domains of the 70-kilodalton subunit of human replication protein A. *Biochemistry* **35**, 10558–10568.
37. Kim, C., Snyder, R.O., and Wold, M.S. (1992) Binding properties of replication protein A from human and yeast cells. *Mol Cell Biol* **12**, 3050–3059.
38. Kim, C., Paulus, B.F., and Wold, M.S. (1994) Interactions of human replication protein A with oligonucleotides. *Biochemistry* **33**, 14197–14206.
39. Parsons, C.A., Baumann, P., Van Dyck, E., and West, S.C. (2000) Precise binding of single-stranded DNA termini by human RAD52 protein. *EMBO J* **19**, 4175–4181.
40. Fischer, C.J., Maluf, N.K., and Lohman, T.M. (2004) Mechanism of ATP-dependent translocation of E. coli UvrD monomers along single-stranded DNA. *J Mol Biol* **344**, 1287–1309.
41. Luo, G., Wang, M., Konigsberg, W.H., and Xie, X.S. (2007) Single-molecule and ensemble fluorescence assays for a functionally important conformational change in T7 DNA polymerase. *Proc Natl Acad Sci USA* **104**, 12610–12615.

Chapter 28

Visualization of Human Dmc1 Presynaptic Filaments

Michael G. Sehorn and Hilarie A. Sehorn

Abstract

Meiosis is initiated by the programmed formation of DNA double-strand breaks (DSBs). These DSBs are repaired by homologous recombination to promote crossover formation that ensures proper chromosomal segregation in meiosis. hRad51 and hDmc1 are two human recombinases present during meiosis that are homologous to the RecA recombinase from *Escherichia coli*. The hRad51 and hDmc1 recombinases bind the nucleolytically processed ends of the DSB forming a presynaptic filament. Formation of the presynaptic filament is necessary for the search for homology and the progression of recombination. In this chapter, we provide a method to purify hDmc1 and prepare samples for visualizing hDmc1 nucleoprotein presynaptic filaments via transmission electron microscopy.

Key words: Meiosis, homologous recombination, presynaptic filament, transmission electron microscopy, protein purification, human Dmc1.

Abbreviations

HR	homologous recombination
ss	single stranded
DSB	DNA double-strand break
NTA	nitro triacetic acid

1. Introduction

Homologous recombination (HR) is a ubiquitous DNA repair pathway utilized to repair the most detrimental form of DNA damage, the DNA double-strand break (DSB). DSBs arise from exogenous and endogenous events such as exposure to ionizing radiation and collapsed replication forks. Meiotic programmed DSBs formed by the Spo11 topoisomerase (1) are another strong inducer of HR. After the introduction of a DSB, the ends of

H. Tsubouchi (ed.), *DNA Recombination*, Methods in Molecular Biology 745,
DOI 10.1007/978-1-61779-129-1_28,

the DSB are processed to expose regions of single-strand DNA (ssDNA). These regions of ssDNA are bound by the eukaryotic homologs of the *Escherichia coli* RecA recombinase, hRad51 and hDmc1, to form a filamentous structure known as a presynaptic filament. Formation of the presynaptic filament is critical for the search for homology necessary to repair the DSB by homologous recombination. In the absence of DNA, hDmc1 exists as an octameric protein ring (2). Upon the addition of ssDNA, hDmc1 was shown to form stacked rings where ssDNA threads the hole of the octameric ring of hDmc1 (**Fig. 28.1b**) (2–4). This is a characteristic of hDmc1 proteins that RecA and hRad51 do not possess. The stacked protein rings on ssDNA appear to be unable to conduct HR reactions (4). In the presence of ATP, hDmc1 is capable of forming short presynaptic filaments on ssDNA (**Fig. 28.1a**) (4, 5). Similar to RecA and hRad51, the hDmc1 nucleoprotein filament was shown to be the active form of the recombinase (4). The method described herein provides a protocol for the chromatographic purification of hDmc1 from insect cells and the preparation of hDmc1 nucleoprotein filaments to be visualized by transmission electron microscopy.

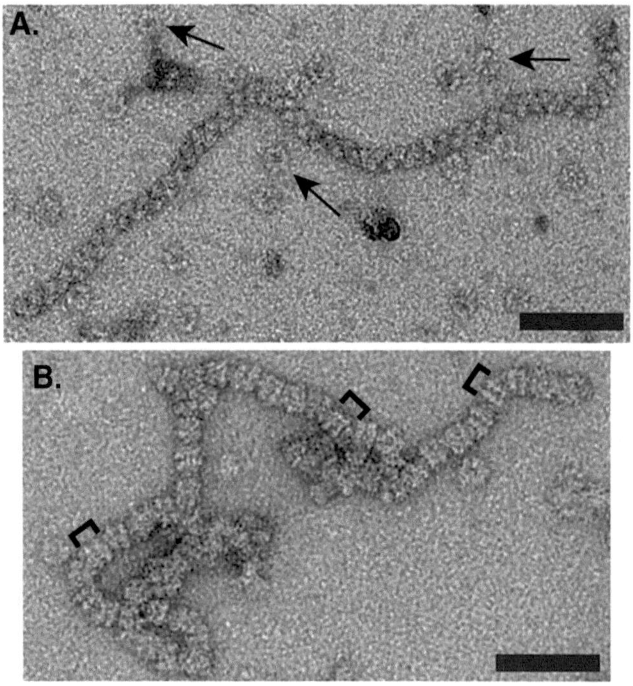

Fig. 28.1. Electron microscopy of hDmc1 nucleoprotein complexes. (**a**) hDmc1 incubated with ssDNA in the presence of ATP. A short hDmc1 helical nucleoprotein filament is shown traversing the micrograph. This is the catalytically active form of hDmc1 (4). Three *black arrows* indicate octameric ring structures of free hDmc1 not bound to ssDNA. (**b**) hDmc1 incubated with ssDNA in the absence of ATP (2). The stacked hDmc1 rings are unable to promote HR (4). Three *black brackets* indicate examples of stacked octameric ring structures hDmc1. A *black bar* denotes 50 nm.

2. Materials

2.1. Amplification of 6XHIS-hDmc1 Baculovirus

1. 25 cm^2 cell culture flasks
2. SF-900 II SFM (1×) medium (Invitrogen) supplemented with 50 μg/ml gentamicin sulfate
3. Gentamicin sulfate solution 10 mg/ml (Invitrogen)
4. Steriflip filtration apparatus (Millipore)

2.2. Determination of 6XHIS-hDmc1 Baculovirus Titer

1. SF-900 II SFM (1×) medium (Invitrogen) supplemented with 50 μg/ml gentamicin sulfate
2. SF-900 II SFM (1.3×) medium (Invitrogen) supplemented with 50 μg/ml gentamicin sulfate
3. Gentamicin sulfate solution 10 mg/ml (Invitrogen)
4. 4% agarose solution (Invitrogen)
5. 6-Well cell culture plates

2.3. Expression of hDmc1 in Insect Cells

1. The recombinant baculovirus for 6XHIS-hDmc1 that contains a 6Xhistidine tag fused to the amino terminal end of the hDmc1 (4)
2. High Five insect cells (Invitrogen)
3. Express Five SFM (Invitrogen) supplemented with 50 μg/ml gentamicin sulfate
4. Gentamicin sulfate solution 10 mg/ml (Invitrogen)

2.4. Purification of hDmc1

1. Q Sepharose (GE Healthcare Life Sciences)
2. Mono S (GE Healthcare Life Sciences)
3. Nickel-NTA (nitro triacetic acid) agarose (Qiagen)
4. 1.0 cm diameter Econo-column (Bio-Rad Laboratories)
5. 1.5 cm diameter Econo-column (Bio-Rad Laboratories)
6. 50 ml Superloop (GE Healthcare Life Sciences)
7. Centricon-30 microconcentrator (Millipore)
8. Cell breakage buffer: 50 mM Tris–HCl, pH 7.5, 2 mM EDTA, 10% sucrose, 150 mM KCl, 1 mM dithiothreitol, 1 mM PMSF (phenylmethylsulfonyl fluoride) containing the following protease inhibitors at 3 μg/ml each: aprotinin, chymostatin, leupeptin, and pepstatin
9. K buffer: 20 mM KH_2PO_4 at pH 7.4, 0.5 mM EDTA, 1 mM β-mercaptoethanol, and 10% glycerol

2.5. Preparation of Electron Microscopy Samples

1. 2% uranyl acetate (Electron Microscopy Sciences) in water filtered through a 0.22 μm syringe filter prior to use
2. Whatman filter paper

3. Carbon-coated 400 mesh copper grids (Ted Pella, Inc.)
4. 25 mM ATP
5. 30 mM $MgCl_2$
6. 3 M KCl
7. TE buffer: 10 mM Tris–HCl, pH 8.0, 1 mM EDTA
8. ϕX174 virion (+) DNA 1 mg/ml diluted to 0.25 mg/ml in TE buffer (New England Biolabs)
9. 5× reaction buffer: 175 mM Tris–HCl, pH 7.6, 5 mM DTT, 12 mM $MgCl_2$
10. Dilution buffer: 35 mM Tris–HCl, pH 7.6, 1 mM DTT, 2.4 mM $MgCl_2$, 2 mM ATP

3. Methods

The detection of a right-handed, helical nucleoprotein filament Rad51 or Dmc1 presynaptic filament ultimately depends on visualization by electron microscopy. The use of electron microscopy has yielded significant information about hRad51 and hDmc1 nucleoprotein filaments (2–4, 6–13). The electron microscope has also provided insight into the function of recombination mediators (14–19) that promote the formation of presynaptic filaments on RPA-coated ssDNA substrates and other proteins that affect presynaptic filament dynamics (20–25). As novel recombination mediators and accessory factors for hDmc1 are elucidated, the use of electron microscopy will no doubt continue to be a valuable tool in the analysis of their role in hDmc1 presynaptic filament dynamics.

3.1. Amplification of Baculovirus in Sf9 Insect Cells

1. Remove SF9 cells from liquid nitrogen and place in a 37°C water bath.
2. Gently agitate cells until almost thawed.
3. As soon as the cells are thawed, place the cells on ice.
4. Add 4 ml of SF-900 II SFM (1×) medium to a sterile 25 cm^2 flask.
5. Decontaminate the cell vial with 70% ethanol and dry the vial with a Kimwipe.
6. Transfer the 1 ml cell suspension directly into the 4 ml of medium in the 25 cm^2 flask.
7. Transfer the 25 cm^2 flask to a 27°C incubator and allow the cells to attach for 1 h.
8. Gently remove the medium.

9. Add 5 ml of fresh SF-900 II SFM (1×) medium and incubate at 27°C for 24 h.
10. Gently remove the medium.
11. Continue to incubate the cells until a confluent monolayer has formed (*see* **Note 1**).
12. Add 2 ml of the cells from the 25 cm^2 flask at a density of 1×10^6 cells/ml to each well of a 6-well plate.
13. Incubate the cells at room temperature for 1 h.
14. After the cells have attached, add 100 μl of the 6XHIS-hDmc1 P1 viral stock to each well.
15. Incubate the cells for 72 h at 27°C.
16. Collect the 2 ml of medium containing virus from each well and transfer to a 15 ml conical tube.
17. Centrifuge the tubes at 1,000×*g* for 10 min.
18. Filter the clarified supernatant containing virus using a Millipore Steriflip apparatus.
19. Store the filtered supernatant called the P2 viral stock at 4°C in the dark.
20. Transfer the appropriate amount of cells from Step 11 in **Section 3.1** to an Erlenmeyer flask to seed a cell suspension culture of 25 ml at a cell density of 1×10^6 cells/ml of medium.
21. Incubate the Erlenmeyer flask at 27°C with constant shaking at 80 rpm for 24 h or until the cell density of the suspension culture reaches 2×10^6 cells/ml of medium.
22. Dilute the cells with enough medium to lower the density of the cell suspension to 1×10^6 cells/ml of medium.
23. Repeat Steps 21 and 22 until the cell culture is at least 50 ml with a cell density of 2×10^6 cells/ml of medium.
24. Transfer 50 ml of the cell suspension at a density of 2×10^6 cells/ml of medium to a new Erlenmeyer flask.
25. Add enough medium to lower the cell density of the suspension culture to 1×10^6 cells/ml of medium.
26. Add the 1 ml of the P2 viral stock and continue to incubate with shaking at 80 rpm at 27°C for 7 days.
27. Centrifuge the cell culture at 1,000×*g* for 10 min.
28. Filter-sterilize the supernatant containing the P3 virus into a new sterile bottle and store at 4°C in the dark.

3.2. Determination of 6XHIS-hDmc1 Baculovirus Titer

1. From the suspension culture in Step 11 in **Section 3.1**, transfer cells at a density of 5×10^5 cells/ml into each well of a 6-well plate.

2. Incubate the cells at room temperature for 1 h to allow the cells to attach to the plate.
3. Place the bottle of agarose gel in the 70°C water bath.
4. Place an empty 100 ml bottle and the bottle of SF-900 II SFM (1.3×) insect medium in the 37°C water bath.
5. Allow the cells to continue to incubate until the cells achieve at least 50% confluence.
6. Prepare a 10^{-1} to 10^{-8} serial dilution of the P3 virus supernatant by sequentially diluting 100 μl of the previous dilution into 900 μl of SF-900 II SFM cell culture medium.
7. Remove the supernatant from each of the wells of the 6-well plate.
8. Transfer 1 ml of the virus dilution to the appropriate well.
9. Incubate for 1 h at room temperature.
10. Prepare the SF-900 plaque overlay medium by dispensing 30 ml of the SF-900 II SFM (1.3×) medium and 10 ml of the melted 4% agarose into the empty bottle.
11. Remove the supernatant from the wells and gently replace with 2 ml of the SF-900 plaque overlay medium.
12. After the agarose has solidified, place the plates in an incubator at 27°C.
13. Monitor the plates until the plaque count does not change for two consecutive days.
14. Calculate the pfu (plaque forming units)/ml of P3 viral stock using the following equation:

$$[(1/\text{dilution factor}) \times (\text{number of plaques})] \times [1/(\text{ml of inoculum per plate})].$$

3.3. Expression of hDmc1 in Insect Cells

1. Remove High Five cells from liquid nitrogen and place in a 37°C water bath.
2. Gently agitate cells until almost thawed.
3. As soon as the cells are thawed, place the cells on ice.
4. Add 4 ml of complete Express Five SFM to a 25 cm^2 flask.
5. Decontaminate the cell vial with 70% ethanol and dry the vial with a Kimwipe.
6. Transfer the 1 ml cell suspension directly into the 4 ml of Express Five SFM medium in the flask.
7. Transfer the flask to a 27°C incubator and allow the cells to attach for 1 h.
8. Gently remove medium.

9. Add 5 ml of fresh Express Five SFM medium and incubate at 27°C for 24 h.
10. Continue to incubate the cells until a confluent monolayer has formed.
11. Subculture the cells to obtain enough cells to start a Erlenmeyer flask containing 25 ml of medium with 1×10^6 cells/ml of Express Five SFM medium.
12. Incubate the Erlenmeyer flask at 27°C with constant shaking at 80 rpm for 24 h or until the cells reach 2×10^6 cells/ml of medium.
13. Dilute the cells with enough medium to bring the density of the cells to 1×10^6 cells/ml medium.
14. Repeat Steps 11–13 until the desired number of flasks contains cell cultures of 100 ml at a cell density of 2×10^6 cells/ml of medium.
15. Add enough medium (~100 ml) to dilute the cell suspension culture in each flask to a cell density of 1×10^6 cells/ml of medium.
16. For each flask with ~200 ml cell suspension at a cell density of 1×10^6 cells/ml, transfer 25 ml of cell suspension to each of eight Erlenmeyer flasks.
17. Add the 6XHIS-hDmc1 baculovirus at a MOI (multiplicity of infection) of 10 to each of the flasks to infect the High Five insect cells with the recombinant hDmc1 virus (*see* **Note 2**).
18. Allow the cells to continue growth for 72 h with slow shaking (80 rpm) at 27°C.
19. Pour the 25 ml culture into a 50 ml conical tube and centrifuge at 1,500×*g* for 8 min to pellet the cells.
20. Remove the media and place the cell pellet on ice for protein purification or store at –80°C.

3.4. hDmc1 Protein Purification

1. Prior to starting the purification of hDmc1, pour approximately 40 ml of resuspended Q Sepharose media into a 1.5 cm diameter Econo-column and allow the media to settle to ~20 ml.
2. Connect the flow adaptor and equilibrate the column with 200 ml of K buffer containing 150 mM KCl using an ÄKTA FPLC (GE Healthcare Life Sciences) at 2 ml/min.
3. All the steps involved with purification of hDmc1 (*see* **Note 3**) are to be performed at 4°C.
4. Resuspend each cell pellet in 4 ml of cell breakage buffer.
5. Combine the resuspended cells and add cell breakage buffer to a final volume of 50 ml.

6. Lyse the resuspended cells using a French Press (Thermo Scientific) at 20,000 psi. A single pass through the pressure cell is sufficient.
7. Sonicate the extract twice at a power setting of 8 for 15 s each using a Branson Sonifier.
8. Centrifuge the extract in a Beckman Type Ti45 rotor at 40,000 rpm (125,000×*g*) for 90 min to clarify the extract.
9. Load the clarified extract onto the equilibrated ~20 ml Q Sepharose column.
10. Fractionate the protein bound to the Q Sepharose column using a 100 ml gradient of K buffer containing 150–700 mM KCl.
11. Pool the fractions (~375 mM KCl) containing hDmc1.
12. Pour 6 ml of nickel-NTA slurry into an Econo-column with 1.0 cm diameter and allow the liquid to drain leaving a packed column. Do not allow the column dry completely as this will cause channels to form in your column.
13. Using a disposable Pasteur pipette, apply 30 ml of K buffer containing 150 mM KCl to the top of the packed column. Continue until all 30 ml is added to the column. Allow the column to drain. Again, do not allow the column run dry.
14. Using a disposable Pasteur pipette, resuspend the washed nickel-NTA agarose beads in 6 ml of K buffer containing 150 mM KCl and add them to the pooled hDmc1 fractions and rotate or rock for 2 h.
15. Pour the nickel-NTA agarose with bound hDmc1 protein into a 1 cm diameter Econo-column.
16. Add 3 ml of K buffer containing 150 mM KCl and 10 mM imidazole. Take care to not disturb the NTA nickel resin. Let the buffer drain from the column.
17. Add 3 ml of K buffer containing 150 mM KCl and 20 mM imidazole. Again, take care to not disturb the NTA nickel resin. Let the buffer drain from the column. Repeat this step two more times.
18. Elute the protein by gently adding 30 ml of buffer K containing 150 mM KCl and 300 mM imidazole and collect 3 ml fractions. Take care to not disturb the resin bed.
19. Pool the fractions containing hDmc1.
20. Equilibrate a 1 ml Mono S column with 10 ml of K buffer containing 150 mM KCl using an ÄKTA FPLC at 0.5 ml/min.
21. Dilute the pooled fractions of hDmc1 with K buffer to match conductivity of K buffer containing 150 mM KCl (*see* **Note 4**).

22. Load the conductivity-matched hDmc1 fractions onto the equilibrated 1 ml Mono S column using a 50 ml Superloop connected to an ÄKTA FPLC.
23. Fractionate the protein bound to the Mono S column using a 20 ml gradient of K buffer containing 150–700 mM KCl.
24. Pool the peak fractions containing hDmc1 (~350 mM KCl) and concentrate to 7 mg/ml using a Centricon-30 microconcentrator.
25. Aliquot 5–10 μl of the concentrated protein into 0.6 ml microcentrifuge tubes.
26. Snap-freeze the protein in liquid nitrogen and store at −70°C.

3.5. Preparation of Electron Microscopy Samples

1. Add 1.25 μl of 5× reaction buffer to a 1.7 ml microcentrifuge tube.
2. Add 0.5 μl of 25 mM stock solution of ATP and mix.
3. Add 0.5 μl of 30 mM stock solution of $MgCl_2$ and mix.
4. Add 1.0 μl of the diluted ϕX174 virion (+) DNA and mix.
5. Add 0.38 μl of 3 M KCl and mix.
6. Add 1.62 μl of water and mix.
7. Add 1.0 μl of hDmc1 protein at 7 mg/ml.
8. Mix the reaction and incubate for 1 h at 37°C.
9. Using Dumon #5 tweezers, pick up a 400-mesh copper grid coated with carbon film as close to the edge as possible. Take care to not bend or punch holes through the carbon-coated grid.
10. Place the grid carbon side up in a small glass petri dish and place the dish in the Pelco easiGlow unit.
11. Turn on the vacuum pump to achieve a pressure within the range 0.45–1.1 mbar.
12. Glow discharge the carbon-coated grid at 15 mA for 15 s (*see* **Note 5**).
13. Turn off the vacuum and use the Dumon #5 tweezers to remove the carbon-coated grid from the dish.
14. Lock the tweezers using the accompanying O-ring and place the tweezers holding the grid on a bench with the carbon side of the grid up.
15. Remove 1 μl of the reaction mix and add it to 39 μl of dilution buffer and mix (*see* **Note 6**).
16. Remove 3 μl of the diluted reaction mix and apply to the carbon surface of the freshly glow-discharged carbon-coated grid (*see* **Note 7**).

17. After adsorption for 30 s, blot excess sample away with Whatman paper by holding the tweezers so that the grid is perpendicular to the surface of the Whatman paper. Gently touch the edge of the grid to the Whatman paper to allow the liquid to be wicked away.
18. Place the tweezers back on the bench with the carbon side of the grid up. Apply 5 μl of 2% uranyl acetate solution to the surface of the grid.
19. After 30 s, blot excess stain with Whatman paper as described in Step 15. Allow the sample to air-dry.
20. Use a transmission electron microscope equipped with a LaB6 filament operated at 120 keV at a nominal magnification of 26,000× to visualize hDmc1 filaments (*see* **Note 8**).

4. Notes

1. Always keep a 25 cm^2 flask with attached cells and an Erlenmeyer flask with a suspension of SF9 and High Five cells at 1×10^6 cells/ml of medium. Continued passage of SF9 and High Five cells saves time and serves as a backup if something happens to the cells.
2. The amount of P3 viral stock to infect cells is determined using the following equation:

$$\text{Volume of virus} = (\text{MOI} \times \text{number of cells})/\text{pfu/ml}$$

3. hDmc1 has a molecular weight of 37,681 kDa; however, with the addition of a 6Xhistidine tag the molecular weight of the recombinant hDmc1 is 38,504 kDa.
4. Dialysis against 2 l of K buffer containing 150 mM KCl can also be used.
5. Glow discharge makes the carbon surface temporarily hydrophilic improving sample adsorption.
6. For a higher density of hDmc1 filaments on the carbon-coated grid, reduce the dilution to 10-, 20-, or 30-fold as desired.
7. Grids coated with silicon dioxide can also be used in the absence of glow discharge.
8. The hDmc1 filaments tend to be short (100–550 nm). There will be some “ring” structures on the grid. These rings are hDmc1 in its octameric ring form.

Acknowledgments

This work was supported by National Science Foundation/ EPSCoR grant 2004 RII-EPS-0447660 and Clemson University.

References

1. Keeney, S., Giroux, C.N., and Kleckner, N. (1997) Meiosis-specific DNA double-strand breaks are catalyzed by Spo11, a member of a widely conserved protein family. *Cell* **88**, 375–384.
2. Passy, S.I., Yu, X., Li, Z., Radding, C.M., Masson, J.Y., West, S.C., and Egelman, E.H. (1999) Human Dmc1 protein binds DNA as an octameric ring. *Proc Natl Acad Sci USA* **96**, 10684–10688.
3. Masson, J.Y., Davies, A.A., Hajibagheri, N., Van Dyck, E., Benson, F.E., Stasiak, A.Z., Stasiak, A., and West, S.C. (1999) The meiosis-specific recombinase hDmc1 forms ring structures and interacts with hRad51. *EMBO J* **18**, 6552–6560.
4. Sehorn, M.G., Sigurdsson, S., Bussen, W., Unger, V.M., and Sung, P. (2004) Human meiotic recombinase Dmc1 promotes ATP-dependent homologous DNA strand exchange. *Nature* **429**, 433–437.
5. Bugreev, D.V., Golub, E.I., Stasiak, A.Z., Stasiak, A., and Mazin, A.V. (2005) Activation of human meiosis-specific recombinase Dmc1 by Ca^{2+}. *J Biol Chem* **280**, 26886–26895.
6. Benson, F.E., Stasiak, A., and West, S.C. (1994) Purification and characterization of the human Rad51 protein, an analogue of E. coli RecA. *EMBO J* **13**, 5764–5771.
7. Sung, P., and Robberson, D.L. (1995) DNA strand exchange mediated by a RAD51-ssDNA nucleoprotein filament with polarity opposite to that of RecA. *Cell* **82**, 453–461.
8. Baumann, P., Benson, F.E., Hajibagheri, N., and West, S.C. (1997) Purification of human Rad51 protein by selective spermidine precipitation. *Mutat Res* **384**, 65–72.
9. Yu, X., Jacobs, S.A., West, S.C., Ogawa, T., and Egelman, E.H. (2001) Domain structure and dynamics in the helical filaments formed by RecA and Rad51 on DNA. *Proc Natl Acad Sci USA* **98**, 8419–8424.
10. Yang, S., VanLoock, M.S., Yu, X., and Egelman, E.H. (2001) Comparison of bacteriophage T4 UvsX and human Rad51 filaments suggests that RecA-like polymers may have evolved independently. *J Mol Biol* **312**, 999–1009.
11. Chi, P., Van Komen, S., Sehorn, M.G., Sigurdsson, S., and Sung, P. (2006) Roles of ATP binding and ATP hydrolysis in human Rad51 recombinase function. *DNA Repair (Amst)* **5**, 381–391.
12. Sheridan, S.D., Yu, X., Roth, R., Heuser, J.E., Sehorn, M.G., Sung, P., Egelman, E.H., and Bishop, D.K. (2008) A comparative analysis of Dmc1 and Rad51 nucleoprotein filaments. *Nucleic Acids Res* **36**, 4057–4066.
13. Dupaigne, P., Lavelle, C., Justome, A., Lafosse, S., Mirambeau, G., Lipinski, M., Pietrement, O., and Le Cam, E. (2008) Rad51 polymerization reveals a new chromatin remodeling mechanism. *PLoS One* **3**, e3643.
14. Davies, A.A., Masson, J.Y., McIlwraith, M.J., Stasiak, A.Z., Stasiak, A., Venkitaraman, A.R., and West, S.C. (2001) Role of BRCA2 in control of the RAD51 recombination and DNA repair protein. *Mol Cell* 7, 273–282.
15. Galkin, V.E., Esashi, F., Yu, X., Yang, S., West, S.C., and Egelman, E.H. (2005) BRCA2 BRC motifs bind RAD51-DNA filaments. *Proc Natl Acad Sci USA* **102**, 8537–8542.
16. San Filippo, J., Chi, P., Sehorn, M.G., Etchin, J., Krejci, L., and Sung, P. (2006) Recombination mediator and Rad51 targeting activities of a human BRCA2 polypeptide. *J Biol Chem* **281**, 11649–11657.
17. Esashi, F., Galkin, V.E., Yu, X., Egelman, E.H., and West, S.C. (2007) Stabilization of RAD51 nucleoprotein filaments by the C-terminal region of BRCA2. *Nat Struct Mol Biol* **14**, 468–474.
18. Davies, O.R., and Pellegrini, L. (2007) Interaction with the BRCA2 C terminus protects RAD51-DNA filaments from disassembly by BRC repeats. *Nat Struct Mol Biol* **14**, 475–483.
19. Shivji, M.K., Mukund, S.R., Rajendra, E., Chen, S., Short, J.M., Savill, J., Klenerman, D., and Venkitaraman, A.R. (2009) The BRC repeats of human BRCA2 differentially regulate RAD51 binding on single- versus

double-stranded DNA to stimulate strand exchange. *Proc Natl Acad Sci USA* **106**(32), 13254–13259.
20. Van Dyck, E., Hajibagheri, N.M., Stasiak, A., and West, S.C. (1998) Visualisation of human rad52 protein and its complexes with hRad51 and DNA. *J Mol Biol* **284**, 1027–1038.
21. McIlwraith, M.J., Van Dyck, E., Masson, J.Y., Stasiak, A.Z., Stasiak, A., and West, S.C. (2000) Reconstitution of the strand invasion step of double-strand break repair using human Rad51 Rad52 and RPA proteins. *J Mol Biol* **304**, 151–164.
22. Kiianitsa, K., Solinger, J.A., and Heyer, W.D. (2002) Rad54 protein exerts diverse modes of ATPase activity on duplex DNA partially and fully covered with Rad51 protein. *J Biol Chem* **277**, 46205–46215.
23. Li, X., Zhang, X.P., Solinger, J.A., Kiianitsa, K., Yu, X., Egelman, E.H., and Heyer, W.D. (2007) Rad51 and Rad54 ATPase activities are both required to modulate Rad51-dsDNA filament dynamics. *Nucleic Acids Res* **35**, 4124–4140.
24. Hu, Y., Raynard, S., Sehorn, M.G., Lu, X., Bussen, W., Zheng, L., Stark, J.M., Barnes, E.L., Chi, P., Janscak, P., Jasin, M., Vogel, H., Sung, P., and Luo, G. (2007) RECQL5/Recql5 helicase regulates homologous recombination and suppresses tumor formation via disruption of Rad51 presynaptic filaments. *Genes Dev* **21**, 3073–3084.
25. Bugreev, D.V., Yu, X., Egelman, E.H., and Mazin, A.V. (2007) Novel pro- and anti-recombination activities of the Bloom's syndrome helicase. *Genes Dev* **21**, 3085–3094.

Section IV

Cell Biological Approaches to Study the In Vivo Behavior of Homologous Recombination

Chapter 29

Tracking of Single and Multiple Genomic Loci in Living Yeast Cells

Imen Lassadi and Kerstin Bystricky

Abstract

Nuclear organization is involved in numerous aspects of cellular function. In yeast, analysis of the nuclear position and dynamics of the silent and active mating-type loci has allowed to gain insight into the mechanisms involved in directing mating-type switching. The fluorescent repressor operator systems (FROS) have proven to be a powerful technique to tag DNA sequences to investigate chromosome position and dynamics in living cells. FROS rely on the transgenic expression of a bacterial repressor fused to a fluorescent protein which can bind to its respective operator DNA sequence integrated as multicopy tandem arrays at a specific genomic site. Different FROS exist which facilitate the tagging of up to three different loci simultaneously. This chapter describes detailed protocols for FROS usage and analysis in the yeast *Saccharomyces cerevisiae*.

Key words: *Saccharomyces cerevisiae*, chromosome dynamics, fluorescent proteins, live-cell microscopy, DNA, nuclear organization, LacO, TetO.

1. Introduction

Large-scale genome organization has been recognized as a major player in genome function and regulation. Its complexity asks for a variety of sophisticated tools to determine the interaction between different loci and to study the positioning and dynamic behavior of chromosomes in the nucleus. However, the highly dynamic genome renders real-time studies very difficult. The fluorescent repressor operator systems (FROS) described here are ideally suited to tag DNA sequences and to study genome organization in living cells.

H. Tsubouchi (ed.), *DNA Recombination*, Methods in Molecular Biology 745,
DOI 10.1007/978-1-61779-129-1_29,

Visualization of specific DNA sequences, achieved through FROS, offers unprecedented possibilities to study chromatin organization and dynamics at high resolution in living cells. Two systems currently in use are the tetracycline (Tet) and the lactose (Lac) operator/repressor systems (1, 2, 4, 5). Both FROS rely on the transgenic expression of a bacterial repressor fused to a fluorescent protein. The fusion protein can bind to the respective operator DNA sequences which are integrated as multicopy tandem arrays at specific chromosomal locations. Accumulation of the fluorescent proteins near tagged DNA regions is then visible as a single fluorescent dot under conventional fluorescent microscopes. The FROS was successfully adapted for tagging chromatin in living organisms from bacteria to human cells. In yeast, *Saccharomyces cerevisiae,* the FROS has been used to evaluate the impact of nuclear organization on different molecular mechanisms including DNA repair, DNA replication and transcription.

Several studies have examined the positioning of telomeres within the yeast nucleus. The mechanisms by which telomeres interact and are anchored to the nuclear periphery are sensitive to transcriptional activity and to chromatin conformation (6–11). In particular, silent chromatin is anchored via the Ku70/80 complex, the Sir4 and Esc1 proteins and several integral membrane components. While most transcriptionally active loci localize in the nuclear lumen, galactose-inducible genes can also be found near the periphery, in particular near the nuclear pores (12–15). The 3D architecture of the nucleus thus provides an additional layer of information, which is potentially involved in epigenetic control in addition to regulation conferred by transcription factor binding sites and local chromatin structure (16). In addition, frequent and large movements of chromatin have been described at defined stages of the cell cycle. During G1 phase, large movements have been shown to tightly correlate with energy levels within the cell, being ATP- but not microtubule dependent. Chromosomal motion was significantly constrained in S phase presumably due to the association of multiple origins in large replication factories (17). In contrast, telomeres and centromeres impose replication-independent constraint on chromatin movement in both G1 and S phases.

DNA repair studies also benefitted from FROS to study the behavior of a double-strand break (DSB) (18–21) during the search for a homologous donor sequence. Persistent associations between donor and recipient loci following formation of a DSB during the yeast mating-type switch were shown to follow frequent transient pairing between donors and template (20). Frequent interaction of the silent mating-type loci depends on Sir proteins and chromatin conformation, but is independent of the Ku70/80 complex and of the position of *HML* and *HMR* relative to the nuclear periphery (9). The frequency of interaction between *HML* and *HMR* as determined using chromosome conformation capture (3C) correlates with the percentage of cells in which these two loci tagged by two different FROS colocalize (<250 nm separation between the center of the fluorescent spots detected) in 3D (9). The 3D nuclear positioning of the three mating-type loci in the budding yeast is also thought to play a role in the directionality of donor selection. Prior to recombination, both *HML* and *HMR* appear rather confined near the nuclear periphery in both cell types, while the actively transcribed *MAT* locus moves rapidly inside the lumen of the nucleus. Although HM and *MAT* loci were not aligned in the absence of damage and their relative nuclear positioning was reported to be mating-type independent, the dynamic behavior of the silent loci varied in a- vs α-cells and upon induction of the DSB break at *MAT* (18, 19). In the case of a non-repairable DSB in the absence of both HM loci, the broken chromosomal locus was found to eventually relocalize to the nuclear periphery where unrepaired DSBs were generally grouped into foci near the nuclear pores (22).

The combination of several distinct FROS has also permitted to gain insight into chromatin folding. Physical distance measurements between differentially labeled sites at varying genomic distances or near telomeres of different chromosome arms provide experimental data to extract physical parameters based on polymer models of the compaction and folding behavior of chromatin in yeast (23, 24). In conclusion, these systems are valuable tools in yeast, which readily performs homologous recombination. The availability of two different operators allows for tagging of multiple loci simultaneously when using two fluorophores with different emission spectra. A third FROS based on the lambda operator has been recently adapted from bacteria (25) to yeast in our group with the aim to track the movement of three distinct loci simultaneously which will further expand the possibilities of investigation (Lassadi, Goiffon, Kangoué and Bystricky, in preparation).

Here we describe the experimental procedures, and discuss applications and shortcomings, for efficient usage of the FROS in the budding yeast *S. cerevisiae*.

2. Materials

2.1. Cell Culture and Transformation for Insertion of the FROS/Cell Culture and Strain Construction

1. Yeast minimal and rich media (SC and YPD) are described by Rose et al. (3).
2. 10× TEL: 0.1 M Tris–Cl, pH 7.5 + 0.01 M EDTA + 1 M LiAc.
3. 40% PEG: MW = 6,000.

4. 1× TEL: 0.1 M LiAc in TE or 10× TEL diluted 10× in TE. (Make fresh as required.)
5. DNA carrier (10 mg/ml).
6. 10× Colony PCR buffer: 0.125 M Tris–Cl (pH 8.5) + 0.56 M KCl.
7. 25 mM $MgCl_2$.
8. Taq polymerase.
9. dNTPs.
10. 100% DMSO.
11. Lysis buffer: 2% of Triton X-100, 1% SDS, 100 mM NaCl, 10 mM Tris–Cl (pH 8), 1 mM Na_2EDTA.
12. Phenol/chloroform/isoamyl alcohol, 25:24:1.
13. Acid-washed glass beads (0.2–1 mm size).
14. 3 M NaAc.
15. 100% Cold EtOH.
16. 70% EtOH.
17. TE/2.5 μg/μl RNase A (store at –20°C).

2.2. Live-Cell Fluorescence Microscopy

1. Well concavity slide (1.4–1.6 mm thick) from Electron Microscopy Sciences.
2. SC + 2% glucose + 3% agarose media (aliquots stored at 4°C can be kept several months).
3. Coverslips.
4. VaLap (1/3 vaseline, 1/3 lanoline, 1/3 paraffin).
5. Microscopes, filters, sources of illuminations (*see* **Section 3.3.3**).

3. Methods

3.1. Strain Constructions

The strain construction procedure consists of two steps: the first one involves the integration of the repressor followed by the integration of the operators (*see* **Note 1**).

The integration of plasmid sequences encoding the repressor–fluorescent fusion proteins is realized by simple transformation (*see* **Note 2**). Homologous sequences for recombination can be found within many genomic selection marker genes containing a point mutation or a small deletion. Recombination will then lead to duplication of the marker sequence. Transformants are screened by microscopy (*see* **Notes 3** and **4**).

The insertion of the operator repeats can be performed in two different ways:

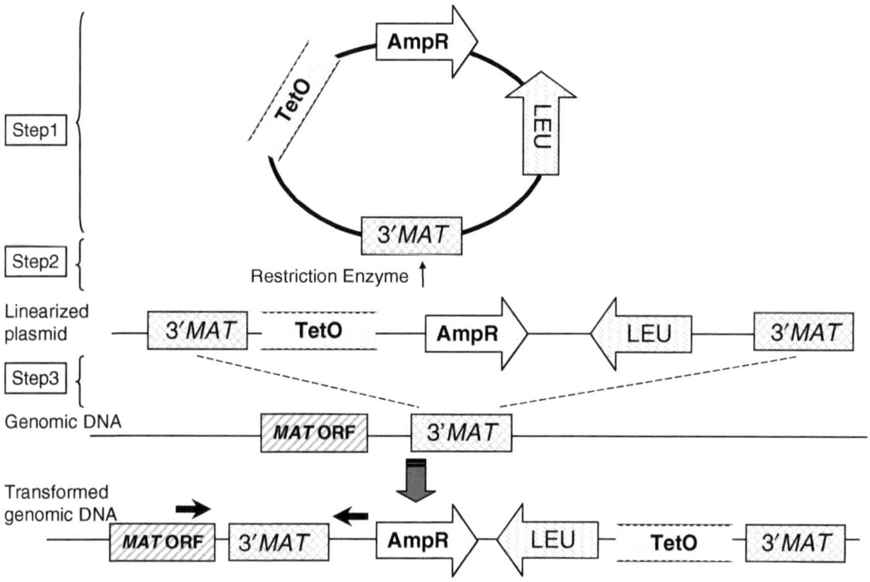

Fig. 29.1. Outline of the operator-bearing plasmid method. *Step 1*: Cloning of a 200-bp sequence which will be used as homology to integrate the plasmid by recombination within a locus of interest at the 3′ of *MAT*. *Step 2*: Linearization of the plasmid at the *MAT* locus by a unique restriction enzyme. *Step 3*: Transformation of the yeast with the linearized plasmid. The *arrows* represent the primers for testing correct integration by PCR.

The first technique consists of insertion of a multimerized lacO, tetO, or lambdaO into the chromosome by standard transformation techniques (**Fig. 29.1**). The integration will be targeted to the selected locus via homologous sequences cloned into the plasmid next to the operator repeats (*see* **Note 5**). Homologous sequences can be obtained by PCR.

The second technique is a cloning-free technique (13, 26) (**Fig. 29.2**). It is a straightforward two-step, PCR-based method to insert arrays of lacO or tetO into specific loci in the budding yeast genome. The method entails insertion of a "marker" generated by PCR with classical long primers (27) (optimal size of the locus-specific primer tails varies from 60 to 80 bp) near the locus of interest (*see* **Note 6**), followed by the replacement of this marker by a linearized "tagging" plasmid bearing an array of lacI or tetR-binding motifs (*see* **Note 7**). The repressor fusion protein will bind to the integrated repeats and result in the appearance of a bright focal spot (**Fig. 29.3d–f**) (*see* **Note 8**).

3.1.1. Yeast Transformation

1. Inoculate 5 ml YPAD with yeast. Leave overnight at 30°C with shaking (200 rpm).
2. Dilute the fresh pre-culture in 50 ml YPAD to OD of 0.1–0.15.
3. Incubate at 30°C with shaking until OD = 0.5–0.7.

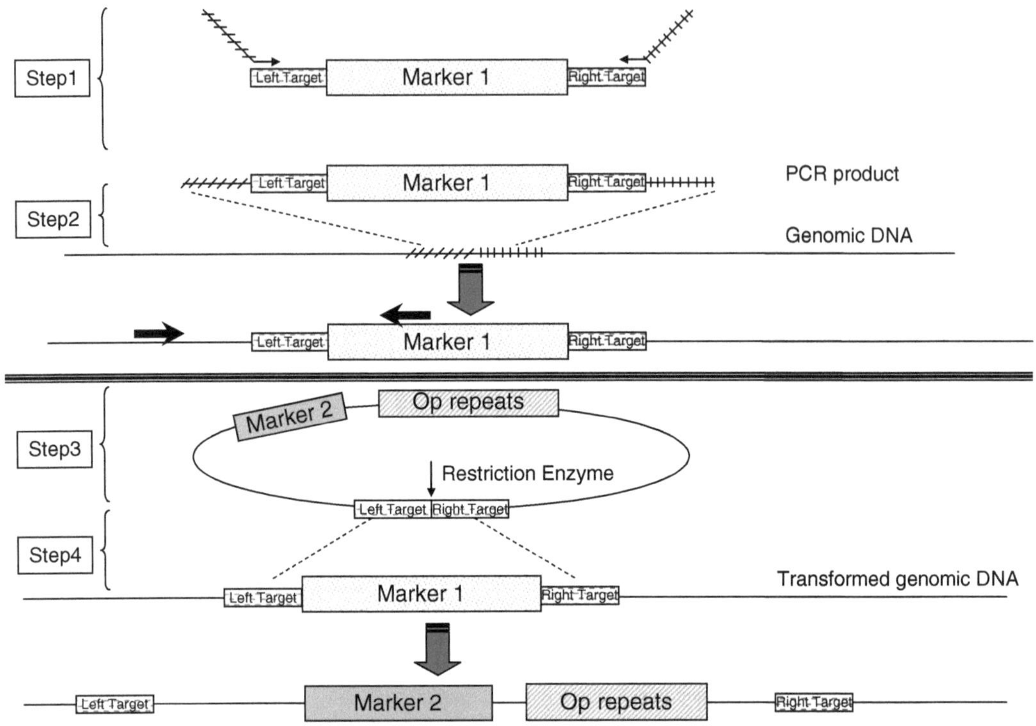

Fig. 29.2. Outline of the cloning-free chromatin tagging method. *Step 1*: Creation by PCR of a selective marker flanked by homologous sequences to the locus of interest. *Step 2*: Integration of the PCR product into the genome and selection of the transformant by the *marker 1*. *Step 3*: Linearization of a plasmid containing the operators. *Step 4*: Exchange by recombination of the *marker 1* by the *marker 2* adjacent to the operators. The *arrows* represent the primers for testing the correct integration by PCR.

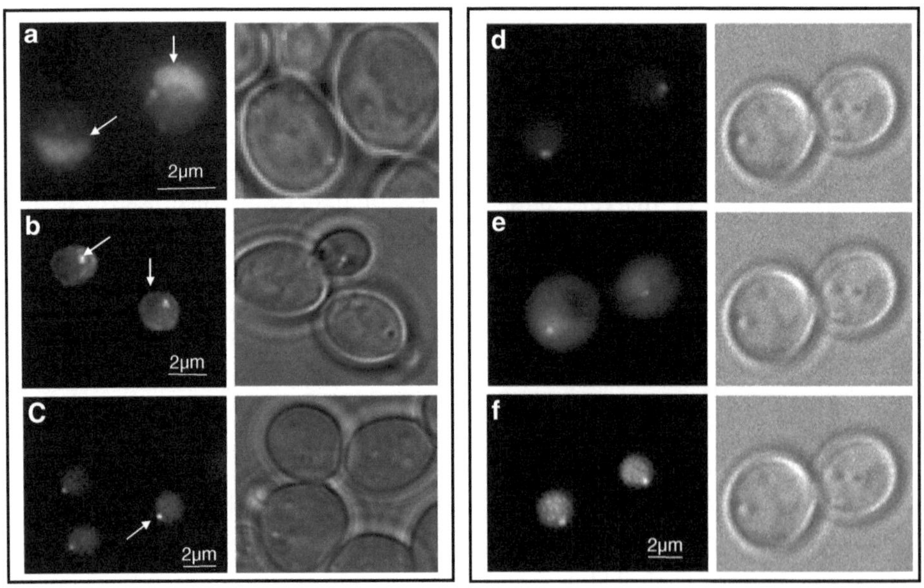

Fig. 29.3. Nuclear landmarks and FROS-tagged sites. (**a**) Nucleolus stained using Nop1-CFP; (**b**) the nuclear envelope stained using Nup49-GFP; (**c**) SPB tagged using Spc42-CFP; (**d**) LacI-CFP, lacOp::*HMR;* (**e**) λcI-YFP, λOp::*MAT;* (**f**) mch-TetR, tetOp::*HML*.

4. Centrifuge for 3 min at 4,000 rpm at 4°C in a 50-ml Falcon tube.
5. Wash the pellet *twice* with 10 ml sterile H_2O.
6. Resuspend the pellet with 1 ml of 1× TEL (freshly made). Transfer to microtube.
7. Centrifuge for 1 min at 7,000 rpm at room temperature.
8. Resuspend the pellet with (*X*+1) × (110 μl of 1× TEL, *X* being the number of plasmids that need to be transformed and "+1" for the negative control (i.e., the strain alone, tube which will contain everything except plasmid DNA).
9. Distribute 110 μl per transformation (i.e., *X*+1) in a microtube.
10. Per tube, add the following:
    i. 8 μl boiled SS-DNA carrier (at 10 mg/ml ⇔ 80 μg). *For the first use, boil DNA carrier for 5 min, then always keep on ice (no need to repeat it each time but repeat it if transformation efficiency is decreasing*).
    ii. 10 μl column-purified PCR or linearized plasmid (0.1–10 μg) (not purified) *or* 1 μl of plasmid (0.3–1 μg).
11. Pipet up and down or gently tap the tube with finger.
12. Per tube, add the following:
    i. 570 μl of 40% PEG – gently tap the tube with finger.
    ii. 70 μl of 10× TEL – gently tap the tube with finger (do *not* vortex!).
13. Incubate for 30 min at 30°C.
14. Heat shock for 10 min at 42°C.
15. Centrifuge for 2 min at 7,000 rpm at room temperature. Discard the supernatant. Let the PEG settle down by gravity and remove the remaining supernatant.
16. Wash the pellet with 1 ml sterile H_2O. Do not resuspend cells.
17. Centrifuge for 2 min at 7,000 rpm at room temperature.
18. Resuspend in 100 μl sterile H_2O (use 1 ml pipette tips not to damage the sensitive cells).
19. Plate all on selective plates (dropout plates) (*see* **Note 9**).
20. Incubate for 2 days at 30°C.
21. Restreak eight clones on dropout plates to confirm the gain of the selection gene before further analysis (*see* **Note 10**).

3.1.2. Classical PCR

PCR mix:

20 μl	Polymerase buffer (5×)
8 μl	dNTP at 10 mM each

1 μl	Primer 1 at 100 μM
1 μl	Primer 2 at 100 μM
63.5 μl	H_2O ultrafiltrated
5 μl	100% DMSO
0.5 μl	Phusion polymerase (Finnzyme) or other polymerase of your choice
1 μl	Plasmid (50–150 ng)/genomic DNA (0.2–2 μg)
100 μl	

PCR cycle:

94°C	4 min	
94°C	30 s	× 30 cycles
55°C	30 s	× 30 cycles
72°C	1 min	
72°C	5 min	
10°C	hold	

Adapt the elongation time to the size of fragment you are expecting (~1 min elongation for 1 kb fragment).

3.2. Verification of Sequence Integration

Integration of the operators at the targeted genomic position has to be verified. If using the cloning-free method, integration of the marker 1 can be tested by PCR on colonies or classical PCR on genomic DNA using one primer located on the genome (adjacent to the selected sequence for targeting the repeats) and one on the marker 1 (**Fig. 29.2**). The second step cannot be verified by PCR, since integrated repeats are very difficult to amplify. Loss of marker 1, though, as well as the gain of the marker 2 can be used as a test of the correct integration by analyzing cell growth on appropriate dropout plates. If using the operator-bearing plasmid method, PCR on colonies can be used to test correct integration of the plasmidic sequences. Primers used should amplify a plasmid fragment which does not contain the repeats (**Fig. 29.1**). To determine the number of integrated repeats, Southern blotting can be performed. In addition, the transformed yeast's karyotype can be analyzed by pulsed-field electrophoresis (for tagging subtelomeres and the rDNA-containing chromosome XII in particular).

3.2.1. Colony PCR

1. Take a 1-mm round-shaped yeast colony from a fresh plate (not older than 1 week), isolate with a sterile tip, and inoculate 20 μl zymolyase at 5 mg/ml (*see* **Note 11**).
2. Incubate for 20 min at 37°C.
3. Inactivate the enzyme for 5 min at 95°C. This step is not necessary if you plan to start the PCR immediately.
4. Centrifuge to pellet cell debris.
5. PCR mix:

2 μl	10× colony PCR buffer.
1.2 μl	25 mM $MgCl_2$.
1 μl	dNTP at 2.5 mM each.
1 μl	Primer 1 at 10 μM.
1 μl	Primer 2 at 10 μM.
12.5 μl	H_2O ultrafiltrated.
0.3 μl	GoTaq polymerase.
1 μl	Prepared yeast.
20 μl.	

6. PCR cycle:

94°C	4 min	
94°C	30 s	× 30
55°C	30 s	× 30
72°C	1 min	
72°C	5 min	
10°C	hold	

Adapt the elongation time to the size of fragment you are expecting. After PCR and visual verification under the microscope, several transformants should be frozen immediately (in 50% glycerol) from a fresh overnight culture from isolated colonies (*see* **Note 12**).

3.2.2. Genomic DNA Extraction

1. Grow 5 ml yeast cultures to saturation in YPAD at 30°C overnight.
2. Centrifuge for 5 min at max. speed in a clinical centrifuge. Pour off the supernatant.
3. Resuspend in 500 μl ultrafiltrated H_2O. Transfer to microfuge tube.
4. Centrifuge for 5 s, pour off the supernatant, and resuspend the pellet in the remaining droplet.
5. Add the following:
   - 200 μl of lysis buffer.
   - 200 μl of phenol/chloroform/isoamyl alcohol 25:24:1.
   - Around 40 μl acid-washed glass beads.
6. Cover the tube top with parafilm.
7. Vortex for 4 min at maximum speed under the hood.
8. Centrifuge for 5 min.
9. Transfer the aqueous phase into new microfuge tube.
10. Add the following:
    - 0.1 volume of the aqueous phase of 3 M NaAc
    - 3 volumes of the aqueous phase of 100% cold EtOH.
11. Leave for >1 h at –20°C.
12. Centrifuge for 10 min at max. speed. Take out the supernatant.

13. Wash with 1 ml of 70% EtOH.
14. Air-dry the pellet or speed-vac.
15. Resuspend in 50 μl of TE /7.5 μl RNase A.
16. Incubate for 15 min at 37°C in water bath.
17. Check purity and quantify.
18. Store at −20°C.

3.2.3. Classical PCR

See Section **3.1.2**.

3.3. Live-Cell Fluorescence Microscopy

To immobilize cells for acquisition, centrifuge 1 ml cell culture in a microfuge tube and spread the well-resuspended cell pellet on a SD–3% agarose patch. Concave slides are commercially available to prepare the agarose patch. Specific chambers in which yeasts are covered with transparent medium during observation are also available for time-lapse imaging on an inverted microscope (Ludin chambers; Lab-Teks). To avoid cells from moving, microscopy slides or coverslips can also be pre-coated with polylysine or concanavalin A.

3.3.1. Cell Culture

1. Pre-culture from a single colony in YPAD and incubate overnight at 30°C (*see* **Note 13**).
2. Dilute the pre-culture to an OD for the culture to reach an early exponential phase of growth (0.5–1 × 10^7 cells/ml) in synthetic transparent medium (*see* **Note 14**).
3. When the OD is around 0.2–0.5, 1 ml of the culture is pelleted, washed in water, and then resuspended in 3.5 μl of SC medium (*see* **Note 15**).
4. The cells should then be spotted onto SD–agarose-filled slides (YNB + 2% sugar/carbon source + 3% (w/v) agarose) and immobilized.
5. For time-lapse acquisition, the slide can be sealed to avoid any liquid evaporation.
6. Use the VaLap (1/3 vaseline, 1/3 lanoline, and 1/3 paraffin) as a sealing medium. This mounting protocol has been shown to prevent both rotation and other movements of the entire yeast nuclei during image acquisition (7, 13).

3.3.2. Slide Preparation

1. Melt the SD–agarose medium at 95°C for 5 min.
2. Transfer 150 μl onto cleaned and heated concave slides. Immediately, slip on heated normal slide on to it to remove any excess agarose.
3. Harvest 1 ml of the culture by centrifugation for 1 min at 13,000 rpm.
4. Wash the cell pellet twice with water.

5. Resuspend the pellet in 3.5 μl of SC medium.
6. Once the agarose has solidified, gently remove the upper slide. Spot 3.5 μl of the concentrated cells and cover with a coverslip.

3.3.3. Image Acquisition

Live-cell microscopy of fluorescently tagged loci can be performed using a large range of commercially available wide-field or confocal, upright or inversed microscopes. Budding yeasts are best imaged with 100× or 63× objectives. There is no golden rule for the choice of the microscope. In fact, a compromise between optimal spatial and temporal resolution depending on the intensity and the number of fluorophores you plan to visualize and on the objectives of your project will be necessary (**Fig. 29.4**).

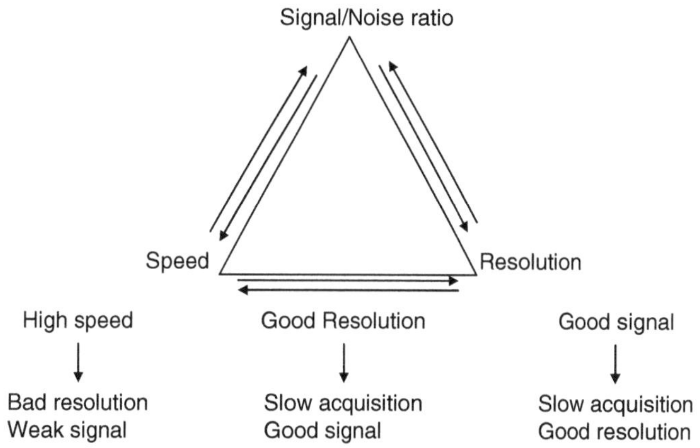

Fig. 29.4. The parameters limiting the realization of an optimal device. Optimal case: Strong excitation means good signal/good resolution/high speed *but* photobleaching and phototoxicity. Compromise is the key word for successful microscopy use.

Tracking of chromatin movements in yeast requires very rapid imaging rates (subsecond range) which are not always compatible with 3D acquisition. In addition, the main limitations of time-lapse microscopy are photobleaching and toxicity of prolonged exposure of the cells. Cells that complete a full cell cycle after imaging are usually considered to be intact. In addition, position and mobility of a chromosomal locus can vary with stages of the cell cycle. Using a transmission-phase image, cell cycle stage of budding yeast can be determined based on bud presence and size, as well as the shape and position of the nucleus (**Fig. 29.5**).

3.3.3.1. Wide-Field Microscopy

The wide-field microscope, as its name indicates, illuminates a wide field of the sample which allows analysis of 20–200 yeast cells per image when using a 100× objective and no further magnification. The microscope should be equipped with a xenon or mercury light source or a monochromator and a high-speed charged

Fig. 29.5. Cell cycle phases of a haploid cell of *S. cerevisiae*. The different cell cycle phases can be identified according to bud size and the nuclear orientation and size.

coupled device (CCD) camera (at least 6 × 6 μm). Due to the limited focal depth, acquisition in 3D is usually necessary to avoid a sampling bias in favor of the nuclei for which the locus lies in the focal plane. Three-dimensional acquisition is possible by acquiring stacks of images by varying the position of the objective coupled to a piezo. For yeast nuclei, *z*-steps corresponding to half the point spread function (PSF) (28), which is typically around 200–250 nm, are reasonable.

Applications: Quantification of positioning of specific loci relative to nuclear landmarks (for example, a tagged chromosomal site near a telomere relative to the nuclear pore) or distance measurements between two tagged loci.

Time lapse of the trajectories of one or several loci in 2D (high risk of "losing" the signal if the locus moves out of the focal plane) or in 3D (the analyzed dynamics has to be slower than the acquisition time for a stack of images) is useful to determine changes in positioning relative to nuclear landmarks or other tagged loci over several minutes or hours.

Advantages: Maximal fluorescent signal, large sampling, several fluorophores, and reasonable cost.

Limitations: Poor resolution in *z*, time lapse in 3D at best every 2–4 s for a single fluorophore, one wavelength/image (single camera setup).

3.3.3.2. Laser Scanning Confocal Microscopy (LSCM)

The confocal microscope uses a scanning point of light instead of full sample illumination. The focal plane for detection and illumination paths goes through a pinhole. The use of a pinhole allows to eliminate out-of-focus light, which increases the optical resolution and contrast at the cost of decreased signal intensity. LSCM is well suited for the study of single nuclei and avoids bleaching of neighboring cells on the same slide. Different fluorochromes can be excited and detected simultaneously (single track) or in successive scans (multiple track) using krypton/argon and helium/neon mixed gas lasers. As a consequence of the pinhole arrangement, light arriving at the detector comes predominantly from a narrow focal plane, which improves the *z*-resolution significantly compared to conventional microscopy. Emitted light from the sample is detected on photomultiplicators by a scanning laser beam. Thus the number of pixels acquired is directly proportional to the time of acquisition. Acquisition of stacks of images for detection in 3D is also possible – with the same limitations as described above for the wide-field microscope. If acquisition in 2D is judged sufficient for image analysis and interpretation, which is the case for tracking of trajectories of single or multiple loci over time (*see* **Note 16**) (7, 29), the optimal focal plane of the fluorescent signal can be manually adjusted during acquisition. For 3D time-lapse imaging, stacks should be taken of ROI as small as possible (*see* **Note 17**) (30).

Application: Rapid time-lapse acquisition of trajectories of one or several fluorescent sites in 2D and 3D (7, 17).

Advantages: Simultaneous excitation and detection of several fluorophores; single-cell tracking; phototoxicity and photobleaching effects limited to the illuminated field.

Limitations: Limited sample size due to imaging of single cells; limited resolution; difficult if emission intensities of fluorophores are distinct (>2 times) (*see* **Note 18**).

3.3.3.3. Nipkow Spinning Disk or Rapid Confocal Microscopy

Spinning disk microscopy is based on a microlens-enhanced Nipkow disk which is a spinning disk with a spiral pattern of (pin) holes arranged to raster scan a fluorescent sample. The illumination light from a laser source scans the specimen as numerous small points simultaneously. The historically low-light efficiency of the Nipkow disk has been dramatically improved in recent years. The addition of a microlens now makes rapid image acquisition of fluorescent labels in yeast possible.

Application: Rapid acquisition of FROS signals and fluorescent fusion proteins in 3D on large sample sizes – especially suited for high-throughput acquisition of 3D positioning of tagged genomic sites relative to nuclear landmarks (12, 24).

Advantages: The spinning disk technology combines certain advantages of the wide-field microscope (sample size, CCD camera acquisition) and the classic confocal microscope (focal acquisition through pinholes; laser source). Simultaneous acquisition of several fluorophores is possible (dual camera setup).

Limitations: Reduced light efficiency, fixed pinhole, and pixel size.

3.3.3.4. New Developments for Imaging in 3D

Confocal fluorescence microscopy occupies a special place because it is well suited for 3D imaging, allowing to cut the sample observed in regularly spaced optical "slices" which are then reunited to form 3D images. This method, however, requires the acquisition of a succession of pictures, and each time exposing the object of interest with a laser. In living cells, the dose of illumination is associated with photobleaching and phototoxicity, resulting in damage to cells. Thus, exposure should be minimized at the cost of signal quality. Such compromises also include reduction of the acquisition rate: to obtain a high-resolution 3D image, as many views as possible must be acquired. It is also a technique requiring expensive instrumentation and 3D reconstruction software. To meet the requirements of speed and precision, new microscopy techniques have been developed. For example, SPIM for "selective plane illumination microscopy" (31) illuminates a thick sample slice (mm) with a planar beam of light and the sample is observed at 90° to the beam. The 3D reconstruction is achieved by rotating the sample around an axis. This method is suitable for observation of thick fabrics, but it is slow because the reconstruction requires the acquisition of successive images involving mechanical rotation of the sample. In order to acquire a sufficient number of 3D images to establish laws of chromosome movements in vivo, especially using multiple fluorophores simultaneously, we have recently reported a new technique based on a stereovision device including micromirrors (32). This technique allows to reconstruct a 3D scene from multiple views of a single scene and thus to track trajectories of chromosomal sites in 3D on very short timescales. The micromirrors are assembled above the sample to be observed on a standard microscopy slide and placed on the support of a right or an inverted microscope. The device can thus be directly adapted to conventional wide-field microscopes. Fluorescent samples can be observed using both the direct and the reflected images that are obtained by each facet of the mirror. Determination of the coordinates of the images is sufficient to reconstruct the object in 3D using stereovision algorithms. In addition, the performance of the device allows us to observe the movement of genes with an acquisition speed of 20 ms with an accuracy of 20 nm in *xy*, these conditions being virtually unattainable with confocal microscopy. The functional use of micromirrors has been validated for the imaging of living yeast cells (32). Yeast can be introduced in large numbers in microgrooves, which makes the method suitable for a wide sampling, which is often necessary for imaging.

3.3.4. General Parameters and Microscope Settings

The setup for acquisition will depend on the application and corresponds to a compromise between speed, resolution, and signal-to-noise ratio (**Fig. 29.4**).

3.3.4.1. Wide-Field Microscope

In our laboratory, yeast nuclear organization and chromatin behavior are routinely studied by wide-field fluorescence microscopy using an Olympus IX-81 microscope, equipped with a CoolSNAP HQ camera (Roper Scientific) and a Polychrome V (Till Photonics), electric piezo with accuracy of 10 nm and imaged through an Olympus oil immersion objective 100× PLANAPO NA1.4. The optimum acquisition parameters are a stack of 21 images with a step size of 0.2 μm and an exposure time of 100–250 ms for GFP and RFP and 300–500 ms for CFP and YFP. Mono or dual filters can be used. These filters from

Chroma Technologies are a good choice: GFP Ex (470BP40) Em (525BP50); YFP/CFP Ex (422-42/457-80+492-509/521-551);CFP Em (457-482);YFP Em (520-548); and mch Ex: (573/24) Em (630BP60).

3.3.4.2. Confocal Microscope

For time-lapse microscopy, the LSM510 or 710 scanning confocal microscope is particularly well adapted, especially when using CFP–YFP. For GFP and RFP, LEICA SP2 or SP5 AOBS is also suitable. To reduce the risk of damage by illumination, the laser transmission is kept as low as possible, and the cells are imaged as rapidly as possible within a minimal region of interest (ROI). Useful settings for the Zeiss LSM510 are as follows: The argon lasers 458 nm (5 mW), 488 nm (25 mW), or 514 nm (25 mW) are used with an output of 25%. Available filters for channel 1 include the following: Lp 505 for GFP alone; Lp 530 for YFP; and for channel 3: Bp 470-500 for YFP/CFP single-track acquisition. The channel setting, pinhole 1–1.2 airy unit (corresponding to optical slice of 700–900 nm); detector gain, 930–999; amplifier gain, 1–1.5; amplifier offset, 0.2–0.1 V; laser transmission AOTF=0.1–4% for GFP, 1–25% for YFP; and 20–60% for CFP in single-track acquisition. Scan setting: maximum speed 10 (0.88 μs/pixel); 8 bits in one scan direction; zoom 1.8 for a 100× objective, 3 for a 63× objective to reach a pixel size of 100×100 nm. Imaging intervals can be as short as 1 s in 2D for a ROI encompassing a single nucleus.

3.3.4.3. Spinning Disk Microscope

In our laboratory, we routinely use an Andor revolution Nipkow disk confocal system installed on a Zeiss inverted microscope (Axiovert 200 M), featuring a Yokogawa CSU22 confocal spinning disk unit and cooled Andor EMCCD camera (DV885). The system is controlled using Andor revolution IQ software (mode "revolution fast"). Images are acquired using a Zeiss 100× objective (Plan APOCHROMAT, 1.4 NA, oil immersion). Single laser lines used for excitation are diode-pumped, solid-state lasers (DPSSL) exciting GFP fluorescence at 488 nm (25 mW, coherent) and mCherry fluorescence at 560 nm (20 mW; Melles Griot) (YFP and CFP). A Semrock biband-pass emission filter (Em01-R488/568-15) allows collection of green and red fluorescence. Routine 3D static analysis, Z-stacks of 41 images with a 250 nm *z*-step, with an exposure time of 200 ms per fluorophore, and an EM consist of acquisition of the EM-CCD camera set to 150 (pre-EM gain 2.0).

3.4. Image Analysis

Here we describe several image analysis and quantification tools that are available to the community. It is important to note that these programs have been developed for a precise application. Some reprogramming is likely necessary to enable analysis of new structures and of images with significantly different acquisition parameters.

3.4.1. ImageJ Applications

3.4.1.1. Pointpicker

Principle: Determination of the sub-nuclear position of a XFP-tagged locus within the nucleus.

The relative position of the tagged locus is calculated using two parameters: the spot distance from the nuclear envelope identified by using the Nup49-XFP ring (14) and the nuclear diameter (D) (*see* **Note 19**). The precise relative radial position is x/r with $r = D/2$. Each spot is then classified with respect to three concentric zones of equal surface: the peripheral zone (zone 1) is a ring of a width=0.184 × the nuclear radius (r), the zone 2 lies between 0.184 and 0.422r, and zone 3 is the center of the nucleus with a radius of 0.578r (6, 8, 19, 29, 33). The nuclei in which the tagged locus is positioned at the very top or bottom of the nucleus are not scored. The identification of each parameter is manual.

3.4.1.2. Spot Distance

Principle: Determination of the distance between two loci of different colors (*see* **Note 20**).

Cells or nuclei visible in an image are first segmented based on the background fluorescence of the unbound repressor–XFP. Then, each spot is identified as the brightest pixel in the cell. Its relative x, y coordinates are identified. The Z-coordinate corresponds to the Z-plane number of the acquired image. The distance (nm) between two spots of different "colors" is then calculated following the formula $d = \sqrt{(xa)^2+(yb)^2+(zc)^2}$ with $x = x_1-x_2$; $y= y_1-y_2$; $z = z_1-z_2$.

This plug-in has been developed for two (10) or three different fluorescent spots (D. Sage; unpublished). The signals are scored on 3D stacks using at least 100 nuclei, monitoring nuclear integrity and cell cycle stage through nucleolar shape and nuclear diameter (**Fig. 29.6**) (9, 10) (I.L. and K.B., unpublished).

3.4.1.3. Spot Tracker

Principle: Determination of the trajectory of a locus over time.

Characterization of the parameters of a locus' movement requires determination of its coordinates at every time point. Coordinates within the nuclear volume depend on the assignment of the nuclear center as a reference. The spot tracker plug-in of ImageJ allows for automatic detection of the nucleus (based on diffuse tet repressor fluorescence or the Nup49 staining) and determination of the x, y coordinates of the fluorescent spot in each time frame (2D acquisition or maximal projection of a 3D acquisition) (34, 35). Observation of the movement of a locus over time gives information about its velocity, track length, and the subvolume of the nucleus that this locus occupies during a given period of time. This movement can be quantified by the mean square displacement (MSD) analysis ($d(\Delta t)^2 = <(r(t+\Delta t) - r(t))^2>$), assuming that the movement of the spot follows a random walk. It describes a linear relationship between different time intervals and the square of the distance traveled by a particle

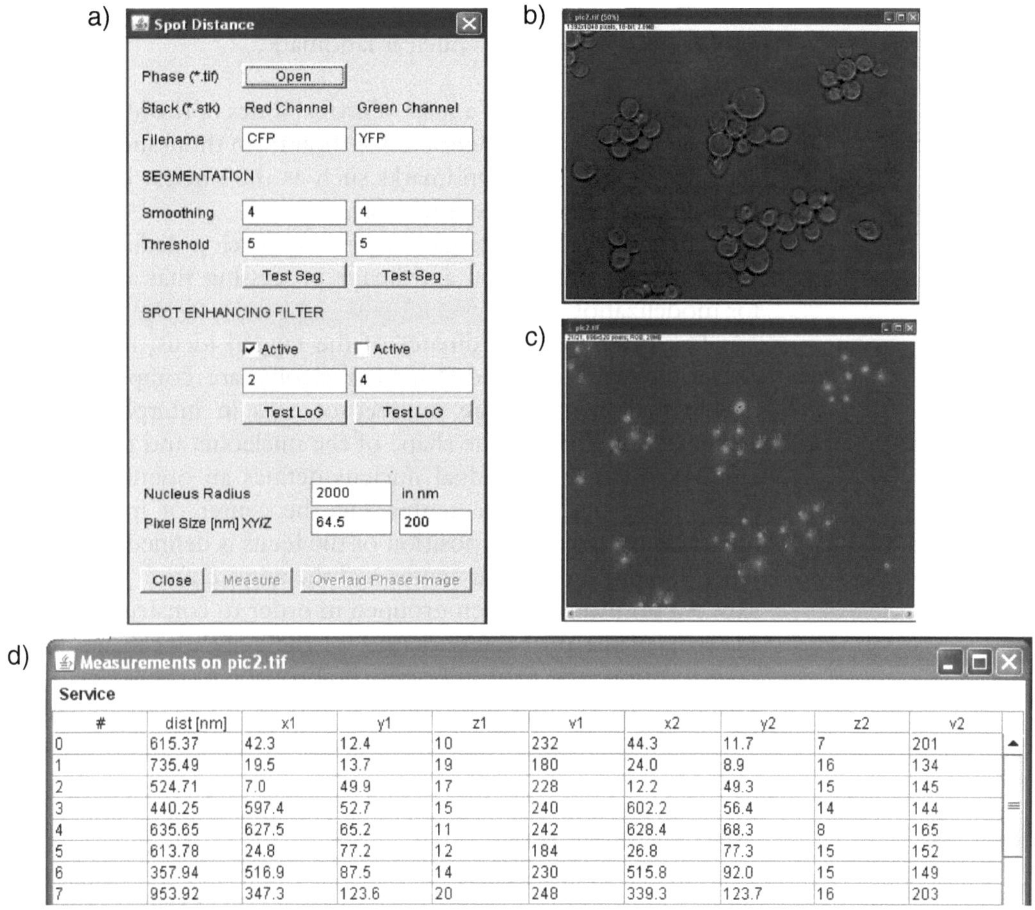

Fig. 29.6. Spot distance plug-in on ImageJ. Summary of the main windows of the analysis. (**a**) Start window where the choice of number or colors analyzed is specified. (**b**) and (**c**) Transmission picture opened by the user will allow the automated opening of the corresponding fluorescent emission pictures which will be merged together. (**d**) Analysis step with the spot identification in each cell and the result table.

during this period of time. The distance traveled by the spot for each time interval is calculated and plotted as the square of the mean against increasing time intervals. The slope of the curve reflects the diffusion coefficient of the locus. The linearity of the curve is usually lost at larger time intervals due to spatial constraint impacting the freedom of movement of the locus. The value at which the curve reaches a plateau is related to the volume to which the locus' movement is restricted. For chromosomal loci in yeast, the maximal diffusion coefficient is in the range of 1×10^{-4} to 1×10^{-3} $\mu m^2/s$ (7, 17, 36).

Spatial constraints are determined based on measurements that reflect the actual distances *d* of the tagged locus covered from any one time point to all others after an alignment of nuclear centers in all frames, on measurements of the distance between two distinct fluorescent spots, or on measurements of the position of a single spot relative to a nuclear landmark.

3.4.2. Matlab Applications: Nucloc (www.nucloc.org)

Principle: Generation of a map of probabilities of the location of a tagged locus at subdiffraction resolution from thousands of living cells by using nuclear landmarks such as the nuclear center, the nuclear envelope and the nucleolus.

The existing tool box can be used to develop dedicated algorithms for deconvolution and image processing that are suitable for modelization.

The positional coordinates of the tagged locus, the nuclear center, and the centroid-shaped nucleolus are computationally determined from a large number of cells in interphase ($N >$ 1,000). Estimation of the shape of the nucleolus and the nuclear envelope in each individual nucleus defines an oriented central axis joining the nuclear center and the center of mass of the nucleolar centroid. The position of the locus is defined by its distance R from the nuclear center and the angle α from the central axis. All positions are then grouped in order to construct a single 2D map of the probability density of the gene's localization. At each location, the map indicates the probability ("heat map" in 10% increments) that the locus lies inside a 3D volume obtained by rotation of a small surface element around the central axis (12, 13).

3.4.3. Nemo

Principle: Nemo is capable of automatically determining the 3D coordinates of the center of mass of point-like structures and other nonuniform structures (for example, DAPI-stained nuclei and chromosome territories in vertebrate cells). Distances can be recorded between all identified structures. Parameters for setup, processing, and validation can be adapted by the user. Nemo is a stand-alone Java application available for Windows and Linux platforms (37). The program is distributed under the Creative Commons License and can be freely downloaded from https://www-lgc.toulouse.inra.fr/nemo.

4. Notes

1. Plasmids containing operator repeats should be amplified in recombination-deficient bacterial strains (RecA-, Stbl2, and SURE) grown at 30°C rather than 37°C. In addition, multiple clones should be checked by restriction enzyme digestion after amplification to ascertain the number of operator repeats.

2. The lambda, Lac, or Tet repressors fused in frame to a fluorescent protein (GFP, CFP, YFP, mCherry, or RFP = XFP) should be integrated into the yeast genome before integrating the operator repeats. The repressor will immediately bind to and stabilize the repeats. Indeed, during each additional transformation, the number of repeats may be reduced by recombination.
3. Correctly integrated repressor–XFP chimera will be expressed by the cell and can thus be visualized easily under the fluorescent microscope. Distribution of the free chimera may vary from cell to cell depending on cell cycle stage and age of the cell.
4. Unbound repressor protein displays different cellular distributions depending on the repressor system and the fluorescent protein used. TetR is nuclear thanks to a nuclear localization signal, while lambdacI usually diffuse in the entire cell with the exception of the vacuole (**Fig. 29.3d–f**).
5. For the operator integration using the plasmid technique, the extent of homology should be at least 200 bp. The choice of the homologous sequence can be difficult given the fact that the plasmid has to be linearized, using a unique restriction enzyme site near the center of this 200-bp sequence. It is also possible to introduce a specific restriction enzyme site within the 200-bp sequence by PCR.
6. For both techniques, the choice of the genomic site of integration is a crucial step. While the integration of operators does not seem to change the physiology of the yeast cell, it is still important to consider that insertion of 5–15-kb plasmid sequences including several kilobases of bacterial tandem repeats may not be neutral. The integration site has to be appropriately chosen in order to avoid potential modifications of the chromatin environment and deregulation of nearby coding sequences. Thus, due to the high content of coding sequence, it is nearly impossible to insert sequences without modifying the neighboring genomic elements. At the least, one should preserve all regulatory elements involved in the studied mechanism and compare the functionality of the tagged locus with the untagged strain. For example, to study the position of a transcribed gene, its promoters, terminators, or other known regulatory sequences have to remain intact and normal gene activation should be ascertained after integration. Similarly, normal origin firing should be checked when tagging ARS loci. In addition, operator insertions near the mating loci should not alter the mating behavior. These

constraints, in addition to the difficulty of finding unique sequences in some areas of the genome (for example, in subtelomeric regions), may greatly reduce the choice of available sites to which operators can be targeted. In addition, one has to be careful to not insert the repeats too far from the site of interest. The behavior of a distant labeled site may be uncoupled and independent of the locus of interest. Thus, we strongly recommend integrating operator repeats at < 5 kb away from the locus of interest.

7. The same operator system is usually used only once in the same strain albeit the fact that the commonly used repressors cannot homodimerize. The main concern is the instability of the operators which can readily recombine with each other, in addition to the fact that two identical operators will necessarily be detected with the same fluorophore.
8. Different clones can display different intensities of the fluorescent spot because the size of the integrated arrays may vary considerably. This variability has to be taken into consideration during the screening for the best clone. The spot intensity can also vary according to the genomic region of investigation and to the fluorophore fused to the repressor.
9. The number of different fluorescent proteins and operator systems to be integrated into the same strain can be limited by the availability of specific selective markers. To circumvent this limitation, different repressor fusion proteins can be cloned into the same plasmid under a single selection marker. We further suggest the use of the *URA3* selection marker for labeling of nuclear landmarks which can recombine out when yeast is grown on 5′-FOA-containing plates. To this end, the *URA3* gene flanked by homologous sequences can be lost by recombination when the cell is under stress in presence of 5′-FOA (38). Hence this marker can be used repeatedly.
10. During transformation, in the case where the selective marker is an antibiotic (such as hygromycin and kanamycin), the strains should be incubated at least 3 h in YPD after the heat shock at 30°C to allow phenotype expression and then plated on the selective plate. The transformants can also be plated overnight on YPD plates and then patched on selective plates the day after.
11. When performing PCR on colony, it is very important to pick one single colony in order to avoid mixing genomic DNA from two different transformants.
12. When strains are recovered from frozen stocks, the preservation of the fluorescent signal should be ascertained and

their genotype should be checked on selective medium. Indeed, recombination events frequently lead to loss of operator repeats.

13. Ade– strains should always be cultured in YPAD media especially for microscopy use. In fact, these strains accumulate a compound which is auto-fluorescent in red.

14. For microscopy analysis, all strains have to be cultured under the same conditions. Imaging should be performed on cells at an early exponential phase of growth. A preculture of stationary phase cells should be diluted in fresh media and cells should be analyzed after not less than two doubling times (OD 600 nm around 0.2–0.4). Generation rates change according to the carbon source used. In fact, doubling times in galactose, raffinose, or lactate-containing media are longer than those in glucose media. Mutant strains usually also display a slower growth compared to the wild-type strain.

15. Reproducible analyses can be obtained only by using identical culture and observation conditions including OD, incubation time, and temperature. The temperature of the room should be carefully controlled (±2°C). Alternatively, the entire imaging part of the microscope can be enclosed in a commercially available temperature-regulated box or a heated stage can be used.

16. *When to use 2D or 3D analysis*: 3D acquisition does not bring additional information in all cases. In an isotropic volume, it is assumed that the movement of a molecule has an equal probability in each direction, thus analysis in 2D is recommended. In addition, quantitative analysis of the displacement of a locus over time (MSD) at very short time intervals (<50 ms) can be performed in 2D.

    The position of a tagged site relative to a nuclear landmark can be determined in 2D, although acquisition should be realized in 3D. The focal plane in which the spot is the brightest is selected to extract positional information (6, 8, 11, 16, 19, 29, 33).

    Time-lapse experiments in 3D (4D) based on stacks of images are inherently slow and useful if the dynamics of the studied process is meaningful over minutes rather than seconds or milliseconds. The number of focal planes can be reduced at the cost of precision to gain time. Subsequent image analysis and interpretation are still difficult, although important progress has been made in recent years. The realization and analysis of time-lapse experiments in 3D using more than one fluorophore (5D, 6D, . . .) remains challenging, in particular when simultaneous multicolor images are desired.

As long as approximations and experimental conditions are clearly defined, analyses can be performed in 2D or in 3D. Relative differences between locus position and dynamics can be compared in different strains or under varying conditions.

17. For time-lapse imaging, reducing the number of pixels of the detected section is very useful. The software driving the acquisition should allow selecting a region of interest (ROI), ideally limited to the dimensions of a single nucleus.
18. The use of larger *z*-steps (400–500 nm) will reduce the precision in determination of the coordinates of the fluorescent spot. However, when speed and photobleaching are a concern, larger *z*-steps allow reduction of the number of planes and thus the total acquisition time per 3D stack.
19. The position and dynamics of specific loci in the nucleus can be analyzed using one or multiple nuclear structures as references. Useful architectural elements include the nuclear envelope (Nup49-XFP), the spindle pole body (SPB; Spc42-XFP or Spc27-XFP), or the nucleolus (Nop1-XFP) (**Fig. 29.3a–c**). Sir protein, Rap1, or replication foci also represent suitable reference points. The position and behavior of a chromosomal site can be determined relative to not only these nuclear landmarks but also inferred positions of the nuclear center (from the envelope or the diffuse repressor fluorescence) and to other chromosomal sites.
20. *When to use one or several fluorophores*: If you can, use only one GFP (the enhanced S65T mutant or EGFP) (39). The spectral characteristics, in particular fluorescence intensity and photostability, will give you higher resolution and the possibility to track fluorescence over long acquisition periods. Excitation times can be as low as 20 ms per image with an acceptable signal/noise ratio making GFP ideally suited for 3D stacks and time-lapse acquisitions of chromosomal loci which are highly dynamic. The same fluorophore can be used several times as long as each fluorescent structure can be identified. A good example is the use of Nup49-GFP to identify the nuclear envelope together with a GFP-tagged locus (6, 8, 11, 16, 19, 29, 33). The fluorescent spot has to be brighter than the nucleopores to be able to distinguish the two during image analysis (**Fig. 29.3b**). The utilization of the same fluorophore to tag two distinct loci is less obvious. Different clues are used to differentiate the two spots. First, two operator arrays with a difference in the number of repeats (128 vs 256) display different size dots, an approach that was successfully used to follow the mating-type loci (18, 21). However, if the two loci are frequently within close proximity to each other or colocalize,

the use of the same fluorophore is prone to artifacts and not recommended. The use of other fluorophores is limited in living yeast, due to the toxicity of some of the available ones. Emission and excitation spectra of the combined fluorophores should be separate to avoid "bleeding" of the emission spectrum of one into the excitation range of the second. Useful combinations are CFP–YFP (7, 9, 10) and GFP–mCherry or mRFP (24).

References

1. Belmont, A.S. (2001) Visualizing chromosome dynamics with GFP. *Trends Cell Biol* **11**, 250–257.
2. Belmont, A.S., and Straight, A.F. (1998) In vivo visualization of chromosomes using lac operator-repressor binding. *Trends Cell Biol* **8**, 121–124.
3. Rose, M.D., Winston, F., and Hieter, P. (1990) Methods in yeast genetics: A laboratory course manual. Cold Spring Harbor Laboratory Press, Cold Spring Harbor, NY.
4. Michaelis, C., Ciosk, R., and Nasmyth, K. (1997) Cohesins: chromosomal proteins that prevent premature separation of sister chromatids. *Cell* **91**, 35–45.
5. Straight, A.F., Belmont, A.S., Robinett, C.C., and Murray, A.W. (1996) GFP tagging of budding yeast chromosomes reveals that protein–protein interactions can mediate sister chromatid cohesion. *Curr Biol* **6**, 1599–1608.
6. Bupp, J.M., Martin, A.E., Stensrud, E.S., and Jaspersen, S.L. (2007) Telomere anchoring at the nuclear periphery requires the budding yeast Sad1-UNC-84 domain protein Mps3. *J Cell Biol* **179**, 845–854.
7. Bystricky, K., Laroche, T., van Houwe, G., Blaszczyk, M., and Gasser, S.M. (2005) Chromosome looping in yeast: telomere pairing and coordinated movement reflect anchoring efficiency and territorial organization. *J Cell Biol* **168**, 375–387.
8. Ebrahimi, H., and Donaldson, A.D. (2008) Release of yeast telomeres from the nuclear periphery is triggered by replication and maintained by suppression of Ku-mediated anchoring. *Genes Dev* **22**, 3363–3374.
9. Miele, A., Bystricky, K., and Dekker, J. (2009) Yeast silent mating type loci form heterochromatic clusters through silencer protein-dependent long-range interactions. *PLoS Genet* **5**, e1000478.
10. Schober, H., Kalck, V., Vega-Palas, M.A., Van Houwe, G., Sage, D., Unser, M., Gartenberg, M.R., and Gasser, S.M. (2008) Controlled exchange of chromosomal arms reveals principles driving telomere interactions in yeast. *Genome Res* **18**, 261–271.
11. Taddei, A., Hediger, F., Neumann, F.R., Bauer, C., and Gasser, S.M. (2004) Separation of silencing from perinuclear anchoring functions in yeast Ku80, Sir4 and Esc1 proteins. *EMBO J* **23**, 1301–1312.
12. Berger, A.B., Cabal, G.G., Fabre, E., Duong, T., Buc, H., Nehrbass, U., Olivo-Marin, J.C., Gadal, O., and Zimmer, C. (2008) High-resolution statistical mapping reveals gene territories in live yeast. *Nat Methods* **5**, 1031–1037.
13. Cabal, G.G., Genovesio, A., Rodriguez-Navarro, S., Zimmer, C., Gadal, O., Lesne, A., Buc, H., Feuerbach-Fournier, F., Olivo-Marin, J.C., Hurt, E.C., and Nehrbass, U. (2006) SAGA interacting factors confine subdiffusion of transcribed genes to the nuclear envelope. *Nature* **441**, 770–773.
14. Gard, S., Light, W., Xiong, B., Bose, T., McNairn, A.J., Harris, B., Fleharty, B., Seidel, C., Brickner, J.H., and Gerton, J.L. (2009) Cohesinopathy mutations disrupt the subnuclear organization of chromatin. *J Cell Biol* **187**, 455–462.
15. Taddei, A., Van Houwe, G., Hediger, F., Kalck, V., Cubizolles, F., Schober, H., and Gasser, S.M. (2006) Nuclear pore association confers optimal expression levels for an inducible yeast gene. *Nature* **441**, 774–778.
16. Gartenberg, M.R., Neumann, F.R., Laroche, T., Blaszczyk, M., and Gasser, S.M. (2004) Sir-mediated repression can occur independently of chromosomal and subnuclear contexts. *Cell* **119**, 955–967.
17. Heun, P., Laroche, T., Shimada, K., Furrer, P., and Gasser, S.M. (2001) Chromosome dynamics in the yeast interphase nucleus. *Science* **294**, 2181–2186.
18. Bressan, D.A., Vazquez, J., and Haber, J.E. (2004) Mating type-dependent constraints on the mobility of the left arm of yeast chromosome III. *J Cell Biol* **164**, 361–371.

19. Bystricky, K., Van Attikum, H., Montiel, M.D., Dion, V., Gehlen, L., and Gasser, S.M. (2009) Regulation of nuclear positioning and dynamics of the silent mating type loci by the yeast Ku70/Ku80 complex. *Mol Cell Biol* **29**, 835–848.
20. Houston, P.L., and Broach, J.R. (2006) The dynamics of homologous pairing during mating type interconversion in budding yeast. *PLoS Genet* **2**, e98.
21. Simon, P., Houston, P., and Broach, J. (2002) Directional bias during mating type switching in *Saccharomyces* is independent of chromosomal architecture. *EMBO J* **21**, 2282–2291.
22. Nagai, S., Dubrana, K., Tsai-Pflugfelder, M., Davidson, M.B., Roberts, T.M., Brown, G.W., Varela, E., Hediger, F., Gasser, S.M., and Krogan, N.J. (2008) Functional targeting of DNA damage to a nuclear pore-associated SUMO-dependent ubiquitin ligase. *Science* **322**, 597–602.
23. Bystricky, K., Heun, P., Gehlen, L., Langowski, J., and Gasser, S.M. (2004) Long-range compaction and flexibility of interphase chromatin in budding yeast analyzed by high-resolution imaging techniques. *Proc Natl Acad Sci USA* **101**, 16495–16500.
24. Therizols, P., Duong, T., Dujon, B., Zimmer, C., and Fabre, E. (2010) Chromosome arm length and nuclear constraints determine the dynamic relationship of yeast subtelomeres. *Proc Natl Acad Sci USA* **107**, 2025–2030.
25. Fekete, R.A., and Chattoraj, D.K. (2005) A cis-acting sequence involved in chromosome segregation in *Escherichia coli*. *Mol Microbiol* **55**, 175–183.
26. Rohner, S., Gasser, S.M., and Meister, P. (2008) Modules for cloning-free chromatin tagging in *Saccharomyces cerevisiae*. *Yeast* **25**, 235–239.
27. Baudin, A., Ozier-Kalogeropoulos, O., Denouel, A., Lacroute, F., and Cullin, C. (1993) A simple and efficient method for direct gene deletion in *Saccharomyces cerevisiae*. *Nucleic Acids Res* **21**, 3329–3330.
28. Dufour, A., Shinin, V., Tajbakhsh, S., Guillen-Aghion, N., Olivo-Marin, J.C., and Zimmer, C. (2005) Segmenting and tracking fluorescent cells in dynamic 3-D microscopy with coupled active surfaces. *IEEE Trans Image Process* **14**, 1396–1410.
29. Heun, P., Laroche, T., Raghuraman, M.K., and Gasser, S.M. (2001) The positioning and dynamics of origins of replication in the budding yeast nucleus. *J Cell Biol* **152**, 385–400.
30. Rosa, A., Maddocks, J.H., Neumann, F.R., Gasser, S.M., and Stasiak, A. (2006) Measuring limits of telomere movement on nuclear envelope. *Biophys J* **90**, L24–6.
31. Huisken, J., Swoger, J., Del Bene, F., Wittbrodt, J., and Stelzer, E.H. (2004) Optical sectioning deep inside live embryos by selective plane illumination microscopy. *Science* **305**, 1007–1009.
32. Hajjoul, H., Kocanova, S., Lassadi, I., Bystricky, K., and Bancaud, A. (2009) Lab-on-Chip for fast 3D particle tracking in living cells. *Lab Chip* **9**, 3054–3058.
33. Hediger, F., Taddei, A., Neumann, F.R., and Gasser, S.M. (2004) Methods for visualizing chromatin dynamics in living yeast. *Methods Enzymol* **375**, 345–365.
34. Meister, P., Gehlen, L.R., Varela, E., Kalck, V., and Gasser, S.M. (2010) Visualizing yeast chromosomes and nuclear architecture. Methods enzymology, Guide to yeast genetics, J. Abelson and M. Simon, eds.,Vol. 470 (New York, NY: Academic Press), 535–567.
35. Sage, D., Neumann, F.R., Hediger, F., Gasser, S.M., and Unser, M. (2005) Automatic tracking of individual fluorescence particles: application to the study of chromosome dynamics. *IEEE Trans Image Process* **14**, 1372–1383.
36. Vazquez, J., Belmont, A.S., and Sedat, J.W. (2001) Multiple regimes of constrained chromosome motion are regulated in the interphase Drosophila nucleus. *Curr Biol* **11**, 1227–1239.
37. Iannuccelli, E., Mompart, F., Gellin, J., Lahbib-Mansais, Y., Yerle, M., and Boudier, T. (2010) NEMO: a tool for analyzing gene and chromosome territory distributions from 3D-FISH experiments. *Bioinformatics* **26**, 696–697.
38. Boeke, J.D., LaCroute, F., and Fink, G.R. (1984) A positive selection for mutants lacking orotidine-5′-phosphate decarboxylase activity in yeast: 5-fluoro-orotic acid resistance. *Mol Gen Genet* **197**, 345–346.
39. Heim, R., Cubitt, A.B., and Tsien, R.Y. (1995) Improved green fluorescence. *Nature* **373**, 663–664.

Chapter 30

Cell Biology of Homologous Recombination in Yeast

Nadine Eckert-Boulet, Rodney Rothstein, and Michael Lisby

Abstract

Homologous recombination is an important pathway for error-free repair of DNA lesions, such as single- and double-strand breaks, and for rescue of collapsed replication forks. Here, we describe protocols for live cell imaging of single-lesion recombination events in the yeast *Saccharomyces cerevisiae* using fluorescence microscopy.

Key words: Homologous recombination, fluorescence microscopy, DNA damage, DNA double-strand break repair.

1. Introduction

In the budding yeast *Saccharomyces cerevisiae*, homologous recombination (HR) is catalyzed by proteins encoded by the *RAD52* epistasis group of genes including *RAD50-59, XRS2, MRE11,* and *RFA1-3* (1). However, many additional proteins regulate and coordinate HR according to the molecular nature of the DNA lesion, cell cycle, and developmental phase. Most of these proteins are expressed constitutively, but their copy number per cell varies greatly from <50 molecules of Tel1 to >5,000 molecules of Rfa1. During HR, these proteins are assembled in a coordinated manner into dynamic giga-Dalton complexes at the site of the DNA lesion (2, 3) (**Fig. 30.1a**). These assemblies of high local concentration of HR and other DNA damage response proteins appear as cytological foci (**Fig. 30.1b**). Remarkably, the appearance of DNA damage-induced foci is highly conserved from yeast to human (4). Therefore, insight into the molecular principles that govern these DNA repair factories in yeast is likely

H. Tsubouchi (ed.), *DNA Recombination*, Methods in Molecular Biology 745,
DOI 10.1007/978-1-61779-129-1_30, © Springer Science+Business Media, LLC 2011

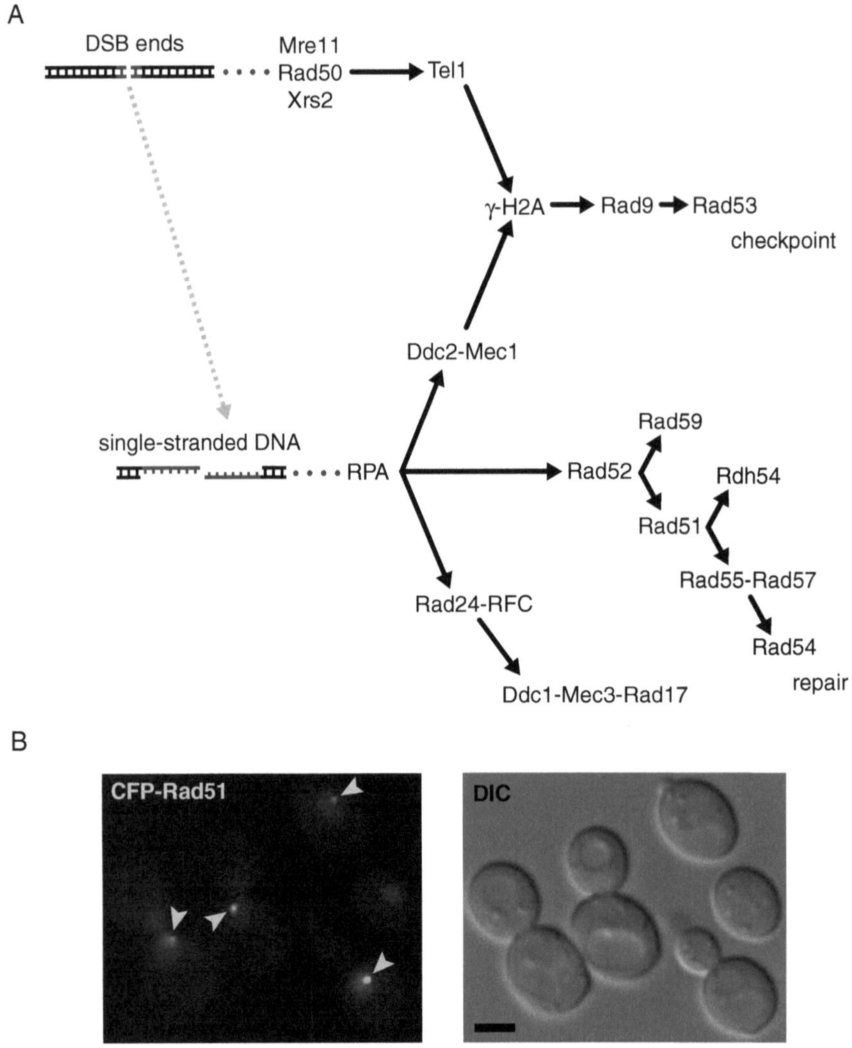

Fig. 30.1. Assembly of checkpoint and recombination proteins in response to DNA double-strand breaks. (**a**) Sequential assembly of cytological foci. *Arrows* indicate the sequential assembly of DNA repair proteins at a DNA double-strand break as described (20). DSB, double-strand break. (**b**) Formation of Rad51 foci in response to DNA damage. Exponentially growing cells (strain ML494-15C) expressing CFP-Rad51 from the endogenous locus were examined for Rad51 foci. In brief, cells were grown by shaking in liquid SC medium containing 100 μg/ml adenine at 25°C to an OD_{600} of 0.2–0.3 before addition of 200 μg/ml zeocin. After continued shaking for 1 h, cells were harvested by centrifugation at 3,000 rpm and processed for fluorescence microscopy as described (**Section 3**). The CFP fluorophore was visualized on a Zeiss AxioImager Z1 wide-field microscope using a Zeiss Plan-Apo 100×/1.40 objective (Carl Zeiss, Jena, Germany), a band-pass CFP filter set from Chroma (Brattleboro, VT), and an ORCA C4742-95-12ER CCD camera (Hamamatsu, Japan). Images were acquired using Volocity software (Improvision, Coventry, UK). DIC, differential interference contrast. *Arrows* indicate Rad51 foci. Scale bar, 3 μm.

to be extendible to higher eukaryotes. For single-lesion analysis, yeast has the advantage over higher eukaryotes that its relatively small genome experiences fewer spontaneous DNA lesions per cell. As a consequence, it is easier to discern the signal from a single specific lesion from the background of spontaneous random lesions.

2. Materials

1. 10× Pfu reaction buffer: 200 mM Tris–HCl (pH 8.8 at 25°C), 100 mM $(NH_4)_2SO_4$, 100 mM KCl, 1% (v/v) Triton X-100, 1 mg/ml BSA, and 20 mM $MgSO_4$.
2. 20× PBS (phosphate buffered saline): 160 g NaCl, 4 g KCl, 28.8 g NA_2HPO_4, 4.8 g KH_2PO_4 in 1,000 ml. Adjust pH to 7.4 at 25°C with either HCl or NaOH as appropriate.
3. YPD (yeast extract peptone dextrose) liquid medium: 10 g yeast extract, 20 g peptone, 20 g glucose in 1,000 ml. Autoclave to sterilize.
4. SC (synthetic complete) powder: 100 g yeast nitrogen base without amino acids and without NH_4SO_4 (MP Biomedicals, Solon, OH), 293.11 g NH_4SO_4, 1.2 g adenine sulfate, 1.2 g L-arginine sulfate, 1.2 g L-histidine-HCl, 1.8 g L-isoleucine, 3.6 g L-leucine, 1.8 g L-lysine-HCl, 1.2 g L-methionine, 3 g L-phenylalanine, 1.2 g L-tryptophan, 1.8 g L-tyrosine, 1.2 g uracil, 9 g L-valine (5). Grind in ball mill overnight.
5. SC liquid medium: 7.25 g SC powder and 20 g glucose in 1,000 ml. Adjust pH to 5.8 with NaOH/HCl.
6. 5-FOA (5-fluoroorotic acid) solid medium: Autoclave solution I (12 g agar in 255 ml H_2O). Filter sterilize solution II (4.35 g SC powder, 18 mg uracil, 450 mg 5-FOA, and 12 g glucose in 345 ml). Cool solution I to 60°C after autoclaving and heat solution II to 60°C. Gently mix the two solutions and cool to 45°C before pouring into the plates.
7. All PCR fragments are agarose gel purified using GeneJET™ Gel Extraction Kit (Fermentas, Germany).
8. The following chemicals were used: bleomycin (Sigma-Aldrich) (6), zeocin (Invitrogen), camptothecin (Sigma-Aldrich), methyl methanesulfonate (Sigma-Aldrich), raffinose (Sigma-Aldrich), galactose (Sigma-Aldrich), petroleum jelly (Vaseline; VWR Scientific Products), beeswax (Sigma-Aldrich), lanolin (Sigma-Aldrich), NuSieve® GTG® low-melt agarose (Lonza Rockland, Inc., Rockland, ME), and 4′,6-diamidino-2-phenylindole dihydrochloride (DAPI; Sigma-Aldrich).
9. Yeast strains ML193-3B (*MAT*a *ADE2 RAD5 TetI-mRFP1::iYGL119W*) and ML494-15C (*MAT*a *ADE2 RAD5 YFP-RAD55 CFP-RAD51*) are derivatives of W303-1A (7, 8).

10. The sequences of plasmids pWJ716, pWJ1108, pWJ1162, pWJ1163, pWJ1164, pWJ1165, pWJ1350, pWJ1351, and pWJ1379 are available upon request (9–12). Plasmid pJH132 is described previously (13).
11. Microscope hardware. For microscopy, yeast cells are relatively small and the number of proteins per cell is low compared to vertebrate systems. For this reason, the essential elements of the microscope hardware are a high-sensitivity cooled CCD or EM-CCD camera and a high-magnification objective (100×) with high numerical aperture (≥1.4).

3. Methods

3.1. Fluorescence Tagging of Endogenously Expressed HR Proteins

We have previously developed a PCR-based method for fluorescence tagging of endogenously expressed HR proteins with cyan, yellow, or red fluorescent protein (CFP, YFP, and RFP, respectively, or XFP, collectively) (11). In brief, approximately 300 bp on either side of the genomic integration site is amplified by PCR and fused in a second PCR to a cassette containing the gene encoding a fluorescent protein and two-thirds of either the 5′- or 3′-end of the *Kluyveromyces lactis URA3* marker (**Fig. 30.2**). The sequence of the *K. lactis URA3* gene is sufficiently divergent from the *URA3* gene of *S. cerevisiae* to prevent targeting to this locus and thus preventing a potential source of false-positive Ura$^+$ transformants. The split marker approach may ease the fusion PCR by allowing for shorter PCR fragments compared to generating one fusion PCR with the full *URA3* marker and both sequences of homology. The resulting PCR fragments are co-transformed into the target strain and transformants selected on synthetic complete medium lacking uracil (SC-Ura). The integration generates a direct repeat of the gene encoding the fluorescent protein, which allows for subsequent popout of the *URA3* marker by genetic recombination. The protocol for monomeric red fluorescent protein (mRFP) tagging is shown here:

1. Template genomic DNA is isolated from the target strain as described (14).
2. The mRFP-5′-K.l.URA3 fragment is amplified by PCR from pWJ1350 using a high-fidelity polymerase such as Pfu and primers Kli3′ and mRFPstart-F. The 3′-K.l.URA3-mRFP fragment is amplified by PCR from pWJ1351 using primers Kli5′ and mRFPend-R (**Fig. 30.2a**). PCR conditions: 95°C for 2 min, 30 cycles of 95°C for 30 s, 52°C for 30 s, and 72°C for 4 min, followed by 72°C for 1 min and subsequent cooling to 4°C. The PCR fragments are agarose gel purified.

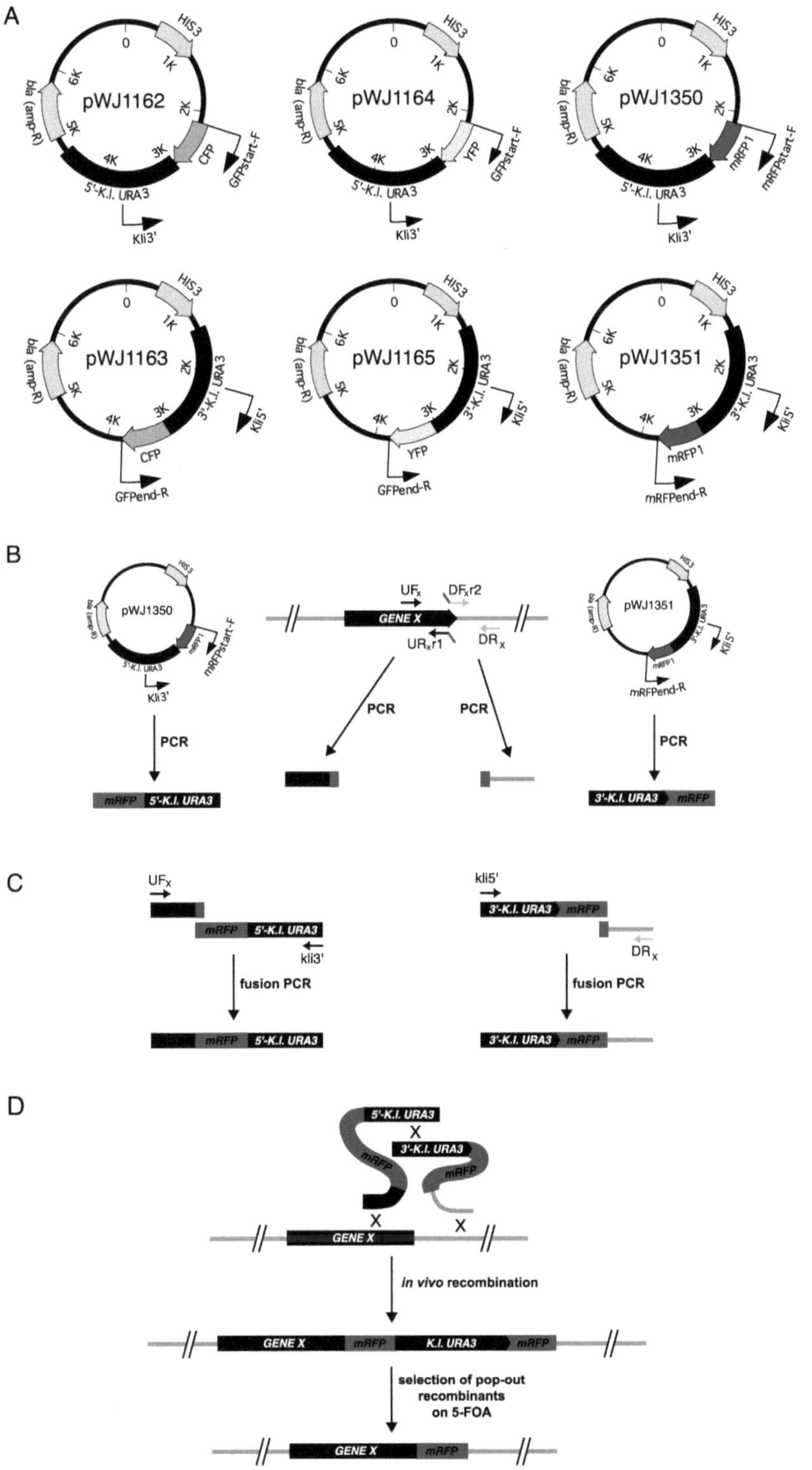

Fig. 30.2. Cloning-free fluorescence tagging of endogenous genes. (**a**) Vectors pWJ1162, pWJ1163, pWJ1164, pWJ1165, pWJ1350, and pWJ1350 harboring XFP-K.l.URA3 cassettes for CFP, YFP, or mRFP1 tagging (21–23). (**b**) PCR amplification of targeting sequences. (**c**) Adaptamer-mediated fusion PCR. A sequence overlap of 18–22 nucleotides between the two PCR products facilitates their fusion in a second PCR. (**d**) Gene targeting and marker elimination by popout recombination.

3. A region of approximately 300 bp immediately upstream of the mRFP integration site is amplified from the genomic DNA using a high-fidelity polymerase such as Pfu and primers UF_X and UR_Xr1 (**Table 30.1** and **Fig. 30.2b**). Similarly, a region of approximately 300 bp immediately downstream of the mRFP integration site is amplified using primers DF_Xr2 and DR_X. *See* **Note 1** for primer design. The PCR fragments are agarose gel purified.
4. Approximately 200 ng of the mRFP-5′-K.l.URA3 fragment is fused by PCR to an equimolar amount of the gene-specific upstream fragment using a high-fidelity polymerase such as Pfu and primers Kli3′ and UF_X. Similarly, 200 ng of the 3′-K.l.URA3-mRFP fragment is fused by PCR to an equimolar amount of the gene-specific downstream fragment using primers Kli5′ and DR_X. PCR conditions: 95°C for 2 min, 30 cycles of 95°C for 30 s, 52°C for 30 s, and 72°C for 4.5 min, followed by 72°C for 2 min and subsequent cooling to 4°C. The PCR fragments are gel purified.
5. To integrate mRFP into the genome, 0.3–1 μg of each of the fusion fragments is co-transformed into the target strain by the LiAc method (15). Transformants are selected on

Table 30.1
Primers used in these protocols

Name	Sequence (5′ to 3′)
DF_Xr2	GCATGGATGAACTATACAAATGAXXXXXXXXXX
Down-e	CGATCTTCTACCCAGAATCACXXXXXXXXXXXXXX
DR_X	XXXXXXXXXXXXXXXXXXX
E-Kl	GTGATTCTGGGTAGAAGATCG
F-Kl	CGATGATGTAGTTTCTGGTT
GFPend-R	TTTGTATAGTTCATCCATGC
GFPstart-F	ATGAGTAAAGGAGAAGAAC
I-SceI-Down-F	TTACGCTAGGGATAACAGGGTAATATAGCGXXXXXXXX
I-SceI-Up-R	CGCTATATTACCCTGTTATCCCTAGCGTAAXXXXXXXXXX
Kli3′	GAGCAATGAACCCAATAACGAAATC
Kli5′	CTTGACGTTCGTTCGACTGATGAGC
mRFPend-R	GGCGCCGGTGGAGTGG
mRFPstart-F	ATGGCCTCCTCCGAGGAC
UF_X	XXXXXXXXXXXXXXXXXXX
Up-f	AACCAGAAACTACATCATCGXXXXXXXXXXXX
UR_Xr1	GTTCTTCTCCTTTACTCATnnnnnnXXXXXXXXXXX

Linker sequences are indicated by nnnnnn and gene-specific sequences are indicated by XXXXXX.

synthetic complete medium lacking uracil (SC-Ura). Targeting efficiency varies with the genomic locus by at least a factor 10.

6. The integration generates a direct repeat of the mRFP sequence flanking the *K.l.URA3* marker (**Fig. 30.2d**). This configuration allows for subsequent popout of the *URA3* marker. Efficient popout is achieved by growing cells overnight in 2 ml of yeast extract peptone dextrose (YPD) medium before plating 200 μl of the culture on plates containing 5-fluoroorotic acid (5-FOA) (5) (*see* **Note 2**).

3.2. Fluorescence Visualization of an Inducible DSB Site

Specific genomic mega-endonuclease restriction sites such as I-*Sce*I and *HO* can be marked fluorescently by integration of a tandem array of 100–200 copies of a recognition sequence for a DNA-binding protein fused to XFP. We have good experience with Tet and Lac repressor binding sites (16, 17) (*see* **Note 3**). The protocol for genomic integration of the I-*Sce*I cut site and the *tetO* tandem array is shown here:

1. Template genomic DNA is isolated from the target strain as described (14).
2. The I-*Sce*I cut site is integrated into the genome essentially as described (9). In brief, a region of approximately 300 bp immediately upstream of the I-*Sce*I integration site is amplified from the genomic DNA using a high-fidelity polymerase such as Pfu and primers Up-f and I-SceI-Up-R (**Table 30.1**). Similarly, a region of approximately 300 bp immediately downstream of the I-*Sce*I integration site is amplified using primers I-SceI-Down-F and Down-e (**Fig. 30.3a**). The PCR fragments are agarose gel purified and fused in a second PCR using primers Up-f and Down-e. The PCR fusion is agarose gel purified.
3. The 5′-K.l.URA3 fragment is amplified by PCR from pWJ716 using primers Kli3′ and E-Kl. The 3′-K.l.URA3 fragment is amplified by PCR from pWJ716 using primers Kli5′ and F-Kl. PCR conditions: 95°C for 2 min, 30 cycles of 95°C for 30 s, 52°C for 30 s, and 72°C for 2 min, followed by 72°C for 1 min and subsequent cooling to 4°C. The PCR fragments are agarose gel purified.
4. Approximately 200 ng of the 5′-K.l.URA3 fragment is fused by PCR to an equimolar amount of the fusion fragment containing the I-*Sce*I cut site (step 2) using primers Kli3′ and Up-f. Similarly, 200 ng of the 3′-K.l.URA3 fragment is fused by PCR to an equimolar amount of the fusion fragment containing the I-*Sce*I cut site using primers Kli5′ and Down-e. PCR conditions: 95°C for 2 min, 30 cycles of 95°C for 30 s, 52°C for 30 s, and 72°C for 4.5 min, followed by 72°C for 2 min and subsequent cooling to 4°C. The PCR fragments are agarose gel purified.

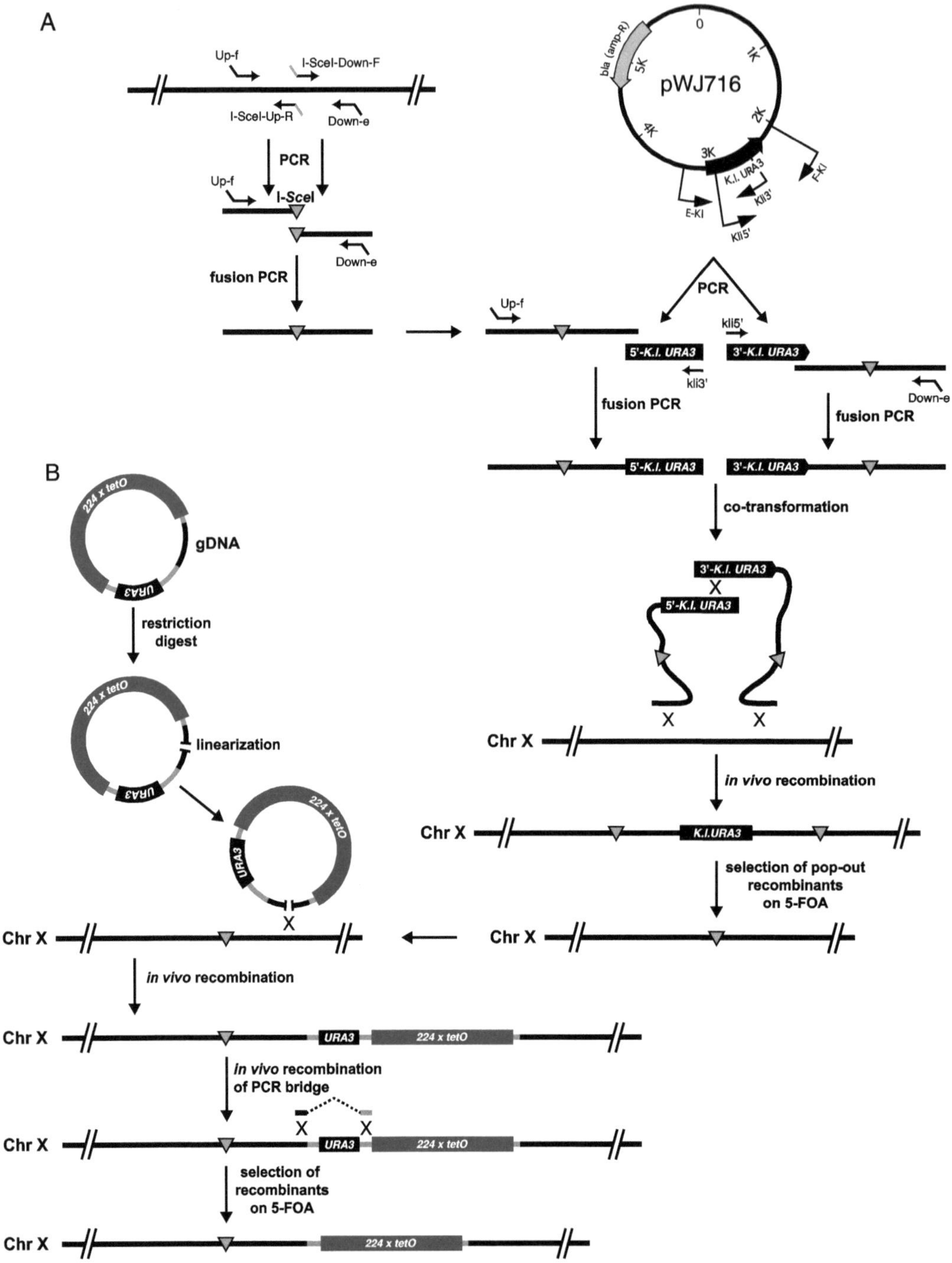

Fig. 30.3. Construction of a fluorescently marked DSB site. (**a**) PCR-based insertion of a unique site-specific I-*Sce*I site in the genome. (**b**) Genomic integration of a *tetO* array in a strain constitutively expressing TetR-mRFP (e.g., strain ML193-3B) gDNA, genomic target DNA.

5. To integrate the I-*Sce*I cut site into the genome, 0.3–1 μg of the fusion fragment is co-transformed into the target strain by the LiAc method (15). Transformants are selected on synthetic complete medium lacking uracil (SC-Ura).
6. The integration generates a direct repeat of the I-*Sce*I cut site containing sequence flanking the *K.l. URA3* marker (**Fig. 30.3a**). This configuration allows for popout of the *URA3* marker. Efficient popout is achieved by growing cells overnight in 2 ml of YPD medium before plating 200 μl of the culture on plates containing 5-FOA (5).
7. To insert an array of Tet repressor binding sites adjacent to the I-*Sce*I cut site, a 1-kb fragment of genomic target DNA (gDNA) is cloned into an integrative plasmid pWJ1379 containing the *tetO* array and the *URA3* selectable marker (**Fig. 30.3b**). The gDNA is selected to contain a unique centrally located restriction site for subsequent linearization of the plasmid before transformation into yeast (*see* **Note 4** for plasmid construction).
8. To integrate the *tetO* array into the genome, 1 μg of linearized plasmid is transformed into a target strain ML193-3B expressing TetR-mRFP by the LiAc method (15). Transformants are selected on synthetic complete medium lacking uracil (SC-Ura). Initially, transformants are screened visually by fluorescence microscopy to select the clones that harbor the *tetO* array as determined by the appearance of a red dot in each cell. Subsequently, correct genomic integration is verified by PCR or Southern blot analysis.
9. If desired, the co-integrated vector backbone of the *tetO* plasmid can be deleted by transformation of a PCR fusion bridging the region to be deleted (**Fig. 30.3b**). The PCR bridge should have >300 bp of homology to the genomic target on each side of the region to be deleted. The PCR bridge is constructed by fusion PCR of regions upstream and downstream of the sequence to be deleted using relevant primers essentially as described in step 2. Since the *URA3* marker can also be deleted by loss of the entire *tetO*-containing region, it advisable to use 1–2 μg of PCR bridge in the LiAc transformation (15). Transformants are incubated for 3 h in YPD in order to allow for the *URA3* gene product to disappear before plating on solid medium containing 5-FOA (5) (*see* **Note 5**).

3.3. Cell Culture for Fluorescence Microscopy

For yeast live cell imaging, the best results are obtained by culturing and mounting cells in identical medium. YPD medium should be avoided for imaging, because it quenches a broad range of wavelengths. By contrast, filter-sterilized minimal medium has excellent optical properties (5). A protocol for imaging exponentially growing cells is shown here:

1. Inoculate 2 ml of SC medium containing 100 μg/ml adenine from a single colony (*see* **Note 6**).
2. Grow by shaking overnight at 25°C (*see* **Note** 7).
3. Dilute culture to $OD_{600} = 0.2$ and grow for one cell cycle (2.5 h for wild type) prior to microscopy.

3.4. Induction of HR Foci by DNA-Damaging Agents

Foci of HR proteins can be induced by a number of agents or enzymes that cause DNA double-strand breaks (DSBs) and/or collapsed replication forks. Examples of agents and doses are 20 Gy of γ-irradiation (1 DSB per haploid cell) (18), 1 μg/ml bleomycin for 1 h (0.5 DSB per haploid cell) (6), 200 μg/ml zeocin for 1 h, 5 μg/ml camptothecin for 1 h, and 0.03% (v/v) methyl methanesulfonate (MMS) for 1 h.

3.5. Induction of HR Foci by Mega-endonucleases

Strains in which a single site-specific endonuclease restriction site such as the *HO* or the I-*Sce*I site in the genome has been marked fluorescently as described above (**Section 3.2**) are transformed with a plasmid expressing the appropriate endonuclease from an inducible promoter, e.g., pJH132 (pGAL-HO) or pWJ1108 (pGAL-I-SceI), respectively:

1. Transform the appropriate plasmid (expressing HO or I-*Sce*I) into the assay strain by the LiAc method (15). Select transformants on the appropriate glucose-containing minimal medium.
2. Inoculate fresh transformants in 2 ml of the appropriate minimal, raffinose-based medium (2% (w/v) raffinose, autoclaved) (*see* **Note 8**).
3. Grow by shaking at 25°C for 24 h.
4. Dilute culture to $OD_{600} = 0.2$ and grow for one cell cycle (3.5 h for wild type) prior to endonuclease induction.
5. Add galactose (sterile-filtered, stock solution, 30% (w/v)) to a final concentration of 2% (w/v). This induces expression of the endonuclease.
6. Grow by shaking at 25°C for another 90 min prior to microscopy.

3.6. Sample Preparation

Live cell imaging is preferred over fixed cells, because fixation may generate artifacts or otherwise decrease the quality of the obtained data. This protocol describes preparation of both types of samples:

1. Harvest 1.5 ml of cells at $OD_{600} = 0.4$–0.6 by centrifugation at 1,500×*g*.

2. For live cell imaging, resuspend the pellet of cells in 50 μl of medium by vortexing and proceed to mounting the cells (**Section 3.7**).
3. For fixed cell imaging, resuspend the pellet of cells in PBS containing 4% paraformaldehyde (wear gloves).
4. Incubate shaking for 30 min at room temperature.
5. Wash twice for 30 min in PBS at room temperature. Store in the dark at 4°C in PBS containing 0.02% sodium azide (optional for long-term storage).

3.7. Mounting of Live Cells

Cells are mounted on standard glass slides and covered by cover glass appropriate to the optics of the microscope. Issues to consider before mounting the cells are the cell density and immobilization. A high cell density is desired to maximize data acquisition. However, at high cell density and growth rate (glucose), the medium quickly becomes saturated with carbon dioxide, which precipitates as gas bubbles that displace cells. Therefore, the optimal cell density must often be determined empirically. For most strains, the cells are immobilized on the slide simply by adjusting the volume applied (usually 2–3 μl) so that the cells settle in a monolayer with the cells touching both the slide and the cover glass. However, for mutant strains that exhibit heterogeneous cell size, it can be necessary to immobilize cells in low-melt agarose. For long-term imaging (>30 min), the edges of the cover glass are sealed to prevent evaporation by a mixture of 1 volume of petroleum jelly (Vaseline), 1 volume of beeswax, and 1 volume of lanolin. The protocol is as follows:

1. Prepare a solution of 1.2% (w/v) low-melt agarose (gelling at 36°C) in appropriate medium (e.g., SC). Aliquot in microcentrifuge tubes and use each aliquot only once. Melt by boiling and keep at 42°C prior to use. Melt the wax solution (e.g., in a 65°C incubator).
2. Add 2 μl cell suspension to the slide and mix with 2 μl of agarose solution by pipetting. Apply cover glass as fast as possible.
3. Seal with melted wax using a flat metal spatula. Heating of the spatula over a gas burner may be necessary to facilitate dispersion of wax along the sides of the cover glass.

3.8. DAPI Staining

Yeast cells can be DAPI stained to visualize DNA content without fixation:

1. Add DAPI to the liquid medium at 10 μg/ml and grow by shaking for 30 min.
2. Wash cells in SC without DAPI before microscopy.

Due to the intense DAPI staining of mitochondria, it can be difficult to discern the nuclear compartment. To visualize staining of nuclear DNA, it is often necessary to eliminate mitochondrial DNA. This is achieved by the following protocol:

1. Inoculate an overnight culture in YPD containing 25 μg/ml ethidium bromide at 30°C.
2. The next morning, dilute 100-fold into YPD containing 25 μg/ml ethidium bromide. Grow again overnight at 30°C.
3. Plate for individual colonies on YPD.
4. To test for loss of mitochondrial DNA, pick five candidates and examine them under the microscope after DAPI staining. Additionally, candidates can be tested for lack of growth on yeast extract peptone medium containing 2% (w/v) galactose due to loss of mitochondria.

3.9. Time-Lapse Microscopy

Imaging of cells over time requires that phototoxicity is minimized and that favorable growth conditions can be maintained. Phototoxicity can be reduced by decreasing the fluorescence exposure time or intensity, and by reducing the number of optical sections acquired. The easiest way to maintain favorable growth conditions is to reduce the cell density of the slide so that only a single cell is present in the field of view, thereby prolonging the time that the cell can grow before nutrients are exhausted locally. For this reason it is not recommended to concentrate the cells by centrifugation before mounting:

1. Cells are cultured and mounted essentially as described above (**Sections 3.3** and **3.7**).
2. The appropriate acquisition parameters for time-lapse microscopy should be determined empirically. In our case, we were able to image Rad52-YFP with minimal phototoxicity for 20 time points over 4 h by reducing the number of optical sections from 11 to 9 (0.5 μm between sections), reducing the neutral density filter for the fluorescence path from 25 to 10% transmission, and by reducing the exposure time from 1 to 0.5 s.

4. Notes

1. Usually, primers are designed to insert a small flexible linker between mRFP and the protein of interest. Obviously the stop codon from the gene to be fused to *mRFP* should not be included in the primers for tagging at the N terminus. For XFP fused to the N terminus, we have good experience with a $Gly_2ProGly_2$ linker and for XFP fused to the C terminus we routinely use an Ala_4 linker.

2. Generally, correct genomic targeting is achieved for >90% of Ura^+ colonies. However, if the genomic target sequence contains repetitive or ARS-like sequences, the transformed PCR fragments can circularize to a self-replicating extrachromosomal circle. These incorrect clones are easily identified during the popout step, because they give rise to a confluent layer of Ura^- cells on 5-FOA due to the instability of centromere-less extrachromosomal circles.
3. When using the Lac repressor, a mutant LacI-R197K protein is used, which binds to DNA in the presence of both glucose and galactose.
4. The *tetO* array should be positioned 3–5 kb away from the I-*Sce*I cut site, in order to prevent or delay its conversion into single-stranded DNA during resection of the DSB generated at I-*Sce*I, which will abolish binding by the Tet repressor. Since the *tetO* array is unstable in *E. coli*, it is essential that the *tetO* plasmid is cloned in a rec^- strain at 30°C. For the *tetO* array, we have good experience with the SURE2 strain (Stratagene, La Jolla, CA), and for the *lacO* array the STBL2 strain (Invitrogen, Paisley, UK) gives good stability.
5. Generally, correct deletion of the plasmid sequences by the PCR bridge is achieved for 10–30% in the 5-FOA-resistant colonies.
6. The extra adenine is necessary only for *ade2* mutant strains, which will otherwise accumulate a red pigment that is strongly autofluorescent. However, SC medium containing 100 μg/ml adenine can also be used for *ADE2* cells.
7. Some variants of GFP and RFP mature better at 25°C than at 30 or 37°C (19).
8. It is important to use fresh transformants in order to avoid cells in which mutations have accumulated in the endonuclease recognition site. Such mutations tend to appear in old transformants due to leakiness of the *GAL1-10* promoter.

Acknowledgments

The LacI-R197K mutant protein was engineered by Christian Müller. This work was supported by The Danish Agency for Science, Technology and Innovation (ML), the Villum Kann Rasmussen Foundation (ML), GM67055 (RR), and the Lundbeck Foundation (NEB).

References

1. Krogh, B. and Symington, L. (2004) Recombination proteins in yeast. *Annu Rev Genet* **38**, 233–271.
2. Lisby, M. and Rothstein, R. (2004) DNA damage checkpoint and repair centers. *Curr Opin Cell Biol* **16**, 328–334.
3. Lisby, M. and Rothstein, R. (2005) Localization of checkpoint and repair proteins in eukaryotes. *Biochimie* **87**, 579–589.
4. Lisby, M. and Rothstein, R. (2009) Choreography of recombination proteins during the DNA damage response. *DNA Repair (Amst)* **8**, 1068–1076.
5. Sherman, F. Fink, G.R., and Hicks, J.B. (1986) Methods in yeast genetics (Cold Spring Harbor, NY: Cold Spring Harbor Laboratory).
6. Moore, C.W., McKoy, J., Dardalhon, M., Davermann, D., Martinez, M., and Averbeck, D. (2000) DNA damage-inducible and *RAD52*-independent repair of DNA double-strand breaks in *Saccharomyces cerevisiae*. *Genetics* **154**, 1085–1099.
7. Thomas, B.J. and Rothstein, R. (1989) Elevated recombination rates in transcriptionally active DNA. *Cell* **56**, 619–630.
8. Zhao, X., Muller, E.G., and Rothstein, R. (1998) A suppressor of two essential checkpoint genes identifies a novel protein that negatively affects dNTP pools. *Mol Cell* **2**, 329–340.
9. Erdeniz, N., Mortensen, U.H., and Rothstein, R. (1997) Cloning-free PCR-based allele replacement methods. *Genome Res* **7**, 1174–1183.
10. Torres-Rosell, J., Sunjevaric, I., De Piccoli, G., Sacher, M., Eckert-Boulet, N., Reid, R., Jentsch, S., Rothstein, R., Aragon, L., and Lisby, M. (2007) The Smc5–Smc6 complex and SUMO modification of Rad52 regulates recombinational repair at the ribosomal gene locus. *Nat Cell Biol* **9**, 923–931.
11. Reid, R., Lisby, M., and Rothstein, R. (2002) Cloning-free genome alterations in *Saccharomyces cerevisiae* using adaptamer-mediated PCR. *Methods Enzymol* **350**, 258–277.
12. Lisby, M., Mortensen, U.H., and Rothstein, R. (2003) Colocalization of multiple DNA double-strand breaks at a single Rad52 repair centre. *Nat Cell Biol* **5**, 572–577.
13. Jensen, R.E. and Herskowitz, I. (1984) Directionality and regulation of cassette substitution in yeast. *Cold Spring Harb Symp Quant Biol* **49**, 97–104.
14. Hoffman, C.S. and Winston, F. (1987) A ten-minute DNA preparation efficiently releases autonomous plasmids for transformation of *Escherichia coli*. *Gene* **57**, 267–272.
15. Gietz, D., St Jean, A., Woods, R.A., and Schiestl, R.H. (1992) Improved method for high efficiency transformation of intact yeast cells. *Nucleic Acids Res* **20**, 1425.
16. Straight, A.F., Belmont, A.S., Robinett, C.C., and Murray, A.W. (1996) GFP tagging of budding yeast chromosomes reveals that protein–protein interactions can mediate sister chromatid cohesion. *Curr Biol* **6**, 1599–1608.
17. Michaelis, C., Ciosk, R., and Nasmyth, K. (1997) Cohesins: chromosomal proteins that prevent premature separation of sister chromatids. *Cell* **91**, 35–45.
18. Lisby, M., Rothstein, R., and Mortensen, U.H. (2001) Rad52 forms DNA repair and recombination centers during S phase. *Proc Natl Acad Sci USA* **98**, 8276–8282.
19. Lim, C.R., Kimata, Y., Oka, M., Nomaguchi, K., and Kohno, K. (1995) Thermosensitivity of green fluorescent protein fluorescence utilized to reveal novel nuclear-like compartments in a mutant nucleoporin NSP1. *J Biochem (Tokyo)* **118**, 13–17.
20. Lisby, M., Barlow, J.H., Burgess, R.C., and Rothstein, R. (2004) Choreography of the DNA damage response; spatiotemporal relationships among checkpoint and repair proteins. *Cell* **118**, 699–713.
21. Ormo, M., Cubitt, A.B., Kallio, K., Gross, L.A., Tsien, R.Y., and Remington, S.J. (1996) Crystal structure of the *Aequorea victoria* green fluorescent protein. *Science* **273**, 1392–1395.
22. Campbell, R.E., Tour, O., Palmer, A.E., Steinbach, P.A., Baird, G.S., Zacharias, D.A., and Tsien, R.Y. (2002) A monomeric red fluorescent protein. *Proc Natl Acad Sci USA* **99**, 7877–7882.
23. Heim, R. and Tsien, R.Y. (1996) Engineering green fluorescent protein for improved brightness, longer wavelengths and fluorescence resonance energy transfer. *Curr Biol* **6**, 178–182.

Chapter 31

Live Cell Imaging of Meiotic Chromosome Dynamics in Yeast

Harry Scherthan and Caroline Adelfalk

Abstract

Recombination in first meiotic prophase is initiated by endogenous breaks in double-stranded DNA (DSBs) which occurs during a time when chromosomes are remodeled and proteinaceous cores (axes) are assembled along their length. DSBs are instrumental in homologue recognition and underlie the crossovers that form between parental chromosomes to ensure genome haploidization during the following two successive meiotic divisions. Advances in fluorescence microscopy and genetic engineering of GFP-tagged fusion proteins have made it possible to observe the behavior of entire chromosomes and specific subregions in live cells of the yeast *Saccharomyces cerevisiae*. In meiosis we observed that telomeres are dynamic and move about the entire nuclear periphery, only interrupted by their fleeting clustering at the spindle pole body (the centrosome equivalent), known as bouquet formation. This mobility translates to whole chromosomes and nuclei during the entire prophase I. Here we describe a simple setup for live cell microscopy that we used to observe chromosome movements during a time when DSBs are formed and transform into crossover and non-crossover products.

Key words: Chromosome dynamics, live cell microscopy, meiosis, recombination, *Saccharomyces cerevisiae*, telomere.

1. Introduction

Homologous recombination is a means by which dsDNA breaks (DSBs) are repaired. DSBs can occur endogenously during DNA replication or by exposure to genotoxic agents such as reactive oxygen species, ionizing radiation, or chemicals. Programmed DSBs and successive recombination also occur during antigen class switching in lymphocyte maturation and are a virtual hallmark of the meiotic process. In meiosis of most species programmed DSBs provide the substrate for the homologous recombination (HR) machinery that generates two outcomes:

H. Tsubouchi (ed.), *DNA Recombination*, Methods in Molecular Biology 745,
DOI 10.1007/978-1-61779-129-1_31, © Springer Science+Business Media, LLC 2011

crossovers (CO) and non-crossovers (NCO) (1). There are indications from yeast that crossover regulation occurs before or during synapsis (2, 3). CO create a physical link between homologous chromosomes, which are required for their reductional segregation at the meiosis I division. For CO to occur, homologous chromosomes, which are usually not aligned, need to encounter and pair lengthwise (synapse) during the first meiotic prophase. Homologue pairing occurs after completion of premeiotic DNA replication and involves a number of topological events in the nucleus such as attachment of meiotic telomeres to the nuclear envelope and their clustering in a limited nuclear sector (bouquet formation) that is thought to instigate homologue recognition and synaptic pairing. The chromosomal bouquet is usually dissolved when recombination commences and homologous chromosomes become stably connected by the synaptonemal complex (4).

Classically, chromosome (re)positioning has been inferred from the analysis of spread (thus partially disrupted) nuclei, albeit at a high resolution. Such information, when deduced in a developmental series of cells or tissues, has for decades been used to reconstruct cellular and chromosomal mobility. However, such information will always remain a series of snapshots frozen in time. Early attempts to capture chromosome movements in living cells disclosed that chromosome pairing is accompanied by vigorous rotation of nuclei and cellular contents (5). Since then, ever-speeding developments in image acquisition hardware and software have allowed to obtain extended image series of living cells at high speed and resolution (for an overview *see* (6)). Dynamic imaging is now performed in organelles as minute as the yeast nucleus, extending down to dynamic single molecule imaging. Introduction of the green fluorescent protein (GFP)-tagged lac repressor–operator system for spot tagging of chromosome subregions (7) and the growing use of fluorescence-tagged (GFP or its derivatives) versions of fusion proteins have instigated the application of live cell analysis in genetics and cell biology. These tools have provided deep insights into the dynamics of entire chromosomes or specific parts thereof in nuclei of live cells undergoing the vegetative cell cycle or meiotic differentiation and recombination at high speed and resolution (for an overview *see* (8)). Due to the ease of handling and genetic manipulation, such work has particularly been carried out in yeasts (e.g., (9–13)). Live cell analysis of the synaptic meiosis of *S. cerevisiae* has led to the detection of rapid and continuous telomere and chromosome movements (10, 14) that depend on actin polymerization, an intact telomere complex, and may interfere to some extent with recombination (10, 14–17). Here we describe the setup of a live cell microscope system for time-lapse observations of GFP-tagged meiotic telomeres and chromosomes during sporulation and meiotic recombination.

2. Materials

2.1. Strains

Replacement of endogenous proteins with fluorescence-tagged versions of proteins has emerged as a valuable tool to study chromosomes and their subregions in live cells. Chromosome dynamics during meiotic recombination can be investigated, for instance, with a diploid strain stably expressing a GFP-tagged version of the ZIP1 synaptonemal complex protein (18) or with strains where specific DNA loci are tagged by the use of GFP fusion proteins that bind to sub-chromosomal regions, like RAP1 to telomeres (10, 19) (*see* **Note 1**). We successfully used a ZIP1–GFP fusion protein to monitor the progress of synapsis and movements of entire meiotic chromosomes, and a RAP1–GFP fusion to follow telomere movements in meiotic cells (**Table 31.1**, **Fig. 31.1**). Furthermore, the GFP-tagged lac repressor–operator system (7) can be applied to tag and follow particular chromosome regions of interest in live cells.

Table 31.1
Yeast strains used in live cell imaging experiments as described in this chapter. For a start the SK1 strain background may be used because of its rapid sporulation and high degree of synchrony (24, 25), which significantly differs in sporulation time from the standard W303 laboratory strain (21).

Strains		References
ZIP1-GFP^{700} (HW122)	*MATa/MATα lys2/lys2 ho::LYS2/ho::LYS2 ura3/ura3 ZIP1::GFP^{700}/ZIP1::GFP^{700}*	(14)
RAP1-GFP	*MATa/MATα ho::LYS2/ho::LYS2 ade2::hisG/ade2::hisG ura3/ura3 leu2::hisG/leu2::hisG trp1::hisG/trp1::hisG his3::hisG/ his3::hisG RAP1::RAP1-GFP-LEU/RAP1::RAP1-GFP-LEU*	(10)

2.2. Culture Media and Drugs

1. Rich medium agar plates (YPDA): 1% yeast extract, 2% peptone, 2% glucose, 2% agar, in distilled water, autoclave and pour in plates (20) (*see* **Note 2**).
2. YPA (presporulation medium): 1% yeast extract, 2% peptone, 1% potassium acetate, in distilled water, autoclave (*see* **Note 2**).
3. SPM (sporulation medium): 2% potassium acetate (0.025% adenine; *see* **Note 2**) in distilled water, autoclave.
4. Trolox™ (Hoffman-La Roche), a water-soluble form of vitamin E (6-hydroxy-2,5,7,8-tetramethylchroman-2-

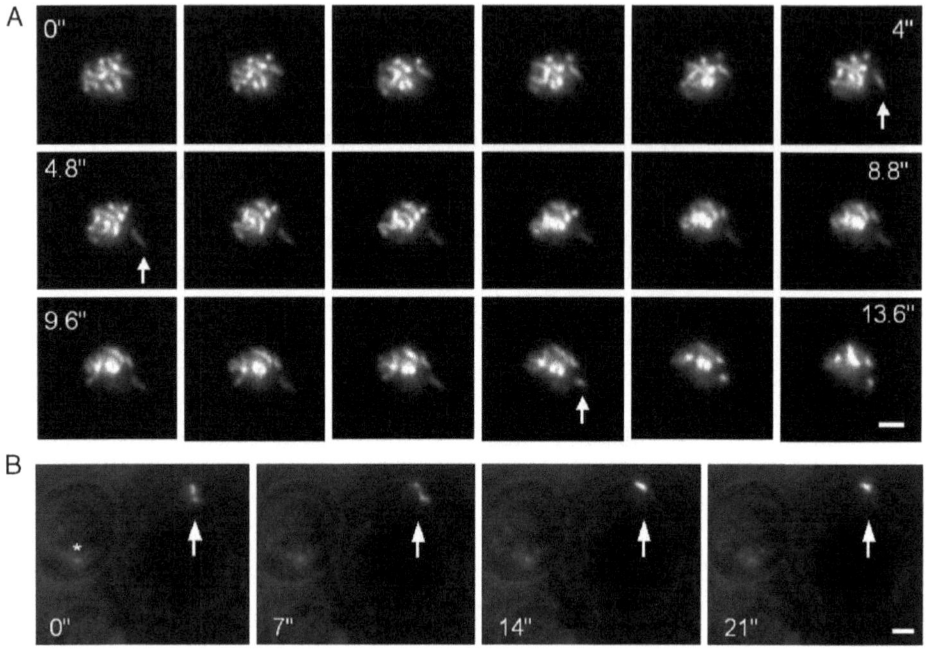

Fig. 31.1. (**a**) A time series of ZIP1–GFP-labeled bivalents of an SK1 pachytene cell displaying rapid chromosome movements recorded at 1.25 Hz (elapsed time indicated in seconds). While there is mobility of bivalents relative to each other in the chromosome mass there is a "maverick" bivalent (*arrows*) leaving the chromosome mass at 4 s and returning into it at 12 s. Focal plane at nuclear equator; bar: 1 μm. For details and movies *see* (14). (**b**) Time series of RAP1–GFP-labeled telomeres in a bouquet stage meiocyte of W303 background with telomere clustering recoded at 0.14 Hz (elapsed time indicated in seconds). Two telomere clusters are in contact at 7″ and fuse to form one strongly fluorescent cluster thereafter. The *dark center* of the cell represents the vacuole. Note the smaller vegetative cell to the left with its nucleus at the lower part of the cell (*) and three relatively immobile telomere clusters. Focal plane at nuclear equator; bar: 1 μm. For details and movies *see* (10).

carboxylic acid; www.sigmaaldrich.com), an antioxidant that can reduce light-induced cell stress.

5. SPM-T: SPM containing 5 μg/ml Trolox™ (*see* **Section 2.2.4**).
6. Microtubule disrupting drugs: nocodazole or benomyl (www.sigmaaldrich.com) (*see* **Note 3**). Dissolve in DMSO as a stock solution: nocodazole 30 mg/ml; benomyl [methyl 1-(butylcarbamoyl)-2-benzimidazolecarbamate] 20 mg/ml.
7. Actin-interfering drug latrunculin B (LatB) used at 30 μM according to the instructions of the manufacturer (www.sigmaaldrich.com). Dissolve in DMSO at 10 mM as a stock solution. Store at –20°C or below.
8. Concanavalin A (ConA; www.sigmaaldrich.com). ConA treatment of glass surfaces supports cell attachment, limiting cell mobility during microscopy. Dissolve in water at 5 mg/ml as a stock solution. Store at −20°C.
9. Hoechst 33342 (www.sigmaaldrich.com), a DNA-binding live cell dye that is used at 0.5 μg/ml (*see* **Note 4**).

2.3. Setup for Live Cell Imaging

1. A temperature-controlled microscope system should be used, since sporulation and recombination are affected by temperature (2, 21). Commercial temperature-controlled microscope stages or hoods are available from several companies/microscope manufacturers. We run a small temperature-controlled microscope room equipped with heating element and air circulation to minimize temperature fluctuations, which avoids focus-instability that can be a problem to high-resolution time-lapse microscopy when temperature gradients are building up in a microscope system.
2. Vibration-free microscope table. Since vibrations emitted by the environment can disturb focus maintenance in high-resolution time-lapse microscopy, the use of a sturdy table is advised. Vibration-damping microscope equipment and tables in desktop or stand-alone format are available from commercial suppliers (e.g., www.speirsrobertson.co.uk or microscope manufacturers). We use a home-made version consisting of an 8 cm thick, 75×75 cm granite plate floating on six low-pressure inflated motor scooter tubes that are placed on a solid table of the same size. If you require details for assembly please contact H. Scherthan.
3. Fast image recording system. Several commercial systems are available for sampling a large number of image frames. The recording of small, light-sensitive cellular objects like yeast nuclei and their sub-nuclear structures requires sensitive high-resolution microscope systems equipped with a high-speed CCD camera (e.g., PCO Sensicam or alike).
4. Low light illumination device. Because yeast and most cells are sensitive to light exposure and since time-lapse analysis most often involves recording several hundred image frames, imaging can induce phototoxic effects, especially at short wavelengths. As a light source of low intensity we use a polychrome IV monochromator (TILL Photonics) in combination with a GFP filter set. The monochromator allows to select an adequate excitation wavelength from its computerized control allowing it to be used with a quadruple band-pass beam splitter and an appropriate barrier filter (www.chroma.com) for analysis of several fluorophors in fixed samples. We prefer a 100 × plan neofluar oil-immersion lens with a high numerical aperture, which depends on the strength of the GFP fluorescence of the tagged fusion protein(s) under investigation. We usually use an α-Plan-Fluar 100× NA 1.45 lens (www.zeiss.com), which transmits less light than the generally used 63× NA 1.3 objective. Since

this also reduces the intensity of the excitation light there will be less phototoxic effects. An increased magnification relative to the 63× objective is furthermore beneficial for the analysis of sub-micrometer objects such as yeast chromosomes.

5. To capture 3D and 4D information a piezoelectric objective nanopositioning system (PIFOC, www.physikinstrumente.com) controlled by TILLvisION v4.0 software (www.till-photonics.com) is applied in our lab. Alternatively, confocal microscopes (e.g., standard or spinning disk) may be used; for an elaboration on this matter *see* (11, 12).
6. *Live cell chamber.* During image recording cells have to be kept in an environment that allows them to proceed through meiosis on the microscope stage. It is important to provide aeration for sporulation to proceed; this is usually achieved by placing cells in medium in either a glass bottom culture dish (inverted microscope) (13) or a Lundin Chamber (www.lis.ch) (11). Usually, sporulation on the microscope is a good indicator of cell viability. For the standard upright microscope yeast cells may be placed between coverslips or on top of an agarose pad hanging from a coverslip for extended imaging in a standard upright microscope (12). In our lab we apply an inverted microscope with two variant cell chambers: the first one was home-made by H. Scherthan and consists of a stainless steel chamber of the size of a standard microscope slide (**Fig. 31.2**) into which a round hole with

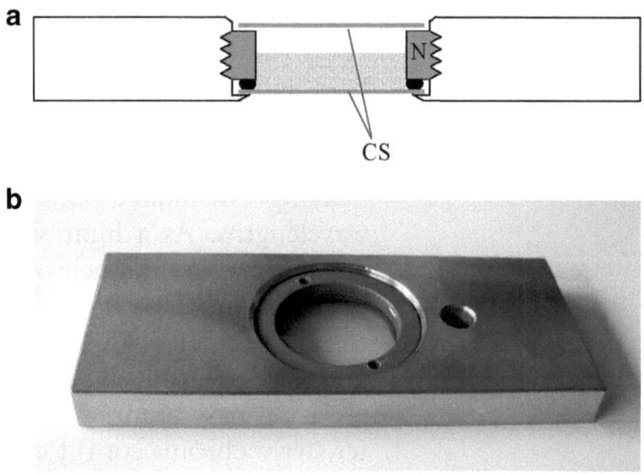

Fig. 31.2. Stainless steel live cell chamber (*see* **Section 2.2**, point 6). (**a**) Outline of the chamber with the body, coverslip (*CS*), nut (*N*), and O-ring seal (*black*). The cells are layered on the lower ConA-coated coverslip and are covered with medium (*light gray*). A second coverslip is loosely placed on top to prevent evaporation. (**b**) Picture of the assembled chamber. Showing medium (*in the center*) residing on the coverslip followed by a seal and plastic nut (*with two opposing holes*). In operating mode the nut is covered by a second coverslip. The hole to the *right* can be used to mount a temperature sensor.

a thread has been milled to accommodate a round coverslip of 25 mm diameter (e.g., www.hecht-assistent.com) at the rim at the bottom of the chamber (*see* **Note 5**). On top of the coverslip there follows a Sykes-Moore gasket (e.g., www.dunnlab.de) and a threaded plastic ring that is screwed on top, so there is no contact between the medium and metal (**Fig. 31.2**). Finally, a second coverslip is loosely placed on top of the plastic ring to allow for aeration of the chamber (*see* **Note 5**). This chamber has the advantage that it can also be used on an upright microscope when the second coverslip is placed between the gasket and the plastic nut. Since this limits aeration, it is only appropriate for short-term imaging in yeast or for imaging of mammalian cells (Scherthan, unpublished).

Alternatively to the custom-made cell chamber, we sometimes use a commercial 8-well coverslip (Lab-Tek®II Chamber #1.5 German Coverglass System; www.nuncbrand.com/NAG/). In this case four different cell cultures can be placed next to each other and the outer 4 wells of the 8-well coverglass system can be filled with water to compensate for evaporation. However, this is only advisory for recording of short movies, e.g., in comparative drug experiments, since sporulation will proceed in the non-imaged cultures.

3. Methods

3.1. Cell Culture

1. To initiate a meiotic cell culture with sufficient cell number, inoculate 50 ml YPA with 5–6 colonies of a diploid SKl strain (*see* **Note 6**) grown for 36–48 h on a YPD plate. Grow the cells in a large Erlenmeyer flask with constant agitation at 30°C overnight (13–16 h) to a concentration of $\sim 2\times 10^7$ cells/ml.
2. Centrifuge the cell suspension for 4 min at 2,500 rpm.
3. Resuspend cells in 10 ml SPM and repeat Steps 2 and 3.
4. Spin down the cells again and resuspend them at a density of about 4×10^7 cells/ml in SPM-T (approx. 10 ml) and incubate for 3 h at 30°C. That will be the time when most cells are in leptonema.

3.2. Sample Preparation

Before the experiments begin, the temperature-controlled stage or microscope room should have reached the desired temperature (in this case $\leq$ 30°C). The required time has to be empirically determined and implemented in the particular experimental design.

1. Live cell chamber: Coverslips should be cleaned and sterilized with 70% ethanol before coating. Coat a round coverslip by application of 10 μl Concanavalin A to the glass surface in the center of the coverslip. Spread the liquid carefully with the pipette tip. Let the solution sit for 5 min and remove surplus liquid with a micropipette. Dry the coverslip in a dust-free environment. Add the glass and gasket and tighten the screw (**Fig. 31.2**). Allow the chamber to sit in the preheated room at 30°C.
2. Remove 150 μl of the cell suspension from the shaker and apply to the live cell chamber (*see* step 1 of this **section**).
3. Allow the cells to adhere for several seconds during which they will form a monolayer on the coverslip. Slowly add 400–500 μl of SPM-T medium and cover loosely with a second coverslip (*see* **Notes 5** and 7).

3.3. Observation of ZIP1–GFP-Tagged Chromosomes in Live Meiocytes

1. Start up a computer-controlled fluorescence microscope system (*see* **Note 8**).
2. Place the chamber on the microscope and monitor for appearance of ZIP1–GFP-expressing cells with dimly green nuclei (preleptotene/leptotene cells). Nuclei that display distinct green spots are in zygotene as synapsis is commencing. Pachytene cells exhibit green rods (bivalents) due to ZIP1–GFP fluorescence at the SC (**Fig. 31.1a**). The chromosome movements at this stage are so rapid that they can be seen by eye.
3. Choose appropriate settings for time-lapse microcopy. As a rule of thumb adjust exposure time to obtain sufficient signal-to-noise ratio but as short-as-possible exposure time (e.g., 200 ms for ZIP1–GFP-expressing cells). This can be obtained by reducing the size of the field that is recorded (region of interest, ROI) or by binning. First record a time series at a fixed focal plane, since this minimizes light exposure and thus maximizes viability of cells (*see* **Note 9**).
4. Run image software to obtain time-lapse image data (*see* **Note 10**).
5. Control experiments with wild-type cells exposed to the actin drug latrunculin B or fixed with formaldehyde should be performed to control the setup (*see* **Note 11**).

4. Notes

1. RAP1 binds not only to *S. cerevisiae* telomeres but also to interstitial loci (22); therefore, when using mutants controls by, e.g., telomere, FISH should be performed to

ensure that the right target structures are followed in the time-lapse analyses.

2. Presporulation growth in YPA (23) improves the synchrony of sporulation. To avoid accumulation of the red metabolic product phosphoribosyl-amino-imidazol in *ade* mutants, add 0.05% adenine to the media.
3. Drugs are diluted with an appropriate solvent (usually DMSO for lipophilic drugs); in such cases a control culture with the appropriate solvent has to be investigated as well.
4. Cells can be incubated with the live cell stain Hoechst 33342 at 0.5 μg/ml in medium for 15 min and then briefly washed and mounted on the cell chamber. However, Hoechst staining is only appropriate for short-term time-lapse videos. *S. cerevisiae* is generally sensitive to light at intensities and wavelengths used in live cell imaging. Phototoxicity is encountered particularly when cells have been stained by the DNA minor groove-binding dye Hoechst 33342, which is also due to the short wavelength used for its excitation. Therefore, it is advisory to avoid Hoechst in long-term experiments or when a large number of image frames have to be recorded, e.g., in 3D and 4D (XYZT) image recording. Live cell stains with other properties can be obtained from commercial suppliers (e.g., Molecular Probes at www.invitrogen.com).
5. The stainless steel chamber has the advantage that it will facilitate temperature maintenance once it has been brought to the desired temperature. Make sure that the top coverslip has no solution between the glass and the nut, since this will block aeration.
6. Sporulation time varies considerably between different strains (21). Meiotic studies usually apply the budding yeast strain SK1 (24) that can be induced to sporulate at high synchrony (25) but is sensitive to temperature variation (2). Strain SK1 (24) which is widely used for meiotic studies shows a maximum of pachytene cells ~4–5 h after transfer to sporulation medium.
7. When the 8-well coverslip chamber is used instead of the custom-made cell chamber only 100 μl of cell suspension is applied per well and overlaid with 200 μl of SPM-T. The inverted microscope stage has the advantage that a cell culture that has been observed can be treated by addition of drugs on stage. In this case the 8-well coverslip chamber is advantageous, since drugs and control can be treated next to each other under similar conditions.

8. Time-lapse microscopy may be performed using commercial computer-controlled microscope systems (e.g., Carl Zeiss, Leica, or Applied Precision Inc.). We attached a TILL Photonics Imaging System driven by TILLvisION software to a Carl Zeiss Axiovert inverted microscope equipped with a PIFOC Piezo drive for 3D recording and a polychrome IV light source (www.till-photonics.com). Generally, the use of wide-field fluorescence microscopy with or without image deconvolution is advisory, since these systems show a higher sensitivity and allow for shorter exposure times, as compared to laser scanning microscopes that use mechanical pin hole systems that block out-of-focus blur.

   Three-dimensional focal stacks can be recorded along a time line and later displayed as a 4D series (e.g., by using commercial software packages such as SVI Huygens, www.svi.nl). Alternatively, open source image processing solutions (ImageJ) can be found at http://rsbweb.nih.gov/ij/index.html. Four-dimensional series with the TILLvisION software may be obtained by maximum projection of the 3D image stacks and assemblage of these in a movie using a TILLvisION macro operation.
9. As a start, ZIP1–GFP-expressing cells may be recorded in 4D by six image stacks spaced 0.5 μm with 200 ms exposure time (×100 lens) and image stacks taken every 10 s over 10 min. For fixed focal plane recording (2D) at nuclear equator 200 ms exposure time and image recording in 5 s intervals for 30 min have proven amenable using our TILL setup. These times may vary, however, with different microscope systems. To determine how many exposures are possible with a certain cell strain and mic system run, e.g., a 2 Hz images series over 7 min, in the resulting movie compare the image quality in the first and last frames. Determine the number of frames up to the point when the signal fades, which will correspond to the max image number that can be recorded. Use this number to set up the frame rate and time span to be recorded. For example, 400 frames may be recorded over 40 min at 1 frame/6 s (0.16 Hz).
10. Since chromosome dynamics may be variable with time and some movements saltatory in nature, i.e., detectable only in a few adjacent image frames/stacks, it is recommended to record several cells at different frame rates (e.g., 2 and 0.2 Hz frame rate). Data should be averaged from different cells and image series recorded with the same frame rate.
11. It is always a good control of the successful setup that cells complete sporulation on the microscope system after

imaging. Furthermore, include controls with cells treated with the actin drug latrunculin B that reversibly abolishes telomere clustering and paralyzes chromosome motion and nuclear deformations, except for Brownian movements (10, 14). To reveal non-biological noise in the experiments, cells are fixed with formaldehyde on stage by addition of 0.027 volumes of 37% formaldehyde into the medium of the chamber during recording. Formaldehyde fixation will not only rapidly arrest chromosome and cellular mobility but also leads to rapid fading of GFP fluorescence.

Acknowledgments

We thank David B. Kaback, UMDNJ, New Jersey, USA, and Edgar Trelles-Sticken, previously MPI-MG, Berlin, for fruitful collaboration, and J. Loidl, Univ. of Vienna, Austria, for critical comments on the manuscript. The work in the lab of HS was supported in part by H.-H. Ropers, Max-Planck-Institute for Molecular Genetics, Berlin, Germany, and the DFG (SCHE 350/10-1, SPP 1384).

References

1. Martinez-Perez, E., and Colaiacovo, M.P. (2009) Distribution of meiotic recombination events: talking to your neighbors. *Curr Opin Genet Dev* **19**, 105–112.
2. Borner, G.V., Kleckner, N., and Hunter, N. (2004) Crossover/noncrossover differentiation, synaptonemal complex formation, and regulatory surveillance at the leptotene/zygotene transition of meiosis. *Cell* **117**, 29–45.
3. Fung, J.C., Rockmill, B., Odell, M., and Roeder, G.S. (2004) Imposition of crossover interference through the nonrandom distribution of synapsis initiation complexes. *Cell* **116**, 795–802.
4. Scherthan, H. (2007) Telomere attachment and clustering during meiosis. *Cell Mol Life Sci* **64**, 117–124.
5. Parvinen, M., and Soderstrom, K.O. (1976) Chromosome rotation and formation of synapsis. *Nature* **260**, 534–535.
6. Frigault, M.M., Lacoste, J., Swift, J.L., and Brown, C.M. (2009) Live-cell microscopy – tips and tools. *J Cell Sci* **122**, 753–767.
7. Robinett, C.C., Straight, A., Li, G., Willhelm, C., Sudlow, G., Murray, A., and Belmont, A.S. (1996) In vivo localization of DNA sequences and visualization of large-scale chromatin organization using lac operator/repressor recognition. *J Cell Biol* **135**, 1685–1700.
8. Swedlow, J., Goldman, R., and Spector, D. (2009) Live cell imaging: a laboratory manual, 2nd Edition (Cold Spring Harbor, NY: Cold Spring Harbor Laboratory Press).
9. Straight, A.F., Belmont, A.S., Robinett, C.C., and Murray, A.W. (1996) GFP tagging of budding yeast chromosomes reveals that protein–protein interactions can mediate sister chromatid cohesion. *Curr Biol* **6**, 1599–1608.
10. Trelles-Sticken, E., Adelfalk, C., Loidl, J., and Scherthan, H. (2005) Meiotic telomere clustering requires actin for its formation and cohesin for its resolution. *J Cell Biol* **170**, 213–223.
11. Hediger, F., Taddei, A., Neumann, F.R., and Gasser, S.M. (2004) Methods for visualizing chromatin dynamics in living yeast. *Methods Enzymol* **375**, 345–365.
12. Dresser, M.E. (2009) Time-lapse fluorescence microscopy of Saccharomyces cerevisiae in meiosis. *Methods Mol Biol* **558**, 65–79.

13. Asakawa, H., and Hiraoka, Y. (2009) Live-cell fluorescence imaging of meiotic chromosome dynamics in Schizosaccharomyces pombe. *Methods Mol Biol* **558**, 53–64.
14. Scherthan, H., Wang, H., Adelfalk, C., White, E.J., Cowan, C., Cande, W.Z., and Kaback, D.B. (2007) Chromosome mobility during meiotic prophase in Saccharomyces cerevisiae. *Proc Natl Acad Sci USA* **104**, 16934–16939.
15. Wanat, J.J., Kim, K.P., Koszul, R., Zanders, S., Weiner, B., Kleckner, N., and Alani, E. (2008) Csm4, in collaboration with Ndj1, mediates telomere-led chromosome dynamics and recombination during yeast meiosis. *PLoS Genet* **4**, e1000188.
16. Koszul, R., Kim, K.P., Prentiss, M., Kleckner, N., and Kameoka, S. (2008) Meiotic chromosomes move by linkage to dynamic actin cables with transduction of force through the nuclear envelope. *Cell* **133**, 1188–1201.
17. Conrad, M.N., Lee, C.Y., Chao, G., Shinohara, M., Kosaka, H., Shinohara, A., Conchello, J.A., and Dresser, M.E. (2008) Rapid telomere movement in meiotic prophase is promoted by NDJ1, MPS3, and CSM4 and is modulated by recombination. *Cell* **133**, 1175–1187.
18. White, E.J., Cowan, C., Cande, W.Z., and Kaback, D.B. (2004) In vivo analysis of synaptonemal complex formation during yeast meiosis. *Genetics* **167**, 51–63.
19. Hayashi, A., Ogawa, H., Kohno, K., Gasser, S.M., and Hiraoka, Y. (1998) Meiotic behaviours of chromosomes and microtubules in budding yeast: relocalization of centromeres and telomeres during meiotic prophase. *Genes Cells* **3**, 587–601.
20. Pringle, J.R., Adams, A.E., Drubin, D.G., and Haarer, B.K. (1991) Immunofluorescence methods for yeast. *Methods Enzymol* **194**, 565–602.
21. Primig, M., Williams, R.M., Winzeler, E.A., Tevzadze, G.G., Conway, A.R., Hwang, S.Y., Davis, R.W., and Esposito, R.E. (2000) The core meiotic transcriptome in budding yeasts. *Nat Genet* **26**, 415–423.
22. Gotta, M., Laroche, T., and Gasser, S.M. (1999) Analysis of nuclear organization in *Saccharomyces cerevisiae. Methods Enzymol* **304**, 663–672.
23. Roth, R., and Halvorson, H.O. (1969) Sporulation of yeast harvested during logarithmic growth. *J Bacteriol* **98**, 831–832.
24. Kane, S.M., and Roth, R. (1974) Carbohydrate metabolism during ascospore development in yeast. *J Bacteriol* **118**, 8–14.
25. Padmore, R., Cao, L., and Kleckner, N. (1991) Temporal comparison of recombination and synaptonemal complex formation during meiosis in *S. cerevisiae*. *Cell* **66**, 1239–1256.

Chapter 32

Chromosome Structure and Homologous Chromosome Association During Meiotic Prophase in *Caenorhabditis elegans*

Kentaro Nabeshima

Abstract

Successful meiotic recombination is driven by a series of programmed chromosome dynamics that include changes in the protein composition of meiotic chromosomes and the juxtaposition of homologous chromosomes. The simultaneous visualization of both chromosome-bound proteins and the status of homologous association is an important experimental approach to analyze the mechanisms supporting proper meiotic chromosome association. One of a number of model organisms used for meiosis research, the nematode *Caenorhabditis elegans* offers an excellent environment to study meiotic chromosome dynamics. Here I will describe how to visualize both chromosome structure and specific chromosomal loci simultaneously, in a whole-mount *C. elegans* germ line. It combines immunofluorescent (IF) staining for a meiotic chromosome structural component with fluorescent in situ hybridization (FISH).

Key words: *C. elegans*, chromosome axis, FISH, germ line, homologous pairing, immunofluorescence, meiosis, synaptonemal complex, whole-mount gonad.

1. Introduction

For most sexually reproducing organisms, the pairwise alignment and juxtaposition of homologous chromosomes during meiotic prophase are stabilized by a protein structure called the synaptonemal complex (SC). In this structure, chromosomal DNA is organized into a linear array of chromatin loops with the base of each loop attached to an axial structure (1). The two homologous axes are joined together by the SC central region (or transverse filament) (2). In *Caenorhabditis elegans*, a number of proteins have been identified as a component of the SC structure. The axis structure includes proteins such as HIM-3 (3) and HIM-3-related

H. Tsubouchi (ed.), *DNA Recombination*, Methods in Molecular Biology 745,
DOI 10.1007/978-1-61779-129-1_32,

proteins, HTP-1 (4, 5) and HTP-3 (6), as well as cohesion proteins, such as REC-8 (7). Proteins enriched in helical coiled-coil domains, SYP-1 (8), SYP-2 (9), SYP-3 (10), and SYP-4 (11) are known to constitute the SC central region. IF staining for these proteins is regularly used to investigate the structural integrity of the SC. An important biological problem is the functional correlation between the integrity of the SC and proper homolog association. To investigate this, it is useful to access the status of homologous association and SC formation at the same time, by the combined use of IF and FISH. For example, cytological demonstration of the loading of a SC central region component at unpaired loci in a mutant has been widely used to show defects in this coordination and resultant SC formation between non-homologous chromosomes (4, 12, 13). In the first part I will describe how to prepare a FISH probe from a YAC (yeast artificial chromosome) containing *C. elegans* genomic DNA (14). In the second part, I will describe how to carry out IF combined with FISH.

2. Materials

2.1. FISH Probe Preparation

1. *C. elegans* YAC clone (Wellcome Trust Sanger Institute, Cambridge, UK)
2. GELase (EPICENTRE Biotechnologies, Madison, WI)
3. Illustra GenomiPhi V2 DNA Amplification Kit (GE Healthcare, Piscataway, NJ)
4. MinElute Reaction Cleanup kit (Qiagen, Valencia, CA)
5. ULYSIS Alexa Fluor Nucleic Acid Labeling Kit (Invitrogen, Carlsbad, CA)
6. Restriction enzymes: *Alu*I, *Hae*III, *Mse*I, *Msp*I, *Rsa*I, and *Sau*3AI (New England Biolabs, Ipswich, MA)
7. Performa DTR Gel Filtration Cartridges (EdgeBio, Gaithersburg, MD)
8. Heating block

2.2. IF and FISH

1. 10× egg salt buffer: 118 mM NaCl, 48 mM KCl, 2 mM $CaCl_2$, 2 mM $MgCl_2$, and 25 mM HEPES [pH 7.5] (15). I usually use Milli-Q water. Regular distilled water is also fine to use for the method described here.
2. Dissection buffer: 1× egg salt buffer, 1% (v/v) Tween-20, 15 mM sodium azide.
3. 2× permeabilization buffer: 1× egg salt buffer, 20% Tween-20.
4. Fixation buffer: 1× egg salt buffer, 4% paraformaldehyde.

5. Paraformaldehyde 16% solution, EM grade (Electron Microscopy Sciences, Hatfield, PA). Content of an ampoule is stored at 4°C in an airtight and opaque container for up to 1 week.
6. 10× phosphate-buffered saline (PBS): Dissolve 80 g of NaCl, 2 g of KCl, 14.4 g of Na_2HPO_4, and 2.4 g of KH_2PO_4 in 1 l of H_2O. Adjust the pH to 7.4 with HCl if necessary. Autoclave to sterilize.
7. Post-fixation buffer: 1× PBS containing 4% paraformaldehyde.
8. 20× SSC: 3 M NaCl (175.3 g NaCl/l), 0.3 M $C_6H_5O_7Na_3$ (88.23 g trisodium citrate dihydrate/l).
9. 2x SSCT: 2x SSC containing 0.1% Tween-20.
10. IF washing buffer (PBST): 1× PBS containing 0.1% Tween-20.
11. Blocking buffer: 1× PBST containing 0.5% BSA.
12. BSA: bovine serum albumin from further purified Fraction V, 98% (Sigma Aldrich, St Louis, MO).
13. Formamide, deionized OmniPur (EMD Chemicals, Gibbstown, NJ).
14. Dextran sulfate: dextran sulfate sodium salt from Leuconostoc spp. for molecular biology (Sigma Aldrich).
15. Hybridization buffer: Dissolve 0.5 g dextran sulfate in 2.5 ml formamide and 0.75 ml 20× SSC in a 14 ml capped conical tube that is gently agitated for several hours on a Nutator mixer (BD Diagnostic Systems, Sparks, MD). Add H_2O to bring total volume to 4.3 ml.
16. 95% ethanol.
17. Secondary antibody: Alexa Fluor labeled (Invitrogen).
18. ProLong Gold Antifade Reagent (Invitrogen).
19. DAPI (4′,6-diamidino-2-phenylindole) dissolved in water, 50–100 μg/ml.
20. Parafilm M (American National Can Company, Neenah, WI).
21. Microscope cover slips, 22×22 mm No. 1 and 22×40 mm No. 1.5 (Fisher, Pittsburgh, PA).
22. Microscope slide glass, SuperFrost Plus (Fisher, Pittsburgh, PA).
23. Surgical blade and a handle: sterile scalpel blade #11 and scalpel handle #7 (Electron Microscopy Sciences, Hatfield, PA).
24. Omni Slide, flat bed thermal cycler (Thermo Fisher Scientific Inc., Waltham, MA).
25. Dissection microscope with a transmitted light base.

3. Methods

In the first part I will describe how to prepare a FISH probe from a YAC carrying *C. elegans* genomic DNA. Using the same principle, BACs (bacterial artificial chromosomes) have also been reported to be a source for DNA amplification (16). In this protocol, chemical reaction is used to label probe DNA, instead of commonly used enzymatic reactions (e.g., tail labeling with terminal deoxynucleotidyl transferase). This chemical method is more reproducible compared to enzymatic methods because there is no enzymatic component whose activity could vary depending on the manufacturers and product age and batches. In addition, this method is also more cost-effective. The probes produced by this method usually generate lower background noise and higher *S/N* ratio compared to probes generated by other methods. In the second part, I will describe how to carry out IF combined with FISH. In this protocol, IF is carried out before FISH procedure, which helps to preserve intact chromosome structure for IF because heat denature and formamide treatment of FISH would disrupt the chromosome structure. On the other hand, Alexa Fluor dyes used for IF are largely resistant to heat denature and formamide treatment, and they retain sufficient fluorescence after FISH. For FISH, a single YAC clone is usually used to prepare a probe. Since the size of genomic DNA cloned into a YAC varies, it might be necessary to combine several YACs or select a YAC that has longer genomic sequence. If multiple YACs are combined, a total of several hundred thousand bases of DNA can be amplified in one amplification reaction. A probe covering 200–300 K bases of genomic sequence usually produces sufficient FISH signal to detect using a conventional compound fluorescent microscope. I have tested five fluorophores: ULYSIS Alexa Fluor-488, -546, -532, -594, and -647. All of them produced excellent results. Of these, ULYSIS Alexa Fluor-594 produced the strongest signal.

3.1. Preparing FISH Probe from a YAC

3.1.1. Digestion of an Agarose Gel Slice Containing a YAC

1. Each YAC should first be purified by pulsed-field gel electrophoresis. The agarose band containing the YAC, excised from pulsed-field gel, can be stored in a micro-centrifuge tube and kept at −20°C for several months before processing (*see* **Note 1**).
2. Thoroughly melt the gel slice by boiling for 3 min.
3. Transfer 100–200 μl of the molten gel slice solution (it is necessary to know the exact volume of the solution for digestion and you may not transfer all the solution) by a pipette to a new micro-centrifuge tube that is pre-heated to 45°C on a heating block. Leave molten agarose solution for 2 min at 45°C to equilibrate its temperature.

4. Add 50× GELase digestion buffer (2 μl/100 μl of molten agarose solution) and GELase (0.5 unit/100 μl of molten agarose), mix well, and incubate at 45°C for at least 1 h (*see* **Note 2**).
5. After digestion, leave at room temperature (RT) for 10 min, and check if the solution is completely clear. If there is white turbidity on top of the solution, boil it again and digest it with additional 0.2–0.4 unit of GELase for 30–60 min.
6. After digestion, store digested agarose gel solution at −20°C. There is no need to purify DNA from this solution for later steps.

3.1.2. Amplify DNA and Digestion of Amplified DNA

1. Mix 2 μl of molten YAC gel slice solution with 18 μl GenomiPhi (*see* **Note 3**) sample buffer. Boil this mixture for 3 min.
2. Cool the mixture to 4°C on ice and centrifuge the tube briefly at 4°C to redeposit the sample to the bottom of the tube. Put the tube back on ice.
3. For each amplification reaction, combine 18 μl of GenomiPhi reaction buffer with 2 μl of GenomiPhi enzyme mix on ice. Add this to the cooled sample. Total volume of the reaction is 40 μl and this size of reaction usually yields 4–6 μg amplified DNA.
4. Incubate the sample at 30°C for 16–18 h.
5. Heat the sample at 65°C for 10 min. Cool the reaction to RT and centrifuge the tube briefly to redeposit the sample to the bottom of the tube (*see* **Note 4**).
6. Add 5.8 μl 10x NE Buffer 2 and 2 μl of each of following six restriction enzymes (*Alu*I, *Hae*III, *Mse*I, *Msp*I, *Rsa*I, and *Sau*3AI) and mix well (*see* **Note 5**).
7. Incubate the reaction at 37°C for more than 4 h (to overnight) to digest DNA.
8. Incubate the reaction at 65°C for 10 min for heat inactivation of restriction enzymes.
9. Clean up the reaction with a MinElute Reaction Cleanup kit (Qiagen). Divide one sample between two MinElute columns (i.e., ~29 μl × 2). Elute DNA in 11 μl (per column) component C (50 mM Tris–HCl, pH 7.4) of ULYSIS buffer (*see* **Note 6**).
10. Combine two eluted samples in one tube.
11. Take 1 μl of DNA solution and dilute DNA solution 50- to 100-fold to measure OD_{A260} with a spectrophotometer (*see* **Note 7**).
12. Calculate DNA concentration (μg/ml) based on the equation: $OD_{A260} \times 50 \times$ dilution factor. If dilution factor

is 100 then DNA concentration is OD $_{A260}$ × 5 (μg/μl). If amplification is successful, it is usually around 0.2–0.3 μg/μl.

3.1.3. Labeling DNA

1. Prepare the ULS labeling reagent stock solution following the manufacturer's instruction. Add 100 μl of component B (fluorescent dye solvent) to component A (fluorescent dye) and thoroughly mix them, except for Alexa Fluor-488 ULS reagent that is in a smaller aliquot (20 μl/tube). Refer to the instructions in the kit.
2. Make 20 μl (for Alexa Flour-488 make 24 μl) DNA solution containing 2 μg of DNA (*see* **Note 8**) with DNA solution (from Step 9 in the previous section) and component C.
3. Denature the DNA by boiling for 5 min and then cool the sample on ice. Centrifuge the tube briefly at 4°C to redeposit the sample to the bottom of the tube. Place the sample on ice.
4. Add 5 μl (1 μl for Alexa Fluor-488) of ULS labeling reagent stock solution to the denatured sample DNA solution. The final volume of reaction is 25 μl.
5. Incubate the reaction on a heating block at 80°C for 15 min. Use aluminum foil to cover tubes during the labeling reaction. Stop the reaction by putting the reaction tube on ice. Centrifuge the tube briefly to redeposit the sample to the bottom of the tube.
6. Purify the labeled DNA by using a column prepared following the manufacturer's instruction (*see* **Note 9**).
7. The purified sample is ready to use for FISH experiment. Usually, there is no need to concentrate the labeled DNA; 1–2 μl of this solution contains 75–150 ng of labeled DNA that is enough for one hybridization experiment. Store the purified DNA at −20°C.

3.2. Immunofluorescent Staining Combined with Fluorescent In Situ Hybridization for Whole-Mount C. elegans Germ Line

3.2.1. Immunofluorescent Staining for Chromosome Structural Proteins

1. Place a drop of 50 μl of dissection buffer on a cover slip (22×22) and suspend 20–30 worms in it. Quickly dissect worms on a cover slip under a dissection microscope (*see* **Note 10**).
2. Collect all the dissected gonads with micro-pipetter in 30 μl of suspension and transfer onto a new cover slip. Add an equal volume (30 μl) of permeabilization buffer and mix well. Leave for 5 min in a humid chamber at RT.
3. Collect all the permeabilized gonads with micro-pipetter in 30 μl of suspension and transfer onto a new cover slip. Add an equal volume (30 μl) of 2× fixation buffer and mix

well. Leave for 1 min. During this fixation step, carefully remove 45 μl from the fixation mixture using a micropipetter. Take extra care not to remove any of the worm tissues.

4. Put a SuperFrost Plus slide on the cover slip (to attach specimen to a positively charged surface). Turn the slide glass upside down (now the cover slip is on the slide glass) and wick away excess liquid with a piece of paper (e.g., kimwipe, *see* **Note 11**).
5. Put the slide glass with the cover slip on it into liquid nitrogen and wait for about 30 s until temperature is equilibrated. Take it out from liquid nitrogen, immediately take the cover slip off with a razor blade (*see* **Note 12**), and put the slide into cold (−20°C) 95% ethanol. Leave it for 5 min in −20°C freezer.
6. Take the slide out from 95% ethanol and quickly wick away the liquid from the area without specimen (both left and right sides of the area where the specimens are present and the back side of a slide glass). Do not completely dry it. Put the slides into PBST in a Coplin jar and leave for 10 min.
7. Repeat the washing of the slide glass in PBST in a Coplin jar for 10 min two more times.
8. Put the slide glass into blocking solution in a Coplin jar and leave for 30 min.
9. Take the slide out from blocking solution, wick away liquid from the area without specimen (as in Step 6). Using aspirator, take almost all the liquid from the surface of the slide glass but leave a very thin layer of liquid in the area where the specimens are present. Before the slide glass is completely dried up, apply 50 μl of a solution of primary antibody diluted in blocking buffer (*see* **Note 13**). Use a piece of a hand-cut parafilm (20×20 mm) to cover the specimen and place the slide into humid chamber (*see* **Note 14**). Leave overnight at 4°C.
10. Wash the slides in PBST in a Coplin jar for 10 min. Repeat washing two more times.
11. As in Step 9, remove liquid from the surface of the slide glass and apply 50 μl solution of secondary antibody (labeled with Alexa Fluor) diluted in blocking buffer (1:400). Incubate the slide in a humid chamber at RT for 2 h.
12. Wash the slide glass in PBST in a Coplin jar for 10 min. Repeat washing two more times.
13. As in Step 9, remove liquid from the surface of the slide glass and apply 400 μl of post-fixation solution over the

specimen. Leave it in a humid chamber in the dark for 10 min at RT.

14. Remove post-fixation solution from the slide glass and put the slide glass into 2× SSCT in a Coplin jar in the dark. Leave for 5 min at RT. Repeat this washing once. Hereafter, all the washing steps need to be done in the dark.

3.2.2. Fluorescent In Situ Hybridization

1. Put the slide glass into 2x SSCT containing 5% formamide in a Coplin jar and leave it for 5 min at RT.
2. Put the slide glass into 2x SSCT containing 10% formamide in a Coplin jar and leave it for 5 min at RT.
3. Put the slide glass into 2x SSCT containing 25% formamide in a Coplin jar and leave it for 5 min at RT.
4. Put the slide glass into 2x SSCT containing 50% formamide in a Coplin jar and leave it for 3 min at RT (*see* **Note 15**).
5. Put the slide glass into 2x SSCT containing 50% formamide in a Coplin jar and put it into a 37°C water bath. Leave it for 1 (up to 4) h.
6. Take a slide glass out from the Coplin jar, wipe the area that does not contain any specimen, and put it into 95% ethanol (at RT). Leave it for 5–10 min at RT.
7. Mix 2 μl probe solution (*see* **Note 16**) with 13 μl hybridization solution (*see* **Note 17**) and put it onto a cover slip (22×22 mm).
8. Take the slide glass out from the Coplin jar and remove liquid on the surface of the slide glass (as in Step 9 of **Section 3.2.1**). Put the slide glass onto a cover slip so that the area with specimen is covered by probe/hybridization buffer mix as well as by a cover slip. Quickly turn them upside down so that the cover slip is on the slide glass.
9. Put water into a channel surrounding a heating block of Omni Slide thermal cycler to keep the hybridization chamber humid (*see* **Note 18**). Put the slide glass onto the flat bed of Omni Slide thermal cycler and put a lid on the machine to close the hybridization chamber. Start the program described below (*see* **Note 19**).

Temp (°C)	Time (min)	Ramp	Inc.
80	10:00	0.5	0
50	1:00	0.5	0
45	1:00	0.5	0
40	1:00	0.5	0
38	1:00	0.5	0
37	hold		

10. Leave it overnight (in the dark if probe is directly labeled with fluorescent dye).
11. Take the slide out of Omni Slide thermal cycler and put into pre-heated (37°C) 2x SSCT containing 50% formamide in a Coplin jar. Leave it for 30 min in a water bath. Check that the cover slip falls off by itself in 2–3 min (*see* **Note 20**). Make sure to wash slides for at least 30 min after the cover slip comes off.
12. Put the slide into another pre-heated (at 37°C) 2x SSCT containing 50% formamide in a Coplin jar. Leave it for 30 min in a 37°C water bath.

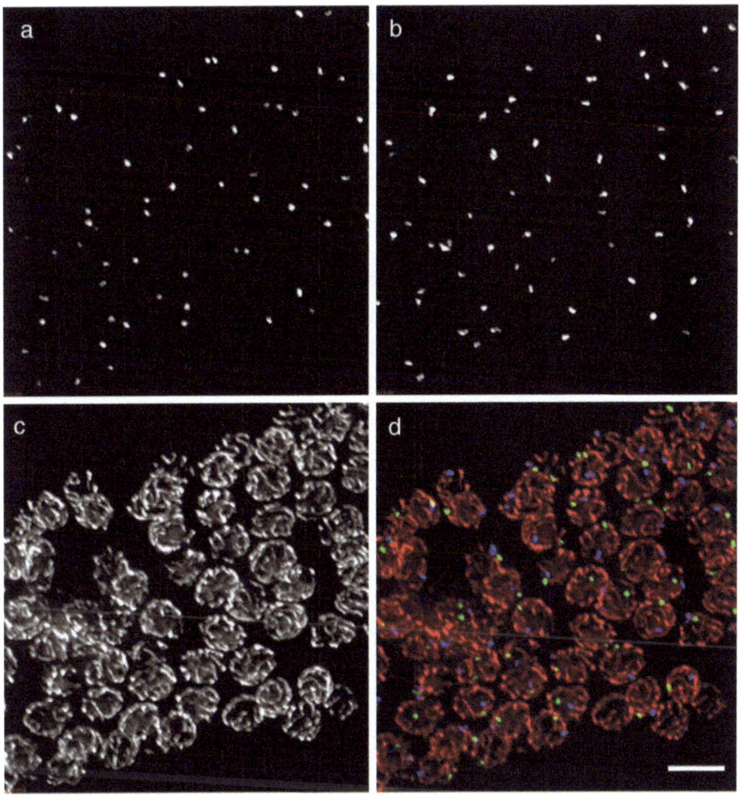

Fig. 32.1. Visualization of both chromosomal loci and SC central region structure in *C. elegans* germ line. (**a** and **b**) The left and the right end of chromosome IV is visualized by FISH using two probes generated from a cocktail of YACs: Y41H10, Y59E4, and Y47B5 labeled with Alexa-488 for the left end and Y51H4 and Y43D4 labeled with Alexa-647 for the right end. (**c**) SC central region structure, SYP-1 is visualized by IF using guinea pig anti-SYP-1 antibody (8) and Alexa-555-labeled goat anti-guinea pig IgG antibody. (**d**) A merge of the left end of chromosome IV (*green*), the right end of chromosome IV (*blue*), and SYP-1 (*red*). A part of the pachytene region containing about 60 nuclei of germ line in a wild-type animal is shown. The image is a projection generated from deconvoluted optical sections covering an entire thickness of nuclei taken with a wide-field fluorescent microscope (60× objective); the projection depth is 5 μm and the thickness of optical section is 0.2 μm. Bar = 5 μm.

13. Put the slide glass into pre-heated (at 37°C) 2x SSCT containing 25% formamide in a Coplin jar and leave it for 10 min at RT.
14. Put the slide glass into 2x SSCT containing 10% formamide in a Coplin jar and leave it for 10 min at RT.
15. Put the slide glass into 2x SSCT containing 5% formamide in a Coplin jar and leave it for 10 min at RT.
16. Put the slide glass into 2x SSCT in a Coplin jar and leave it for 10 min at RT.
17. Put the slide glass into 2x SSCT containing DAPI (0.5–1 μg/ml) in a Coplin jar and leave it for 10 min at RT.
18. Put the slide glass into 2x SSCT in a Coplin jar and leave it for 3 min at RT.
19. Put the slide glass into 2x SSCT in a Coplin jar and leave it for 40–60 min at RT.
20. Take a slide glass out from the Coplin jar and remove liquid from the surface of the slide glass (as in Step 9 of **Section 3.2.1**). Apply 15 μl of an appropriate mounting medium (e.g., Prolong Gold). Put a cover glass (22×45 mm No. 1.5) on top slowly not to make any bubble on a specimen. Cure mounting medium if necessary. Seal the cover slip by applying a thin layer of clear nail polish to the area surrounding it. Store the slide at 4°C or –20°C. *See* **Fig. 32.1** for example images.

4. Notes

1. There is no need to use low-melting agarose for PFG electrophoresis. I use SeaKem GTG Agarose (Cambrex, East Rutherford, NJ) to obtain better resolution. Limit exposure of the gel slice to UV light and use longer wavelength UV light. The presence of ethidium bromide will not affect later steps.
2. GELase is a product used to digest molten low-melting agarose, according to the manufacturer's instruction, but this enzyme can also digest molten standard agarose including SeaKem GTG agarose.
3. This kit utilizes bacteriophage phi29 DNA polymerase that has strong strand displacement and DNA synthesis activity. Compared to PCR amplification using degenerated primers, this method has two major advantages: (1) DNA synthesis primed with random hexamers avoids

biased amplification of DNA that often accompanies with degenerated PCR-based amplification and (2) high yield of this polymerase enables to amplify large amount of DNA in a small reaction. Although the manufacturer does not recommend the amplification of YACs, this kit is capable of amplifying YAC DNA. In cases where more amplified DNA is required, increase the reaction volume but not the amount of input DNA. Yield of amplified DNA is proportional mainly to reaction volume, but not to the amount of input DNA. When reaction volume is increased, the capacity of purification media needs to be taken into consideration. MinElute, used in this protocol, has 5 μg maximum binding capacity per column.

4. There might be white turbidity after overnight incubation. It apparently does not affect later processes. Just proceed to the next step.
5. Fragmenting DNA to less than 1,000 bp is essential for both subsequent labeling reaction and probe penetration into gonads. This digestion usually yields ~300-bp fragments.
6. Purity of DNA seems to be important: MinElute Cleanup kit is convenient to get pure and concentrated DNA. I also tested desalting with G-25 column and subsequent ethanol precipitation (EtOH ppt) to purify digested DNA. There was no difference between MinElute purification and G-25/EtOH ppt for labeling efficiency.
7. In the next labeling step, a non-enzymatic method for chemically labeling DNA is used. Because of this (purely chemical reaction), it is critical to accurately measure DNA concentration and put an exact amount of DNA into the reaction. Use spectrophotometer. Do *not* rely on a gel electrophoresis to estimate DNA concentration, though it is more convenient. Ethidium bromide staining for these shorter and broad-range fragments leads to a completely false estimate.
8. The kit instructions recommend that 1 μg DNA is used for a 25 μl reaction. I saw better results when 2 μg instead of 1 μg DNA was used. But do *not* increase the amount of DNA further. When I used 4 μg DNA per 25 μl reaction, I did not see any significant improvement.
9. For purification of labeled DNA from unincorporated fluorophore, the manufacturer recommends a gel filtration-based spin column for purification of sequencing sample. I use Performa DTR Gel Filtration Cartridges (EdgeBio). I also used Centri-Sep columns (Applied Biosystems Inc., Foster City, CA), which worked equally well.

10. In order to synchronize the age of worms, pick L4 worms 24 h before dissection. Use a surgical blade (Sterile scalpel blade, #11) or an injection needle (Needle 25G $1^1/_2$, Becton Dickinson, Franklin Lakes, NJ) to dissect worms. Make a complete cut at the position of the vulva before worms are completely paralyzed. Check most of the gonad arms are extruded from the body. Gonad parts remaining inside the body will not be stained. Do not leave worms for longer than 7 min in dissection buffer before dissection as it gets difficult to extrude entire gonad arms.
11. Remove excess liquid by absorbing it from the side of the cover slip with a piece of paper. If there is too much liquid left, it causes a considerable loss of specimen from a surface of slide glass in later washing steps.
12. Put a razor blade between the cover slip and the slide glass, and quickly flick the cover slip off. A careful and complete removal of cover slip is necessary because it tends to stick back to a slide glass by static.
13. I use 1:400 dilution for both rabbit anti-HIM-3 and guinea pig anti-SYP-1 antibodies.
14. Put a spacer (e.g., broken plastic pipettes) on paper towels dampened with plenty of water lining the bottom of a shallow plastic container. Put slide glasses on the spacer so as not to directly touch the damp paper towels and put on a lid to seal the container.
15. This four-step change in formamide concentration helps to preserve gonad structure. With conventional two-step change (i.e., 25%→50%), gonads are often damaged and burst.
16. 75–150 ng of labeled probe from one YAC is enough for a slide. This corresponds to 1–2 μl of probe solution from standard labeling reaction with 2 μg DNA in 25 μl reaction mixture (see also the protocol for making FISH probes from YACs). If you use multiple probes, just combine them and reduce the volume to less than 2 μl with speed vacuum or you can even completely dry them. Then adjust the final volume to (or resuspend the pellet in) 2 μl water. Do *not* increase the concentration of probe too much (twofold to threefold increase is ok, but usually there is no need). It drastically increases background noise and makes signal–noise ratio worse.
17. Hybridization solution does not need to contain a reagent to out-compete non-specific DNA binding, such as salmon sperm DNA. When I included salmon sperm DNA (5 μg/ml, Invitrogen, ready to use for hybridization) in hybridization solution, it did not reduce background but disrupted chromosome morphology.

18. Putting additional paper towels dampened with water next to the slide glass on a flat bed helps to maintain humidity. Make sure no water spills into inside of thermal cycler.
19. Lowering denaturing temperature to 80°C from conventional 89°C helps preservation of chromosome morphology and gonad integrity without any apparent decrease in hybridization efficiency for this preparation. If you do not have access to a Omni Slide machine, use a regular heating block set at 80°C (humidified using a shallow plastic box (e.g., the lid from racks of P2 Pipetman tips) and kimwipes placed around the interior rim of the lid, moistened with water for the denaturing step, and a humid chamber at 37°C for the hybridization step (17)).
20. Do not try to directly remove the cover slip, since it will break and remove specimens from the slide. If the cover slip is stuck on the slide glass and does not come off easily by itself in the first wash (in 2× SSCT containing 50% formamide), gently shake the slide glass in the first wash until the cover slip comes off by itself.

Acknowledgments

The author would like to thank Dr. Anne M. Villeneuve for extensive support and helpful suggestions during part of the development of this method as well as for providing anti-SYP-1 antibody and Dr. Raymond Chan for critical reading of the manuscript and insightful comments. This work was supported by March of Dimes, Basil O'Conner Starter Scholar Award (#5-FY07-666).

References

1. Moens, P.B., and Pearlman, R.E. (1988) Chromatin organization at meiosis. *Bioessays* **9**, 151–153.
2. von Wettstein, D., Rasmussen, S.W., and Holm, P.B. (1984) The synaptonemal complex in genetic segregation. *Ann Rev Genet* **18**, 331–413.
3. Zetka, M.C., Kawasaki, I., Strome, S., and Muller, F. (1999) Synapsis and chiasma formation in *Caenorhabditis elegans* require HIM-3, a meiotic chromosome core component that functions in chromosome segregation. *Genes Dev* **13**, 2258–2270.
4. Martinez-Perez, E., and Villeneuve, A.M. (2005) HTP-1-dependent constraints coordinate homolog pairing and synapsis and promote chiasma formation during *C. elegans* meiosis. *Genes Dev* **19**, 2727–2743.
5. Couteau, F., and Zetka, M. (2005) HTP-1 coordinates synaptonemal complex assembly with homolog alignment during meiosis in *C. elegans*. *Genes Dev* **19**, 2744–2756.
6. Goodyer, W., Kaitna, S., Couteau, F., Ward, J.D., Boulton, S.J., and Zetka, M. (2008) HTP-3 Links DSB formation with homolog pairing and crossing over during *C. elegans* meiosis. *Dev Cell* **14**, 263–274.
7. Pasierbek, P., Jantsch, M., Melcher, M., Schleiffer, A., Schweizer, D., and Loidl, J. (2001) A *Caenorhabditis elegans* cohesion

protein with functions in meiotic chromosome pairing and disjunction. *Genes Dev* **15**, 1349–1360.

8. MacQueen, A.J., Colaiacovo, M.P., McDonald, K., and Villeneuve, A.M. (2002) Synapsis-dependent and -independent mechanisms stabilize homolog pairing during meiotic prophase in *C. elegans*. *Genes Dev* **16**, 2428–2442.
9. Colaiacovo, M.P., MacQueen, A.J., Martinez-Perez, E., McDonald, K., Adamo, A., La Volpe, A., and Villeneuve, A.M. (2003) Synaptonemal complex assembly in *C. elegans* is dispensable for loading strand-exchange proteins but critical for proper completion of recombination. *Dev Cell* **5**, 463–474.
10. Smolikov, S., Eizinger, A., Schild-Prufert, K., Hurlburt, A., McDonald, K., Engebrecht, J., Villeneuve, A.M., and Colaiacovo, M.P. (2007) SYP-3 restricts synaptonemal complex assembly to bridge paired chromosome axes during meiosis in *Caenorhabditis elegans*. *Genetics* **176**, 2015–2025.
11. Smolikov, S., Schild-Prufert, K., and Colaiacovo, M.P. (2009) A yeast two-hybrid screen for SYP-3 interactors identifies SYP-4, a component required for synaptonemal complex assembly and chiasma formation in *Caenorhabditis elegans* meiosis. *PLoS Genet* **5**, e1000669.
12. Couteau, F., Nabeshima, K., Villeneuve, A., and Zetka, M. (2004) A component of *C. elegans* meiotic chromosome axes at the interface of homolog alignment, synapsis, nuclear reorganization, and recombination. *Curr Biol* **14**, 585–592.
13. Nabeshima, K., Villeneuve, A.M., and Hillers, K.J. (2004) Chromosome-wide regulation of meiotic crossover formation in *Caenorhabditis elegans* requires properly assembled chromosome axes. *Genetics* **168**, 1275–1292.
14. Coulson, A., Waterston, R., Kiff, J., Sulston, J., and Kohara, Y. (1988) Genome linking with yeast artificial chromosomes. *Nature* **335**, 184–186.
15. Edgar, L.G. (1995) Blastomere culture and analysis. In Methods in cell biology: *Caenorhabditis elegans*, modern biological analysis of an organism, H.F. Epstein and D.C. Shakes, eds. (New York, NY: Academic Press), p. 317.
16. Roohi, J., Cammer, M., Montagna, C., and Hatchwell, E. (2008) An improved method for generating BAC DNA suitable for FISH. *Cytogenetic Genome Res* **121**, 7–9.
17. Dernburg, A.F. (1999) Fluorescence in situ hybridization in whole-mount tissues. In Chromosome structural analysis, a practical approach, W.A. Bickmore, ed. (Oxford: Oxford University Press), p. 142.

INDEX

H. Tsubouchi (ed.), *DNA Recombination*, Methods in Molecular Biology 745,
DOI 10.1007/978-1-61779-129-1, © Springer Science+Business Media, LLC 2011

MIX
Papier aus verantwortungsvollen Quellen
Paper from responsible sources
FSC® C105338

If you have any concerns about our products,
you can contact us on
ProductSafety@springernature.com

In case Publisher is established outside the EU,
the EU authorized representative is:
Springer Nature Customer Service Center GmbH
Europaplatz 3, 69115 Heidelberg, Germany

Printed by Libri Plureos GmbH
in Hamburg, Germany